11 Springer Series in Solid-State Sciences
Edited by Hans-Joachim Queisser

Springer Series in Solid-State Sciences

Editors: M. Cardona P. Fulde H.-J. Queisser

B. R. Nag

Electron Transport in Compound Semiconductors

With 148 Figures

Springer-Verlag Berlin Heidelberg New York 1980

Professor Dr. Biswaranjan Nag

Institute of Radiophysics and Electronics
University College of Science and Technology
92, Acharya Prafulla Chandra Road, 700 009 Calcutta, India

Series Editors:

Professor Dr. Manuel Cardona
Professor Dr. Peter Fulde
Professor Dr. Hans-Joachim Queisser

Max-Planck-Institut für Festkörperforschung
Heisenbergstrasse 1, D-7000 Stuttgart 80, Fed. Rep. of Germany

ISBN-13:978-3-642-81418-1 e-ISBN-13:978-3-642-81416-7
DOI: 10.1007/978-3-642-81416-7

Library of Congress in Publication Data. Nag, B. R. Electron transport in compound semiconductors. (Springer series in solid-state sciences ; v. 11) Bibliography : p. Includes index. 1. Compound semiconductors. 2. Electron transport. I. Title. II. Series. QC611.8.C64N34 537.6'22 79-26983

Softcover reprint of the hardcover 1st edition 1980

2153/3130-5 4 3 2 1 0

To Mriduchandra,
Biswadeep, and Mridula

Preface

Discovery of new transport phenomena and invention of electron devices through exploitation of these phenomena have caused a great deal of interest in the properties of compound semiconductors in recent years. Extensive research has been devoted to the accumulation of experimental results, particularly about the artificially synthesised compounds. Significant advances have also been made in the improvement of the related theory so that the values of the various transport coefficients may be calculated with sufficient accuracy by taking into account all the complexities of energy band structure and electron scattering mechanisms. Knowledge about these developments may, however, be gathered only from original research contributions, scattered in scientific journals and conference proceedings. Review articles have been published from time to time, but they deal with one particular material or a particular phenomenon and are written at an advanced level. Available text books on semiconductor physics, do not cover the subject in any detail since many of them were written decades ago. There is, therefore, a definite need for a book, giving a comprehensive account of electron transport in compound semiconductors and covering the introductory material as well as the current work. The present book is an attempt to fill this gap in the literature.

The first chapter briefly reviews the history of the developement of compound semiconductors and their applications. It is also an introduction to the contents of the book. The fundamentals of crystal structure, energy band structure and electron statistics are discussed in Chaps.2-5. Care has been taken to include all the recent developments when presenting these introductory subjects. The $\underline{k}\cdot\underline{p}$ perturbation theory and the statistics related to nonparabolic band structure and quantizing magnetic fields have been discussed at great length.

The scattering theory and the Boltzmann equation are presented, respectively, in Chaps.6,7. Special attention has been given to those methods which were developed in the past few years for calculating the transport coefficients including the effects of the nonparabolicity of the band structure and the polar character of the materials. Low-field DC transport coefficients, i.e., drift mobility, Hall mobility, magnetoresistance, diffusivity, thermoelectric power and electronic thermal conductivity are covered in Chap.8. A major part of the discussion in this chapter is based on the author's own research and is not available in the literature. Chapter 9 deals with the AC transport coefficients. Low-frequency AC conductivity, AC magnetoconductivity, and the phenomena of cyclotron resonance, Faraday rotation and free-carrier absorption are all discussed in detail.

Electron transport in quantizing magnetising fields has been a subject of special interest in relation to compound semiconductors. Phenomena associated with this aspect of transport are presented in Chap.10. Hot-electron transport is discussed in Chap.11. The study of this aspect of transport has been particularly fruitful for the realisation of new devices and it is a subject of continued interest. The discussion in this chapter has been aimed at acquainting the reader with early research and at the same time making him aware of current trends.

Chapter 12 gives a review of the properties of some important compound semiconductors. Chapter 13 is the concluding chapter and it deals with problems of current interest.

The book has been written at an introductory level giving all the mathematical details and computer programs to suit the needs of students and newcomers in the field. Enough current information on the properties and physical constants of all kinds of semiconducting compounds and up-to-date surveys of research results are also included to make the book useful for specialists. It is hoped that the book will be of use to all physicists and engineers who have a broad interest in the properties and applications of compound semiconductors and that it will stimulate further research in the area.

Calcutta, February 1980 *Biswaranjan Nag*

Acknowledgments

It is almost impossible to name and acknowledge individually the contributions of all the colleagues and students whose discussions and exchange of ideas helped the author in writing the book. He would, however, like to record his indebtedness to P.N. Robson, P.N. Butcher, R.A. Stradling, J.C. Woolley, J. Frey, P.J. Price, A.K. Das, P.K. Basu, A.N. Chakravarti, D. Mukherji and P. Fentem with whom he had discussed parts of the book at different times. He is particularly indebted to D. Chattopadhyay for carefully reading the manuscript, suggesting improvements, and sharing with the author the labour of preparing the final version. He is also indebted to A. Choudhury, N. Purkait, S. Dhar, S. Sen, N.R. Saha, D. Roy Choudhury, P.C. Rakshit, G. Ghosh, D. Das Gupta and a few other colleagues for their help in the preparation of the book. Thanks are also due to A.K. Chakrabarty, A. Saha and A.M. Banerjee for carefully typing the manuscript and for drawing the figures.

Some figures presented in the book are reproduced from earlier publications and the original sources are mentioned in each case. The author takes this opportunity to acknowledge the generosity of the authors and publishers for permitting him to use this material.

The book was partly written when the author was a Jawaharlal Nehru Fellow and he takes this opportunity to record his appreciation of the support provided by the Jawaharlal Nehru Memorial Fund for this work.

Biswaranjan Nag

Contents

1. Introduction

1.1 Historical Note

Electron devices using the rectifying properties of compound semiconductors have been utilised from the very early stages of the development of electromagnetic waves as a means of communication. Natural crystals of galena (lead sulphide) were used in point-contact detectors as early as in 1904 [1.1]. However, as the technology of preparation and purification of semiconductors advanced, interest was shifted to the elemental semiconductors, silicon and germanium. Subsequent point-contact detectors, which continued to be used as microwave mixers and detectors even after the advent of vacuum tubes, were mostly made with silicon. Interest in elemental semiconductors was also greatly enhanced when transistor action was discovered in germanium in 1948 [1.2]. Silicon soon replaced germanium in transistors because of its greater thermal stability. As a result of the invention of the transistor, germanium and silicon became the two most intensely studied semiconductors in course of the next few years.

Work on compound semiconductors, however, was continued as there were applications for some of these materials in infrared detectors and in thermoelectric generators and as phosphors and photoconductors. But, the development of the transistor and the subsequent growth of its use in electronic appliances had also inaugurated an era of intensive study of the properties of semiconductors in general. Side by side with the detailed study of elemental semiconductors, work was also continued on the growth and purification of crystals of other well-known as well as less-well-known compound semiconductors. Around 1950 [1.3], due to these efforts, the properties of a new class of synthetic semiconductors, the so-called III-V compounds, became known. It had been pointed out in 1926 that elements of group IIIb and Vb when melted together and resolidified should crystallise with structures similar to those of germanium and silicon [1.4]. Although the prediction of their crystal structure was confirmed in 1929 [1.5] and the semiconducting property of InSb was reported in 1950, it was WELKER's [1.6] work which drew particular attention to the importance of these semiconductors by discussing their special properties, for example, the high value of electron mobility and the low value of the energy band gap. Subsequent research on this new class of semiconductors has vastly extended the field of application of compound semiconductors.

The developments in III-V compounds have also ushered in a new era of materials research. It was soon established that compound semiconductors may be made not only by combining one element of group III with one of group V, but also by combining elements of other groups or by making solid solutions of different compounds. Some of the mixed compounds have definite band structure and behave exactly like the elemental or binary semiconductors. Due to the development of III-V compounds, semiconductors with many different band structures were available. Even then, the diversity of the structures is limited by the inherent characteristics of the compounds, which cannot be controlled. With mixed compounds [1.7], however, it is possible to prepare a semiconductor almost with a specified band structure. The entire band structure in between the two limits determined by the two compounds may be obtained by adjusting the proportion of the individual compounds. This particular characteristic of mixed compounds has greatly increased the potential of compound semiconductors, and only the future can tell how much of this potential will be put to good use.

The potential of compound semiconductors has been further extended, very recently, by the introduction of the idea of semiconductor superlattices [1.8]. It has been suggested that when two semiconductors are grown in alternate layers of thicknesses comparable with the atomic spacing, the band structure will have microstructures which may be controlled by choosing the individual semiconductors and by varying the layer thicknesses. The predicted properties of these materials are very interesting. We may hope that radically new kinds of devices will be invented when the technology of preparation of these materials is fully developed. Mixed compounds have provided us with a means of tailoring the band structure by varying the composition. Superlattice semiconductors will make possible the alteration of the band structure by partially controlling the crystal structure.

1.2 Applications

The full range of possibilities of compound semiconductors have not yet been exploited in pratical electronic devices. But, even at present, compound semiconductors have many applications, which make them no less important

Table 1.1. Applications of compound semiconductors

Device	Commonly used materials	Reasons for the choice
Tunnel diode	GaAs	Tunnelling efficiency
Laser (injection and beam pumped)	GaAs, $GaAs_x P_{1-x}$, ZnS ZnO, CdS, CdTe, CdSe, $CdS_x Se_{1-x}$	Direct nature and specific values of energy band gap
Light emitting diodes (LED's)	GaP, $GaAs_{1-x} P_x$	Values of energy band gap, quantum efficiency
Infrared detectors	InSb, $Hg_{1-x} Cd_x Te$ PbS, PbSe, $Pb_x Sn_{1-x} Te$	Values of energy band gap, noise characteristics
Galvanomagnetic devices	InSb, InAs, HgTe, HgSe	High value of electron mobility
Thermoelectric devices	Bi_2Te_3, PbS, $InAs_x P_{1-x}$ $GaAs_x Sb_{1-x}$	High values of thermoelectric coefficients and allowable operating temperature
Gunn-effect devices [1.11]	GaAs, InP, CdTe	Specific band structure
Acoustoelectric devices	CdS, ZnO	Strong piezoelectric interaction

than elemental semiconductors. We briefly review these applications in this section.

Compound semiconductors are not widely used in the construction of diodes or transistors, although it is possible to use them [1.9,10], and for extreme ambient conditions of temperature, devices made with them may be more convenient. Only gallium arsenide has been considered a serious competitor of silicon for applications in some types of transistors and diodes. The relatively sparse use of compound semiconductors in the common junction devices is partly due to the nonavailability of simple and routine technology for the growth of the single crystals and for the processing of these materials. In spite of the rapid increase in the use of gallium arsenide in recent years there are very few sources of good crystals, and consequently, there is very little possibility of using compound semiconductors in the major section of the semiconductor industry, unless radically new methods are developed for the growth of single crystals, or unless devices utilising these materials are shown to have some definite advantages.

However, compound semiconductors are widely used in various other semiconductor devices, which are listed in Table 1.1. Also shown are the compounds used and the properties which make them suitable for a particular device.

The utility of the compound semiconductors in the various devices listed above is due to their special properties. These properties are the direct and controllable energy band gap, the high value of the electron mobility and of the thermoelectric coefficients, the structure of the energy bands required for the Gunn effect, and a strong piezoelectric interaction.

In materials with a direct energy band gap, the tunnelling process and the absorption and emission of light occur efficiently, as there is no change of the crystal momentum in the electron transitions, and phonons are not involved. Tunnel diodes, injection lasers, and light-emitting diodes (often abbreviated to LED's) are therefore made with compound semiconductors having a direct energy band gap. An additional criterion for the selection of materials for lasers and LED's is the value of the band gap, which determines the wavelength of emission. The variability of the energy band gap in compound semiconductors and in mixed compounds has made it possible to construct these devices for particular wavelengths.

An essential requirement for the detection of infrared radiation is that the energy band gap of the material used for detection should be less than the quantum energy of the radiation to be detected. Also, the efficiency is a maximum if the energy band gap is only slightly smaller than the threshold value for absorption. Another requirement is that the energy band gap should be direct so that the absorption coefficient for the radiation may be high. As the interest is mostly in wavelengths larger than 2-3 micrometres materials with values of energy band gap lower than 0.5 eV are required. Only some particular compounds satisfy these requirements.

The major consideration for the selection of materials for galvanomagnetic [1.12] and thermoelectric [1.13] devices is the sensitivity. A high value of sensitivity is obtained if the relevant transport coefficient has a large value. As mentioned earlier, the electron mobility in some compound semiconductors has a high value and this property is the basis of the choice of these compounds for use in Hall-effect and magnetoresistance devices for magnetic-field meters, for analog multipliers and for power meters and amplifiers. The use of some other compounds in thermoelectric devices is similarly due to the high value of their thermoelectric coefficients.

In the selection of materials for Gunn-effect devices [1.11], the band structure is the deciding factor. The effect is observed in materials with a special band structure. It is required that the electrons should occupy at low fields low effective-mass minima. But, when a high electric field is applied, these electrons should be transferred through the hot-electron effect [1.14] to higher effective-mass valleys. The Gunn effect has been observed in many semiconductors, including CdTe, InSb, $Ga_x In_{1-x}Sb$, InAs, germanium, and silicon [1.15]. But, in many of these materials, the effect is observed only for particular orientations of the electric field and for particular conditions of stress and temperature. Even when the effect is produced, the instabilities are weak. Only in some compound semiconductors is the instability strong and is it produced at room temperature with the material unstressed. These materials are therefore the natural choice for the construction of Gunn-effect devices.

The piezoelectric interaction in semiconductors has been used to realize various devices useful for signal processing [1.16]. Suitability of a particular material for these devices is mainly determined by the electromechanical coupling factor, which is again related to the crystal structure. Electrons are strongly coupled with lattice vibrations in semiconductors that have an asymmetric crystal structure. Since II-VI compounds have hexagonal crystal structure, these are most suitable for use in acoustoelectric devices, in which acoustic signals are amplified and processed using their interaction with accelerated electrons. Most favoured among these compounds is CdS, which has a strong acoustoelectric coupling and at the same time may be grown easily.

Table 1.1 shows that the field of application of compound semiconductors is fairly wide. In some areas, the peculiarities of their properties make them the only usable materials. With the rapidly increasing applications of electronics in various fields, compound semiconductors have, therefore, gained considerable technological importance, and scientific curiosity in these materials is rapidly being transformed into commercial interest.

1.3 Transport Coefficients of Interest

It is evident from the discussion of the previous section that the applications of compound semiconductors are based on transport properties. For the construction of diodes and transistors the properties of interest are mobility, diffusivity, minority carrier lifetime, and surface effects. Other transport properties are used as diagnostic tools for the development of good materials. But, for the full range of applications of compound semiconductors, detailed knowledge of all transport coefficients including those mentioned above is necessary, as devices exploiting all these properties directly are used.

The properties of interest for devices involving electron-light interaction, e.g. laser diodes, LED's, and infrared detectors, are related to interband transitions as well as to free-carrier absorption of infrared radiation or of light. The diffusion and recombination processes are also of interest as the detailed characteristics of the devices also depend on the behaviour of the carriers introduced by injection or generated by excitation.

Galvanomagnetic or thermoelectric devices are based on the bulk transport coefficients of materials. For these devices, a detailed knowledge of the behaviour of the majority carriers in bulk and often homogeneous materials is necessary. The properties of interest are the electron mobility, Hall coefficient, magnetoresistance coefficient, and the thermoelectric transport coefficients.

Gunn-effect devices rely on the transport characteristics in the presence of a high electric field. In the presence of very high fields, the carriers do not remain in thermal equilibrium with the lattice and the relative importance of the various scattering mechanisms is altered. The energy distribution function for the carriers is also radically changed so that for an understanding of the Gunn-effect devices acquaintance with altogether new aspects of transport theory is needed. The low-field theory based on the equilibrium distribution function, which is quite sufficient for the devices used for so long, is of little help.

In some of the devices, e.g. tunable Raman lasers [1.17], the operation is controlled by a magnetic field, high enough to produce quantisation of the energy bands. Hence, transport properties of compound semiconductors in high magnetic fields are also of practical interest. Such properties, in addition, have been of immense help in the elucidation of the energy band structure of these materials.

We conclude that a number of transport properties are of interest for the development of devices using compound semiconductors, namely, (i) drift mobility, (ii) Hall mobility, (iii) magnetoresistance, (iv) diffusivity, (v) thermoelectric coefficients, (vi) free-carrier absorption, (vii) interband excitation, and (viii) recombination. We are interested in these properties both for direct and alternating currents and also for low as well as high values of both electric and magnetic fields. This book is concerned with all these properties. Only interband transition characteristics and recombination mechanisms are excluded as these may best be dealt with in a book solely devoted to the subject [1.18].

1.4 Scope of the Book

The transport theory for compound semiconductors is based on the same principles as the theory for elemental semiconductors. It relies on the effective mass approximation, Bloch wave functions, first- and second-order perturbation theories, and the Boltzmann transport equation. Hence the theory may be considered to be very much a classical one, and is discussed in many available text books. However, there are some important differences between elemental and compound semiconductors which make the details of the transport theory for compound semiconductions somewhat involved. The complexities arise from the low values of the energy band gaps and of the effective masses, and from the ionicity of these materials.

Due to the low band gap, the simplifying plane-wave approximation for the electron wave functions is mostly inapplicable for the experimental conditions. More complicated wave functions have to be considered and the nonparabolicity of the energy-crystal-momentum relations [1.19] has to be taken into account to obtain results even to a first-order approximation. The theory is thus already complicated for compound semiconductors with simple, single-valley, isotropic band structures.

The ionicity of the materials introduces an additional complexity in the theory. Due to the partially polar nature of the materials, the polar optic phonon scattering is the predominant scattering process. The relaxation-time approximation is invalid for this scattering mechanism, and hence theoretical procedures have not been very successful in the past for the evaluation of the transport coefficients of compound semiconductors. In spite of the rapid advances in materials technology, progress with the theory has been rather slow. Only in the last few years have some new methods [1.20-22] been developed enabling the accurate computation of transport coefficients for the predominant polar optic phonon scattering. These methods are undergoing further development [1.23], but the progress achieved so far is worthy of consideration.

In most of the commonly used diodes and transistors, the carrier concentration is such that nondegenerate statistics may be employed for the analysis. However, in devices using compound semiconductors the doping level is often very high. Since the density of states is low due to the low value of the effective mass of the electrons, the materials easily become degenerate; therefore, degenerate statistics should be used in the theory. Progress in this direction has been made only recently [1.24].

This book is concerned with these recent developments in the theory of electron transport in compound semiconductors. The theory related to the evaluation of the various transport coefficients is discussed. In particular, drift mobility, Hall mobility, magnetoresistance, diffusion, Faraday rotation, hot-electron transport, and quantum-magnetic-field effects receive detailed attention.

Along with the theory, experimental results are also discussed. But, it should be evident from Table 1.1 that in spite of the availability of a large number of compound semiconductors, only a few of them are useful in practice. Of these, it is mostly the n-type material which is commonly utilised in devices, and so the detailed discussion has been confined to n-type materials.

2. Crystal Structure

The transport characteristics of a material are intimately related to the energy band structure and the nature of the collision processes. For the development of the theory of electron transport, therefore, the form of the energy bands and the scattering mechanisms are required to be known in detail. The characteristic features of the energy bands are, however, intimately related to the crystal structure. The common forms of the crystals of compound semiconductors are described in this chapter.

The useful and well-known compound semiconductor crystals have one of the three structures: that of zinc blende (also known as sphalerite), the wurtzite (sometimes called zincite), or rock salt (also called sodium chloride). We describe these structures and the associated nomenclature below. We shall also describe briefly the chalcopyrite crystal structure, as a class of compound semiconductors having this structure is receiving intense attention at present [2.1]. For other possible structures, the reader may consult the references cited at the end of this chapter.

2.1 Zinc-Blende Structure

An atom of one kind constituting the binary compound is surrounded in zinc-blende structure by four equidistant atoms of the second kind, which occupy the apexes of a tetrahedron, the atom of the first kind being at its centre. The orientation of the neighbouring atoms is such that the atoms also occupy the sites of two interpenetrating face-centred-cubic (fcc) lattices. Each cubic lattice is formed by atoms of the same kind. The two cubic lattices are oriented parallel to each other and the corner of one cube is placed on the body diagonal of the other cube at a distance of a quarter of the length of the diagonal. Figure 2.1 shows a face-centred-cubic lattice, and Fig.2.2 shows the sphalerite structure. Atoms connected tetrahedrally are also indicated in the figure. It is also evident that an atom has 12 atoms of the same kind as its second nearest neighbours.

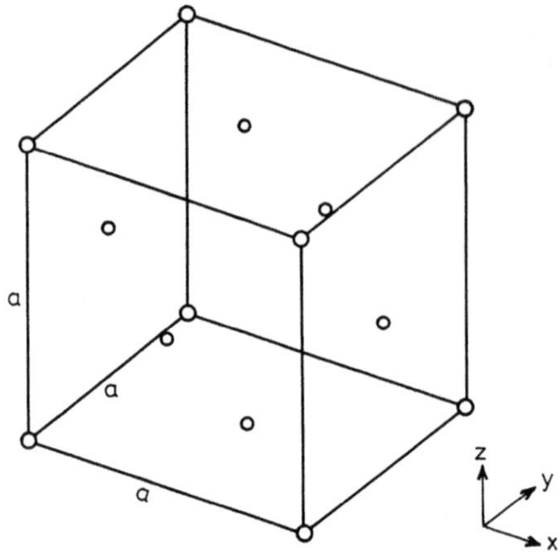

Fig. 2.1. Face-centered-cubic lattice

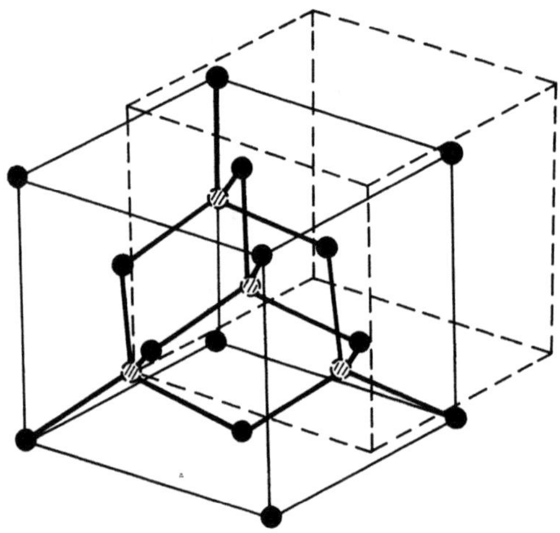

Fig. 2.2. Sphalerite crystal structure

The space lattice for the sphalerite crystal structure may be considered as an fcc lattice with a basis of two atoms of the two kinds [Ref.2.2, p.5]. The lattice constant is the length of the edge of the cubic lattice and the basis vector, the vector connecting the two atoms forming the basis, is (a/4,a/4,a/4), a being the lattice constant.

The primitive translation vectors for the fcc lattice are usually taken as

$$\underline{a}_1 = (a/2)(\hat{\underline{x}}+\hat{\underline{y}}) \quad , \quad \underline{a}_2 = (a/2)(\hat{\underline{y}}+\hat{\underline{z}}) \quad , \quad \underline{a}_3 = (a/2)(\hat{\underline{z}}+\hat{\underline{x}}) \quad , \tag{2.1}$$

where $\hat{\underline{x}}$, $\hat{\underline{y}}$, $\hat{\underline{z}}$ are the unit vectors along the directions of the Cartesian coordinates, which are chosen in the directions of the cube edges, with the origin at the corner of a cube.

The primitve unit cell shown in Fig.2.3 is trigonal and has a volume of $a^3/4$. The eight atoms at the corners of the cube are shared with the neighbouring eight cubes, and the six atoms on the faces are shared between two neighbouring cubes. The total number of atoms which can be associated with a cubic cell is thus $8/8+6/2=4$. Each primitive trigonal unit cell has, therefore, one atom. Considering the two interpenetrating lattices formed by the two kinds of atoms, we find that the total number of atoms in a primitive unit cell is two. The number of atoms per unit volume of the crystal is thus $8/a^3$.

The basis vectors $\underline{b}_i$ of the reciprocal lattice are obtained from the definition [2.3,4]

$$\underline{a}_i \cdot \underline{b}_j = 2\pi\delta_{ij} \quad , \tag{2.2}$$

where δ_{ij} is the Kronecker delta function.

These vectors for the fcc cubic lattice are given by

$$\underline{b}_1 = (2\pi/a)(\hat{\underline{x}}+\hat{\underline{y}}-\hat{\underline{z}}) \quad , \tag{2.3a}$$

$$\underline{b}_2 = (2\pi/a)(\hat{\underline{y}}+\hat{\underline{z}}-\hat{\underline{x}}) \quad , \tag{2.3b}$$

$$\underline{b}_3 = (2\pi/a)(\hat{\underline{z}}+\hat{\underline{x}}-\hat{\underline{y}}) \quad . \tag{2.3c}$$

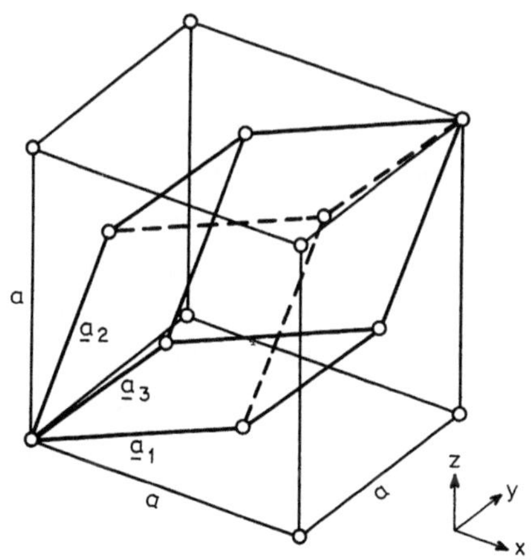

Fig. 2.3. Primitive unit cell for an fcc lattice

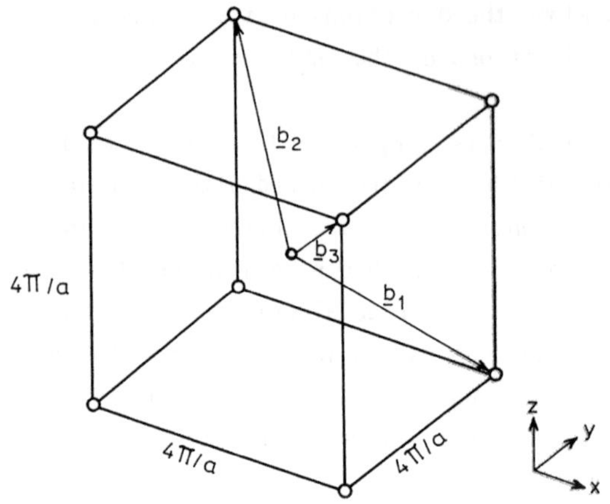

Fig. 2.4. Reciprocal lattice for the fcc structure

The reciprocal lattice vectors and the reciprocal lattice are shown in Fig.2.4. The x, y, z directions are chosen parallel to the cube edge and the origin is at the centre of the cube. It is seen that the reciprocal lattice is a body-centred-cubic lattice with a primitive cell of volume $4(2\pi/a)^3$.

It is common practice to indicate the planes for this lattice by using Miller indices [Ref.2.3, p.23]. A direction is indicated as $[\ell mn]$, whereas a class of directions having the same symmetry is indicated as $\langle \ell mn \rangle$; ℓ, m, and n are the smallest integers having the same ratio as the components of a vector in the same direction. A particular plane normal to this direction is written as (ℓmn) and a class of crystallographically equivalent planes is denoted by $\{\ell mn\}$. A negative index is written with a bar on top. In Fig.2.5 some of the symmetry directions and planes are shown.

The arsenides, antimonides, and phosphides of aluminium, indium, and gallium among the III-V compounds and sulphides, selenides, and tellurides of cadmium, zinc, and mercury among the II-VI compounds have the zinc-blende crystal structure. It should be noted that silicon and germanium have essentially the same kind of crystal structure, but the two atoms forming the basis being similar, their structure, the so-called diamond structure, has inversion symmetry, that is the atoms are symmetrically located about a point midway between two atoms forming a basis. On the other hand, in the sphalerite structure, the two basis atoms being of different kinds, there is no inversion symmetry [2.5]. The basic similarity of the diamond and sphalerite crystal structures gives rise to some similarities in the energy band levels of the materials with these structures. The similarity is so strong that the band structure for III-V compounds with the sphalerite structure may be qua-

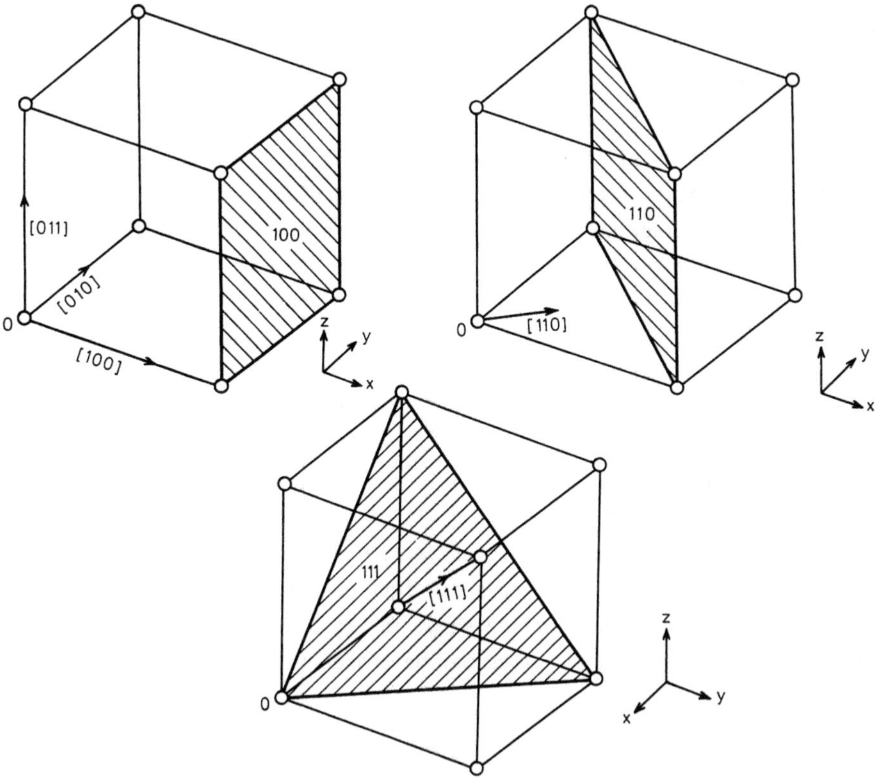

Fig. 2.5. Symmetry directions and symmetry planes for cubic lattice

litatively derived from the band structure of group IV elements by establishing a correspondence between the two kinds of materials [2.6]. However, the lack of inversion symmetry in the sphalerite structure also leads to some peculiarities. These are mostly noticeable in the structure of the valence band and the position of conduction band minima. Further, since the atoms along the <111> axes are of different kinds, and since the atoms have usually some ionicity, they form dipole pairs. As the length of the dipoles is smaller than the spacing between them, materials with the sphalerite structure are polar along the <111> axes. These materials, in contrast to those having the diamond structure, are consequently piezoelectric along the polar axis. The piezoelectric tensor is, however, of the simplest form and has only one independent component [2.7].

2.2 Wurtzite Structure

The basic arrangement of the atoms in the wurtzite structure is similar to that in the sphalerite structure. An atom of one kind is surrounded tetrahedrally by four atoms of the second kind. But the tetrahedrons are so oriented that the location of the atoms fit two interpenetrating close-packed hexagonal lattices. The two lattices have the same axis, but one of them is displaced with respect to the other. The wurtzite structure may, therefore, be considered as the close-packed hexagonal structure with a basis of two atoms. The close-packed hexagonal and the wurtzite structure are shown respectively in Figs.2.6,7. The basis vector is oriented along the axis of the hexagons and has a length of 3/8 times the separation c between the two hexagonal faces.

The close-packed hexagonal structure is not a space lattice. The lattice is the hexagonal space lattice with a basis of two atoms, which are located at (0,0,0) and at the tip of the vector $2\underline{a}_1/3 + \underline{a}_2/3 + \underline{c}/2$, the $\underline{a}_1$, $\underline{a}_2$, and $\underline{c}$ vectors being shown in Fig.2.6. The space lattice for the wurtzite structure is, therefore, the hexagonal lattice shown in Fig.2.8.

The primitive translation vectors may be chosen as

$$\underline{a}_1 = (a/2)(\hat{\underline{x}}-\sqrt{3}\,\hat{\underline{y}}) \quad , \tag{2.4a}$$

$$\underline{a}_2 = (a/2)(\hat{\underline{x}}+\sqrt{3}\,\hat{\underline{y}}) \quad , \tag{2.4b}$$

$$\underline{c} = c\hat{\underline{z}} \quad . \tag{2.4c}$$

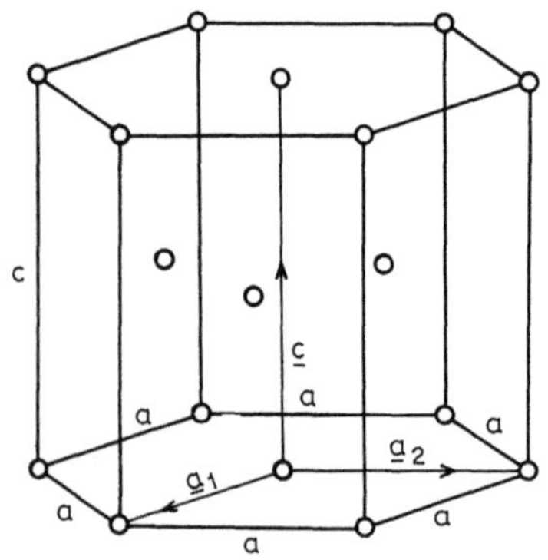

Fig. 2.6. Hexagonal close-packed crystal structure

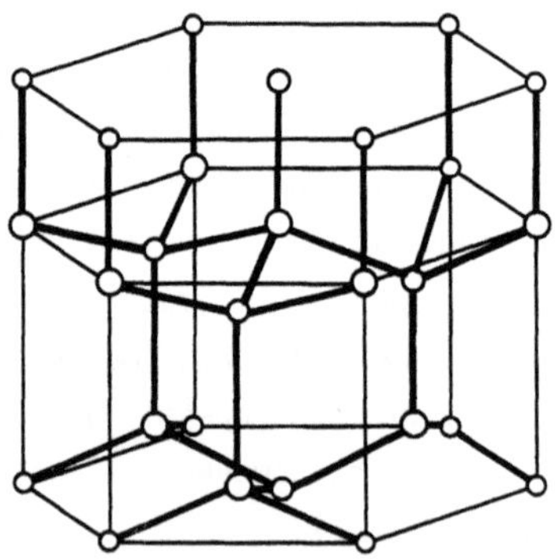

Fig. 2.7. Wurtzite crystal structure

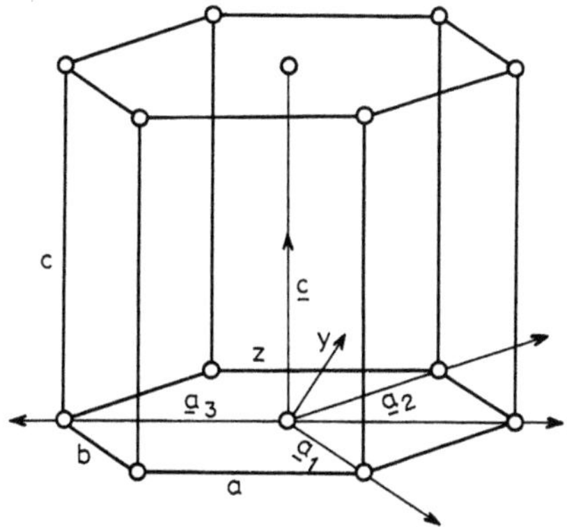

Fig. 2.8. Hexagonal lattice and primitive translation vectors

The reference coordinate system is chosen as Cartesian with $\hat{\underline{x}}$, $\hat{\underline{y}}$, $\hat{\underline{z}}$ vectors as shown in Fig.2.8, the origin being chosen at the centre of the hexagonal face. The length of the sides of the hexagonal bases is a. It should be noted that there are two lattice constants a and c for the wurtzite structure. The ratio of c to a is ideally $(8/3)^{\frac{1}{2}} = 1.63$.

The atoms at the corners of the cell formed by the primitive translation vectors are shared with the neighbouring eight cells. Hence, each such cell in the hexagonal lattice has one atom; the corresponding number for the close-packed hexagonal structure is two, and that for the wurtzite structure is four. The volume of a cell is $\sqrt{3}a^2c/2$, and the number of atoms per unit volume for the wurtzite structure is therefore $8(\sqrt{3}a^2c)^{-1}$.

As mentioned earlier, the atoms are arranged with tetrahedral symmetry both in wurtzite and sphalerite structures. The environment of an atom in the two structures is thus similar. As in the sphalerite structure, an atom in the wurtzite structure has four atoms of the other kind as nearest neighbours and twelve atoms of the same kind as second nearest neighbours. In the complete structure, the atomic arrangement along the c axis (the axis of the hexagon) in the wurtzite structure is similar to the arrangement along the <111> axes in the sphalerite structure. Photographs of the crystal models shown in Fig.2.9 and 2.10 illustrate this similarity. Crystals with wurtzite structure are therefore polar along the c axis. However, the wurtzite structure has uniaxial symmetry. Hence, in addition to being piezoelectric, crystals with wurtzite structure are also pyroelectric [2.8]. The piezoelectric tensor also has a number of independent components [2.9].

The basis vectors for the reciprocal lattice of the hexagonal space lattice are given by

Fig. 2.9

Fig. 2.10

Fig. 2.9. Photograph of sphalerite structure along [111] direction

Fig. 2.10. Photograph of wurtzite structure along [0001] direction

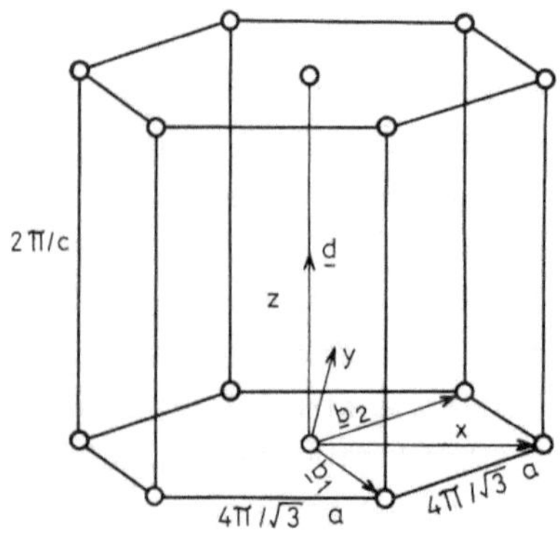

Fig. 2.11. Reciprocal lattice for hexagonal direct lattice

$$\underline{b}_1 = (2\pi/a)(\hat{\underline{x}}-\hat{\underline{y}}/\sqrt{3}) \quad , \tag{2.5a}$$

$$\underline{b}_2 = (2\pi/a)(\hat{\underline{x}}+\hat{\underline{y}}/\sqrt{3}) \quad , \tag{2.5b}$$

$$\underline{d} = (2\pi/c)\hat{\underline{z}} \quad . \tag{2.5c}$$

The reciprocal lattice is also hexagonal, as shown in Fig.2.11.

The planes and directions for the hexagonal lattice are often denoted by hexagonal indices in which $\underline{a}_1$, $\underline{a}_2$, another redundant direction $\underline{a}_3$, and $\underline{c}$ are taken as references [2.10]. These vectors are shown in Fig.2.8. In this notation, the c axis is indicated by [0001].

Among the III-V compounds, the relatively unknown crystals of nitrides of aluminium, indium, and gallium have wurtzite structure. Most of the II-VI compounds have both a sphalerite and a wurtzite modification. The lattice spacing in the two forms is very nearly the same, but the properties, particularly the piezoelectric and pyroelectric properties, are different. Due to the similarity between the sphalerite and the wurtzite structure, energy band structure of materials with these two kinds of structures is similar. The uni axial nature of the wurtzite structure, however, introduces some finer dissimilarities, which appear in the shape of the constant energy surfaces for both the conduction and the valence bands and in the degeneracy of the valence band.

2.3 Rock-Salt Structure

The arrangement of atoms (shown in Fig.2.12) in the rock-salt structure is such that atoms of the two kinds occupy alternate positions on a face-centred cubic lattice. It may also be considered as two interpenetrating parallel face-centred lattices, the corner of one of which is positioned on the body diagonal of the other at a distance of half a diagonal length. The crystal structure may therefore be described also as face-centred-cubic lattice with a basis of two atoms of two kinds, the basis vector being (a/2,a/2,a/2). The space lattice is thus the face-centred-cubic lattice; the lattice constant

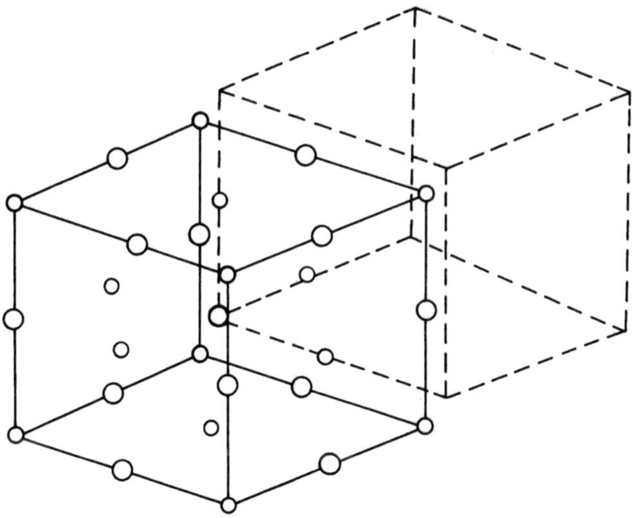

Fig. 2.12. Rock-salt crystal structure

is a. The edge of the cube and the reference axes are taken parallel to the cube faces, with the origin at a corner. The primitive translation vectors, reciprocal lattice, the number of atoms per unit cell, and the nomenclature for the rock-salt structure are therefore identical to those for the sphalerite structure. However, due to the difference in the lengths of the basis vectors in the two systems, there are some important differences in their properties. Firstly, there is inversion symmetry in the rock-salt structure, although the lattice is formed by atoms of two kinds, which are often ionic. As mentioned earlier, sphalerite structure lacks inversion symmetry. Secondly, an atom in the rock-salt structure is surrounded by six atoms and has twelve atoms as next nearest neighbours. The corresponding figures for sphalerite structure are four and twelve.

The important compound semiconductors having the rock-salt structure are the sulphides, selenides, and tellurides of lead. The band structure of these materials have broad similarity with those of the sphalerite and wurtzite structure, but, due to the differences mentioned above, there are also significant differences, the most important being the position of the valence band maxima, and the shape of the associated constant energy surfaces.

2.4 Chalcopyrite Structure

The crystal structures described above are relatively simple as they may be derived from the cubic and hexagonal structure with a basis of two atoms. The crystal structure of the mixed compounds consisting of group II, IV, and V elements and group I, III, and VI elements are, however, relatively complex. Both these groups of compounds crystallise with the structure of chalco-

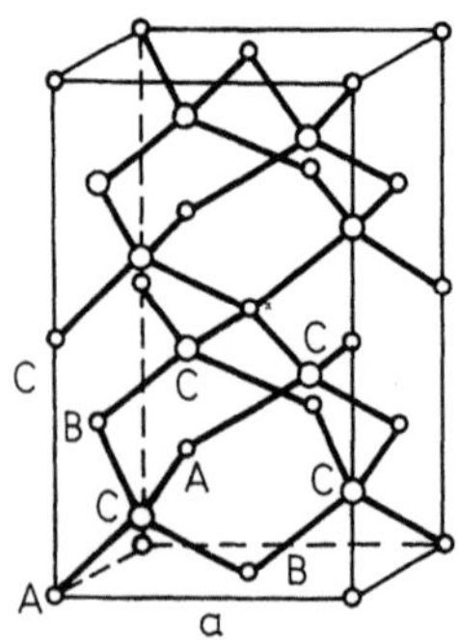

Fig. 2.13. Chalcopyrite crystal structure

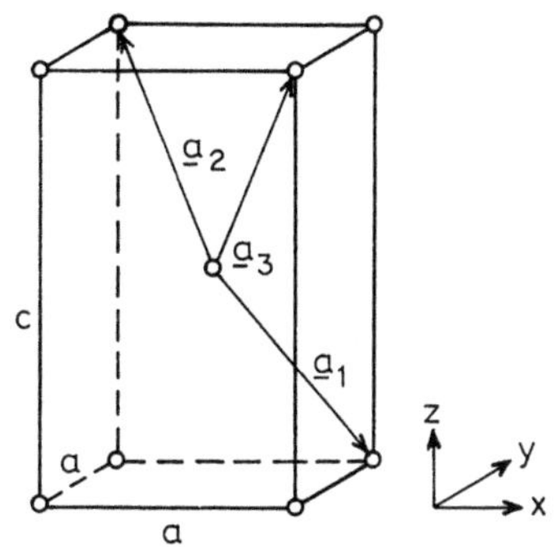

Fig. 2.14

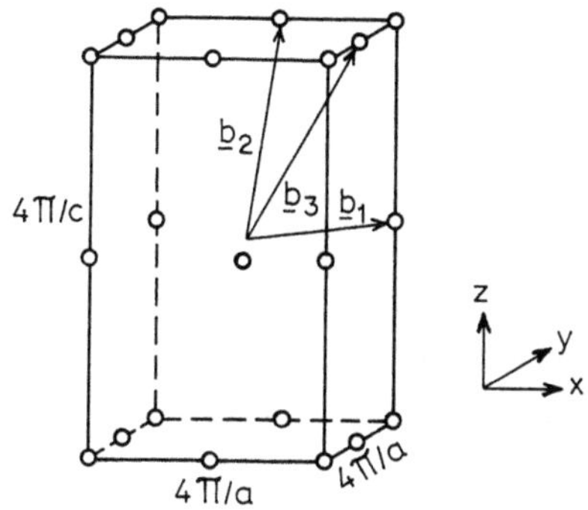

Fig. 2.15

Fig. 2.14. Body-centred tetragonal lattice

Fig. 2.15. Reciprocal lattice of body-centred tetragonal direct lattice

pyrite (the mineral $CuFeS_2$). The arrangement of the atoms in this structure is illustrated in Fig.2.13. The lattice is body-centred tetragonal (Fig.2.14) and the structure has a basis of eight atoms (two A, two B, and four C atoms for the ABC_2 compound) with the following coordinates for the A, B, and C atoms:

A atoms at (0,0,0) and (a/2,0,c/4),
B atoms at (a/2,a/2,0) and (0,a/2,c/4),
C atoms at (xa,a/4,c/8), (a-xa,3a/4,c/8), (xa,3a/4,3c/8) and (a-xa,a/4,3c/8).
Here a and c are respectively the length of the base and the height of the tetragon. The value of x is slightly larger than 0.25 (the value for the zinc-blende structure) and ranges between 0.25 and 0.30.

The primitive translation vectors for the lattice may be chosen as

$$\underline{a}_1 = (a/2)\hat{\underline{x}} + (a/2)\hat{\underline{y}} - (c/2)\hat{\underline{z}} \quad , \tag{2.6a}$$

$$\underline{a}_2 = -(a/2)\hat{\underline{x}} + (a/2)\hat{\underline{y}} + (c/2)\hat{\underline{z}} \quad , \tag{2.6b}$$

$$\underline{a}_3 = (a/2)\hat{\underline{x}} - (a/2)\hat{\underline{y}} + (c/2)\hat{\underline{z}} \quad . \tag{2.6c}$$

The volume of the unit cell is $a^2c/2$ and the number of atoms per unit cell is 8.

The reciprocal lattice vectors are

$$\underline{b}_1 = (2\pi/a)(\hat{\underline{x}}+\hat{\underline{y}}) \quad , \tag{2.7a}$$

$$\underline{b}_2 = 2\pi(\hat{\underline{y}}/a+\hat{\underline{z}}/c) \quad , \tag{2.7b}$$

$$\underline{b}_3 = 2\pi(\hat{\underline{x}}/a+\hat{\underline{z}}/c) \quad . \tag{2.7c}$$

The reciprocal lattice (Fig.2.15) is thus face-centred tetragonal with a square base of sides $4\pi/a$ long and with height $4\pi/c$.

When the difference between A and B atoms are ignored and c is equal to 2a, the chalcopyrite structure is the same as the zinc-blende structure, i.e. two interpenetrating face-centred-cubic lattices. However, c is not usually equal to 2a and x is different from 0.25. The bonding is, however, tetrahedral and perfectly tetrahedral between B and C atoms. Hence, there is similarity between the properties of zinc-blende and chalcopyrite semiconductors. The properties of chalcopyrite materials may be understood in terms of the properties of zinc-blende materials provided the effects arising from the following differences are kept in mind. In chalcopyrite crystals atoms are more compressed in the c direction; the spacing between C and A atoms is larger than between C and B atoms; and the unit cell is doubled in the c direction. We shall see later that these differences cause some significant differences between the band structure of chalcopyrite materials and that of other compound semiconductors.

We have discussed above the main crystal structures in which most of the important compound semiconductors crystallise. There are, however, some other structures like the cinnabar, rhombohedral, or multilayer hexagonal structures which are relevant to the study of some compound semiconductors. However, this book is concerned only with the compound semiconductors for which the knowledge of the crystal structures discussed above would be sufficient.

3. Energy Band Structure

The compound semiconductors crystallize mostly in one of the four crystal structures, described in the preceding chapter. The forms of the energy bands, being determined by the crystal structure, are also basically of four types and these are described in this chapter. Information relating to the energy band structure is usually presented by plotting the energy of the electron, E, for values of the wave vector, $\underline{k}$, limited to within the first Brillouin zone. We first introduce the concept of the Brillouin zone and describe its shape for the different crystal structures. The salient features of the energy bands for each structure are then discussed. It should be borne in mind that we are concerned with the extrema of energy bands when dealing with problems of transport. Hence although the detailed energy band diagrams are given for some of the important compounds, the discussion is mostly confined to the E-$\underline{k}$ relations for the extrema.

3.1 Electron Wave Vector and Brillouin Zone

The electrostatic potentials in a solid vary sharply with distance as the atoms are spaced at distances of the order of a few angstroms. The electrons being thereby subjected to very large forces, their motion cannot be described properly by classical mechanics, and wave mechanics must be used. The Schrödinger equation giving the time-dependent wave function ψ for the electron is written as

$$-(\hbar^2/2m_0)\nabla^2\psi + V\psi = -(\hbar/i)(\partial\psi/\partial t) \quad , \tag{3.1}$$

where V is the potential energy operator which has the periodicity of the lattice, $\hbar$ is Planck's constant divided by 2π (also called Dirac's constant), m_0 is the free electron mass, $i = \sqrt{-1}$, t is the time, and ∇^2 is the Laplacian operator. The stationary states of the electron for a time-dependent V are obtained by replacing the right hand side by $E\psi$, where E is the total energy of the electron, independent of time for stationary states. The resultant equation is then time independent. The solution of the time-independent Schrödinger equation for a periodic V has the form

$$\psi(\underline{r}) = U_{\underline{k}}(\underline{r}) \exp(i\underline{k}\cdot\underline{r}) \quad , \tag{3.2}$$

and the complete wave function is

$$\psi(\underline{r},t) = U_{\underline{k}}(\underline{r}) \exp[i(\underline{k}\cdot\underline{r}-Et/\hbar)] \quad .$$

Here, $U_{\underline{k}}(\underline{r})$ is a periodic function having the same periodicity as the lattice.

The expectation values of the momentum and energy of the electron are, in general, given by

$$\langle\underline{p}\rangle = \int_{\Omega} \psi^*(\hbar/i)\underline{\nabla}\psi d\Omega \quad , \tag{3.3}$$

and

$$\langle E\rangle = \int_{\Omega} \psi^* H\psi d\Omega \quad , \tag{3.4}$$

where Ω is the volume of the crystal, $(\hbar/i)\underline{\nabla} \equiv \underline{p}$ is the momentum operator and H is the Hamiltonian operator, being given by

$$H \equiv -(\hbar^2/2m_0)\nabla^2 + V \quad . \tag{3.5}$$

ψ^* is the complex conjugate of ψ.

The wave function ψ given by (3.2) is often referred to as the Bloch function. The state of the electron in this formalism is given by the vector $\underline{k}$, and the corresponding energy $E(\underline{k})$. The vector $\underline{k}$ is called the wave vector and $E(\underline{k})$ is the eigenvalue of the eigenfunction $\psi(E,\underline{k})$. In the theory of transport in semiconductors, the electron is treated as a classical particle with momentum $\hbar\underline{k}$ (called the crystal momentum) and mass, called the effective mass [3.1]. The effective mass is given by $(1/m_{ij}) = (1/\hbar^2 k_i)(\partial E/\partial k_j)$, where i, j are respectively the directions of the motion of the electron and of the force on it [3.2]. It is in general a tensor quantity. The only information that is required in this formalism is thus the relation between E and $\underline{k}$, which has, however, to be obtained by solving (3.2) for the particular crystal field.

It is found that the $E(\underline{k})$ are the eigenvalues for wave functions corresponding to $\underline{k}$ as well as $\underline{k}+\ell\underline{g}$, where $\underline{g}$ is a reciprocal lattice vector and ℓ is an integer. Electron states with wave vectors $\underline{k}$ and $\underline{k}+\ell\underline{g}$ cannot be distinguished in experiments and so only electrons having irreducible wave vectors need be considered. The region in $\underline{k}$ space consisting of wave vectors which cannot be further reduced by subtracting or adding multiples of a reciprocal lattice vector is called the first Brillouin zone or just the Brillouin zone. We are concerned only with values of $\underline{k}$ lying within this zone in the interpretation of experimental results on electron transport. The shape of the Brilluoin zone is, therefore, of great significance in transport theory.

The easiest way of constructing a Brillouin zone is to draw in reciprocal space planes normal to the lines joining the origin with the neighbouring lattice sites at their midpoints. It is easily visualised that the region bounded by these planes contains all the irreducible $\underline{k}$ vectors. The shape of Brillouin zone is, thus, determined solely by the crystal structure. Below, we shall describe its shapes for the crystal structures discussed in Chap.2.

3.2 Brillouin Zone for the Sphalerite and the Rock-Salt Crystal Structure

The Brillouin zone for the fcc lattice is shown in Fig.3.1. It has been constructed following the method outlined above. The shape is that of a trunc-acted octahedron or of the so-called tetrakaidecahedron. It has fourteen plane faces; six square faces are along the <100> directions and eight hexagonal faces are along the <111> directions. Since the space lattice for both the sphalerite and the rock-salt crystal structure is the fcc lattice, the Brillouin zone for materials with either of these structure is the same as that shown in Fig.3.1.

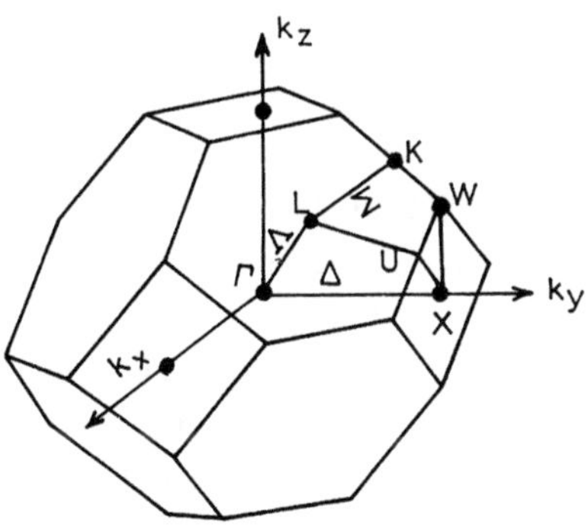

Fig. 3.1. Brillouin zone for the face-centred-cubic lattice

The volume of the Brillouin zone for an fcc lattice can be shown to be $4(2\pi/a)^3$, a being the lattice constant. Considering the wave function (3.2) and the boundary condition it should satisfy, it is found that the density of electron states in $\underline{k}$ space per unit volume of the crystal is $2(1/2\pi)^3$, the factor 2 arising from the spin degeneracy of each state. The total number of states in the Brillouin zone is hence $8/a^3$ for unit volume of the crystal. As indicated earlier the number of atoms per unit volume is $8/a^3$. Hence, the number of states contributed by an atom is 1.

The form of the band structure is determined by the symmetry of the Brillouin zone. The points and directions of symmetry are usually referred to by the letters shown in Fig.3.1 [3.3]. As these symbols are very often used, acquaintance with them is helpful in reading the literature. The zone centre is called the Γ point, the directions <111> and <100> are called respectively Λ and Δ directions and their intersections with the zone boundary are called respectively the L and the X point. Other less important directions and intersection points are indicated in the figure. The symmetry properties of the various points are discussed in the references cited at the end of this chapter. A knowledge of these properties is useful in visualising the shape of the constant energy surfaces in the $\underline{k}$ space. We shall only present these results, omitting the relevant group-theoretical discussion available in the cited references.

3.3 Brillouin Zone for the Wurtzite Structure

The space lattice for the wurtzite structure is the hexagonal lattice. The shape of the Brillouin zone is illustrated in Fig.3.2. It is that of a hexagonal right prism. The volume of the zone is $2(2\pi)^3(\sqrt{3}a^2c)^{-1}$. The number of states per unit volume is $4(\sqrt{3}a^2c)^{-1}$. The number of atoms per unit volume being $8(\sqrt{3}a^2c)^{-1}$ (see Sect.2.2), the number of states per atom is 1/2. To make the picture consistent with that for the sphalerite structure, it is usual to take a double Brillouin zone [3.4] as shown in Fig.3.2, which is obtained by doubling the length of the zone in the [0001] direction.

The directions of symmetry, the symmetry points, and the notation used to denote them are also shown in Fig.3.2. Γ and Γ' are at the centres of the hexagonal faces forming the individual zones. Point A is the zone centre of the upper zone. The [0001] axis is denoted by Δ and [1100] is denoted by Σ. The intersection of the [$\bar{1}$100], the [1100], and the [1$\bar{1}$01] axes with the zone boundaries are denoted respectively by K, M, and L.

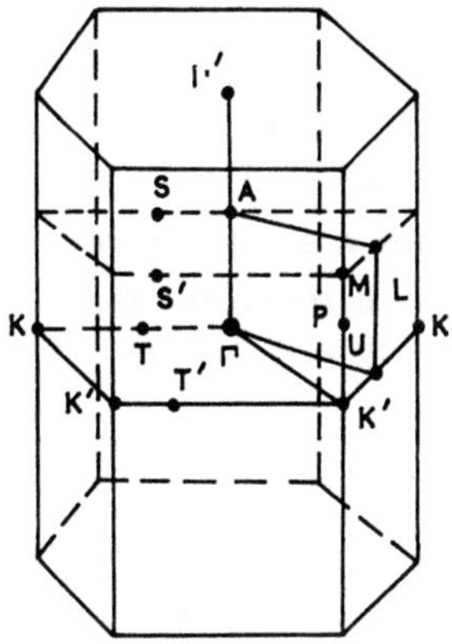

Fig. 3.2. Brillouin zone for the hexagonal lattice

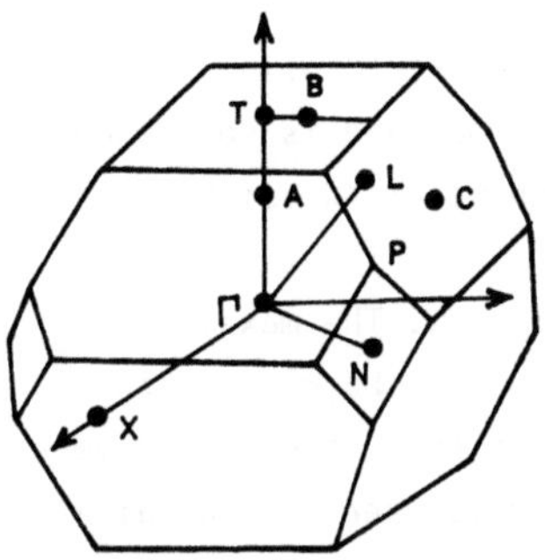

Fig. 3.3. Brillouin zone for the chalcopyrite crystal structure

3.4 Brillouin Zone for the Chalcopyrite Structure

The space lattice for the chalcopyrite structure being body-centred tetragonal, the reciprocal lattice is face-centred tetragonal. The Brillouin zone has the shape shown in Fig.3.3. We note that it is very similar to that for body-centred cubic structure but the six corners are truncated. The volume of the Brillouin zone is $(2a/c)(2\pi/a)^3$. It has also different symmetries in comparison with structures discussed in the preceding sections. The symmetry directions and the nomenclature for the symmetry points are also shown in Fig.3.3.

3.5 E-$\underline{k}$ Diagrams

The characteristics of the energy bands are usually indicated by plotting the energy eigenvalues of the electrons for different values of $\underline{k}$ in the

Brillouin zone. The crystal potential is different in different directions because of the difference in the spacing of the atoms. As a result, the value of E depends both on the magnitude and on the direction of $\underline{k}$. The special features of the diagrams obtained by plotting E versus $\underline{k}$ occur at the points of symmetry or in the directions of symmetry. The diagrams are, therefore, usually given for the zone centre and the <100> and <111> directions. We review in this section the characteristic features of the E-$\underline{k}$ diagrams for the crystal structures discussed earlier.

3.5.1 Energy Bands for the Sphalerite Structure

Two representative E-$\underline{k}$ diagrams for the sphalerite structure are shown in Fig.3.4. We may list the following important characteristics from these diagrams.

i) The conduction band has a minimum at the zone centre (Γ point), minima on the <111> directions and on the <100> directions. The band is not degenerate at the X_1 point, but there is a second set of minima at X_3 points, well separated from the X_1 minima [3.5].

ii) The valence band has three branches V_1, V_2, and V_3. The maxima of V_2 and V_3 are at the zone centre. The maxima of V_1 lie along the <111> directions slightly away from the Γ point [3.6]. The maxima of V_1 and V_2 are very nearly the same, but that of V_3 is separated by a finite energy interval Δ, which is called the spin-orbit splitting as it originates from the spin-orbit interaction [3.7,8].

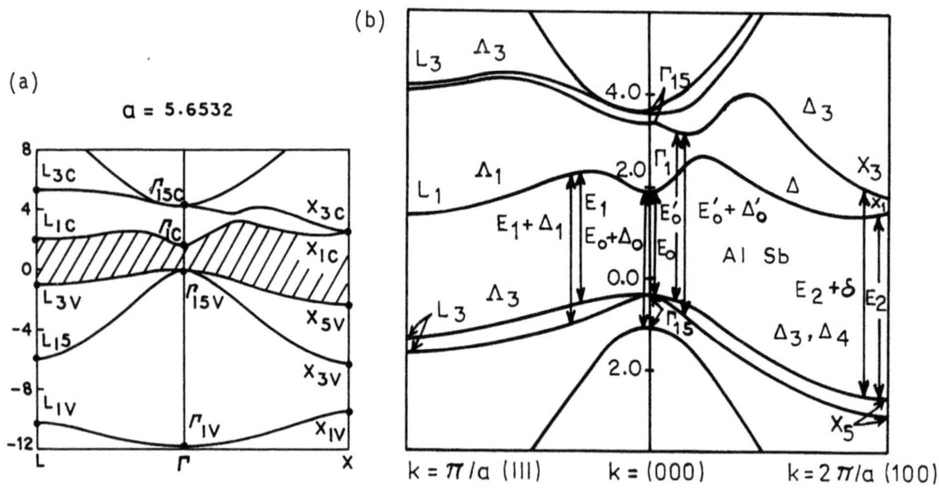

Fig. 3.4. (a) Energy band diagram of gallium arsenide [3.5e]; (b) Energy band diagram of aluminium antimonide [3.5b]

The E-$\underline{k}$ diagrams of silicon and germanium are very similar to the diagram for the spnalerite structure described above. This similarity is due to the close resemblance of the sphalerite structure with the diamond-type crystal structure of silicon and germanium. However, due to the lack of inversion symmetry we get two distinctive features in the E-$\underline{k}$ diagram for sphalerite crystal structure. First, the two conduction band curves are separated at the X points, whereas for the diamond structure, the curves meet at this point. Second, all three valence band maxima lie at the zone centre for the diamond structure, but for the sphalerite structure, one of the valence band curves has maxima away from the zone centre. These differences, though important from theoretical standpoint, are so minute that for all practical purposes they may be neglected. The separation between the X_1 and X_3 points is usually such that X_3-point minima are not normally populated even when the lowest minimum in the conduction band is at X_1. The difference in the energy values for the maxima of V_1 and the value corresponding to the Γ point is so small that in most experiments, this difference is not perceptible. Only in some very special experiments do the differences between the band structure for the sphalerite structure and the diamond structure pointed out above become significant.

It may be noted that, in dealing with transport problems, we are concerned usually with the lowest minima or the highest maxima as these are populated respectively by electrons and holes under normal conditions. Other minima or maxima become important only in high-field transport and optical interaction experiments. We have detailed experimental knowledge only about the lowest minima and highest maxima and our knowledge about other extrema are indirect inferences or in some cases only good guesses. Theoretical calculations of recent years have, however, yielded results confirming these guesses in some cases.

The diagram given in Fig.3.4 shows only the characteristic positions of the extrema. The other required information about the band structure are the algebraic expressions relating E with $\underline{k}$ near the extrema. These relations have some general characteristics, which may be listed as follows.

The energy E varies with $\underline{k}-\underline{k}_0$ parabolically, where $\underline{k}_0$ is the value of $\underline{k}$ at an extremum. The expression relating E with $\underline{k}-\underline{k}_0$ (denoted below by $\underline{k}$ alone) has the simplest form for the Γ-point minimum [Ref.3.9, p.75]. The relation is

$$E = \hbar^2 k^2 / 2m^* \quad . \tag{3.6}$$

Surfaces in $\underline{k}$ space consisting of points on which the energy is the same, i.e. the constant-energy surfaces, are, therefore, spherical for the Γ-point minimum. The constant m^* in (3.6) is the effective mass (Sect.3.1), which is, therefore, a scalar quantity for this minimum.

The E-$\underline{k}$ relation for the minima lying on the Λ or Δ directions is of the form [Ref.3.9, p.80]

$$E = \frac{\hbar^2}{2}\left(\frac{k_{\parallel}^2}{m_{\parallel}} + \frac{k_{\perp}^2}{m_{\perp}}\right) \quad . \tag{3.7}$$

where $k_{\parallel}$ and $k_{\perp}$ represent respectively the components of $\underline{k}$ in a direction parallel to the direction of symmetry, measured from the position of the minimum, and that in a transverse direction. The constant energy surfaces are spheroidal, and the shape of the spheroid is usually prolate. It should also be noted that the effective mass for these minima is a tensor quantity, the tensor being diagonal, with diagonal components $m_{\parallel}$, $m_{\perp}$, and $m_{\perp}$, when the direction of symmetry corresponding to the particular minimum is chosen as a reference axis.

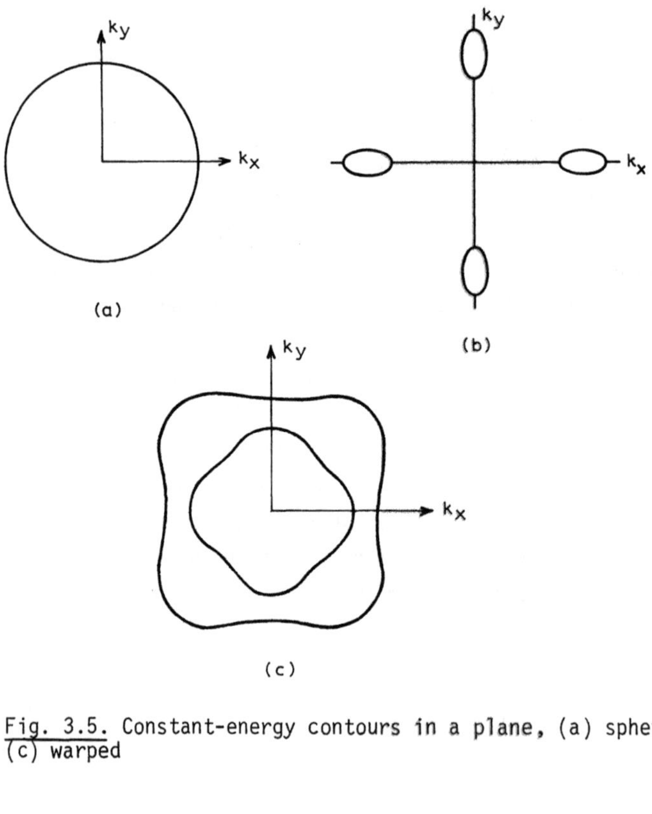

Fig. 3.5. Constant-energy contours in a plane, (a) spherical, (b) ellipsoidal, (c) warped

The E-$\underline{k}$ relation for the valence band maxima is [3.10]

$$E = (\hbar^2/2m_0)\left\{Ak^2 \pm \left[B^2k^4 + C^2\left(k_x^2k_y^2 + k_y^2k_z^2 + k_z^2k_x^2\right)\right]^{\frac{1}{2}}\right\} \quad . \tag{3.8}$$

The constant-energy surfaces are, therefore, warped for the maxima, but in many cases the warping is small enough and may be neglected. With this approximation, (3.8) is simplified to [3.11]

$$E = \frac{\hbar^2 k^2}{2m_0}\left[A \pm \left(B^2 + \frac{C^2}{6}\right)^{\frac{1}{2}}\right] \quad . \tag{3.8a}$$

Constant energy contours of the various kinds discussed above are shown in Fig.3.5.

The energy E for points in $\underline{k}$ space away from the extrema varies nonparabolically with $\underline{k}$. The nature of the nonparabolicity has, however, been investigated in some detail only for the Γ-point minimum. We shall discuss the nature and effect of this nonparabolicity in a later section. Suffice it to mention at this stage that the nonparabolic E-$\underline{k}$ relation may be written for most purposes as [3.6c]

$$\hbar^2k^2/2m^* = E(1+\alpha E) \quad , \tag{3.9}$$

where m* is the effective mass for k tending to zero, and α is a constant which in many materials is approximately equal to $1/E_g$, E_g being the separation between the minimum and the valence band maximum, or the so-called band gap.

All the materials having the sphalerite crystal structure have energy bands with the characteristics discussed above. The difference in the energy bands of the various materials is in the relative position of the conduction band minima and in the values of the effective masses. We have listed in Table 3.1 the energy band parameters of different compound semiconductors having the sphalerite crystal structure.

It is evident from Table 3.1 that the lowest minimum in most of the compound semiconductors with sphalerite crystal structure is at the zone centre. The transport theory for these materials is simpler on this account. The effective mass being a scalar quantity, the treatment of galvanomagnetic effects and of other transport coefficients involves simpler algebraic manipulations.

We also find from the data listed in Table 3.1 that the energy bands of materials with sphalerite structure have three important properties worthy of mention. First, the band gap in most of the materials is direct. The ab-

Table 3.1. Energy band parameters of compound semiconductors with sphalerite structure
A. Material with direct gap at Γ point

Material	Lattice constant [Å]	Energy band gap E_g [eV] at 0 K	dE_g/dT [10^{-4} eV/K]	Spin-orbit splitting [eV]	Electron effective mass in units of electron mass m_0	Valence band parameters A	B	C^2
GaAs	5.6533	1.55	-4.2-4.6	0.35	0.07	-5.8	- 2.43	27.7
InP	5.4505	1.34	-4.2	0.21	0.078	-5.75	- 2.78	27.4
InAs	6.0584	0.36	-4.0	0.43	0.026	-14.5	-12.74	73.09
InSb	6.48	0.24	-3.6	0.9	0.0145	-25	-21	256
ZnS	5.4093	3.84	-4.6	0.27	0.27	-3.06	0.53	2.18
ZnSe	5.6687	2.82	-8	0.43	0.17	-2.12	- 0.48	4.9
ZnTe	6.1037	2.38	-5	0.91	0.17	-3.29	- 0.08	6.6
CdTe	6.481	1.6	-2.3	0.81	0.16	-2.23	- 1.03	6.18
HgS	5.851	2.1						
HgSe	6.084	-0.1						
HgTe	6.460	-0.1						
CdS	5.820							
CdSe	6.05							

B. Materials with conduction band minima at X point and valence band maxima at Γ point

Material	Lattice constant [Å]	Energy gap at 0 K [eV] X	Γ	dE_g/dT [10^{-4} eV/K]	Spin-orbit splitting [eV]	Electron effective mass in units of electron mass m_0 $m_\parallel$	$m_\perp$	Valence band parameters A	B	C^2
GaP	5.4504	5.3	2.87	-2.5	0.8	1.51	0.28	-4.5	-2.1	20
GaSb	6.12			-4.0				-7.46	-4.50	38.7
AlSb		4.2	1.8		0.4	0.95	0.25	-5.96	-3.36	23
AlAs	5.660	4.9	2.9		0.3					
AlP	5.467	5.3			0.06					

sorption of light and infrared radiation due to interband transition of electrons and the phenomenon of tunnelling can occur efficiently. Second, the band gap may be very small so that the effects of nonparabolicity will, in general, be significant. Lastly, the effective mass may also be small; in some cases it is as small as one tenth of that for germanium. The electron mobility has, as a result, a high value, and the materials become degenerate for not-too-large values of the carrier concentration.

3.5.2 Energy Bands for the Wurtzite Structure

The energy bands of materials with wurtzite structure and with sphalerite structure should have similarities due to the similarity of the two struc-

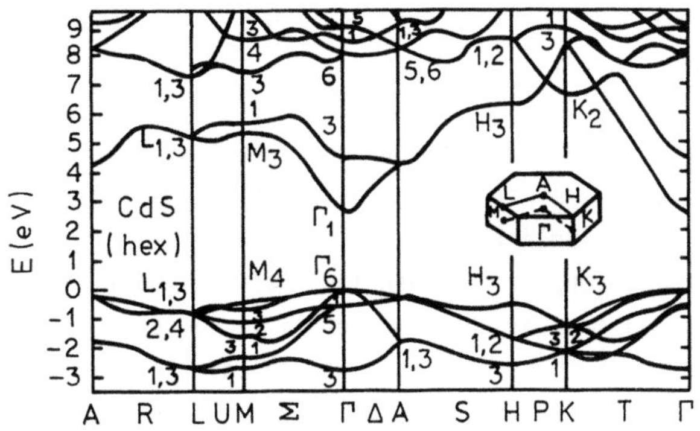

Fig. 3.6. Energy band diagram of cadmium sulphide [3.5g]

tures. It should be remembered that the number and orientation of nearest and next nearest atoms are the same, differences existing only in the position of third nearest atoms. However, the details of the band structure for wurtzite crystals are not yet fully available. Techniques of material preparation have not reached the stage at which the methods for detailed exploration of the band structures may be successfully employed. However, the characteristics of the lowest conduction band minima and of the highest valence band maxima are known with some certainty.

In Fig.3.6 a representative band structure for wurtzite materials is shown. All the materials whose band structure is known have extrema at the Γ point. However, there are some marked differences in the relative position and in the shape of the constant energy surfaces at the extrema for these materials and with sphalerite structure. These differences originate from the uniaxial nature of the wurtzite crystals, the presence of the crystal field and the difference in the shape of the Brillouin zone for the two kinds of materials.

In wurtzite materials, all three valence band maxima are separated from each other. Each of these bands is split by spin-orbit interaction but the splitting is so small that for most experiments it may be neglected. The E-$\underline{k}$ relation for the extrema is [3.6d]

$$E = A\left(k_x^2+k_y^2\right) + Bk_z^2 \pm C\left(k_x^2+k_y^2\right)^{1/2} \quad , \tag{3.10}$$

where z direction is chosen parallel to the c axis. The constant energy surfaces are, therefore, spheroidal with the c axis as the principal axis for both the conduction band minimum and the valence band maximum. However, the anisotropy in the conduction band is small and may be neglected. There is very little information available about the values of A and B for valence bands, although there is evidence showing that these are considerably different [3.12,13].

It may be pointed out at this stage that the apparent effective mass of the carriers in a polar material varies with frequency due to the polaron effect [3.14]. The electrons produce polarisation in the lattice, which has to move with the electron when the coupling between the electron and lattice is strong. In effect, the inertial mass of the electron is increased, the increase being determined by the strength of electron-lattice coupling, and is large for a polar material. Materials with either sphalerite or wurtzite crystal structure are polar, but since materials with the latter kind of structure are more polar, the effect of polarons on the effective mass is more apparent in these materials. It should be noted that this effect is insignificant if the frequency of variation of the fields is high as in light signals, since the polarisation is then unable to follow the field. The effective masses measured in other transport experiments involving low frequencies are, however, the values modified by the polaron effect. The bare mass is related to the polaron mass for weak coupling by the relation $m_p = m^*(1+\alpha/6)$, where α depends on the electron-lattice interaction and is given by [3.15]

$$\alpha = \frac{e^2}{4\pi\varepsilon_0}\left(\frac{1}{K_\infty} - \frac{1}{K_S}\right)\left(\frac{m^*}{2\hbar^3\omega_\ell}\right)^{\frac{1}{2}} , \tag{3.11}$$

where ω_ℓ is the polar optic phonon frequency, and K_∞, K_S are the high-frequency and static dielectric constants. The characteristic parameters of the band structure of compound semiconductors with wurtzite crystal structure are summarised in Table 3.2.

As the materials have direct gaps in the energy band structure, efficient light interaction devices may be made using these materials. The band gap is also such that the light emission due to interband transitions is in the visible region for some of them. These materials are, therefore, extensively used in lasers and light-emitting diodes. It should also be noted that the effective mass of the electrons in these materials is fairly large. Hence, the mobility is relatively small. Due to the large value of band gap, the effects of nonparabolicity are also small.

3.5.3 Energy Bands for the Rock-Salt Structure

The general features of the band structure of crystals with rock-salt structure should be similar to those of crystals with sphalerite structure as the

Table 3.2. Energy band parameters of semiconductors with wurtzite crystal structure

Material	Lattice a [Å]	constants c [Å]	Energy band gap E_g [eV] at 0 K	dE_g/dT [10^{-4}eV/K]	Spin-orbit splitting S_0 [eV]	Crystal field splitting C_r [eV]	Electron effective mass m_e in units of m_0	Hole effective mass in units of m_0 $m_\perp$	$m_\parallel$
ZnO			3.44	-9.5	0.0087		0.27		
ZnS	3.8225	6.2610	3.91	-8.5	0.092	0.055	0.205	0.51	>1
CdS	4.1368	6.7163	2.58	-5.2	0.065	0.027	0.17	0.7	5
ZnSe	4.003	6.540	2.80	-4.6	0.42		0.13	0.7	5
CdSe	4.299	7.010	1.84				0.13	0.45	>1
ZnTe	4.27	6.99					0.17		1
CdTe	4.57	7.47					0.096	0.4	

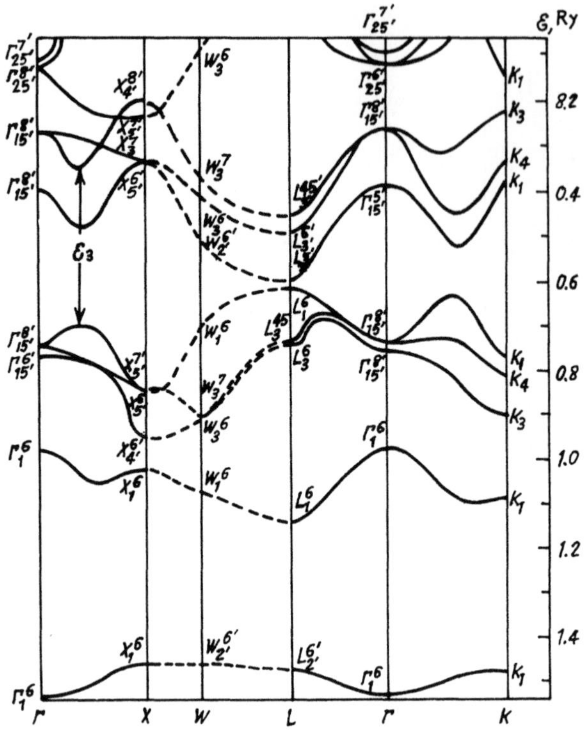

Fig. 3.7. Energy band diagram of lead sulphide [3.17]

space lattice is face-centred cubic in both kinds of crystals. Detailed experimental information is, however, available only for the lowest conduction band minimum and for the highest valence band maximum. In contrast to sphalerite and wurtzite crystals, the valence band maximum and the conduction band minimum are both at the L point (Fig.3.7) and, are almost mirror images of one another. The E-k relation for both the extrema is [3.5,16,17]

$$E = \frac{\hbar^2}{2}\left(\frac{k_\parallel^2}{m_\parallel} + \frac{k_\perp^2}{m_\perp}\right) \quad , \tag{3.12}$$

where $m_\parallel$ is the effective mass for the <111> direction and $m_\perp$, that for a perpendicular direction. The values of $m_\parallel$ and $m_\perp$ are also nearly the same for the two extrema.

In general, $m_\parallel$ is different from $m_\perp$, but the anisotropy in the effective mass varies from material to material. Some compounds have strong anisotropy, whereas in others, it may be neglected and the effective mass may be treated as a scalar quantity. It may also be noted that as in wurtzite crystals, all

Table 3.3. Energy band parameters of IV-VI compounds

Material	Hole effective mass in units of m_0		Electron effective mass in units of m_0		Band gap at 0 K	dE_g/dT
	$m_\parallel$	$m_\perp$	$m_\parallel$	$m_\perp$	[eV]	[10^{-4}eV/K]
PbS	0.105	0.075	0.105	0.080	0.29	4.7
PbSe	0.068	0.034	0.07	0.040	0.15	4.7
PbTe	0.31	0.022	0.24	0.024	0.19	4.7

three valence band maxima are separated from each other. The parameters of the band structures of rock-salt crystals are listed in Table 3.3.

Crystals with rock-salt structure are thus direct band gap materials. The band gap ranges between 0.15 and 0.24 eV for the three compounds listed. The nonparabolicity of the bands is, therefore, very significant and must be taken into account in transport calculations. In contrast to III-V and II-VI semiconductors, the temperature coefficient of the band gap is positive for IV-VI compounds [3.18]. We should note, however, that these are not high-mobility materials like some of the III-V compounds. The important applications of these materials are mostly in infrared detectors, for which the direct nature of band gap is advantageous.

3.5.4 Band Structure of Mixed Compounds

Crystals are being grown from alloys of all kinds. However, alloys having well-defined band structure and good semiconducting properties may be classified into two groups. In one group belong those compounds which are obtained by mixing two compounds with the same crystal structure. The band structure of the resultant compound has, therefore, the same general features as those of the ingredients and the exact structure may be considered as an interpolation between the structures of the two compounds. Very little, however, is known about the fine structure of the bands. The available information is mostly related to the variation of band gap with the composition. For a compound having the composition $A_{1-x}\,B_x\,C$, the band gap is found to be given by [3.19]

$$E_g(A_{1-x}\,B_x\,C) = (1-x)E_g(AC) + x[E_g(BC)-c] + cx^2 \quad . \tag{3.13}$$

The value of c is a measure of the deviation from the linear interpolation curve and is often referred to as the bowing parameter. Values of c for the various mixed compounds are given in Table 3.4.

Table 3.4. Energy band-gap bowing parameter of mixed compounds

Material	Bowing parameter [eV]
GaAs-P	0.21
InAs-P	0.20-0.26
Ga-InSb	0.43
Ga-InAs	0.33-0.56
InAs-Sb	0.58
Ga-AlAs	0.2
Ga-InP	0.88
ZnS-Se	0
ZnSe-Te	1.28
ZnS-Te	2.40

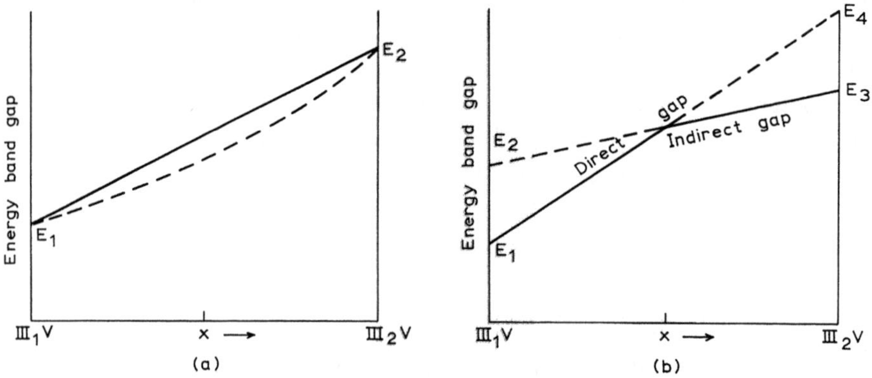

Fig. 3.8. Energy gap in a mixed compound, $III_{1(x)}III_{2(1-x)}V$. (a) Direct band: E_1, band gap in III_1V compound; E_2, band gap in III_2V compound. (b) Indirect band: E_1, direct band gap in III_1V compound; E_2, indirect band gap in III_1V compound; E_3, direct band gap in III_2V compound; E_4, indirect band gap in III_2V compound

If both the compounds have the same kind of band gap, direct or indirect, the approximately linear variation continues from one end to the other. On the other hand, if one compound has a direct gap and the other has an indirect gap, the variation of band gap (shown in Fig.3.8) is given by two straight lines intersecting at the point where there is a transition in the nature of the band gap of the mixed compound.

The effective mass of the electrons has been studied in only a few mixed compounds [3.20]. It appears from these results that the effective mass also varies linearly with the composition. The linear interpolation is often taken for granted as being applicable to almost all the parameters of mixed compounds, although the energy as mentioned above is known to deviate a little. Only future studies may reveal how far this assumption is valid for the vari-

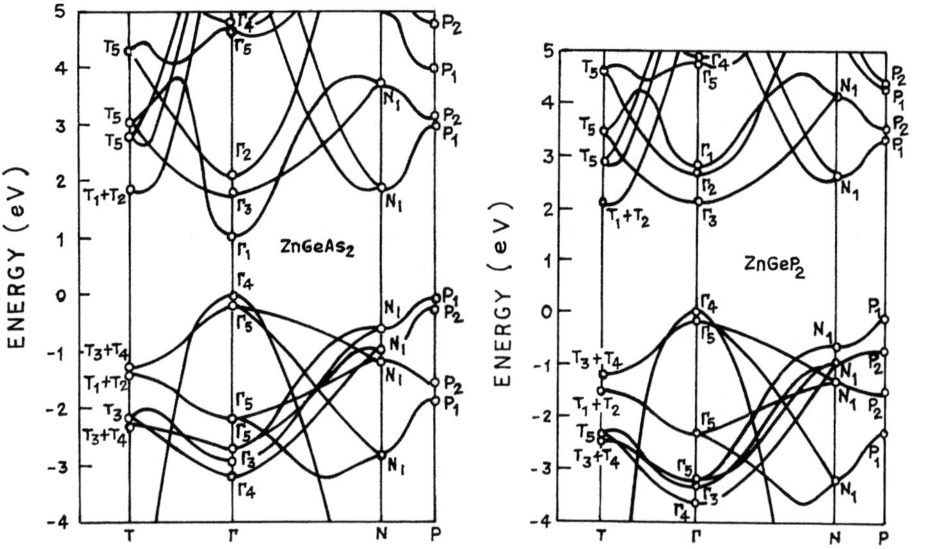

(a) (b)

Fig. 3.9. (a) Energy band diagram of $ZnGeAs_2$; (b) Energy band diagram of $ZnGeP_2$ [3.22]

ous semiconductor parameters. But, in the absence of experimental results we have to rely on the applicability of this interpolation method for the interpretation of the transport behaviour of the semiconductors obtained by mixing two compounds.

The second kind of mixed compound is obtained by mixing more than two elements, chosen so as to form bands similar to those in III-V or II-VI compounds discussed earlier. Examples of such compounds which have been studied [3.21] are $CdSiP_2$, $CdSnAs_2$, $CdGeP_2$, $CdSiAs_2$, $ZnSiAs_2$, $ZnSnAs_2$, $ZnSiP_2$, $ZnGaP_2$, $ZnSnP_2$, and $ZnGeAs_2$; and $CuInSe_2$, $CuGaS_2$, $CuGaSe_2$, $AgInS_2$, $AgGaSe_2$, $AgGaS_2$, and $AgInSe_2$. Of these, the first group are II-IV-V_2 compounds, analogous to III-V binary compounds; while the other group are I-III-VI_2 compounds, analogous to II-VI compounds. The band structures of these ternary compounds have been theoretically calculated [3.22], so that some typical features are known. These may be listed as follows. The conduction band has two classes of lowest minima, one at Γ_1 derived from the Γ_1 of the corresponding binary compound, and the other at Γ_3 or T_1+T_2 derived from the X_1 minimum of the corresponding compound. The degeneracy in the valence band maximum is removed due to the crystal field resulting from the asymmetry in the atomic positions. These features are illustrated in Fig.3.9, which shows the calculated band structure of $ZnGeAs_2$ and $ZnGeP_2$.

Experimental results are available only for the band gap and about the nature of the extrema. The major part of the evidence has come from electroreflectance experiments, which confirm the theoretical predictions noted above. The pressure dependence of band gap also confirms the existence of two kinds of lowest conduction band minima at Γ_1 and at Γ_3 or T_1+T_2. The band gaps for II-IV-V_2 compounds differ by about 0.3 eV from the band gaps in the corresponding III-V compounds, whereas in a I-III-VI_2 compound, the band gap differs by as much as 1.6 eV from that of the corresponding II-VI compounds. The E-$\underline{k}$ relation near the extrema for both the conduction and the valence bands has been shown to be of the form

$$E = \pm\frac{\hbar^2}{2}\left(\frac{k_x^2+k_y^2}{m_\perp}+\frac{k_z^2}{m_\parallel}\right) , \tag{3.14}$$

where the z direction is as shown in Fig.3.7, and plus and minus signs correspond respectively to the conduction and valence band. The energy E is measured from the extremum. The calculated values of $m_\perp$ and $m_\parallel$ [3.23,24] and the energy band parameters are listed in [3.21] for the various chalcopyrite compounds.

3.6 Conclusion

We have presented briefly in this chapter the general features of the energy band structure of the various compound semiconductors. We find that in all the materials except in GaSb and in mixed compounds with a particular composition, the band structure is characterised by a conduction band minimum well separated from other minima. The low-field transport properties are, therefore, related to the properties of this minimum. We note that the effective mass of the electrons for this minimum is isotropic but energy dependent in many compounds. In some compounds, however, the effective mass is a tensor quantity which has only three elements for properly chosen coordinate axes. We should also note that when dealing with high-field transport or low-field transport in GaSb and similar compounds we may have to consider electrons with different masses.

The valence band, on the other hand, is found to have a doubly degenerate maximum in all compounds except in IV-VI compounds. The next maximum is usually separated by a large energy gap. The effective mass of the holes corresponding to the two maxima are very different. We are, therefore, required to consider two varieties of holes (heavy and light) in low-field transport studies. The concentration of heavy holes is, however, much larger than that of the light holes and the properties would be predominantly determined by heavy holes. We also note that the effective mass of the holes is direction dependent to a large extent as the constant energy surfaces are warped. Because of this complexity, transport theory for p-type materials has not yet reached an advanced state and we shall not concern ourselves much with holes.

4. Theory of Energy Band Structure

The information relating to the energy band structure of various materials reviewed in the previous chapter has been obtained partly from experiments and partly from theoretical calculations. Experiments giving information about the band structure will be discussed in later chapters. Here we present some simple models in terms of which common features of the band structure are often discussed.

The basic symmetries in the band structure follow from the crystal structure and may be made evident by applying group-theoretical considerations. This aspect is not discussed here. The interested reader may consult [4.1] for such a discussion.

The detailed nature of the energy bands may be worked out, in principle, by solving the Schrödinger equation provided the crystal potential is known. There are also methods for the calculation of the band structure of solids from first principles which partially utilise the information about the band structure obtainable experimentally. References to papers and books [4.2] discussing these methods are given at the end. The methods involve elaborate computations, and only a few groups have tackled these problems.

4.1 Models of Band Structure

Although the detailed knowledge of the band structure is helpful for the interpretation of some experiments, only an acquaintance with some special features is necessary for the development of a general transport theory. We shall be concerned only with the simple models which are useful for this purpose.

The simplest model for band structure calculations which is discussed in most of the books is the Kronig-Penney model [4.3]. In this model, the periodic potential is approximated by periodic square wells. The Schrödinger equation for this potential distribution can be solved easily, and the result shows that the discrete atomic energy levels broaden into bands in solids.

Two other simple models associated with the free-electron approximation and the tight-binding approximation, are also discussed in many textbooks. We shall review these models briefly as these are often used in the derivation of electron wave functions in semiconductor superlattices. The major part of our discussion will, however, be concerned with the $\underline{k}\cdot\underline{p}$ perturbation method. It is essential to know this method for the development of the transport theory of materials with small band gap, which constitute a sizeable fraction of the compound semiconductors. This method, though it does not proceed from first principles, has been extensively used in the derivation of the E-$\underline{k}$ relation away from the band extrema. It should be remembered that all but a few experiments give direct information only about the extrema. For the analysis of transport problems in which the electron states away from the minima play an important role, we have to rely mostly on the results obtained by the $\underline{k}\cdot\underline{p}$ perturbation method.

4.2 Free-Electron Approximation Model

The band structure or the E-$\underline{k}$ relation for the electrons in a solid is given by the solution of the time-independent Schrödinger equation,

$$(\hbar^2/2m_0)\nabla^2\psi + [E-V(\underline{r})]\psi = 0 \quad , \tag{4.1}$$

where $V(\underline{r})$ is the potential energy of an electron due to the potential produced by the nuclei of the atoms and other electrons. It is a periodic function having the same periodicity as the lattice.

The general solution of the equation is the Bloch function

$$\psi = U_{\underline{k}}(\underline{r})\ \exp(i\underline{k}\cdot\underline{r}) \quad , \tag{4.2}$$

where $U_k(\underline{r}+\underline{d}) = U_k(\underline{r})$, $\underline{d}$ being a primitive translation vector of the lattice.

In the free-electron approximation model we consider only the states for which E is much larger than $V(\underline{r})$. We may then assume that the electron is almost free and its dynamics may be obtained using a wave function slightly altered from the free-electron wave function. The term $V(\underline{r})$ in (4.1) may be dealt with as a perturbation.

In order to simplify the analysis we shall outline the method of solution based on the above approximation for a one-dimensional crystal. $V(\underline{r})$ is, therefore, a function of x alone. The lattice spacing is assumed to be a. Now, the solution of (4.1), as argued above, should be a Bloch function but at the same time should reduce to the free-electron function for $E \gg V(x)$. We may, therefore, try a solution of the form

$$\psi(x) = \left[c_0 + \sum_{n \neq 0} c_n \exp(-ink_L x)\right] \exp(ikx) \quad , \tag{4.3}$$

where the $k_L = 2\pi/a$. c_n are much smaller than c_0. The magnitudes of the c_n would decrease as E increases in comparison with V(x).

Substituting (4.3) in the Schrödinger equation, and replacing V(x) by

$$V(x) = \sum_{n=-\infty}^{\infty} V_n \exp(-ink_L x) \quad , \tag{4.4}$$

we obtain

$$c_0\left(k_0^2-k^2\right) \exp(ikx) + \sum_{n\neq 0}\left[c_n\left(k_0^2-k_n^2\right)-\left(2m_0/\hbar^2\right)c_0 V_n\right] \exp(ik_n x)$$
$$-\left(2m_0/\hbar^2\right)\sum_{n\neq 0}\sum_{n'\neq 0} c_n V'_n \exp[i(k_n-n'k_L)x] = 0 \quad , \tag{4.5}$$

where $k_0^2 = \left(2m_0/\hbar^2\right)E$, $k_n = k - nk_L$. Note that the V_n are the Fourier coefficients of the periodic potential function.

Multiplying (4.5) by e^{-ikx} and integrating between the limits 0 and a, we get

$$c_0\left(k_0^2-k^2\right) - \left(2m_0/\hbar^2\right)\sum_{n\neq 0} c_n V_{-n} = 0 \quad . \tag{4.6}$$

Similarly, multiplying by $e^{-ik_n x}$ and integrating over the same limits, we get

$$c_n\left(k_0^2-k_n^2\right) - \left(2m_0/\hbar^2\right)c_0 V_n - \left(2m_0/\hbar^2\right)\sum_{n'\neq 0}\sum_{n''\neq 0} c_{n'} V_{n''} = 0 \quad . \tag{4.7}$$

The summation in the third term is restricted to values of n' and n" satisfying the relation $n' + n'' = n$.

The set of linear homogeneous equations given by (4.6) and (4.7) may in principle be solved, but in the free-electron approximation we simplify the analysis by neglecting the third term of (4.7) which will be of a lower order as it involves the products of $c_{n'}$ and $V_{n''}$, each of which is assumed small.

The coefficients in the function ψ are thus given by

$$c_n = 2m_0 V_n\left[\hbar^2\left(k_0^2-k_n^2\right)\right]^{-1} c_0 \quad . \tag{4.8}$$

The E-$\underline{k}$ relation as obtained from (4.6) by substituting (4.8) is

$$E = \frac{\hbar^2 k^2}{2m_0} + \sum_{n\neq 0} \frac{2m_0}{\hbar^2} \frac{V_n V_n^*}{\left(k^2-k_n^2\right)} \quad , \tag{4.9}$$

where V_n^* is the complex conjugate of V_n. Equation (4.9) is obtained by replacing k_0^2 by $\left(2m_0/\hbar^2\right)E$ and by noting that $V_{-n} = V_n^*$ and $k_0 \approx k$ [Ref.4.4,p.227].

The approximate relation given above is evidently invalid when $k \approx nk_L/2$ since then c_n is very large and the assumption on which the formula is based is not obeyed. The proper relation for this condition may be derived by retaining only the terms containing c_0 and c_n in the expression for ψ. We then get from (4.6) and (4.7)

$$\left(k_0^2-k^2\right)c_0 - \left(2m_0/\hbar^2\right)c_n V_n^* = 0 \quad , \tag{4.10}$$

$$\left(2m_0/\hbar^2\right)V_n c_0 - c_n\left(k_0^2-k_n^2\right) = 0 \quad . \tag{4.11}$$

A nontrivial solution is obtained only when the determinant of the coefficients vanishes and the E-$\underline{k}$ relation obtained by applying this condition is

$$E = \frac{\hbar^2}{4m_0}\left\{k^2+k_n^2 \pm \left[\left(k^2-k_n^2\right)^2 + 4\left(\frac{2m_0}{\hbar^2}\right)^2 V_n V_n^*\right]^{\frac{1}{2}}\right\} \quad . \tag{4.12}$$

The corresponding ψ function is

$$\psi = c_0\left[1+2V_n m_0 \hbar^{-2}\left(k_0^2-k_n^2\right)^{-1}\right]\exp(-ink_L x) \quad . \tag{4.13}$$

The plus sign in (4.12) should be taken when $k > nk_L/2$ and the minus sign, when $k < nk_L/2$. Also (4.13) reduces to $c_0 e^{ikx}$ when $\left(k^2-k_n^2\right)$ is very large.

The E-$\underline{k}$ relation given by (4.9) and (4.12) is illustrated in Fig.4.1. The figure shows that the continuous energy levels of a free electron are grouped into bands in a solid. We also find that when the free-electron approximation is valid, the widths of the forbidden bands are smaller than those of the allowed bands.

It is also of interest to note that near the band edges, i.e. when $\underline{k} \approx n\underline{k}_L/2$, the E-$\underline{k}$ relation (4.12) simplifies to the parabolic relation

$$E = \pm \frac{\hbar^2}{2m_0} k_1^2\left(1 + \frac{4E_n}{\Delta E}\right) + E_n \pm \frac{\Delta E}{2} \quad , \tag{4.14}$$

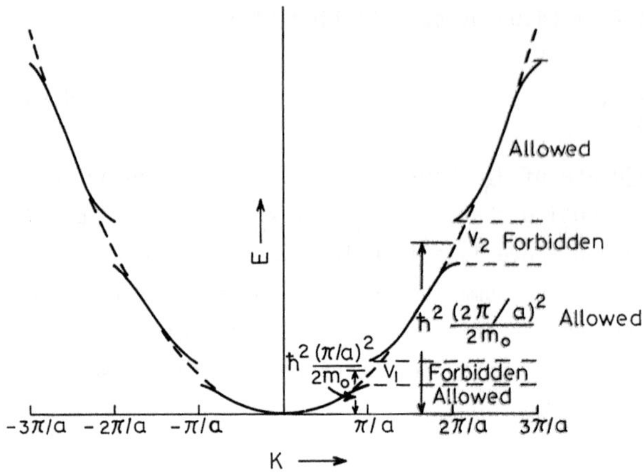

Fig. 4.1. E-k diagram for the free-electron approximation

where $\Delta E = 2|V_n|$, $k_1 = k - n\pi/a$, $E_n = \left(\hbar^2/2m_0\right)(n\pi/a)^2$, ΔE is the band gap and E_n is the energy at the band edge. The band gap for the nth band is thus twice the nth Fourier coefficient of the periodic potential. The effective mass of the electron at the band edge is also given by

$$\frac{1}{m_n^*} = \frac{1}{\hbar^2}\frac{d^2E}{dk_1^2} = \frac{1}{m_0}\left(1 + \frac{4E_n}{\Delta E}\right) \quad . \tag{4.15}$$

The basic assumption in the free-electron approximation as discussed above is that the crystal potential is small in comparison to the electron energy. Evidently, this assumption cannot be obeyed near the atomic cores in any material. In some metals, the assumption is applicable for the outer electrons over a large part of the crystal excluding points very near the atomic cores. The results obtained by using the free-electron model are hence in good agreement with experiments for metals. For the study of common semiconductors, the method is not of much use as the basic assumption is not valid. However, in semiconductor superlattices the magnitude of the potential barrier being of the order of 0.1-0.5 eV, the method is applicable for electron energies larger than this barrier energy. We shall discuss in Sect.4.4 the theory of energy bands in semiconductor superlattices. Suffice it to mention at this stage that for the development of transport theory involving electrons of energy greater than the barrier energy, the electron wave functions and E-k relations are often obtained by using the free-electron approximation method for superlattices.

4.3 Tight-Binding Approximation Model

The tight-binding approximation method is based on assumptions which are just the opposite of those in the free-electron approximation. In this method the energy bands in the solid are assumed to develop as modifications of the energy levels appropriate to the atoms constituting the solid rather than as modifications to the free-electron energy.

The discrete atomic levels transform into bands due to the perturbation in the atomic potentials caused by the neighbouring atoms. If the spacing between the atoms is such that the perturbation in the potential is small, the electron wave function in the solid will have the basic character of the atomic wave functions. This expectation is also likely to be true for the electron states, the atomic wave functions for which do not extend far away from the atoms. In the tight-binding approximation, these conditions are assumed to hold and the electron wave function is taken to be

$$\psi = \sum_n c_n U_0(\underline{r}-\underline{r}_n) \quad , \tag{4.16}$$

where $U_0(\underline{r}-\underline{r}_n)$ represents the atomic wave function for an atom at $\underline{r}_n$.

For nonoverlapping atomic potentials or for nonoverlapping atomic wave functions, c_n would be unity and the energy of the electron corresponding to this wave function will be N-fold degenerate, if there are N atoms in the solid. The energy E_0 will satisfy the equation

$$\left[-\left(\hbar^2/2m_0\right)\nabla^2+V_a(\underline{r}-\underline{r}_n)\right]\psi = E_0\psi \quad , \tag{4.17}$$

where $V_a(\underline{r}-\underline{r}_n)$ is the potential due to the atom at $\underline{r}_n$.

However, when the atoms come together to form a solid, there must be some overlap in the potentials and also in the wave functions since this overlap is the origin of the binding energy. The potential of the individual atoms cannot be considered to be isolated, but should be taken to constitute the periodic potential in the crystal. The electron wave function will have the Bloch form, though near an atom it will approximate the atomic wave function. A wave function having this dual character may be constructed by taking c_n as (see Fig.4.2 for the meaning of $\underline{r}_n$)

$$c_n = \exp(i\underline{k}\cdot\underline{r}_n) \quad . \tag{4.18}$$

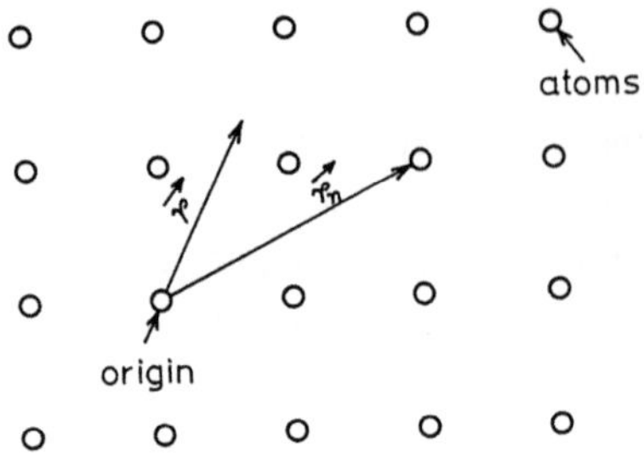

Fig. 4.2. Coordinate system of (4.18)

The resultant function is

$$\psi_{\underline{k}}(\underline{r}) = \sum_n U_a(\underline{r}-\underline{r}_n)\ \exp(i\underline{k}\cdot\underline{r}_n) \quad . \tag{4.19}$$

Since U_a decreases rapidly with distance,

$$|\psi_{\underline{k}}(\underline{r})| \approx |U_a(\underline{r}-\underline{r}_n)| \quad \text{when} \quad \underline{r} = \underline{r}_n .$$

Also, we find that

$$\begin{aligned}\psi_{\underline{k}}(\underline{r}+\underline{r}_m) &= \sum_n U_a(\underline{r}+\underline{r}_m-\underline{r}_n)\ \exp(i\underline{k}\cdot\underline{r}_n)\\ &= \exp(i\underline{k}\cdot\underline{r}_m)\sum_n \exp[i\underline{k}\cdot(\underline{r}_n-\underline{r}_m)]U_a(\underline{r}-\underline{r}_n+\underline{r}_m)\\ &= \exp(i\underline{k}\cdot\underline{r}_m)\psi_{\underline{k}}(\underline{r}) \quad .\end{aligned} \tag{4.20}$$

Hence $\psi_{\underline{k}}(\underline{r})$ as in (4.19) has the necessary periodicity of the Bloch function.

The atomic functions are usually normalised to a unit cell but the wave function $\psi_{\underline{k}}(\underline{r})$ given by (4.19) is not normalised due to the overlap of the atomic functions. We should note that

$$\begin{aligned}\int\psi^*\psi d\underline{r} &= \int\Big[\sum_\ell \exp(-i\underline{k}\cdot\underline{r}_\ell)U_a(\underline{r}-\underline{r}_\ell)\sum_n \exp(i\underline{k}\cdot\underline{r}_n)U_a(\underline{r}-\underline{r}_n)\Big]d\underline{r}\\ &= N + \int\sum_n\sum_{\ell\neq n} \exp[-i\underline{k}\cdot(\underline{r}_\ell-\underline{r}_n)]U_a(\underline{r}-\underline{r}_\ell)U_a(\underline{r}-\underline{r}_n)d\underline{r}\\ &= N + O_c \quad ,\end{aligned} \tag{4.21}$$

where N is the number of unit cells in the crystal, and the additional term arising from the overlap of the atomic functions has been written as O_c.

The expectation value of the energy as obtained by using the wave function of (4.19) is

$$E = \int \psi^* H \psi d\underline{r} \Big/ \int \psi^* \psi d\underline{r} \quad , \tag{4.22}$$

where H is the Hamiltonian for the electrons in the solid, given by

$$H = \left(-\hbar^2/2m_0\right)\nabla^2 + \sum_m V_a(\underline{r}-\underline{r}_m) \quad , \tag{4.23}$$

$\sum V_a(\underline{r}-\underline{r}_m)$ being the potential energy of the electron for the combined field due to all the atoms. Replacing H by (4.23) and using (4.19), we get

$$\begin{aligned}\int \psi^* H \psi d\underline{r} &= \int \Big\{ \sum_\ell \exp(-i\underline{k}\cdot\underline{r}_\ell) U_a(\underline{r}-\underline{r}_\ell) \Big[\left(-\hbar^2/2m_0\right)\nabla^2 + \sum_m V_a(\underline{r}-\underline{r}_m)\Big] \\ &\quad \times \sum_n \exp(i\underline{k}\cdot\underline{r}_n) U_a(\underline{r}-\underline{r}_n) \Big\} d\underline{r} \\ &= E_0 \int \Big[\sum_\ell \exp(-i\underline{k}\cdot\underline{r}_\ell) U_a(\underline{r}-\underline{r}_\ell) \Big] \Big[\sum_n \exp(i\underline{k}\cdot\underline{r}_n) U_a(\underline{r}-\underline{r}_n) \Big] d\underline{r} \\ &\quad + \int \Big[\sum_\ell \exp(-i\underline{k}\cdot\underline{r}_\ell) U_a(\underline{r}-\underline{r}_\ell) \Big] \Big[\sum_{m \neq n} V_a(\underline{r}-\underline{r}_m) \sum_n \exp(i\underline{k}\cdot\underline{r}_n) U_a(\underline{r}-\underline{r}_n) \Big] d\underline{r} \ .\end{aligned} \tag{4.24}$$

The first term of the above expression is the same as (4.21) and in the second term the summation over m for the individual values of n would give identical values of the sums. Hence, (4.22) may be reduced as follows by replacing r_n by 0:

$$\begin{aligned} E = \Big[(N+O_c)E_0 + N \int \sum_\ell \exp(-i\underline{k}\cdot\underline{r}_\ell) U_a(\underline{r}-\underline{r}_\ell) \\ \times \sum_{m \neq 0} V_a(\underline{r}-\underline{r}_m) U_a(\underline{r}) d\underline{r} \Big] \Big/ (N+O_c) \quad . \end{aligned} \tag{4.25}$$

In the tight-binding approximation we assume that the overlapping of the atomic functions U_a are small. Hence $O_c \ll N$, and we may simplify (4.25) to

$$E = E_0 + E_{0c} \quad , \tag{4.26}$$

where E_{0c} is the difference between the atomic energy levels and the energy levels in the solid arising from the overlap of the atomic potentials and of the wave functions. It is given by

$$E_{0c} = \int \sum_\ell \exp(-i\underline{k}\cdot\underline{r}_\ell) U_a(\underline{r}-\underline{r}_\ell) \sum_{m \neq 0} V_a(\underline{r}-\underline{r}_m) U_a(\underline{r}) d\underline{r} \quad . \tag{4.27}$$

Only the terms corresponding to $\underline{r}_\ell = 0$ and $\underline{r}_\ell = \underline{r}_1$ ($\underline{r}_1$ is the position vector of the atom nearest to the origin) would be significant in the expression for E_{0c} since the overlap between the V_a or between the U_a are assumed to be small. E_{0c}, therefore, consists of two significant terms:

$$\Delta E_0 = -\int U_a(\underline{r}) \sum_m V_a(\underline{r}-\underline{r}_m) U_a(\underline{r}) d\underline{r} \quad , \tag{4.28}$$

and

$$E_1 = -\sum \exp(-i\underline{k}\cdot\underline{r}_1)\beta(\underline{r}_1) \quad , \tag{4.29}$$

where

$$\beta(\underline{r}_1) = -\int U_a(\underline{r}-\underline{r}_1) \sum_{m \neq 0} V_a(\underline{r}-\underline{r}_m) U_a(\underline{r}) d\underline{r} \quad .$$

The summation in (4.29) is over all the nearest neighbouring atoms of the atom at the origin.

ΔE_0 is a correction term giving the shift in the electron energy level when the atoms combine to form the solid. E_1, on the other hand, is the change in energy level, dependent on the wave vector $\underline{k}$, and causes a broadening of the levels and leads to bands. Its value is dependent on the crystal structure. For crystal structures, in which the atoms are symmetrically placed about the origin, $\beta(\underline{r}_1)$ will be the same for all neighbouring atoms and writing β for $\beta(\underline{r}_1)$ we get

$$E_1 = -\beta \sum \exp(-i\underline{k}\cdot\underline{r}_1) \quad . \tag{4.30}$$

The summation may be easily carried out for the crystal structures with simple cubic, body-centred-cubic, and face-centred-cubic lattices. The E_1 for these structures are given below:

Simple cubic: $$E_1 = -2\beta(\cos k_x a + \cos k_y a + \cos k_z a) \quad ; \tag{4.31}$$

bcc: $$E_1 = -8\beta\left(\cos\frac{k_x a}{2}\cos\frac{k_y a}{2}\cos\frac{k_z a}{2}\right) \quad ; \tag{4.32}$$

fcc: $$E_1 = -4\beta\left(\cos\frac{k_x a}{2}\cos\frac{k_y a}{2} + \cos\frac{k_y a}{2}\cos\frac{k_z a}{2} + \cos\frac{k_z a}{2}\cos\frac{k_x a}{2}\right) \quad ; \tag{4.33}$$

a is the lattice constant.

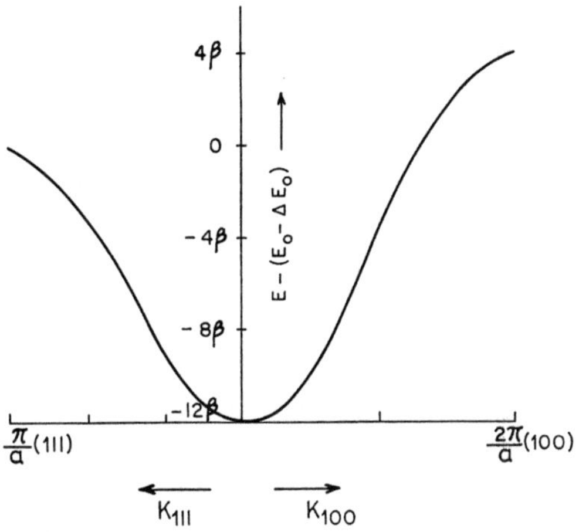

Fig. 4.3. E-k diagram for tight-binding approximation

We have plotted E against k_x for an fcc crystal in Fig.4.3. The energy levels are goruped into bands, the widths of which are determined by the value of β and the crystal structure. It is evident from (4.31-33) that the widths for cubic, bcc, and fcc lattices are given, respectively, by 12β, 16β, and 24β.

As is evident from the above discussion we may compute the E-k relation for crystals in which the assumptions of the tight-binding approximation holds provided β is known. It is possible to relate β to the effective mass of the electrons, since we find that the dispersion relation may be simplified for small values of k to

$$E = E_0 - \Delta E_0 + \beta(k^2a^2+\dots) - n\beta \quad , \tag{4.34}$$

where n = 6, 8, or 12.

Near the energy minima, therefore, the E-k relation is parabolic and the electrons will be characterised by an effective mass given by

$$1/m^* = 2\beta a^2/\hbar^2 \quad . \tag{4.35}$$

Knowing m* we may obtain β, since a may be easily determined. The complete E-k relation, which only involves β and a, may, therefore, be plotted.

It should be kept in mind that the results of the tight-binding approximation apply only when β is small, i.e. when m* is large. In the study of compound semiconductors we are, on the other hand, often concerned with materials for which m* is small, and this method may not be useful. The results presented above, however, make it clear that the atomic wave functions play a significant role in the formation of the electron wave functions for a solid. We also find that the E-$\underline{k}$ relation is, in general, nonparabolic and that the effective mass increases for levels away from the extrema. The results are, however, very useful for the understanding of the behaviour of semiconductor superlattices which we discuss in the next section.

4.4 Energy Bands in Semiconductor Superlattices

Semiconductor superlattice crystals are constructed by growing two similar but different semiconducting compounds in alternate layers with thicknesses two to twenty times the lattice spacing [4.5,6]. Structures have been realised with $GaAs_x P_{1-x}$ and $Ga_{1-x}Al_xAs$, in which the alternate layers have different values of $\underline{x}$ and thicknesses range between 10 and 200 Å. The materials forming the alternate layers have the same kind of band structure but different energy gaps. Part of the difference in band gap is due to the difference in the position of the valence band maximum, but most of the difference is a result of the difference in the position of the conduction band energy minima. The potential distribution in the direction of the growth of the layers (which we take as z direction) is shown in Fig.4.4. This periodic potential is due to the difference in band gaps and is superposed on the intrinsic periodic potential in the crystal. The exact nature of the variation of potential at the interface depends on the condition of growth of the layers and on the exact nature of the band structure of the two compounds.

The band structure of the complete structure depends on the magnitude of the electron mean free path ℓ in relation to the period d. For $\ell \ll d$, the electrons may be considered to belong to the individual layers and the band structure of interest is that of the compound forming the particular layer.

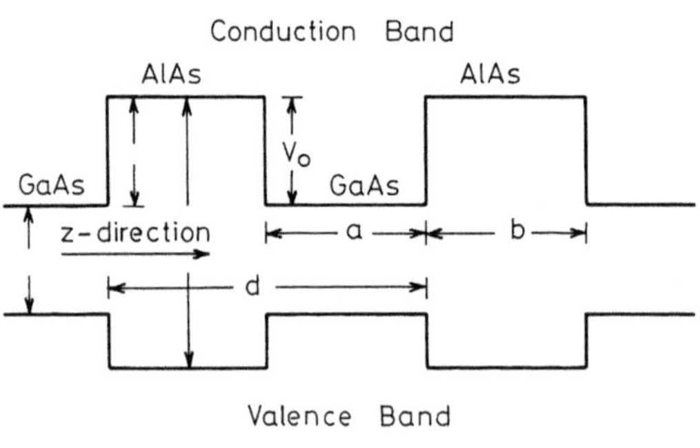

Fig. 4.4. Potential distribution in a semiconductor superlattice

On the other hand, when $\ell \gg d$, an electron may be considered to belong to the complete structure. The energy band diagram for directions transverse to the z direction remain unaffected, but that in the z direction is modified by the superlattice potential. The modified band diagram may be worked out by solving the Schrödinger equation with the extra periodic potential of the superlattice and using for the electron mass the effective mass of the electron in the material forming the superlattice.

In the simple case when the potential distribution is approximated by square wells as in Fig.4.4, the E-$\underline{k}$ relation is similar to that obtained from the Kronig-Penney model for solids and is given by [Ref.4.7, p.4; 4.8]

$$\cos[a(2m^*E)^{\frac{1}{2}}/\hbar] \cosh\left\{b[2m^*(U_0-E)]^{\frac{1}{2}}/\hbar\right\} + (U_0/2E-1)(U_0/E-1)^{-\frac{1}{2}}$$
$$\times \sin[a(2m^*E)^{\frac{1}{2}}/\hbar] \sinh\left\{b[2m^*(U_0-E)]^{\frac{1}{2}}/\hbar\right\} = \cos k(a+b) \quad , \qquad (4.36)$$

where U_0 is the height of the barriers, and a, b are the widths of the layers. Note that m* in (4.36) represents the effective mass of the carriers.

It is evident that as we get bands in solids due to the periodicity of the potential in the solid, in superlattice crystals, the z component of the carrier energy will break into minibands due to the periodic superlattice potential. The band is illustrated schematically in Fig.4.5. The discontinuity in the energy bands occurs for values of $k_z = \pm n\pi/(a+b)$; the widths of the allowed and the forbidden regions are determined by the barrier potential U_0 and the superlattice constants.

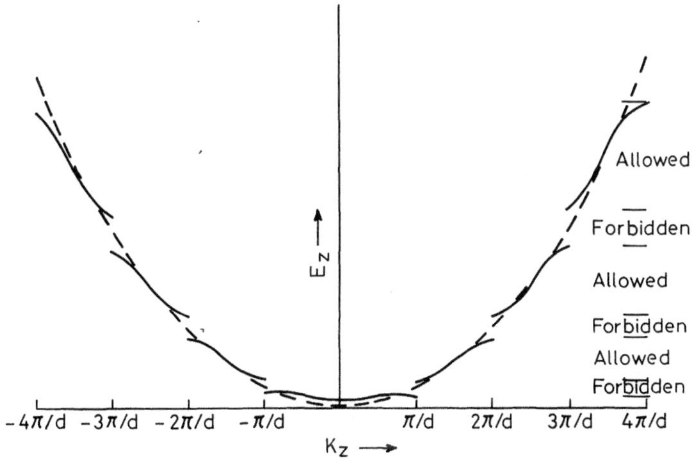

Fig. 4.5. Energy bands in a semiconductor superlattice

Equation (4.36) giving the E-$\underline{k}$ relation is not very convenient for the analysis of experiments or for the development of transport theory. It has been usual to use instead simplified relations, which are applicable for the two extreme conditions, $E \ll U_0$ and $E \gg U_0$. When $E \ll U_0$, we may use the tight-binding approximation. As shown earlier, we may then write for the wave function

$$\psi_s(\underline{r}) = C \exp[i(k_x x + k_y y)] \sum_n \exp[ik_z nd]\psi_s(z-nd) \quad , \tag{4.37}$$

where C is the normalising constant, and ψ_s is the wave function corresponding to a potential well of width d [Ref.4.4, p.88]. It should be noted that in this description, the electron is treated as a free electron for the x and y directions, but its wave function in the z direction is assumed to be modified by the superlattice potential.

The corresponding E-$\underline{k}$ relation may be simply expressed as

$$E_{s\underline{k}} = \hbar^2 k_\perp^2/(2m^*) + \Delta_s \cos k_z d + E_{zs} \quad , \tag{4.38}$$

where $k_\perp$ is the component of the wave vector perpendicular to the superlattice axis (the z axis), E_{zs} is the energy corresponding to the midpoint of the sth forbidden miniband, and Δ_s is the corresponding halfwidth.

We note that E_{zs} and Δ_s may be related to a, b, and U_0 for the square well potential model by simplifying (4.36). We obtain [4.9]

$$E_{zs} = E_s \quad , \quad \Delta_s = 4E_s \exp(-\beta_s b)/[\alpha_s \partial F(\alpha_s)/\partial \alpha_s] \quad , \tag{4.39}$$

where E_s is the value of E corresponding to the solution α_s of the following equation:

$$F(\alpha_s) = \left[\left(\beta_s^2 - \alpha_s^2\right)/(2\alpha_s\beta_s)\right] \sin \alpha_s a + \cos \alpha_s a = 0 \quad ,$$

$$\alpha_s = (2m^* E_s)^{1/2}/\hbar \quad ; \quad \beta_s = \left[2m^*(U_0 - E_s)\right]^{1/2}/\hbar \quad .$$

In real structures, however, the potential distribution is not like square wells and we must treat E_{zs} and Δ_s as experimental parameters.

For the other case, $E \gg U_0$, we may treat the electron as approximately free. The results derived earlier (Sect.3.1) for the periodic potential in the solid would apply for the wave function and the E-$\underline{k}$ relation. Near the extrema of the minibands we get [Ref.4.4, p.229]

$$E = \frac{\hbar^2 k^2}{2m^*} + \frac{\hbar^2}{2m^*}(nk_L)^2 \pm \left[\left(\frac{\hbar^2}{m^*}\right)^2 (nk_L)^2 (k_z - nk_L)^2 + (E_{gn}/2)^2\right]^{\frac{1}{2}} \quad , \tag{4.40}$$

where $k_L = \pi/d$, and

$$E_{gn} = (2U_0/n\pi)\ \sin(nk_L d) \quad . \tag{4.41}$$

We find that the energy bands in superlattice crystals are highly nonparabolic and anisotropic. The transport theory for these structures is, therefore, expected to be very different from that for regular crystals. We shall discuss in a later section the available results, most of which have been obtained by using the simplified wave functions and E-k relations given above.

It is, perhaps, evident that the description of the band structure given above rests on an idealised picture of the superlattice structure. The material forming the individual layers is assumed to be homogeneous and uniform and the width of the layers is also assumed to be uniform. In practical structures, these conditions are rather hard to meet at present, but very rapid progress is being made in realising idealised structures [4.5]. It is, however, more difficult to justify the assumption that the electron mean free path is very large compared with the superlattice period. In actual structures, the mean free path is 3-5 times the lattice period. Further, the mean free path is dependent on the electron energy; hence, the assumption is valid to different extents for electrons of varying energy. The E-k relation which applies to real superlattice crystals may, therefore, be more complex than predicted by the simple theory. In addition, the two compounds often have different kinds of band gap, one direct and the other indirect. Also, the effective mass in one compound may be isotropic, while that in the other compound may be anisotropic and have a very different value. Even the simple theory has to be refined to include the effect of these deviations from the simple picture. Work on these lines is underway [4.10], but real progress may be expected only when superlattice crystals are experimentally proved to be important materials.

4.5 The k·p Perturbation Method for Derivating E-k Relations

The approximation models discussed in Sects.4.1,2 are useful for understanding the formation of energy bands in solids. In some special materials, these models also give acceptable results. As discussed in the previous section, the results obtainable from these models are useful in the theory of semiconductor superlattices. However, the detailed structure of the energy bands can only be obtained by applying more involved methods of analysis.

We are mostly interested in the structure of the energy bands near the extrema for the development of transport theory and the required information is very often obtained from experiments. The experiments, however, give the characteristics of electrons having energies very close to those corresponding to the lowest minimum or highest maximum. In the interpretation of some experiments, on the other hand, the E-$\underline{k}$ relation for larger energies is required to be known. The $\underline{k}\cdot\underline{p}$ perturbation method has been mostly used in such cases.

The $\underline{k}\cdot\underline{p}$ perturbation method [4.11] is based on the fact that the cell periodic functions for the electrons for any $\underline{k}$ but different bands form a complete set and we may express the wave functions for electrons for any $\underline{k}$ in terms of the functions for the minima and maxima. Let us consider the wave function for values of $\underline{k}$ near the minimum located at $\underline{k}$-$\underline{k}_0$ in the nth band. The wave function may be written in terms of the cell periodic functions for $\underline{k}_0$ as

$$\psi = U_{n\underline{k}}(\underline{r})\exp(i\underline{k}\cdot\underline{r}) = \left[\sum c_m U_{m\underline{k}_0}(\underline{r})\right]\exp(i\underline{k}\cdot\underline{r}) \quad . \tag{4.42}$$

Replacing ψ by (4.42) in the Schrödinger equation we get

$$\left[-\left(\hbar^2/2m_0\right)\nabla^2+(\hbar/m_0)\underline{k}\cdot\underline{p}+\left(\hbar^2k^2/2m_0\right)+V(\underline{r})\right]U_{n\underline{k}}(\underline{r}) = E_n(\underline{k})U_{n\underline{k}}(\underline{r}) \quad . \tag{4.43}$$

Since $\exp(i\underline{k}_0\cdot\underline{r})U_{m\underline{k}_0}(\underline{r})$ is the wave function for $\underline{k}=\underline{k}_0$ in the nth band, $U_{m\underline{k}_0}(\underline{r})$ should satisfy the equation

$$\left[-\frac{\hbar^2}{2m_0}\nabla^2+\frac{\hbar}{m_0}\underline{k}_0\cdot\underline{p}+\frac{\hbar^2k_0^2}{2m_0}+V(\underline{r})\right]U_{m\underline{k}_0}(\underline{r}) = E_m(\underline{k}_0)U_{m\underline{k}_0}(\underline{r}) \quad , \tag{4.44}$$

where $E_m(\underline{k}_0)$ is the corresponding energy eigenvalue.

Replacing $U_{nk}(\underline{r})$ by (4.42) and using (4.44), we get from (4.43)

$$\sum c_m\left[E_m(\underline{k}_0)+\frac{\hbar^2}{2m_0}\left(k^2-k_0^2\right)+\frac{\hbar}{m_0}(\underline{k}-\underline{k}_0)\cdot\underline{p}\right]U_{m\underline{k}_0}(\underline{r}) = \sum c_m E_n(\underline{k})U_{m\underline{k}_0}(\underline{r}) \quad . \tag{4.45}$$

Multiplying the two sides of (4.45) by $U^*_{\ell\underline{k}_0}(\underline{r})$ and integrating over the volume of a unit cell (Ω) of the crystal we obtain the following set of homogeneous linear equations giving the c_m

$$c_\ell\left[E_n(\underline{k})-E_\ell(\underline{k}_0)-\frac{\hbar^2}{2m_0}\left(k^2-k_0^2\right)\right] - \sum c_m\frac{\hbar}{m_0}(\underline{k}-\underline{k}_0)\cdot\underline{p}_{\ell m} = 0 \quad , \tag{4.46}$$

where

$$\underline{p}_{\ell m} = \int_\Omega U^*_{\ell\underline{k}_0}(\underline{r})\underline{p}U_{m\underline{k}_0}(\underline{r})d\underline{r} \quad . \tag{4.47}$$

The full set of equations is obtained by giving ℓ successive integral values.

In the general case, the set of equations have a nontrivial solution if the determinant of the coefficients is zero. This condition gives the energy eigenvalues $E_n(k)$ in terms of the quantities $E_m(k_0)$ and $\underline{p}_{\ell m}$. The relative values of the c_m may then be obtained by using these values of $E_n(k)$ and the absolute value determined by applying the condition of normalisation on the resultant ψ. A general solution including the effects of all the bands, however, involves the solution of a determinant having too many terms and this is hardly necessary in practice. The problem may be simplified by introducing approximations and we consider below the results of some simplifying assumptions.

4.5.1 Single-Band Perturbation Theory

We assume that $U_{n\underline{k}}$ is mostly determined by $U_{n\underline{k}_0}$, the contribution of other terms being of a smaller order. In other words, we assume that the c_m are all small compared to c_n. We may then write

$$\psi = \left[c_nU_{n\underline{k}_0}(\underline{r})+\sum_{\substack{m\\m\neq n}} c_mU_{m\underline{k}_0}(\underline{r})\right]\exp(i\underline{k}\cdot\underline{r}) \quad . \tag{4.48}$$

We should note that ψ should be normalised. Hence

$$\sum_m c_m^2 = 1 \quad . \tag{4.49}$$

But, we have assumed that $c_m \ll c_n$. The normalisation condition thus gives $c_n \approx 1$.

We may solve (4.46) by the method of successive approximation. In the first step, neglecting c_m ($m\neq n$) in comparison to c_n in the nth equation we get

$$E_n(\underline{k}) = E_n(\underline{k}_0) + \frac{\hbar^2\left(k^2-k_0^2\right)}{2m_0} + \frac{\hbar}{m_0}(\underline{k}-\underline{k}_0)\cdot\underline{p}_{nn} \quad . \tag{4.50}$$

Further, c_n being approximately equal to unity in the mth equation, all the terms in the summation in (4.46) except the one containing c_n may be neglected. We thus get for c_m

$$c_m \simeq \frac{\hbar}{m_0} \frac{(\underline{k}-\underline{k}_0)\cdot \underline{p}_{mn}}{E_n(\underline{k}_0)-E_m(\underline{k}_0)} c_n \quad , \tag{4.51}$$

neglecting the $\underline{k}\cdot\underline{p}$ perturbation term in the denominator.

We may derive the second order approximate result by replacing c_m in the nth equation by the solutions given by (4.51). The result is

$$E_n(\underline{k}) = E_n(\underline{k}_0) + \frac{\hbar^2}{2m_0}\left(k^2-k_0^2\right) + \frac{\hbar}{m_0}(\underline{k}-\underline{k}_0)\cdot \underline{p}_{nn}$$
$$+\sum_{\substack{m\\ m\neq n}} \frac{\hbar^2}{m_0^2} |(\underline{k}-\underline{k}_0)\cdot\underline{p}_{nm}|^2 \Big/ \left[E_n(\underline{k}_0)-E_m(\underline{k}_0)\right] \quad . \tag{4.52}$$

It should be noted that $\underline{k}_0$ being a point of extremum [see (4.50)],

$$\underline{p}_{nn} = -\hbar\underline{k}_0 \quad . \tag{4.53}$$

We may then express (4.52) in a properly chosen coordinate system as

$$E_n(\underline{k}) = E_n(\underline{k}_0) + \frac{\hbar^2}{2}\sum_i (k_i-k_{0i})^2/m_i \quad , \tag{4.54}$$

where

$$\frac{1}{m_i} = \frac{1}{m_0} + \frac{2}{m_0^2}\sum_{\substack{m\\ m\neq n}} |\hat{\underline{i}}\cdot\underline{p}_{nm}|^2 \Big/ \left[E_n(\underline{k}_0)-E_m(\underline{k}_0)\right] \quad . \tag{4.55}$$

Here $\hat{\underline{i}}$ represents a unit vector in the direction of the ith coordinate axis and the subscript i indicates a component of the vector along the same direction. Evidently, m_i is the effective mass tensor component in the direction of $\hat{\underline{i}}$.

The single band perturbation theory leads us to the important result that near an extremum the E-$\underline{k}$ relation in all materials is parabolic. The approximation should be valid for small values of $\underline{k}-\underline{k}_0$ if the denominators $E_n(\underline{k}_0)-E_m(\underline{k}_0)$ are finite. However, the range of $\underline{k}-\underline{k}_0$ over which it is valid depends on the values of the $E_n(\underline{k}_0)-E_m(\underline{k}_0)$ terms or of m*, m* being related to the inverse of the $E_n(\underline{k}_0)-E_m(\underline{k}_0)$. In materials for which m* is small the range of validity of the parabolic relation would therefore be small. Unless $\underline{k} - \underline{k}_0$ is very small, the perturbation term cannot be considered small in materials with low effective mass, since the c_m, being inversely proportional to the $E_n(\underline{k}_0) - E_m(\underline{k}_0)$, are not small.

The analysis may, however, be improved by combining with the band under consideration those bands which are separated from it by small energies and treating the effects of other bands as small perturbations. The wave function $U_{n\underline{k}}$ should be considered to be constituted by the combination of the functions for the bands close to the band under consideration. The energy eigenvalues may then be worked out following a procedure in which these bands are simultaneously considered. We illustrate this method in the following section considering only two bands, the conduction band and the valence band.

4.5.2 Two-Band Approximation

We shall discuss the method assuming that the conduction band minimum and the valence band maximum are at the zone centre. The effect of the spin-orbit interaction will be neglected for the present and the valence band will be assumed to be triply degenerate.

Let U_c, U_{V1}, U_{V2}, and U_{V3} represent respectively the cell periodic functions for the conduction band and the three degenerate valence bands. Let E_c and E_V be the energies corresponding to the conduction band minimum and the valence band maximum. As we are assuming that the conduction band and the valence bands are strongly coupled, the wave function may be written as

$$\psi = a_{\underline{k}} U_c + b_{\underline{k}} U_{V1} + d_{\underline{k}} U_{V2} + c_{\underline{k}} U_{V3} \quad , \tag{4.56}$$

where the symbols used for the coefficients are chosen to be in conformity with the literature. It may be noted that in contrast to the case considered earlier, all the coefficients $a_{\underline{k}}$, $b_{\underline{k}}$, and $c_{\underline{k}}$ are expected to be of the same order for the case under consideration. Equations giving the energy eigenvalues may now be obtained from (4.46), keeping only the terms corresponding to the conduction band and the degenerate valence bands and neglecting all other terms. We thus get

$$a_{\underline{k}}(E'-E_c) - (\hbar/m_0)\underline{k} \cdot (b_{\underline{k}}\underline{p}_{cV1} + d_{\underline{k}}\underline{p}_{cV2} + c_{\underline{k}}\underline{p}_{cV3}) = 0 \quad ,$$

$$-a_{\underline{k}}(\hbar/m_0)\underline{k} \cdot \underline{p}_{cV1} + b_{\underline{k}}(E'-E_V) - (\hbar/m_0)\underline{k} \cdot (d_{\underline{k}}\underline{p}_{V1V2} + c_{\underline{k}}\underline{p}_{V1V3}) = 0 \quad ,$$

$$-a_{\underline{k}}(\hbar/m_0)\underline{k} \cdot \underline{p}_{cV2} - b_{\underline{k}}(\hbar/m_0)\underline{k} \cdot \underline{p}_{V1V2} + d_{\underline{k}}(E'-E_V) - (\hbar/m_0)\underline{k} \cdot c_{\underline{k}}\underline{p}_{V1V3} = 0 \quad ,$$

$$-a_{\underline{k}}(\hbar/m_0)\underline{k}\cdot\underline{p}_{cV3} - b_{\underline{k}}(\hbar/m_0)\underline{k}\cdot\underline{p}_{V1V3} - d_{\underline{k}}(\hbar/m_0)\underline{k}\cdot\underline{p}_{V2V3} + c_{\underline{k}}(E'-E_V) = 0 \quad , \tag{4.57}$$

where $E' = E - \hbar^2k^2/2m_0$, and E is the energy eigenvalue; $\underline{p}_{cV1}$, $\underline{p}_{V1V2}$, etc., represent the matrix elements for the coupling between the various bands.

The matrix elements may be evaluated if U_c, U_{V1}, U_{V2}, and U_{V3} are known. For the material under consideration, U_c is the atomic s-type function and U_{V1}, U_{V2}, U_{V3} are the p-type functions, but the functions have the periodicity of the lattice. We choose the z axis for the p-type functions to be the same as that of the wave vector $\underline{k}$ for the sake of convenience. On the basis of these considerations and including the spin vector we may choose iS↑, (X+iY)↓, Z↑, (X-iY)↓, iS↓, (X-iY)↑, Z↓, (X+iY)↑ as the base vectors for ψ. Here S is the s-type function for the conduction band, X, Y, Z are the p-type functions for the valence bands; ↑ indicates the spin direction and the directions of the component functions are chosen to be in conformity with the bonding behaviour, for a discussion of which the reader may consult [4.1,2].

It may be readily verified that

$$\underline{k}\cdot\underline{p}_{cV1} = \underline{k}\cdot\underline{p}_{cV2} = \underline{k}\cdot\underline{p}_{V1V2} = \underline{k}\cdot\underline{p}_{V2V3} = \underline{k}\cdot\underline{p}_{V1V3} = 0 \quad . \tag{4.58}$$

The only non-zero $\underline{k}\cdot\underline{p}$ perturbation term is

$$(\hbar/m_0)\underline{k}\cdot\underline{p}_{cV3} = (\hbar/m_0)\underline{k}\cdot(-i\int S\underline{p}Zd\underline{r}) = kP \quad , \tag{4.59}$$

where $P = -(i\hbar/m_0)\int Sp_zZd\underline{r}$.

Using the base vectors mentioned above, we get an 8×8 matrix for the secular equation, which may be written in the form

$$\begin{vmatrix} D & 0 \\ 0 & D \end{vmatrix} = 0 \quad , \tag{4.60}$$

where D is a 4×4 matrix. This form is obtained if the base vectors are combined in the order in which they are written.

Equations (4.57) thus reduce to

$$\begin{aligned} &a_{\underline{k}}(E'-E_c) - c_{\underline{k}}Pk = 0 \quad , \\ &b_{\underline{k}}(E'-E_V) = 0 \quad , \\ &-a_{\underline{k}}Pk + c_{\underline{k}}(E'-E_V) = 0 \quad , \\ &d_{\underline{k}}(E'-E_V) = 0 \quad . \end{aligned} \tag{4.61}$$

Note that the first four vectors, and the second four vectors have the same coefficients and the two sets account for the spin degeneracy.

The energy eigenvalues are thus given by

$$E = E_V\ ,\ E_V\ ;\ (E'-E_c)(E'-E_V) - P^2k^2 = 0\ . \tag{4.62}$$

The above equations give the dispersion relation when the conduction and valence bands are strongly coupled and the extrema are at the zone centre. The dispersion relation for two valence bands having wave functions (X+iY) and (X-iY) is

$$E = E_V + (\hbar^2k^2/2m_0)\ . \tag{4.63}$$

The relation for the conduction band and for the other valence band is to be obtained from (4.62). We find that the relation involves the unknown P, direct evaluation of which requires the knowledge of the wave functions.

We may, however, express P in terms of the effective mass for the band edge and of the energy band gap in the following way. Let E_g be the energy band gap. Equation (4.62) may then be written as

$$(E'-E_c)(E'-E_c+E_g) - P^2k^2 = 0\ . \tag{4.64}$$

When E' tends to E_c, we may neglect $E'-E_c$ in comparison to E_g and simplify (4.64) to

$$(E'-E_c)E_g = P^2k^2\ . \tag{4.65}$$

If the band-edge effective mass is m*, we get

$$P^2 = (E-E_c-\hbar^2k^2/2m_0)E_g/k^2 = \hbar^2(1/m^*-1/m_0)E_g/2\ . \tag{4.66}$$

The dispersion relation as obtained from the two-band approximation is thus

$$E'(1+\alpha E') = \hbar^2(1/m^*-1/m_0)k^2/2\ , \tag{4.67}$$

where $\alpha = 1/E_g$, and E' is measured from the conduction band minimum. We have written the above relation with reference to the conduction band electrons, but similar results for the valence band may also be obtained following the same procedure.

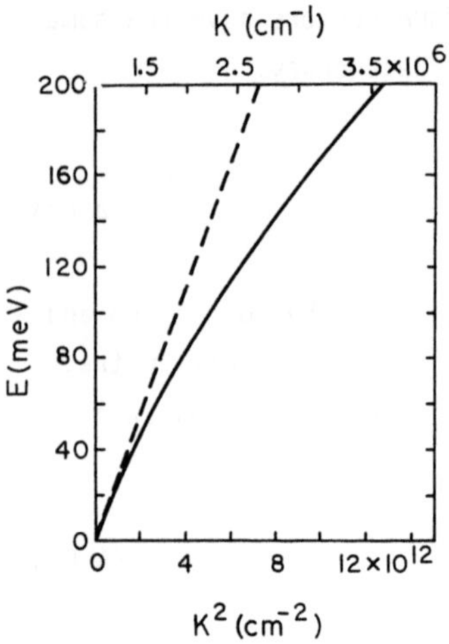

Fig. 4.6. E-$\underline{k}$ diagram for indium antimonide illustrating nonparabolicity [4.33]

Equation (4.67) is often further simplified, noting that $m_0 \gg m^*$ and $\hbar^2k^2/2m_0$ may be neglected in comparison to E, and $1/m_0$ in comparison to $1/m^*$. We then get

$$\hbar^2k^2/2m^* = E(1+\alpha E) \quad . \tag{4.68}$$

The above relation is frequently used to analyse the effects of nonparabolicity of the E-$\underline{k}$ relation on the transport properties of compound semiconductors. We illustrate in Fig.4.6 the E-$\underline{k}$ relation obtained from (4.68) for InSb as an example. It fits the experimental curve [4.33] almost exactly.

We may now evaluate the constants $a_{\underline{k}}$, $b_{\underline{k}}$, and $c_{\underline{k}}$ to obtain the wave function. For the conduction band we get from (4.61)

$$a_{\underline{k}}E - Pkc_{\underline{k}} = 0 \quad . \tag{4.69}$$

In addition, since the wave function should be normalised it is also required that

$$a_{\underline{k}}^2 + c_{\underline{k}}^2 = 1 \quad . \tag{4.70}$$

Solving (4.69) and (4.70) and using (4.68) we get for a_k and c_k

$$a_{\underline{k}} = [(1+\alpha E)/(1+2\alpha E)]^{\frac{1}{2}} \quad ; \quad c_{\underline{k}} = [\alpha E/(1+2\alpha E)]^{\frac{1}{2}} \quad . \tag{4.71}$$

The same expressions also apply to the valence band corresponding to the fourth equation of (4.61). The coefficients $b_{\underline{k}}$ and $d_{\underline{k}}$ for the other valence band function should be chosen as $1/\sqrt{2}$ for normalisation.

4.5.3 Effect of Spin-Orbit Interaction

In the theory illustrated above we have neglected the effect of spin-orbit interaction. The interaction between the spinning and orbital motion of the electrons introduces an extra term in the Hamiltonian which was not included in (4.43). It is given by [4.12]

$$H_{spin} = \frac{\hbar}{4m_0^2c^2}\,[\underline{\nabla}V\times\underline{p}]\cdot[\sigma] \quad , \tag{4.72}$$

where V is the potential energy and the components of $[\sigma]$ are the spin matrices, the axis of quantisation being taken to be that of $\underline{k}$. The effect of this term may be treated as a perturbation and the energy eigenvalues may be worked out following the procedure explained in the earlier section. On including the effect of the spin-orbit interaction we get the following equations:

$$\begin{aligned}
&a_{\underline{k}}(E'-E_c) - c_{\underline{k}}0k = 0 \quad , \\
&b_{\underline{k}}(E'-E_v+2\Delta/3) - c_{\underline{k}}\sqrt{2}\Delta/3 = 0 \quad , \\
&a_{\underline{k}}Pk + b_{\underline{k}}\sqrt{2}\Delta/3 - c_{\underline{k}}(E'-E_v+\Delta/3) = 0 \quad , \\
&d_{\underline{k}}(E'-E_v) = 0 \quad ,
\end{aligned} \tag{4.73}$$

where Δ represents the spin-orbit splitting. It is given by [4.12]

$$\Delta = -3i\left(\frac{\hbar}{4m_0^2c^2}\right)\langle X|\underline{\nabla}V\times\underline{p}|_y Z\rangle \quad .$$

The wave function for the conduction band, light-hole valence band, and the split-off band, which are all coupled in the dispersion relation matrix, has been taken as

$$\psi = a_k iS\uparrow + (b_k/\sqrt{2})(X+iY)\downarrow + c_k Z\uparrow \quad . \tag{4.74}$$

The ψ function for the opposite spin state is obtained by reversing the spins. The function for the heavy-hole band is $\psi = d_{\underline{k}}(X-iY)$ with the two possible spins.

The energy eigenvalues are given by equating the determinant of the coefficients of $a_{\underline{k}}$, $b_{\underline{k}}$, $c_{\underline{k}}$ to zero. The equation giving the values is

$$(E'-E_c)(E'-E_V)(E'-E_V+\Delta) - P^2k^2(E'-E_V+2\Delta/3) = 0 \quad . \tag{4.75}$$

The coefficients $a_{\underline{k}}$, $b_{\underline{k}}$, $c_{\underline{k}}$ for the wave function are obtained by applying the normalisation condition, and (4.73) and (4.75). The constants are given by [4.13]

$$\begin{aligned}
a_{\underline{k}} &= r\left\{\left[E_g-\gamma_k^2(E_g-\delta')\right]/(E_g+\delta')\right\}^{\frac{1}{2}} \quad ; \quad r = \left[6(E_g+2\Delta/3)(E_g+\Delta)/(6E_g^2+9E_g\Delta+4\Delta^2)\right]^{\frac{1}{2}} \; ; \\
\delta' &= E_g^2\Delta/(6E_g^2+9E_g\Delta+4\Delta^2) \quad ; \quad \gamma_k = [(\xi_k-E_g)/2(\xi_k+\delta')]^{\frac{1}{2}} \quad ; \\
\xi_k &= \hbar^2k^2/2m^* \quad ; \quad b_{\underline{k}} = s\gamma_k \quad ; \quad s = \left[4\Delta^2/3(6E_g^2+9E_g\Delta+4\Delta^2)\right]^{\frac{1}{2}} \quad ; \\
c_{\underline{k}} &= t\gamma_k \quad ; \quad t = \left[6(E_g+2\Delta/3)^2/(6E_g^2+9E_g\Delta+4\Delta^2)\right]^{\frac{1}{2}} \quad .
\end{aligned} \tag{4.76}$$

It is of interest to note that the above relations reduce for the conduction band to those derived earlier by neglecting the effect of spin-orbit interaction when $\Delta \ll E_g$ and also when $\Delta \gg E$ and E_g. Since in most of the materials one of these conditions is obeyed, the formulas of the earlier section are mostly used in practice. There are, however, some materials (e.g. InAs) in which Δ is comparable to E_g, and we are required to use the complete expressions of this section.

The matrix element P may be eliminated as in the earlier case and (4.75) may be written in terms of m* (the band edge effective mass) as

$$\frac{\hbar^2k^2}{2}\left(\frac{1}{m^*}-\frac{1}{m_0}\right) = E'\left[\frac{E_g+2\Delta/3}{E_g(E_g+\Delta)}\left(\frac{(E'+E_g)(E'+E_g+\Delta)}{E'+E_g+2\Delta/3}\right)\right] \quad . \tag{4.77}$$

The E-$\underline{k}$ relation given by (4.77) is too complex for computations. If, however, E is small in comparison to E_g, the relation may again be simplified to that given in (4.68). But, α is now given by [4.14]

$$\alpha = \frac{1}{E_g}\left(1-\frac{m^*}{m_0}\right)\left(1-\frac{E_g\Delta}{3(E_g+2\Delta/3)(E_g+\Delta)}\right) \quad . \tag{4.78}$$

The E-$\underline{k}$ relation obtained for the valence band from the above first-order approximate theory is, however, not correct as it predicts negative mass for holes. To obtain the correct relation, the analysis has to be extended to the second-order approximation where we get the E-$\underline{k}$ relation for valence bands [4.12] noted earlier (see Sect.3.4.1).

4.5.4 Nonparabolic Relation for Extrema at Points Other than the Γ Point

The discussion in the preceding sections has been confined to the case where the extrema are located at the Γ point. We saw in Chap.2, where the E-$\underline{k}$ diagrams of various materials are discussed, that this is the case in most of the compound semiconductors. There are, however, some III-V compounds like GaP, GaSb, AlAs, AlSb, AlP in which the conduction band minima are in the <111> or <100> directions away from the Γ points. In IV-VI compounds, on the other hand, the extrema for both the conduction band and the valence band lie at the L point. In these materials, the band gap is also small so that effects of nonparabolicity would be important. It is generally accepted that the E-$\underline{k}$ relation should be similar to (4.67). The relation which has been used in some analyses is [4.15]

$$\frac{\hbar^2}{2}\left(\frac{k_\perp^2}{m_\perp}+\frac{k_\parallel^2}{m_\parallel}\right) = E(1+\alpha E) \quad , \tag{4.79}$$

where $k_\parallel$ and $k_\perp$ represent respectively the components of the wave vector in the direction parallel to and perpendicular to the direction in which the minimum occurs; $m_\parallel$ and $m_\perp$ are the corresponding effective-mass tensor components. The constant α is taken as $1/E_g$; but E_g now represents the direct energy gap between the conduction band and the valence band at the point of the extremum.

The assumption basic to the above relation that the nonparabolicity is independent of the direction of $\underline{k}$ must be considered as simplifying the picture, perhaps beyond justification. The application of the formula to any actual analysis also involves uncertainties, as the direct gap E_g is not experimentally obtainable for extrema away from the zone centre and the values are usually taken from theoretical band-structure calculations. These values often do not agree with those required to fit the experimental data on the basis of (4.79).

In most of the materials having the extrema at the X or L point, the direct band gap is large and the effect of nonparabolicity is small. The formula (4.79) may, therefore, be used to estimate its magnitude in spite of the shortcomings discussed above. However, for the interpretation of hot electron transport or free-carrier absorption results, for which the electrons under consideration penetrate deep into the bands, we need an exact formulation of the nonparabolicity in anisotropic bands. Such a formulation is also necessary for the interpretation of the transport results on lead chalcogenides for which, as mentioned earlier, the bands are anisotropic and the energy band gap is also small. It should be mentioned that for these materials an alternative relation has been given by COHEN [4.16]:

$$\left(E - \frac{\hbar^2 k_{\parallel}^2}{2m_{\parallel n}}\right)\left(1+\alpha E + \frac{\hbar^2 k_{\perp}^2}{2m_{\parallel p}}\right) = \frac{\hbar^2 k_{\perp}^2}{2m_{\perp n}} , \tag{4.80}$$

where $m_{\parallel n}$ and $m_{\parallel p}$ are the effective mass components in the longitudinal direction for electron and hole respectively. Experimental results appear to support this model [4.17].

4.6 External Effects on Energy Bands

We discussed in Chap.3 the energy band structure of semiconductors as found in pure crystals in the absence of externally applied fields. The band structure is altered when the crystals are heavily doped or when very large electric or magnetic fields are applied. Doping alters the band structure as the atoms of the doping material disturbs the periodic potential at random. The alteration becomes significant with increasing concentration of the doping material. In many applications very heavily doped compound semiconductors are used, so knowledge of the band structure of heavily doped materials is often necessary for the analysis of practical devices.

External fields alter the band structure by affecting the dynamics of the electrons. The strength of the fields for which the effects are significant is inversely proportional to the effective mass of the electrons. Effects are produced for moderate values of fields in n-type compound semiconductors as the effective mass of electrons is usually small. The effects lead to many interesting transport phenomena which have been widely studied for the determination of various constants used in the theory of transport. Acquaintance with these effects is often necessary to interpretate other transport results.

We shall discuss in this section the effects of both doping and external fields. The effects of doping are considered first.

4.6.1 Effects of Doping

In lightly doped semiconductors, the band structure of the pure material may be assumed to prevail. The potential wells associated with impurity atoms are isolated from each other and cause perturbations in the motion of the free electrons. Discrete energy levels are also produced, the positions of which may be worked out by applying the Bohr theory of the hydrogen atom. These

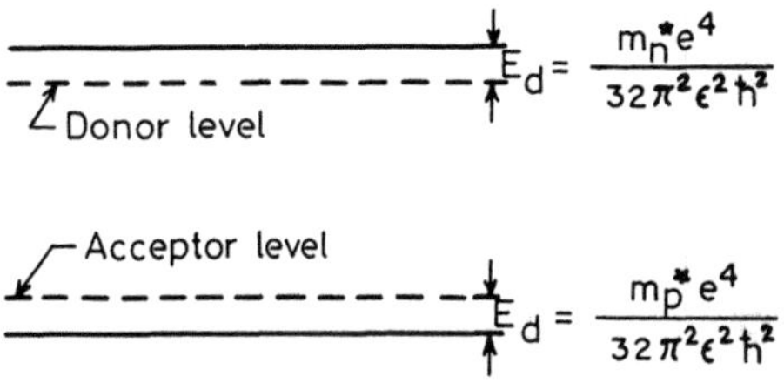

Fig. 4.7. Impurity levels

levels are separated by the ionisation energy from the conduction band, for donor atoms; and from the valence band, for acceptor atoms (see Fig.4.7). As the wave function for the outer electrons of the impurity atoms covers a large number of atoms of the host crystal, the value of the permittivity to be used in the theory is that of the host material. The mass of the electron should also be taken to be the effective mass appropriate to the conduction band for donor atoms, and to the valence band for acceptor atoms. The separation of the impurity levels from the lower edge of the conduction band or the upper edge of the valence band (as the case may be) is therefore given by

$$E_d = m^* e^4 (32\pi^2\varepsilon^2\hbar^2)^{-1} \quad , \tag{4.81}$$

where m^* is the effective mass of the conduction electrons for donor atoms and that of holes for acceptor atoms. The value of E_d is directly proportional to m^* and inversely proportional to the square of the dielectric constant. It should be noted that E_d has the same theoretical value for different atoms contributing the same number of carriers. It is, however, found from experiments that this result, although valid for many impurity atoms, is not true for all.

The total number of energy levels per unit volume of the crystal contributed by the impurity atoms is usually taken to be equal to their concentration as each atom ordinarily contributes one level. Electrons occupying these levels are not available for conduction. But the electron or hole arising from the ionisation of the impurity atoms behaves in the same way as other free electrons or holes.

The potential wells associated with the impurity atoms overlap as the doping level is increased. The overlapping of the potential wells transforms the discrete levels to energy bands as in perfect crystals. It should, however, be borne in mind that the distribution of the impurity atoms is random, and it is almost an impossible task to determine completely and repeatably the characteristic band structure by experiment. The distribution of states has

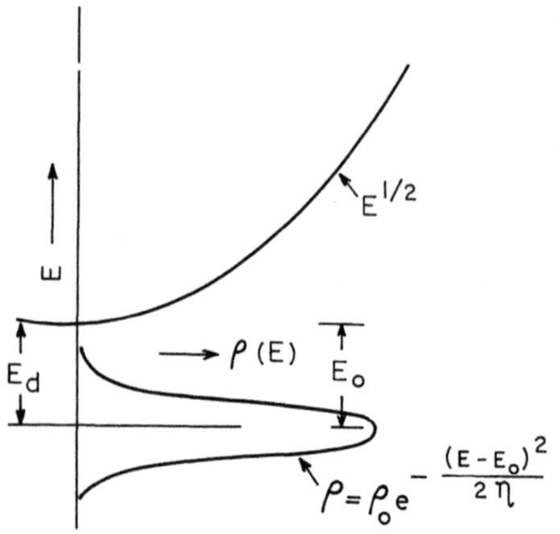

Fig. 4.8. Impurity states in a heavily doped semiconductor

been worked out theoretically by MORGAN [4.18], who shows (as illustrated in Fig.4.8) that the density of the levels should vary as

$$\rho(E) = \rho_0 \exp\left[-(E-E_0)^2/\eta\right] \quad , \tag{4.82}$$

where ρ_0 is the density of states for $E = E_0$, η is the mean deviation, and E_0 is the average of the Gaussian function. The constants ρ_0 and η are related to the impurity concentration, the screening length, and the distribution of the impurity atoms. For detailed expressions for these parameters the reader should consult MORGAN's paper. It may, however, be mentioned that for not-too-heavy doping, E_0 is close to the impurity level discussed earlier.

The impurity levels are thus transformed into Gaussian bands with increase in doping. Electrons occupying these bands take part in transport phenomena, but their nature is not clearly understood. Results of some experiments, particularly at low temperatures, have been explained by assuming that the electrons move by hopping [4.19] from one impurity site to another. But, in some experiments, these electrons have been found to behave very much like conduction electrons with a different effective mass. Though experimental values of the effective mass are available [4.20], no theoretical dispersion relation has yet been worked out.

The nature of impurity levels differs radically in strongly doped and in lightly doped materials. The results reported in the literature are somewhat confusing and we shall describe briefly the results about which there is some agreement. We shall consider for our discussion only n-type materials, there being little difference in the behaviour of the two types of materials. For very heavy doping, the impurity band widens greatly. At the same time, the

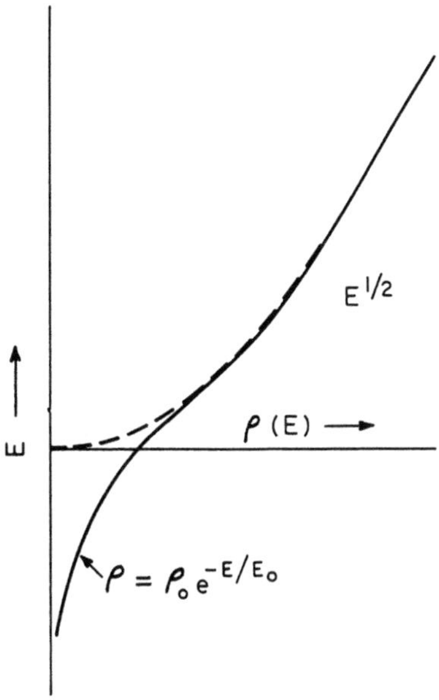

Fig. 4.9. Band tails and impurity states

high concentration of ionised donors provides an attractive potential which lowers the edge of the conduction band. As a result, the distinction between the conduction band and the impurity band vanishes. The edge of the conduction band drops into the forbidden band, forming the so-called band tail.

The electrons suffer frequent collisions under conditions in which the tail is formed so that no stationary state can be visualised for the electrons and it is not very meaningful to talk about a dispersion relation. The dynamical behaviour of the electrons cannot, therefore, be described in the conventional formalism of band theory. We may, however, evaluate the density of states for the tail region, and most of the available results on the behaviour of electrons in the band tail are related to the density of states. Information about the band tail has been obtained experimentally from the study of light absorption, luminescence, junction behaviour, and devices made of degenerate materials [4.21]. It is found that experimental results may be fitted with theory if we assume an exponential density of states in the band tail given by (see Fig.4.9)

$$\rho(E) = \rho_0 \exp(-E/E_0) \quad , \tag{4.83}$$

where ρ_0 is the density of states at E equal to zero, and E_0 is the band tailing parameter often determined from experiments. In the expression given above E is measured from the edge of the conduction band as found in lightly doped materials.

Attempts have been made to derive theoretically the density of states for the conditions of very heavy doping. We shall not discuss these theories, but quote only the results which are often used for the analysis of transport phenomena in heavily doped materials. Specifically, we give below the formulae derived by KANE [4.22], HALPERIN and LAX [4.23], and BONCH-BRUEVICH [4.24].

KANE predicts a tail shape given by

$$\rho(E) = T(E/\eta)m^{*3/2}(2\eta)^{\frac{1}{2}}V\pi^{-2}\hbar^{-3} \quad , \tag{4.84}$$

where $T(x) = \pi^{-\frac{1}{2}}\int_{-\infty}^{x}(x-\zeta)^{\frac{1}{2}}\exp(-\zeta^2)d\zeta$, m^* is the density-of-states effective mass in the bulk, V is the volume of the crystal, $\eta = (4\pi N_i/K_D)^{\frac{1}{2}}(e^2/4\pi\varepsilon_d)$, ε_d is the permittivity, $K_d = 2N_i^{1/6}(3/\pi)^{1/6}a_0^{*-\frac{1}{2}}$, $a_0^* = 4\pi\varepsilon_d\hbar^2/m^*e^2$, and N_i is the impurity concentration.

HALPERIN and LAX, on the other hand, show that the density of states varies with energy very nearly exponentially. The explicit formula derived by them is

$$\rho(E) = \rho_0 \exp(-|E|^n/E_0) \quad . \tag{4.85}$$

The value of n lies between 0.5 and 2 and is dependent on the value of the energy.

BONCH-BRUEVICH has derived two asymptotic formulae of the form

$$\rho(E) = \rho_0\left[1-a_0^2/4E^2\right] \quad , \tag{4.86}$$

where ρ_0 is the density of states for the parabolic band;

$$\rho(E) = (2m^*)^{3/2}a_0^{\frac{1}{2}}\Gamma(3/4)\pi^{-5/2}\hbar^{-3}\left\{1+\Gamma(1/4)a_0/[4\Gamma(3/4)E]\right\} \quad , \tag{4.87}$$

where $\Gamma(n)$ is the gamma function of argument n. The former is valid for E nearly equal to the Fermi energy, and the latter for the tail region.

It is evident that the mathematical expressions given by these authors are quite different from each other. It has not been possible to decide exactly which of the formulae is more appropriate. The particular constants may be chosen for each of them to fit broadly the experimental results.

In recent years, there has also been some discussion in the literature about the density of states for carrier concentrations intermediate between the extreme values for which band tails and Gaussian impurity bands are formed. It has been found that the experimental results may be explained by combining the Gaussian density of states according to MORGAN with the tail states according to BONCH-BRUEVICH [4.25] or by taking the envelope of the combination [4.26].

It appears that we know only broadly the nature of the distortion of the band structure produced by heavy doping. For light doping we find that we may rely fully on our knowledge of the band structure of pure materials. For very heavy doping, however, detailed transport calculations are not possible at the present stage and we shall have to wait till an accepted theory of electron states in heavily doped semiconductors is available.

4.6.2 Effects of Large Magnetic Fields

The band structure may be altered by applying an electric or a magnetic field. The electric field causes a shift in the band edge, the magnitude of which is related to the electric field by the relation

$$\Delta E = (3|e|\mathscr{E}\hbar/4\sqrt{2m^*})^{2/3} \quad . \tag{4.88}$$

This effect is known as Franz-Keldysh effect [4.27]. Calculations show that for the production of a measurable shift a very high value of the electric field is required. The effect is thus observed in few experiments, but there is one example in which the effect is observed in a practical device. The field in the domains of Gunn diodes is often very high, and it has been observed that a light beam may be modulated by the domain field [4.28] through the Franz-Keldysh effect. We shall not, however, discuss the effect any further as it occurs rarely in the usual transport experiments.

The effect of the magnetic field on the band structure is more striking and is easily observed in experiments. A number of interesting transport phenomena originate from the alteration in the band structure by magnetic fields; these have been extensively studied and serve as diagnostic tools for characterising the materials. We shall, therefore, discuss this effect in detail. The basic theory was developed by LANDAU [4.29] to explain the diamagnetic behaviour of quasi-free electrons in a solid, which are expected not to produce such an effect from classical theory. In the presence of a magnetic field, the electrons experience the transverse Lorentz force and curl round the magnetic field and are expected to contribute to the magnetic moment of the system. But for a confined electron system, the total sum when worked out classically turns out to be zero. LANDAU's theory, which is based on the solution of the Schrödinger equation, shows, on the other hand, that the energy of the electrons corresponding to the transverse components of the wave vector are quantised. These quantised levels are called the Landau levels. It is found that the diamagnetic contribution of the quasi-free electrons even in a confined system is finite due to this quantisation. It is also this quantisation which leads to various interesting transport phenomena in the presence of magnetic fields.

Expressions for the energy of the quantised levels are obtained, as mentioned above, by solving the Schrödinger equation, which in the presence of a magnetic field has the form

$$(2m^*)^{-1}(-i\hbar\underline{\nabla}+|e|\underline{A})^2\psi - E\psi = 0 \quad , \tag{4.89}$$

where $-i\hbar\underline{\nabla}$ is the momentum operator, $\underline{A}$ is the vector potential due to the magnetic field, and m^* is the effective mass of the electrons [4.30], which takes into account the effect of the periodic crystal potential. The above formulation would be valid provided the electron orbit extends over a large multiple of lattice spacings. We have neglected the energy arising from the spin of the electrons.

The magnetic vector potential $\underline{A}$ is related to the magnetic induction by the equation

$$\underline{B} = \underline{\nabla} \times \underline{A} \quad . \tag{4.90}$$

In the problem under consideration $\underline{B}$ is specified and we shall assume it to be in the z direction. $\underline{A}$ cannot be uniqually expressed in terms of $\underline{B}$, as we may add to $\underline{A}$ the gradient of any scalar function without affecting (4.90). For the sake of mathematical convenience we choose this function, the so-called Coulomb gauge, such that

$$\underline{\nabla} \cdot \underline{A} = 0 \quad . \tag{4.91}$$

We may then take for $\underline{A}$,

$$\underline{A} = \hat{\underline{j}}Bx \quad . \tag{4.92}$$

It is easy to verify that for this expression for $\underline{A}$, (4.90) and (4.91) are satisfied. We could alternatively choose the function $(\underline{r}\times\underline{B})/2$, but the subsequent analysis then involves a little more algebra.

Replacing $\underline{A}$ in (4.89) by (4.92) and rearranging the terms we get

$$\left[\left(\frac{\partial^2}{\partial x^2}+\frac{\partial^2}{\partial z^2}\right)+\left(\frac{\partial}{\partial y}+\frac{i}{\hbar}|e|\,Bx\right)^2+\frac{2m^*}{\hbar^2}\,E\right]\psi = 0 \quad . \tag{4.93}$$

It is seen that the magnetic field introduces only an additional x-dependent term, so that y and z dependence of the ψ function is unaltered. We may hence try a solution of the form

$$\psi = \psi_1(x)\,\exp\left[i(k_y y+k_z z)\right] \quad . \tag{4.94}$$

Substituting in (4.93) we get

$$\frac{d^2\psi_1}{dx^2} + \left[\frac{2m^*}{\hbar^2}\,E-k_z^2-\left(k_y+\frac{|e|Bx}{\hbar}\right)^2\right]\psi_1 = 0 \quad . \tag{4.95}$$

Putting $x_1 = x - \hbar k_y(|e|B)^{-1}$, we obtain

$$\left(-\frac{\hbar^2}{2m^*}\frac{d^2}{dx_1^2}+\frac{e^2B^2}{2m^*}x_1^2\right)\psi_1(x_1) = \left(E-\frac{\hbar^2k_z^2}{2m^*}\right)\psi_1 \quad . \tag{4.96}$$

Equation (4.96) is identical to the Schrödinger equation for the one-dimensional harmonic oscillator of angular frequency $\omega_B = |e|B/m^*$. The wave function and eigenvalues for (4.96) are, thus, given by [Ref.4.7,p.31]

$$\psi = A\left\{\exp\left[\frac{|e|B}{2\hbar}\left(x-\frac{\hbar k_y}{|e|B}\right)^2\right]\right\}H_n\left(x-\frac{\hbar k_y}{|e|B}\right)\exp\left[i(k_y y+k_z z)\right] \quad ,$$

$$E_\ell = (\ell+1/2)\hbar|e|B/m^* \quad , \tag{4.97}$$

where ℓ is an integer. The energy eigenvalues of quasi-free electrons in the presence of a magnetic field are hence

$$E_{\ell,k_z} = \frac{\hbar^2k_z^2}{2m^*} + \left(\ell+\frac{1}{2}\right)\frac{\hbar|e|B}{m^*} \quad . \tag{4.98}$$

The above E-$\underline{k}$ relation is illustrated in Fig.4.10.

In the derivation given above we have assumed that the effective mass of the particles is isotropic and independent of energy. Also the effect of the

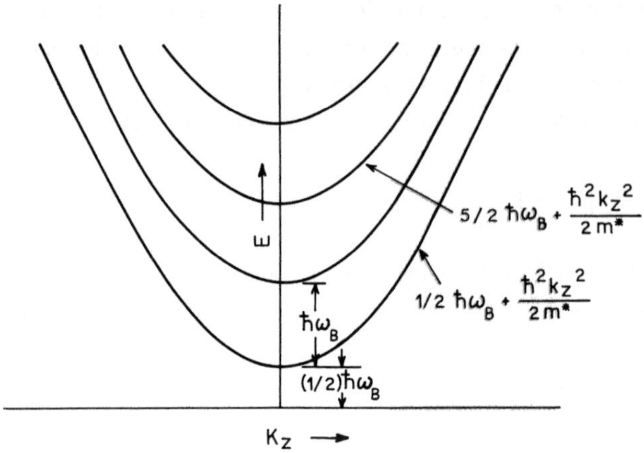

Fig. 4.10. Landau levels

electron spin has been neglected. For the anisotropic effective mass appropriate to ellipsoidal constant energy surfaces, the Landau levels are given by [4.31]

$$E = |e|\hbar(m_1m_2m_3)^{-\frac{1}{2}}\left(\alpha_1^2m_1+\alpha_2^2m_2+\alpha_3^2m_3\right)^{\frac{1}{2}}B(\ell+\frac{1}{2}) \quad , \tag{4.99}$$

where m_1, m_2, and m_3 are the components of the effective mass tensor, and α_1, α_2, and α_3 are the direction cosines of $\underline{B}$ relative to the principal axes of the constant energy ellipsoids.

The effect of the electron spin is to introduce a term $\pm(1/2)g\mu_BB$ to the energy for the Landau levels, where g is the spectroscopic splitting factor and μ_B is the Bohr magneton. The expression for the energy of Landau levels for nonparabolic bands is very involved for a general nonparabolic E-$\underline{k}$ relation. For the dispersion relation of (4.75), the expression for E_ℓ is, when $m_0 \gg m^*$ [4.32],

$$\begin{aligned} E &= \frac{\hbar|e|B}{m^*}\,\frac{E_g(E_g+\Delta)}{3E_g+2\Delta}\,(\ell+\frac{1}{2})\left(\frac{2}{E+E_g}+\frac{1}{E+E_g+\Delta}\right) \\ &\pm\frac{1}{2}g\mu_BB\frac{E_g(E_g+\Delta)}{\Delta}\left(\frac{1}{E+E_g}+\frac{1}{E+E_g+\Delta}\right) \quad . \end{aligned} \tag{4.100}$$

It may be readily shown that for $\Delta \ll E_g$ and also for $E_g \gg \Delta$, when (4.68) applies, equation (4.100) simplifies to

$$E_\ell = \frac{\hbar|e|B}{m^*(1+\alpha E)}\,(\ell+\frac{1}{2}) \pm \frac{1}{2}\,\frac{g\mu_BB}{1+\alpha E} \quad , \tag{4.101}$$

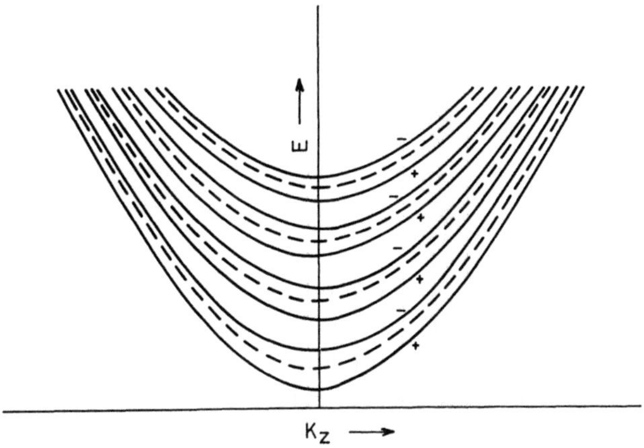

Fig. 4.11. Landau levels in nonparabolic bands including the effect of spin

where $\alpha = 1/E_g$. Equation (4.97) thus applies to this case, provided m^* is interpreted as $(\hbar^2 k^2/2E)$. The energy levels are illustrated in Fig.4.11.

The important point to remember about the effect of the magnetic field is that the general characteristics of the band structure remain unaltered, but in each band the energy of the electron corresponding to the velocity transverse to the magnetic field becomes discrete and the low conduction band edge is shifted upward by $\hbar\omega_B/2$. The discreteness in the levels leads to a host of interesting transport phenomena, important among which are (i) de Haas-Van Alphen oscillations in magnetic susceptibility; (ii) Shubnikov-de Haas oscillations in magnetoresistance; (iii) magnetophonon oscillations in magnetoresistance and thermoelectric power; (iv) oscillations in the absorption coefficient of light; and (v) cyclotron resonance. All these phenomena result from the restrictions on electron motion due to the discretisation of the energy levels. These effects are discussed in Chap.10.

5. Electron Statistics

Electrons in the conduction band and holes in the valence band behave as free particles, but their distribution among the available energy levels, when in thermal equilibrium with the lattice, obeys Fermi-Dirac statistics [5.1]. The function giving the occupancy of the levels, often called the Fermi function, is

$$n_s/g_s = 1/\left\{1+\exp\left[(E_s-E_F)/k_BT\right]\right\} \quad , \tag{5.1}$$

where n_s is the number of electrons occupying levels of energy E_s, the number of which is g_s; k_B is the Boltzmann constant, T is the lattice temperature, and E_F is referred to as the Fermi energy.

To apply the Fermi function in transport calculations, knowledge of g_s and E_F is required. The density of states g_s is essentially determined by the E-$\underline{k}$ relation, whereas E_F is determined by the energy band structure, the doping of the materials, and the operating temperature. We shall discuss in this section the relation of E_F to these parameters for equilibrium conditions. It may be mentioned that the equilibrium may be disturbed by applying high electric fields or by generating excess carriers with light. These nonequilibrium conditions will be discussed in later sections, using the discussion in this section.

We may develop expressions giving the Fermi energy E_F in terms of the band structure parameters and the density and ionisation energy of impurity atoms. However, in most transport experiments, it is the carrier concentration that is directly determined, and the Fermi energy is then calculated from it. We shall first derive the relations used for this purpose. These relations assume different forms of different band structures; we shall derive only those relations which are frequently used.

5.1 Fermi Energy for Parabolic Bands

Equation (5.1) gives the probability of a particular energy level being occupied. To get the number of electrons occupying the level we have to know g_s, the number of such available levels. Using the condition that the wave function at the two ends of a crystal should be the same, the so-called cyclic boundary condition, it may be shown that the number of states per unit volume of the crystal and per unit volume in $\underline{k}$ space is $(1/2\pi)^3$.

If, now, $g(\underline{k})d\underline{k}$ denotes the number of states per unit volume of the crystal in the volume element of $\underline{k}$ space lying between $\underline{k}$ and $\underline{k}+d\underline{k}$, then

$$g(\underline{k})d\underline{k} = (1/2\pi)^3 d\underline{k} \quad . \tag{5.2}$$

The number of electrons occupying these levels is

$$dn = 2(1/2\pi)^3\left\{1+\exp\left[(E-E_F)/k_BT\right]\right\}^{-1} d\underline{k} \quad . \tag{5.3}$$

The factor of 2 is introduced because each level can accommodate two electrons of opposite spin.

The relation between E_F and the electron concentration n is now established by integrating (5.3) over all possible values of $\underline{k}$. Although the maximum value of $\underline{k}$ is limited by the size of the Brillouin zone, the upper limit of integration is taken to be infinity for mathematical convenience. Extension of the limit to infinity does not introduce any error as the integrand reduces to zero for large values of energy due to the exponential function in the denominator. It is this integration where the actual band structure, the E-$\underline{k}$ relation to be more exact, is involved.

For parabolic isotropic bands, the E-$\underline{k}$ relation is

$$E = \hbar^2k^2/2m^* \quad . \tag{5.4}$$

Transforming from $\underline{k}$ space to energy space, we get

$$n = 4\pi\left(\frac{2m^*}{h^2}\right)^{3/2}\int_0^\infty \frac{E^{\frac{1}{2}}dE}{1+\exp[(E-E_F)/k_BT]} \quad , \tag{5.5}$$

where $h = 2\pi\hbar$ is the Planck's constant. It may be noted that the density of states in the energy space varies as $E^{\frac{1}{2}}$, as mentioned earlier in connection with the discussion of the effects of heavy doping.

The integral in (5.5) cannot, in general, be evaluated analytically. However, when E_F is negative and $|E_F|$ is much larger than k_BT, i.e. when the material is nondegenerate, the Fermi function simplifies to the Maxwellian function since we may neglect 1 in comparison to $\exp[(E-E_F)/k_BT]$. We then get

$$n = 4\pi\left(\frac{2m^*}{h^2}\right)^{3/2} \exp(E_F/k_BT) \int_0^\infty E^{\frac{1}{2}} \exp(-E/k_BT)dE$$

$$= 2\left(\frac{2\pi m^* k_BT}{h^2}\right)^{3/2} \exp(E_F/k_BT) = N_c \exp(E_F/k_BT) \quad , \qquad (5.6)$$

where $N_c = 2(2\pi m^* k_BT/h^2)^{3/2} = 2.51 \times 10^{25}(m^*/m_0)^{3/2}(T/300)^{3/2}$ per cubic metre.

When E_F is positive or not much lower than k_BT of band edge, i.e. when the material is degenerate, the integral has to be evaluated numerically. Values of the integral are, however, available in tabulated form. The integrals given in these tables are written as

$$F_j(\eta) = \frac{1}{\Gamma(j+1)} \int_0^\infty \frac{W^j dW}{1+\exp(W-\eta)} \quad , \qquad (5.7)$$

which are often referred to as the Fermi integrals [5.2].

Equation (5.5) may be converted to the form (5.7) by putting $E/k_BT = W$ and $E_F/k_BT = \eta$. We may then write

$$n = N_c \frac{2}{\sqrt{\pi}} \int_0^\infty \frac{W^{\frac{1}{2}}\, dW}{1+\exp(W-\eta)} = N_c F_{\frac{1}{2}}(\eta) \quad . \qquad (5.8)$$

We give in Fig.5.1 the plot of $F_{\frac{1}{2}}(\eta)\exp(-\eta)$. Note that for large negative values of η, n/N_c is equal to $\exp(\eta)$ as has been derived in (5.6). One may calculate η using the values of m^*, T, and n from the above relation. On the other hand, when η is known, one may then obtain N_c and m^*. A table of values of $F_{\frac{1}{2}}(\eta)$ is given in Appendix A for this purpose.

Equation (5.8) may also be simplified when η is very large, i.e. under conditions of extreme degeneracy. Then the Fermi function may be taken as 1 for $W \leq \eta$ and as 0 for $W > \eta$. The integral then reduces to

$$n = N_c \frac{2}{\sqrt{\pi}} \int_0^\eta W^{\frac{1}{2}}\, dW = N_c \frac{4}{3\sqrt{\pi}}\, \eta^{3/2} \quad . \qquad (5.9)$$

The formulae given above have been derived assuming an isotropic E-$\underline{k}$ relation. These may be extended directly to the case of ellipsoidal constant energy surfaces, since we may transform the ellipsoidal surfaces to spherical

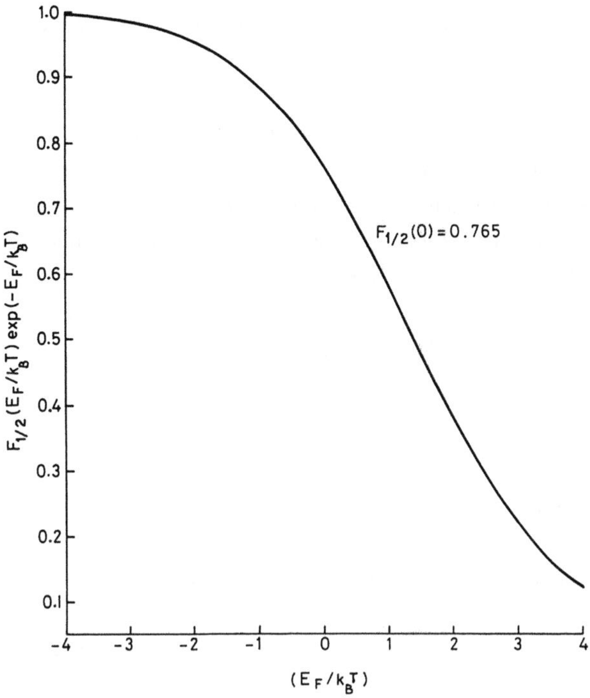

Fig. 5.1. $F_{\frac{1}{2}}(E_F/k_BT)\exp(-E_F/k_BT)$ plotted against E_F/k_BT

surfaces using the transformation (often called the Herring-Vogt transformation) [5.3],

$$\underline{k}^* = \underline{\underline{\alpha}} \cdot \underline{k} \quad ; \quad \underline{\underline{\alpha}} = \begin{bmatrix} (m_0/m_1)^{\frac{1}{2}} & 0 & 0 \\ 0 & (m_0/m_2)^{\frac{1}{2}} & 0 \\ 0 & 0 & (m_0/m_3)^{\frac{1}{2}} \end{bmatrix} \quad . \tag{5.10}$$

Using the above transformation we may write

$$\frac{\hbar^2}{2}\left(\frac{k_1^2}{m_1} + \frac{k_2^2}{m_2} + \frac{k_3^2}{m_3}\right) = \frac{\hbar^2 k^{*2}}{2m_0} = E \quad . \tag{5.11}$$

Following the same procedure as outlined before we find that (5.8) also applies to the case of ellipsoidal constant energy surfaces, but N_c is now given by

$$N_c = 2(2\pi m_d k_B T/h^2)^{3/2} \quad , \tag{5.12}$$

where $m_d = (m_1 m_2 m_3)^{1/3}$; m_d is usually called the density-of-states effective mass.

A similar formula may also be derived for warped constant energy surfaces. However, we shall not discuss it here as warped surfaces are characteristic of p-type material. Compound semiconductors that are p type have not been much studied, mainly because the hole mobility is very low. It may, however, be pointed out that although detailed expressions may be developed taking into account the full complexity introduced by warping of the constant energy surfaces, it is more usual to approximate the warped surfaces by spherical surfaces and use (5.12) with a suitably defined m_d. For a detailed discussion of warped surfaces, see [5.4,5].

5.2 Fermi Energy for Nonparabolic Bands

The E-$\underline{k}$ relation for isotropic nonparabolic bands may be written as

$$\frac{\hbar^2 k^2}{2m^*} = \gamma(E) \quad , \tag{5.13}$$

where m* is the band-edge effective mass, and $\gamma(E)$ is a fairly involved function of the carrier energy E (measured from the band edge) in the general case.

The relation between the concentration of electrons and the Fermi level is now given by

$$\begin{aligned} n &= \frac{1}{4\pi^3} \int \frac{d\underline{k}}{1+\exp[(E-E_F)/k_BT]} \\ &= N_c \frac{2}{\sqrt{\pi}(k_BT)^{3/2}} \int_0^\infty \frac{\gamma^{\frac{1}{2}}(d\gamma/dE)dE}{1+\exp[(E-E_F)/k_BT]} \quad . \end{aligned} \tag{5.14}$$

As mentioned earlier, the commonly used relation for γ is (see Sect.3.5.1)

$$\gamma = \gamma(E) = E(1+\alpha E) \quad ; \quad \alpha \simeq 1/E_g \quad .$$

Equation (5.14) may be simplified if αE_F and αk_BT are small. We may then write

$$\begin{aligned} n &\simeq N_c \frac{2}{\sqrt{\pi}(k_BT)^{3/2}} \left[\int_0^\infty \frac{E^{\frac{1}{2}}dE}{1+\exp[(E-E_F)/k_BT]} + \frac{5\alpha}{2} \int_0^\infty \frac{E^{3/2}dE}{1+\exp[(E-E_F)/k_BT]} \right] \\ &= N_c\left[F_{\frac{1}{2}}(\eta)+\frac{15}{4}\alpha_n F_{3/2}(\eta)\right] \quad ; \quad \alpha_n = \alpha k_BT \quad . \end{aligned} \tag{5.15}$$

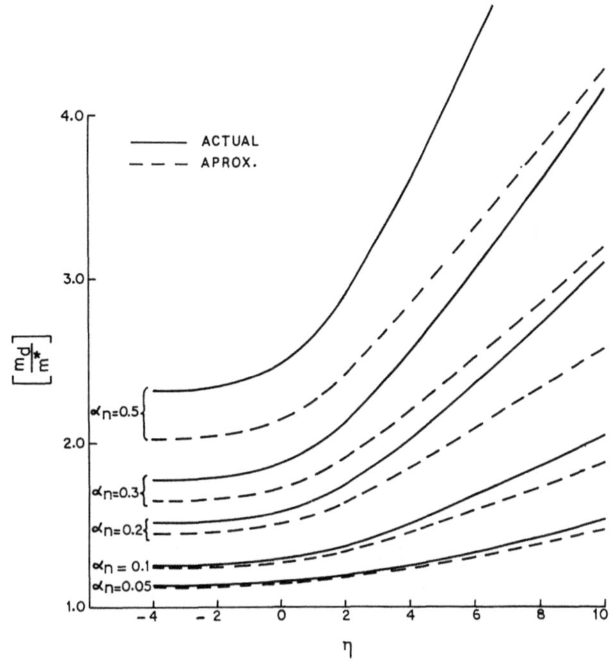

Fig. 5.2. Density of states effective mass for nonparabolic bands

We may rewrite (5.16) in the form of (5.8) by defining a density-of-states effective mass given by

$$m_d = m^*\left[1+(15/4)\alpha_n F_{3/2}(\eta)/F_{\frac{1}{2}}(\eta)\right]^{2/3} \quad . \tag{5.16}$$

The density-of-states effective mass thus varies with η. The variation is illustrated in Fig.5.2 in which m_d is plotted against η for different values of α_n. Equation (5.16) will not be applicable when E_F or k_BT is comparable to E_g. The effective density-of-states mass should then be obtained from the relation

$$m_d = m^*\left(\frac{2}{\sqrt{\pi}}\int\limits_0^\infty \frac{x^{\frac{1}{2}}(1+\alpha_n x)^{\frac{1}{2}}(1+2\alpha_n x)}{\exp(x-\eta)+1}\,dx\right)^{2/3} \Big/ [F_{\frac{1}{2}}(\eta)]^{2/3} \quad .$$

Values obtained from this relation are also shown in Fig.5.2. We find that the approximate relation gives fairly good results upto about $\alpha_n = 0.1$.

It is of interest to note that the study of light absorption in InSb provided the first experimental evidence of the nonparabolicity of energy bands

[5.6]. The absorption coefficient of light signals due to interband transitions in a semiconductor is dependent on the density-of-states effective mass. Study of light absorption for various carrier concentrations showed that the effective mass increases with the carrier concentration indicating a nonparabolic E-$\underline{k}$ relation.

The nonparabolic relation for energy bands with ellipsoidal constant energy surfaces being similar to that considered above (see Sect.4.5.4), (5.16) applies also to such bands, but m^* is replaced by $(m_1m_2m_3)^{1/3}$.

5.3 Fermi Energy in the Presence of a Quantising Magnetic Field

We have seen earlier that the electron states are radically altered in the presence of a strong magnetic field. The alteration in the E-$\underline{k}$ relation and in the electron wave function changes the density of states. Hence (5.3) cannot be applied to this case and, for the derivation of the relation between n and E_F, we must first derive some basic relations giving the density of states under magnetically quantised conditions.

5.3.1 Density of States

The wave function of the electrons in the presence of a quantising magnetic field is given by (see Sect.4.6.2)

$$\psi = A\left\{\exp\left[\frac{|e|B}{2\hbar}\left(x-\frac{\hbar k_y}{|e|B}\right)^2\right]\right\}H_\ell\left(x-\frac{\hbar k_y}{|e|B}\right)\exp[i(k_y y+k_z z)] \quad , \tag{5.17}$$

where A is the normalisation constant, and H_ℓ is the Hermite polynomial of order ℓ. We find that the wave function is periodic in the direction of the applied magnetic field, the z direction, and also in the y direction, which is chosen arbitrarily. The number of states in these two directions are, therefore, determined by the cyclic boundary conditions and are given by $L_3/2\pi$ and $L_2/2\pi$ per unit lengths of k_z and k_y where L_3 and L_2 are the dimensions of the crystal in these two directions. We also note that the energy E is independent of k_y but varies continuously with k_z. The energy levels, which are quasi-continuously distributed in $\underline{k}$ space in the absence of a magnetic field, are redistributed on the surfaces of cylinders of discrete diameters under the action of the quantising magnetic field (Fig.5.3). Since the energies corresponding to values of $\underline{k}$ on these surfaces are independent of k_y, the energy levels should be considered degenerate. The boundary conditions in the x direction for the wave functions give the degeneracy of the energy

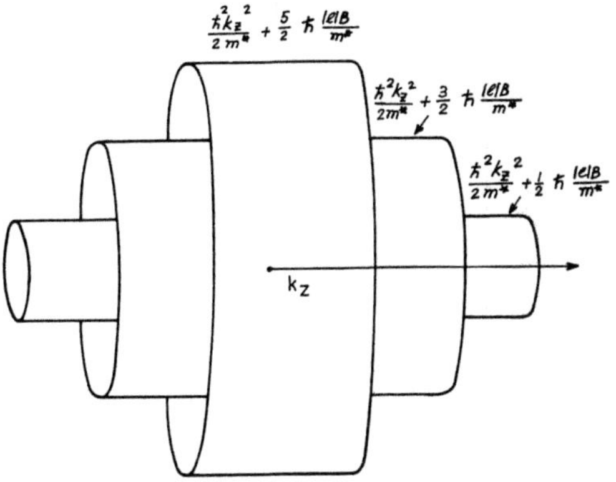

Fig. 5.3. Energy surfaces in a quantising magnetic field

levels. As is evident from (5.17), the wave function is not periodic in the x direction; so that the periodic boundary condition cannot be applied. The function is localised around $x = \hbar k_y/|e|B$, the extent being of the order of $(2\ell\hbar/|e|B)^{\frac{1}{2}}$ [Ref.5.7, p.95], which is a small quantity even for moderately high values of B. The boundary condition that may be applied for the x direction is that $\hbar k_y/|e|B$ should lie within 0 and L_1, where L_1 is the dimension of the crystal in the x direction. These limiting values of k_y would cause ψ to be practically zero at the boundaries of the crystal in the x direction, since ψ extends a very small distance from $\hbar k_y/|e|B$. In thin samples, however, the extent of ψ away from $\hbar k_y/|e|B$ may be comparable to the dimensions of the sample. The results obtained on the basis of the above boundary condition would, therefore, apply strictly to thick samples only. The maximum length of k_y is thus $L_1|e|B/\hbar$ and the degeneracy of the levels is $(L_2/2\pi) \times (L_1|e|B/\hbar)$, as the number of values of k_y per unit length of $\underline{k}$ space along the y direction is $(L_2/2\pi)$.

The number of states per unit volume of the crystal associated with each cylindrical surface of length dk_z, lying between k_z and $k_z + dk_z$ is, therefore,

$$dg(k_z) = |e|B(4\pi^2\hbar)^{-1}dk_z \quad . \tag{5.18}$$

It is of interest to note that the magnetic field does not affect the total number of states but only alters the distribution of the states. In the presence of the magnetic field, the states condense on the circles in the k_x-k_y space defined by

$$(\hbar^2/2m^*)\left(k_x^2+k_y^2\right) = (2\ell+1)\hbar\omega_B/2 \quad . \tag{5.19}$$

The electron states normally distributed in the space extending between the circles of radii $(2m^*/\hbar^2)^{\frac{1}{2}}(\ell\hbar\omega_B)^{\frac{1}{2}}$ and $(2m^*/\hbar^2)^{\frac{1}{2}}(\ell+1)^{\frac{1}{2}}(\hbar\omega_B)^{\frac{1}{2}}$ coalesce on the circle of radius $(2m^*/\hbar^2)^{\frac{1}{2}}[(2\ell+1)/2]^{\frac{1}{2}}(\hbar\omega_B)^{\frac{1}{2}}$. This may be easily demonstrated considering that the number of states per unit cross-sectional area of the crystal in this space in the absence of magnetic field is

$$(4\pi^2)^{-1}\pi\hbar\omega_B(2m^*/\hbar^2) = |e|B(2\pi\hbar)^{-1} \quad . \tag{5.20}$$

This is identical to the degeneracy, as shown before, associated with the cylindrical surfaces per unit cross-sectional area of the crystal. The equivalence discussed here is often useful for the derivation of density of states for complex band structures.

The density of states in the energy space may now be derived by using the dispersion relation in the presence of quantising magnetic field along with (5.18). We get from the dispersion relation (4.98)

$$dk_z = \left(\frac{m^*}{\hbar^2}\right)\frac{dE}{k_z} = \frac{m^*}{\hbar^2}\,\frac{dE}{\left(\frac{2m^*}{\hbar^2}\right)^{\frac{1}{2}}\left[E-(2\ell+1)\,\frac{\hbar\omega_B}{2}\right]^{\frac{1}{2}}} \quad . \tag{5.21}$$

Substituting in (5.18) we obtain

$$g(E)dE = \frac{1}{4\pi^2}\left(\frac{2m^*}{\hbar^2}\right)^{3/2}\frac{\hbar|e|B}{m}\sum_{\ell=0}^{\ell_{max}}\left\{E-(2\ell+1)\frac{\hbar\omega_B}{2}\right\}^{-\frac{1}{2}}dE \quad , \tag{5.22}$$

where $g(E)dE$ gives the number of states lying between E and $E+dE$ and ℓ_{max} is defined by

$$(2\ell_{max}+3)\,\frac{\hbar\omega_B}{2} > E > (2\ell_{max}+1)\,\frac{\hbar\omega_B}{2} \quad . \tag{5.23}$$

In (5.22) a factor of 2 is introduced by the spin degeneracy of the levels. A second factor of 2 occurs because the same value of E corresponds to both $+k_z$ and $-k_z$. Further, the states lying between E and $E+dE$ lie on different subbands and hence for obtaining the total number of states we have to sum over the various subbands. The subband corresponding to ℓ_{max} is such that E is larger than the lower edge of this subband but smaller than the lower edge of the next higher subband.

It may be shown that (5.22) reduces to the expression for the density of states in the absence of magnetic field when B tends to zero. For low magnetic fields we may assume that

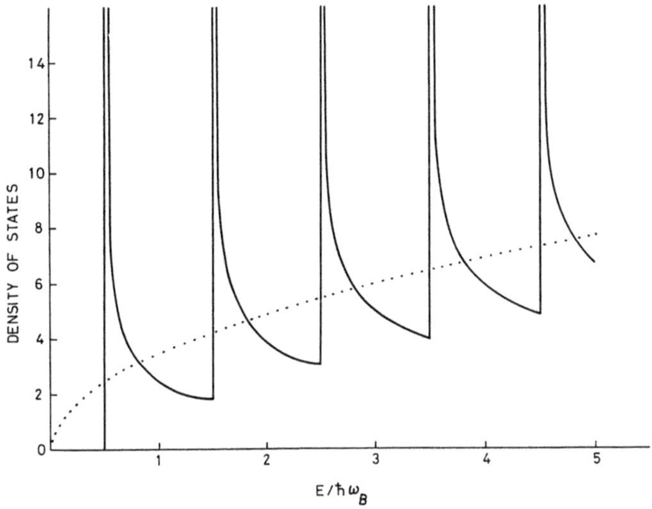

Fig. 5.4. Density of states in a quantising magnetic field for parabolic bands

$$\ell_{max} = E/\hbar\omega_B \quad , \quad \hbar\omega_B(2\ell+1)/2 \simeq \ell\hbar\omega_B \quad ,$$

and replace the summation by an integration. We then get

$$g(E) = \left(\frac{2m^*}{\hbar^2}\right)^{3/2}(4\pi^2)^{-1}\int_0^E \frac{d(\ell\hbar\omega_B)}{(E-\ell\hbar\omega_B)^{\frac{1}{2}}} = 4\pi(2m^*/h^2)^{3/2}E^{\frac{1}{2}} \quad . \tag{5.24}$$

On the other hand, when B is large the density of states departs radically from the $E^{\frac{1}{2}}$ behaviour. The variation of g(E) with E for large values of B is illustrated in Fig.5.4. It is seen that the function is zero up to a value of E of $\hbar\omega_B/2$ and becomes infinity at the points $E=(2\ell+1)\hbar\omega_B/2$. The periodic increase in the value of the density of states is the basis of the various oscillatory phenomena at high magnetic fields.

We should remember that in the preceding discussion we have considered a perfectly crystalline material. In imperfect materials the energy subbands are broadened and cause the density of states to be finite at the points of singularity [5.8]. The peculiar effects of quantisation are not evident unless the broadening is small. Thermal vibrations are one of the main sources of imperfection of the crystals. Hence the oscillatory phenomena characteristic of quantisation are observed only when thermal vibrations do not destroy the discrete nature of the energy levels and usually a fairly low temperature is required for the observation of the effects.

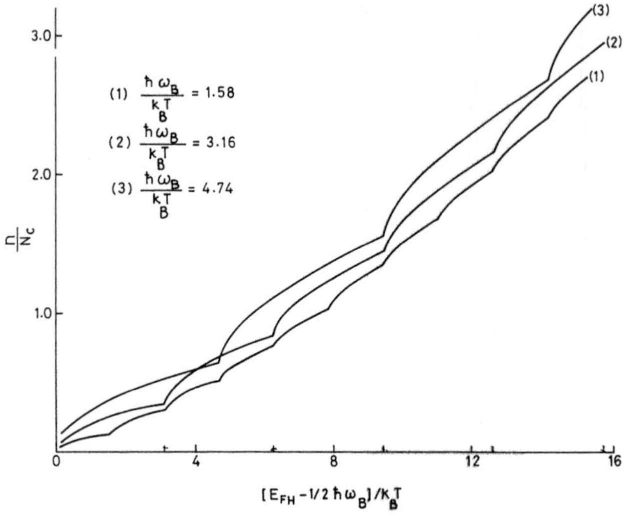

Fig. 5.5. Plot of n/N_C against $\eta'(=\eta-\theta/2)$ showing the effect of magnetic field on the Fermi level for a constant electron concentration

5.3.2 Fermi Level

The relation between carrier concentration and Fermi level is obtained by using (5.22) and (5.1). We get

$$n = N_C \hbar\omega_B \pi^{-\frac{1}{2}} (k_BT)^{-3/2} \int_0^{\infty} \sum_{\ell=0}^{\ell_{max}} \left\{1 + \exp\left[(E-E_F)/k_BT\right]\right\}^{-1} \left\{E-(2\ell+1)\frac{\hbar\omega_B}{2}\right\}^{-\frac{1}{2}} dE \quad . \tag{5.25}$$

We may express (5.25) in terms of the Fermi integrals by changing the order of integration and summation. We obtain

$$n = N_C \theta \pi^{-\frac{1}{2}} \sum_{\ell=0}^{\infty} \int_0^{\infty} W_\ell^{-\frac{1}{2}} \left\{1 + \exp[W_\ell + (\ell+1/2)\theta - \eta]\right\}^{-1} dW_\ell$$

$$= N_C \theta \sum_{\ell=0}^{\infty} F_{-\frac{1}{2}}[\eta-(\ell+1/2)\theta] \quad , \tag{5.26}$$

where $\hbar\omega_B/k_BT = \theta$; $E_F/k_BT = \eta$; $\frac{E}{k_BT} - (\ell+\frac{1}{2})\theta = W_\ell$.

The Fermi level is thus modified by the magnetic field. In Fig.5.5, we have plotted n/N_C as a function of $\eta'(=\eta-\theta/2)$. The quantisation of the bands shifts the conduction band edge by $\hbar\omega_B/2$. It is, therefore, legitimate to compare the value of η with η in the absence of the magnetic field to assess the effect of the magnetic field on the Fermi level. We find that the Fermi level is in general lowered by the magnetic field and the material becomes less degenerate when a magnetic field is applied.

Equation (5.26) is applicable for all values of B. We may simplify the relation for two limiting conditions, nondegenerate and extremely degenerate conditions. For nondegenerate conditions, the relation simplifies to [5.9]

$$n = N_c\theta \sum_{\ell=0}^{\infty} \exp\left[\eta-(\ell+\frac{1}{2})\theta\right] = N_c \exp(\eta)(\theta/2) \operatorname{cosech}(\theta/2) \quad . \tag{5.27}$$

On the other hand, under the conditions of extreme degeneracy, i.e. when η is very large, we get

$$n = N_c(2/\sqrt{\pi})\theta^{3/2} \sum_{\ell=0}^{\ell_{max}} \left[(\eta/\theta)-(\ell+\frac{1}{2})\right]^{\frac{1}{2}} \quad , \tag{5.28}$$

where $(\ell_{max}+3/2)\theta > \eta > \theta(\ell_{max}+1/2)$. Often a third condition becomes applicable to experimental conditions. This is referred to as the extreme quantum limit. Under this condition the Fermi level is below the lower edge of the second subband so that only one subband is occupied by the electrons. This condition occurs when the magnetic field is very large so that $\hbar\omega_B$ is larger than E_F. Since only one band is occupied we get only one term on the right hand side of (5.26) and the relation simplifies to

$$n = N_c\theta F_{-\frac{1}{2}}(\eta-\theta/2) \quad . \tag{5.28a}$$

The formulae given above have been derived assuming an isotropic and parabolic E-$\underline{k}$ relation. We may, however, easily extend them to the case of anisotropic and of nonparabolic bands. For ellipsoidal constant energy surfaces, N_c is given by (5.12) and ω_B by [see (4.99)]

$$\omega_B = (m_1m_2m_3)^{-\frac{1}{2}}(\alpha_1^2m_1+\alpha_2^2m_2+\alpha_3^2m_3)^{\frac{1}{2}}|e|B \quad , \tag{5.29}$$

where α_1, α_2, α_3 are the direction cosines of $\underline{B}$ with respect to the principal axes of the constant energy ellipsoids. The above result may be derived by using the transformation relation (5.10).

In the case of nonparabolic bands, although the effective mass varies with electron energy, (5.20) still applies, since it does not involve the effective mass. Further, the dispersion relation for the nonparabolic bands may be written after neglecting the effects of electron spin as [see (4.101)]

$$E(1+\alpha E) = \frac{\hbar|e|B}{m^*}\left(\ell+\frac{1}{2}\right) + \frac{\hbar^2 k_z^2}{2m^*} \quad , \tag{5.30}$$

where m* is the band-edge effective mass, and $\alpha \simeq 1/E_g$.

Using (5.20,30) we get for the density of states

$$g(E) = \sum_{\ell=0}^{\ell_{max}} (1+2\alpha E)\left[E(1+\alpha E)-(\ell+\frac{1}{2})\hbar\omega_B\right]^{-\frac{1}{2}} \frac{N_c\theta}{(\pi k_B T)^{\frac{1}{2}}} \quad , \tag{5.31}$$

where $\omega_B = |e|B/m^*$. The relation between the carrier concentration and the Fermi level is, hence,

$$n = N_c\hbar\omega_B(\pi)^{-\frac{1}{2}}(k_B T)^{-3/2}\int_0^\infty \sum_{\ell=0}^{\ell_{max}}\left[E(1+\alpha E)-(\ell+\frac{1}{2})\hbar\omega_B\right]^{-\frac{1}{2}}$$
$$\left\{1+\exp\left[(E-E_F)/k_B T\right]\right\}^{-1} \times (1+2\alpha E)dE \quad . \tag{5.32}$$

We may write the above expression in terms of Fermi integrals when $E_F \ll E_g$, a condition which is satisfied in most practical cases. Introducing this condition we get after some algebraic manipulations [5.10]

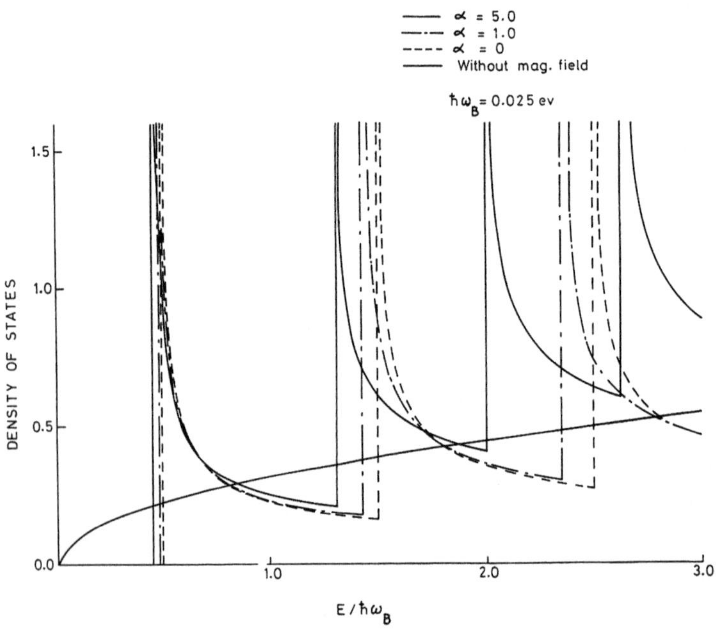

Fig. 5.6. Density of states in a quantising magnetic field for nonparabolic bands

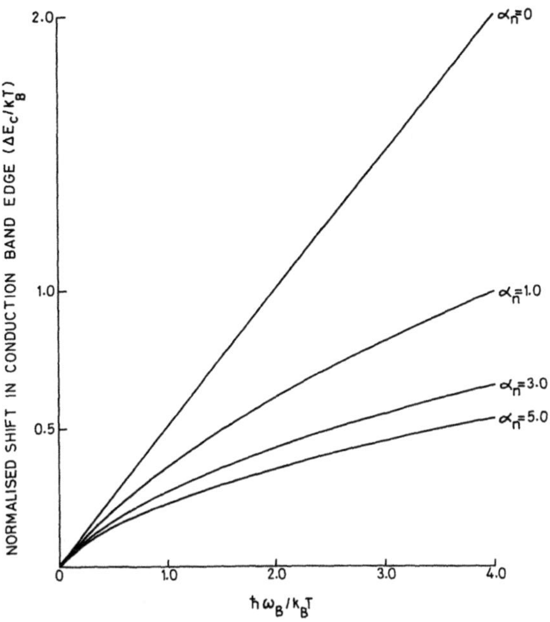

Fig. 5.7. Shift in conduction band edge with magnetic field for nonparabolic bands

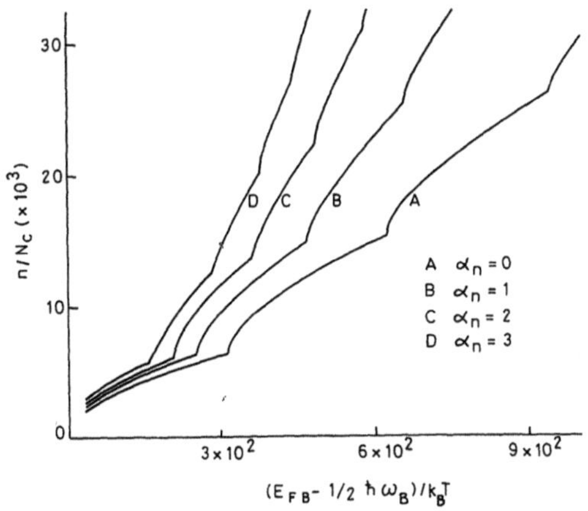

Fig. 5.8. Plot of n/N_C against η' $(=\eta-\theta/2)$ showing the variation of Fermi level with magnetic field for nonparabolic bands

$$n \simeq N_c\theta\sum_{\ell=0}^{\infty}\left[(1+3\alpha b/2)F_{-\frac{1}{2}}(\eta')+(3/4)\alpha_n F_{\frac{1}{2}}(\eta')\right]a^{-\frac{1}{2}}\quad , \tag{5.33}$$

where $\alpha_n = \alpha k_BT$, $\theta = \hbar\omega_B/k_BT$, $a = 1+\alpha(\ell+\frac{1}{2})\hbar\omega_B$, $b = (\ell+\frac{1}{2})\hbar\omega_B/\left[1+\alpha(\ell+\frac{1}{2})\hbar\omega_B\right]$, $\eta' = (E_F-b)/k_BT$.

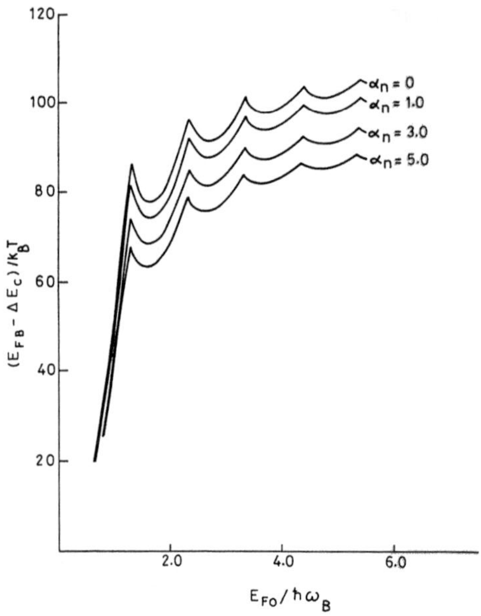

Fig. 5.9. Shift in Fermi level from the conduction band edge with magnetic field for nonparabolic bands

In order to illustrate the effects of nonparabolicity we have plotted in Fig.5.6 the density-of-states function; in Fig.5.7, the shift in conduction band edge; in Fig.5.8, n/N_c against η' $(=\eta-0.5\theta)$; and in Fig.5.9, η' against $E_{F0}/\hbar\omega_B$ for different values of αk_BT. It is seen that the nonparabolicity in the E-$\underline{k}$ relation causes a change in the shapes of all these curves and also that the shift in conduction band edge as well as the Fermi energy is, in general, reduced by the nonparabolicity.

5.4 Fermi Energy and Impurity Density

5.4.1 General Considerations

As mentioned earlier, in a semiconducting material the concentration of carriers under thermal equilibrium conditions is determined by its band structure, the lattice temperature and the concentration of impurity atoms. Although in the analysis of transport phenomena we mostly require the relation between n and E_F, presented in the preceding sections, we often require also the knowledge of the density of the ionised and neutral impurity atoms. The required knowledge may be gained from the relations between the Fermi energy and the ionisation energy and total concentration of impurity atoms. A complete theory giving the relation is, however, complicated by various factors. Each kind of impurity atom in the general case produces a number of energy levels corresponding to different states of excitation and each level is degenerate to a different degree. The positions of the levels are also affected by the magnetic field and by the pressure. The greatest complication is, however, introduced by the fact that a semiconductor, even when prepared carefully, contains a number of species of impurities in addition to those introduced for producing the intended doping level. These extra impurity atoms are more numerous in materials the technology of preparation of which is in the developmental stage. This being the case for most compound semiconductors, the undesired impurity atoms are usually present in large numbers. These extra impurity atoms may be either donors or acceptors and usually produce energy levels varying in position from those contributed by the major impurity level. Theoretical evaluation of the Fermi energy from the knowledge of the impurity concentration is almost an impossible task because of the complications discussed here. However, in the analysis of some transport experiments, as mentioned earlier, we are required to know the concentration of ionised and neutral impurity atoms and the simplified analysis which is used for this purpose is discussed in this section.

5.4.2 General Formula

We shall assume that the majority of the impurity atoms are donors, contributing a single energy level. Some of the other impurity atoms may also be donors, but the energy level contributed by them is assumed to be below that of the major donors. These levels are always occupied unless the Fermi level is lower and we shall assume that the Fermi level always remains higher than the minor donor levels. These donors will therefore behave as neutral atoms and play no role in determining the ionisation of the major donors. We shall leave them out of consideration.

The levels contributed by the acceptor impurity atoms also will always be occupied as the levels are far below the Fermi level. These levels may capture more than one electron. We shall assume that the total number of electrons captured by the acceptor states is N_a, which has to be treated as unknown since the total number of acceptor states and the valence of each state are not usually known. We shall also assume that the number of electrons excited intrinsically from the valence band to the conduction band is negligible in comparison to the donor concentration.

Let the total number of electrons in the conduction band be n. Since all of them have been contributed by the donor atoms (intrinsic excitation being neglected), we have

$$n = N_{Di} - N_a \quad , \tag{5.34}$$

where N_{Di} is the concentration of ionised donors, assumed to be singly ionised. It is also assumed that the hole concentration is negligible in comparison to electron concentration.

However, it follows from statistics [Ref.5.7, p.270] that N_{Di} is related to the density N_{Dn} of the neutral major donors (hereinafter referred to as donors only) by the relation

$$N_{Di}/N_{Dn} = \left\{\beta \exp[(E_F-E_d)/k_BT]\right\}^{-1} \quad , \tag{5.33}$$

where E_d is the energy corresponding to the donor level measured from the bottom of the conduction band. We note that E_d in effect is the donor ionisation energy. The factor β represents the degeneracy of the donor level and multiplies the number of ways in which the donor level may be occupied. The common form of degeneracy is the spin degeneracy of the level and β has a value of 2.

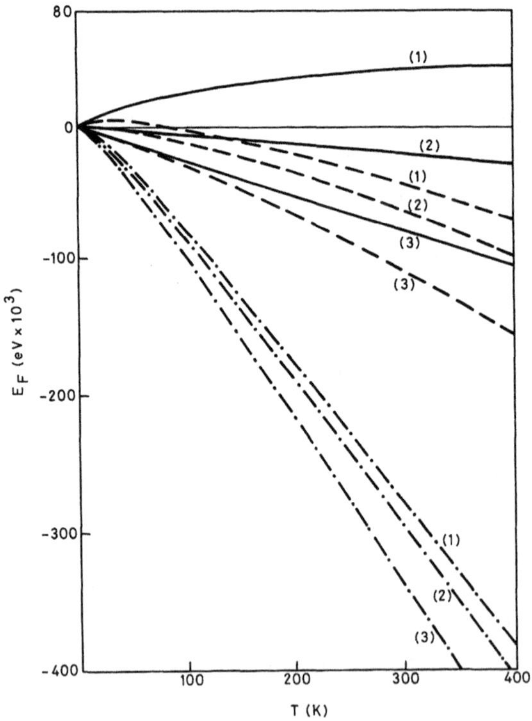

Fig. 5.10a

Equation (5.34) may now be written as

$$n = N_D\left\{1+\beta\ \exp[(E_F-E_d)/k_BT]\right\}^{-1} - N_a \quad , \tag{5.36}$$

where $N_D = N_{Di} + N_{Dn}$.

The above equation may be used to evaluate E_F for any temperature provided the band structure of the material, N_D and N_a are known, since n may be expressed in terms of E_F, m*, and T as discussed in the earlier sections. We shall discuss here the results for a parabolic band.

5.4.3 Discussion of Parabolic Band

Figure 5.10 shows the variation of Fermi energy and electron concentration with temperature for three doping levels at which the sample is nondegenerate, moderately degenerate and degenerate. The ionisation energy has been taken as 5.0 meV. The change in Fermi energy and electron concentration for different values of compensation ratio (K) are also illustrated. The important points of interest are that for temperatures above about 20 K almost all the donors are ionised for the nondegenerate case, but for the degenerate cases, quite

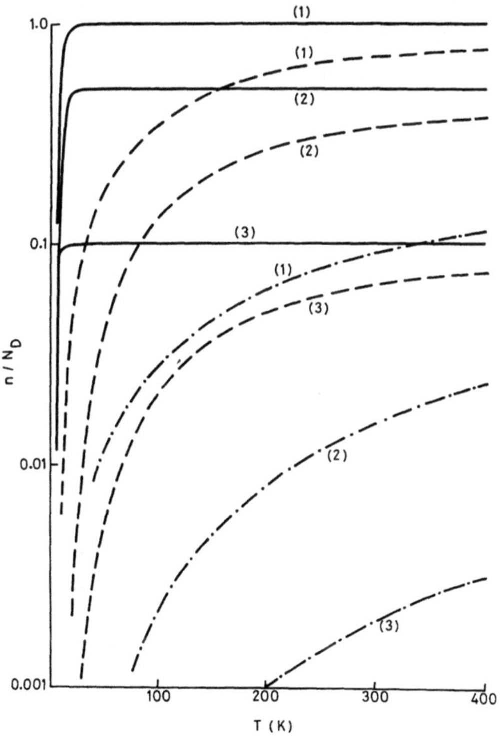

Fig. 5.10b

Fig. 5.10a and b. Fermi level and electron concentration plotted against temperature for $E_d = 5\times10^{-3}$ eV, $m^* = 0.068\ m_0$. Solid lines, $N_D = 10^{13}$ [cm^{-3}]; dotted lines, $N_D = 10^{17}$ [cm^{-3}]; chain line, $N_D = 10^{19}$ [cm^{-3}]; (1) K = 0; (2) K = 0.5; (3) K = 0.9; (a) Fermi level; (b) electron concentration

a large fraction remains neutral. We also note that for nonzero values of the compensation ratio, the Fermi level starts from the donor level and increases or decreases with temperature depending on whether $\beta(N_D-N_a)/N_a$ is larger or smaller than unity. It should be evident that the exact form of the curves will vary from material to material due to variation of E_d and N_c. For a detailed discussion of this variation and for the approximate formulae giving the Fermi level for different ranges of doping, see [Ref.5.9, pp.120-146].

5.4.4 Effect of Magnetic Field

It has been mentioned earlier that the position of the Fermi level is affected by magnetic field. Without entering into a detailed discussion we only note that the magnetic field affects the Fermi level through its effect on N_c as well as on E_d. Donor levels move upward with the magnetic field [5.11], but the shift in the conduction band edge is larger so that the ionisation energy increases in effect. The final result is that electron concentration decreases and at low temperatures electrons may be frozen on the donor states by the application of a magnetic field.

Before concluding this section, we should note that the simple theory discussed above fails completely when the material is heavily doped. The formation of impurity bands and band tails, discussed in the previous chapter introduces complications which cannot be dealt with by the simple theory.

5.5 Conclusions

We have derived in this chapter the formulae which relate the concentration of electrons to the Fermi energy and the density of states. These formulae will be used in later chapters for the calculation of the transport coefficients.

We have also indicated how the Fermi level is determined by impurity concentration. The difficulties encountered in evaluating the Fermi level in terms of impurity concentration have also been discussed. The relation between the impurity concentration and the Fermi energy can rarely be used in practice due to the uncertainty about the degrees of compensation. In transport calculations involving the effects of ionised impurities we have to rely on other experiments [5.12] and often on the transport coefficients [5.13] themselves to estimate their densities. This uncertainty often makes it difficult to arrive at definite conclusions about the agreement between theory and experiments on electron transport in compound semiconductors.

6. Scattering Theory

We have discussed so far the behaviour of an electron in a perfectly crystalline material, in which the wave function of the electron is given by the stationary Bloch functions, which are solutions of the time-independent Schrödinger equation. The motion of an electron is unhampered in a perfect crystal, and the application of an external field would uniformly accelerate the electron causing a linear increase of its drift velocity with time in the direction of the field. But we know that such a linear increase in drift velocity with time does not occur in real crystals. The average electron drift velocity reaches a limiting value, which at low fields is proportional to the magnitude of the field. The limit is set by the interaction of the electron with imperfections in the crystal through processes referred to as scattering or collision processes. If the imperfections are not too frequent we may still work with the stationary wave functions. The electron continues to be in a stationary state until it comes close to an imperfection. As it interacts with the imperfection, the deviation from the periodic potential affects the wave function and when the interaction is complete the electron has a new wave function characterised by a new wave vector and often a different value of energy. The electron continues in the new state till it again interacts with an imperfection. If the deviation from the perfect crystalline potential be small or infrequent, the effects of the imperfections may be worked out by applying perturbation theory.

We shall derive in this chapter the formulae obtained from this theory, taking into account the complete wave functions applicable to nonparabolic bands. As mentioned earlier, since the energy band gap is small in many compound semiconductors, the nonparabolicity cannot be neglected, and in the complete theory of scattering we cannot but include its consequences. The formulae for parabolic bands have been discussed at length in the literature [6.1,2], and we shall base our development of the formulae for nonparabolic bands on these.

6.1 Collision Processes

Imperfections in the periodic potential in a crystal are produced in various ways. The arrangement of atoms in a real crystal differs from that in the perfectly crystalline material due to defects in the structure introduced during growth. These crystal defects assume various forms, e.g. edge dislocations and screw dislocations. As charges collect at the dislocation sites, electrons are scattered at these sites. However, in properly grown crystals, the effects of such dislocations are negligible compared to those of other imperfections [6.3]. Hence we shall exclude the collisions at dislocations from our consideration, but consider the other processes in detail.

The most common form of deviation from the periodic potential is produced by impurity atoms. These atoms remain in the crystals, as complete purification is often not technically possible. These are also sometimes introduced intentionally to provide one type of free carrier of the required concentration. The potential associated with the impurity atoms is different from that of the atoms in the host crystal, so that even when these atoms occupy lattice sites as substitutional atoms, there is a perturbation in the periodic potential. The atoms often occupy also interstitial sites and produce potentials to disturb the periodic potential distribution.

As mentioned earlier (Sect.4.6.1), the impurity atoms provide energy levels in the forbidden gap, and these levels are occupied by electrons (for donor atoms) and are empty (for acceptor atoms) at very low temperature, so that the impurity atoms are mostly neutral at very low temperatures. The collision process is dominated at very low temperatures (near liquid-helium temperature) by collisions with the neutral impurity atoms. We should mention, however, that the theory of scattering by neutral atoms is not yet properly developed [6.4], and we shall present only the available results without entering into a detailed discussion.

The impurity atoms become ionised with increase in lattice temperature (donor atoms contribute the outer electron to the conduction band and acceptor atoms accept an electron from the valence band), and the collision process is then dominated by the collision of electrons with ionised impurity atoms. The importance of this kind of collision process increases as the temperature of the crystal is increased from very low values. However, as the charged impurity atoms cannot scatter high energy electrons, the effect of ionised impurity atoms decrease again as the temperature is increased to higher values. But in compound semiconductors prepared with current technology, the concentration of the impurity atoms being relatively large, the ionised impurities dominate the collision process at liquid-nitrogen temperature or often even at room temperature. The scattering by ionised impurities should therefore be considered as an important collision process for compound semiconductors.

The crystal defects or impurity atoms may be considered as imperfections, the effects of which may be controlled and reduced by improvements in the techniques of crystal preparation. However, in all crystals, even in perfect crystals, there remains an intrinsic collision mechanism. The effect of this cannot be controlled and is determined by the intrinsic physical properties of the crystal. In developing the theory of energy band structures we have assumed that the lattice atoms are frozen in space. But the real situation is that the atoms vibrate about their equilibrium sites, and it is the vibration which accounts for the thermal energy of a crystal. As a result of the vibrations, the periodic potential in the crystal varies with time and causes alterations in the electron states in time. The scattering of the electrons due to the alterations in the periodic potential arising from the vibration of the lattice atoms depends on the nature of these vibrations and also on the kinds of perturbing potentials produced as a result of vibrations. The lattice scattering, i.e. the scattering of electrons due to lattice vibrations, may, therefore, be of different kinds which we discuss below.

The interaction of the electrons with the perturbing potential produced by the lattice vibrations may be dealt with only by adopting the quantum-mechanical description of the vibrations. In this description, the vibrations are said to excite phonons similar to photons associated with electromagnetic waves. The energy and momentum of a phonon associated with a particular mode of vibration are taken respectively as $\hbar\omega_{\underline{q}}$ and $\hbar\underline{q}$, where $\omega_{\underline{q}}$ and $\underline{q}$ are the frequency and wave vector of the vibration. The number of phonons of frequency $\omega_{\underline{q}}$ and wave vector $\underline{q}$ at a particular temperature are given by Bose statistics [6.5], that the occupancy of the states at a temperature T per unit available state is given by

$$n_{\underline{q}} = \left[\exp(\hbar\omega_{\underline{q}}/k_BT)-1\right]^{-1} \quad . \tag{6.1}$$

These phonons are considered as particles and interact with the electrons as particles obeying the principles of conservation of energy and of momentum. We shall discuss the interaction mechanism in the next few sections. Here we shall describe the various kinds of phonons that we have to deal with for the development of the theory of lattice scattering.

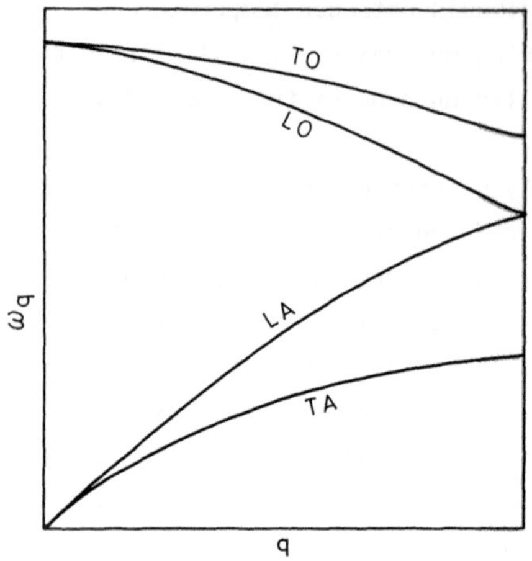

Fig. 6.1. Dispersion curves for lattice vibrations in semiconductors

All semiconducting materials of interest have either two kinds of atoms or double periodicity or both. The dispersion relation for the lattice vibrations (shown schematically in Fig.6.1) has, therefore, four branches, which may be divided into two groups. For one group, ω_q increases with $|\underline{q}|$ approximately linearly at first and attains a limiting value nonlinearly when $|\underline{q}$ approches the Brillouin zone boundary. The neighbouring atoms vibrate in phase as for a propagating acoustic wave. The phonons corresponding to these vibrations are called acoustic phonons. The two curves belonging to this group correspond to the transverse and longitudinal vibrations. For the second group of vibrations, the neighbouring atoms vibrate in opposite phase. The phonons corresponding to these vibrations are called optic phonons.

In scattering theory, the scattering mechanisms are subdivided further in accordance with the nature of perturbing potentials produced by the vibrations. Acoustic phonons produce the perturbing potential in two ways. First, due to the changes in the spacing of the lattice atoms, the energy band gap and the position of conduction and valence band edges vary from point to point (see Fig.6.2). Potential discontinuities are thereby introduced in the conduction band as well as in the valence band. The potential so produced due to the deformation of the crystal is called the deformation potential, the magnitude of which is evidently proportional to the strain produced by the vibrations. The scattering of the carriers through the deformation potential is called the deformation potential scattering. The second kind of perturbation is produced by acoustic vibrations through the piezoelectric effect. If the atoms constituting the crystal are partially ionised, the displacements of the atoms due to the acoustic vibrations produce potentials, the magnitude of which

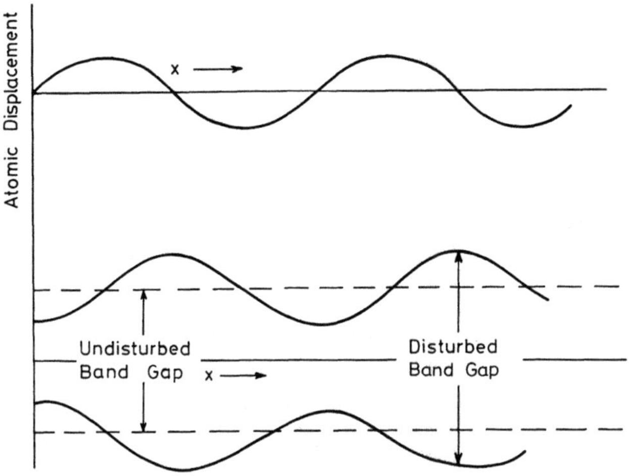

Fig. 6.2. Energy gap in the presence of an acoustic wave

would depend on the arrangement of the ionised atoms in the crystal. The magnitude will be larger in crystals with lower symmetry. The scattering of electrons through the piezoelectric potential will be referred to as piezoelectric scattering. This kind of scattering is important in all compound semiconductors, particularly at low temperatures. It is, however, stronger in materials with wurtzite structure than in materials with sphalerite structure due to the lower symmetry of the former structure.

The optic vibrations also produce perturbing potentials in two ways. The deformation of the crystal due to optic vibrations produces a perturbing potential proportional to the optic strain. The interaction of the electrons with the lattice through the deformation potential produced by optic vibrations is very much dependent on the symmetry of the band structure. This interaction is very weak for electrons at the Γ-point minima or for the <100> minima [6.6]. It is, however, strong for minima along the <111> directions. Since in most compound semiconductors the lowest minimum is at the Γ point, this kind of optic scattering is of little significance. Only in aluminium and gallium and lead compounds is the deformation optic or so-called nonpolar optic scattering of importance. The second kind of optic scattering is more important. This occurs through the polarisation produced by the optic vibrations due to the ionic charges associated with the atoms forming the compounds. The displacement of the neighbouring atoms with the opposite ionic charges results in dipole moments, the potential associated with which scatters the electrons. This kind of scattering is called the polar optic phonon scattering and is often the most important scattering mechanism, particularly at liquid-nitrogen or higher temperatures.

In addition to the four classes discussed above, lattice scattering is often divided into two other classes. In lattice scattering, the electron changes state by sharing its momentum and energy with the phonons. It will be shown later that in the exchange process, the total energy and momentum of the electron and the phonon are conserved. In materials in which the Γ-point minimum is the lowest minimum, the electron after being scattered by the lattice usually remains in the same Γ-point valley in E-$\underline{k}$ space. It may be easily shown, using the conditions of conservation of energy and momentum, that only phonons of low wave vector take part in such interactions. This kind of lattice scattering for which the final and the initial positions of the electron are in the same valley in E-$\underline{k}$ space is classed as intravalley scattering.

In materials with the lowest minima in the <100> or <111> directions, electrons may be scattered by the phonons from one valley to another valley having its energy minimum at the same level. In such transitions, the wave vector of the electron changes by large amounts, so that only phonons with large wave vectors may cause such scattering. Such scattering is classed as intervalley scattering and the corresponding phonons are often referred as intervalley phonons. It should be noted that the intervalley phonons may be of either kind, acoustic or optic.

Intervalley scattering may also occur from a low-valley to a higher non-equivalent valley. This kind of scattering, although mostly unimportant for low fields, assumes great importance at high fields, as discussed in Chap.11.

The different kinds of lattice scattering and impurity atom scattering discussed above are of importance in all semiconductors even when grown with care to ensure uniform crystal structure. In semiconducting alloys, which are alloys of two semiconducting compounds, even when grown with care, we get an additional scattering mechanism, often called alloy scattering. The dissimilarity of the compounds leads to distortions in the band structure at the points in the crystal where the two kinds of compounds meet. Since these boundaries are randomly positioned and randomly oriented, each such boundary acts as an individual scattering centre like the impurity atoms, but the basic mechanism of scattering is similar to that for deformation potential acoustic phonon scattering, as the scattering is due to discontinuities in the band edges.

Lastly, we should remember that in the preceding discussion and in all the theories presented so far we have considered only the states of a single elec-

Table 6.1. Scattering mechanisms

- Scattering Mechanisms
  - Defect scattering
    - Crystal defects
    - Impurity
      - Neutral
      - Ionised
    - Alloy
  - Carrier-Carrier scattering
  - Lattice scattering
    - Intravalley
      - Acoustic
        - Deformation potential
        - Piezo-electric
      - Optic
        - Nonpolar
        - Polar
    - Intervalley
      - Acoustic
      - Optic

tron moving in the periodic field in the crystal, which takes into account the field of the nuclei and all other electrons. In a real crystal we get a host of electrons moving freely inside the crystal. The concentration of these free electrons range typically between 10^{13} and 10^{18} cm^{-3}. Hence, even when we confine our attention to a single electron and assume that the states obtained from this one-electron approximation represent the available states for the free electrons, we should take into consideration the fact that the state of the electron may be changed by collision with another free electron when they come near each other through their Coulomb fields. Such collisions are called interelectronic or carrier-carrier collisions. Evidently the importance of this kind of collisions will increase with increasing electron concentration. The effect is usually negligible compared to the effects of other scattering mechanisms for concentrations of the order of 10^{13}-10^{14} cm^{-3}. At higher concentrations carrier-carrier scattering, however, assume importance.

We have outlined in this section the various kinds of collision processes which we have to consider for the development of the formulae for transport coefficients. The importance of each kind of scattering varies from material to material and depends on the lattice temperature. This variation is of central importance in transport theory and will be discussed in later chapters. For convenience we have collected the various scattering mechanisms in Table 6.1. We shall derive in the following sections the formulae which are used to calculate the effects of various kinds of scattering listed in this table.

6.2 Transition Probability

The effects of collisions, as discussed earlier, are worked out by applying perturbation theory. Let us consider an electron in the state $(E_{\underline{k}},\underline{k})$ with wave function

$$\psi_{\underline{k}} = U_{\underline{k}} \exp\left[i(\underline{k}\cdot\underline{r}-E_{\underline{k}}t/\hbar)\right] \quad . \tag{6.2}$$

The wave function of the electron changes with time under the influence of the perturbing potential. Since the eigenfunctions of the Schrödinger equation form a complete set we may write the wave function in the perturbed state as

$$\psi(t) = \sum_{\underline{k}_i} c_{\underline{k}_i}(t)\psi_{\underline{k}_i}(\underline{r}) \exp(-iE_{\underline{k}_i}t/\hbar) \quad , \tag{6.3}$$

where $\psi_{\underline{k}_i}(\underline{r}) \exp[-i(E_{\underline{k}_i}/\hbar)t]$ is the eigenfunction corresponding to the eigenvalue $E_{\underline{k}_i}$ for the unperturbed Hamiltonian and the $c_{\underline{k}_i}(t)$ are multiplicative constants, to be chosen so as to satisfy the Schrödinger equation with Hamiltonian including the perturbing potential.

Using (6.3), it may be readily shown that the number of transitions to the state $(E'_{\underline{k}},\underline{k}')$ from the state $(E_{\underline{k}},\underline{k})$ per unit time is given by [Ref.6.1, p.30; 6.7]

$$T_{\underline{k}\underline{k}'} = \frac{2\pi}{\hbar}\left| <\psi^*_{\underline{k}'}(\underline{r})|\Delta V|\psi_{\underline{k}}(\underline{r})> \right|^2 \left| \frac{\sin[(E_{\underline{k}}-E_{\underline{k}'})t/\hbar]}{\pi(E_{\underline{k}}-E_{\underline{k}'})} \right| \quad , \tag{6.4}$$

where ΔV represents the perturbing potential energy, and $<\psi^*_{\underline{k}'}(\underline{r})|\Delta V|\psi_{\underline{k}}(\underline{r})>$ is a matrix element giving the $c_{\underline{k}}$, defined as

$$<\psi^*_{\underline{k}'}(\underline{r})|\Delta V|\psi_{\underline{k}}(\underline{r})> = \int\limits_{\substack{\text{unit}\\ \text{cell}}} \psi^*_{\underline{k}'}(\underline{r})[\Delta V\psi_{\underline{k}}(\underline{r})]d\underline{r} \quad . \tag{6.5}$$

The integral is to be evaluated over a unit cell as the wave functions are normalised over the same volume.

For large values of t, $|\{\sin[(E_{\underline{k}}-E_{\underline{k}'})t/\hbar]\}/[\pi(E_{\underline{k}}-E_{\underline{k}'})]|$ behaves as the Dirac delta function $\delta(E_{\underline{k}}-E_{\underline{k}'})$. Hence the transition probability when the interaction is complete may be written as

$$T_{\underline{k}\underline{k}'} = \frac{2\pi}{\hbar}\, |M(\underline{k},\underline{k}')|^2\delta(E_{\underline{k}}-E_{\underline{k}'}) \quad , \tag{6.6}$$

where $M(\underline{k},\underline{k}')$ is written for the matrix element (6.5).

We should note that the above simplification introduces the restriction that the free time between two successive collisions should be large enough. Equation (6.6) is the fundamental equation of scattering theory and is often referred to as the golden rule number 2 [6.8], according to Fermi. The transition probabilities for the various kinds of scattering are evaluated using this theory. The matrix element involved in (6.6) is determined by the perturbing potential and we shall give in the following sections the matrix elements for the different scattering mechanisms discussed before.

In developing the expressions for the matrix elements we should distinguish between lattice scattering and scattering by crystal defects. In the latter case the scattering potentials may be considered isolated from each other, since the spacing between the points of defect is large and irregular. Further, the effect of collisions with defect points appears mostly as a change in the state of the electron; the effect on the impurity atom or the lattice at points of defect is usually very insignificant. We may, therefore, assume for such scattering that $\psi(r)$ in (6.3) is the wave function of the electron alone.

On the other hand, in collisions with the lattice atoms, significant changes are also produced on the lattice atoms. The lattice atoms being strongly coupled to each other, the change in momentum or energy caused by the collisions is shared by all the lattice atoms. In dealing with the collision between the electron and the lattice atoms through the vibration of the latter, we are required to consider ψ in (6.4), so as to represent the combined wave function of the electron and the vibrating lattice atoms. The combined wave function is given by

$$\chi_{\underline{k},n} = \psi_{\underline{k}} \prod_{\underline{q}} \phi_{n,\underline{q}} \quad , \tag{6.7}$$

where $\psi_{\underline{k}}$ is the wave function of the electron, and $\phi_{n,\underline{q}}$ is the wave function for the lattice vibration mode of vector $\underline{q}$ in the nth state of excitation. It should be noted that n is given by (6.1) and the $\phi_{n,\underline{q}}$ are Hermite polynomials [6.9].

Equation (6.6) also indicates that in transitions of the electron from the state $(E_{\underline{k}},\underline{k})$ to $(E_{\underline{k}'},\underline{k}')$, energy is conserved. In the case of defect scattering, since the change in energy of the defect centre (e.g. impurity atom) is considered negligible, the argument of the delta function is $E_{\underline{k}}-E_{\underline{k}'}$. On the other hand, in lattice scattering the change in the electron state occurs by the absorption or the emission of a phonon. Since the argument of the delta function is to represent the total change in the energy of the electron-phonon system, it is given in this case by $E_{k'}-(E_k \pm \hbar\omega_q)$ where the plus and minus signs correspond respectively to the absorption and the emission processes. Note that $\hbar\omega_q$ is the energy of a phonon corresponding to the mode of vibration of wave vector $\underline{q}$, with which the electron interacts.

6.3 Matrix Elements

The perturbing potential produced by crystal defects may be expanded in a Fourier series, written as

$$\Delta V = \sum_{\underline{q}} A(\underline{q}) \exp(i\underline{q}\cdot\underline{r}) \quad , \tag{6.8}$$

where the $A(\underline{q})$ are Fourier coefficients. The matrix element for scattering by crystal defects and impurity atoms is thus given by

$$M(\underline{k},\underline{k}') = \sum_{\underline{q}} \int U^*_{\underline{k}'}(\underline{r}) \exp(-i\underline{k}'\cdot\underline{r}) A(\underline{q}) \exp(i\underline{q}\cdot\underline{r}) U_{\underline{k}}(\underline{r}) \exp(i\underline{k}\cdot\underline{r}) d\underline{r} \quad . \tag{6.9}$$

Since $\exp(i\underline{k}'\cdot\underline{r})$ and $\exp(i\underline{k}\cdot\underline{r})$ are periodic functions, the integral is zero for all values of $\underline{q}$ except $\underline{q} = \underline{k}'-\underline{k}$. The matrix element thus reduces to

$$M(\underline{k},\underline{k}') = A(|\underline{k}'-\underline{k}|) I(\underline{k},\underline{k}') \quad , \tag{6.10}$$

where

$$I(\underline{k},\underline{k}') = \int_{\Omega} U^*_{\underline{k}'}(\underline{r}) U_{\underline{k}}(\underline{r}) d\underline{r} \quad . \tag{6.11}$$

The condition $\underline{q} = \underline{k}'-\underline{k}$ is analogous to the condition of conservation of momentum, as $\hbar\underline{k}$ denotes the crystal momentum before collision and $\hbar\underline{k}'$ that after collision, and $\underline{q}$ may be considered as a vector analogous to the wave vector of the lattice waves.

In the case of lattice scattering, the perturbing potential may be written as

$$\Delta V = \sum_{\underline{q}} A(\underline{q}) \left[a_{\underline{q}} \exp(i\underline{q}\cdot\underline{r}) + a^{\dagger}_{\underline{q}} \exp(-i\underline{q}\cdot\underline{r}) \right] \quad , \tag{6.12}$$

where $A(\underline{q})$ is related to the amplitude of the mode of vibration of wave vector $\underline{q}$, and its magnitude is determined by the various physical parameters like sound velocity, density, ionic charge of the atoms, dielectric constant, deformation potential constant, etc. The symbols $a^{\dagger}_{\underline{q}}$ and $a_{\underline{q}}$ stand for the creation and annihilation operators [6.10].

Substituting (6.12) in the expression for the matrix element and using for the wave function of the electron-phonon system (6.7), we get for the matrix element

$$M(\underline{k},\underline{k}') = \int_{\Omega} \Big\{ U^{*}_{\underline{k}'}(\underline{r}) \exp(-i\underline{k}'\cdot\underline{r}) \prod_{\underline{q}} \phi^{*}_{n',\underline{q}} \sum_{\underline{q}} A(\underline{q}) \left[a_{\underline{q}} \exp(i\underline{q}\cdot\underline{r}) + a^{\dagger}_{\underline{q}} \exp(-i\underline{q}\cdot\underline{r}) \right] \times U_{\underline{k}}(\underline{r}) \exp(i\underline{k}\cdot\underline{r}) \prod_{\underline{q}} \phi_{n,\underline{q}} \Big\} d\underline{r} \quad . \tag{6.13}$$

As $\exp[(\underline{k}-\underline{k}'\pm\underline{q})\cdot\underline{r}]$ is periodic, the first term of the integral has a nonzero value only for $\underline{k}' = \underline{k}+\underline{q}$, and the second term, for $\underline{k}' = \underline{k}-\underline{q}$. Also, since [Ref. 6.1, p.45]

$$a_{\underline{q}}\phi_{n,\underline{q}} = \sqrt{n_{\underline{q}}}\ \phi_{n-1,\underline{q}} \quad ,$$

and

$$a^{\dagger}_{\underline{q}}\phi_{n,\underline{q}} = \sqrt{n_{\underline{q}}+1}\ \phi_{n+1,\underline{q}} \quad ,$$

the integral has nonzero values only for $n' = n-1$ for the first term and $n' = n+1$ for the second term, the $\phi_{n,\underline{q}}$ being orthogonal.

The transition indicated by the first term of (6.13) thus causes a reduction in the number of phonons of mode $\underline{q}$ by unity and corresponds to the phonon absorption process. Similarly, the transition corresponding to the second term is a phonon emission process in which a phonon of mode $\underline{q}$ is emitted. The relation $\underline{k}' = \underline{k} \pm \underline{q}$ may hence be considered to be the condition of conservation of momentum as it indicates that the crystal momentum of the electron after collision is equal to the crystal momentum of the electron before collision plus the pseudomomentum of the absorbed phonon, or minus the pseudomomentum of the emitted phonon.

Using the results discussed above, we get for the matrix element

$$M(\underline{k},\underline{k}') = A(\underline{q})\left(n_{\underline{q}} + \frac{1}{2} \pm \frac{1}{2}\right)^{\frac{1}{2}} I(\underline{k},\underline{k}') \quad . \tag{6.14}$$

The plus and minus signs correspond respectively to the emission and absorption processes. The integral reduces to the above form as the $\phi_{n,\underline{q}}$ are orthonormal functions.

The evaluation of the matrix elements for the various collision processes thus requires the knowledge of $I(\underline{k},\underline{k}')$ and the $A(\underline{q})$. For parabolic energy bands, we may assume that $U_{\underline{k}} \approx U_{\underline{k}'}$, and since the functions are normalised over a unit cell, $I(\underline{k},\underline{k}')$ is equal to unity. For nonparabolic bands, however, $U_{\underline{k}'}$ cannot be taken equal to $U_{\underline{k}}$. The integral $I(\underline{k},\underline{k}')$ then differs significantly from unity and has a complex form. We shall derive in a later section the detailed expression for this integral. Expressions for the $A(\underline{q})$ are the same for parabolic and nonparabolic bands, and as these are discussed thoroughly in other books [6.1,2], we shall only quote the expressions.

The discussion on the scattering of electrons given above is based on the one-electron picture. We cannot treat electron-electron scattering with the same formalism, and we shall discuss some aspects of this case in a later section. The presence of free electrons has, however, an important effect on scattering by charged centres such as ionised impurity atoms, which has to be taken into account even in the one-electron picture. We discuss this effect before further discussion of the scattering probabilities.

6.4 Free-Carrier Screening

In the absence of any potential, we may assume that the free carriers are uniformly distributed in a semiconductor sample. However, in regions where potential discontinuities occur, the free electrons collect or are removed in accordance with the sign of the potential discontinuity. The space charge so produced causes an extra potential and modifies the potential discontinuity, almost screening its effect for large distances. We shall discuss this screening effect with reference to the potential due to an ionised impurity atom. The potential produced by a bare ionised impurity atom is

$$\phi(r) = \frac{Ze}{4\pi\varepsilon r} , \tag{6.15}$$

where ε is the static material permittivity, Ze is the charge on the atom, and r is measured from its centre. The distribution of the free electrons around the charged impurity atom is determined by Poisson's equation. Let $n(\phi)$, $p(\phi)$ denote respectively the concentration of electrons and holes at $\underline{r}$ where the potential is ϕ. Let N_D and N_A denote respectively the concentration of donor and acceptor impurity atoms, and $N_D^+(\phi)$ and $N_A^-(\phi)$ the respective ionised concentrations for a potential of ϕ. The charge density at the point $\underline{r}$ is, then,

$$\rho = |e|\left[p(\phi)-n(\phi)+N_D^+(\phi)-N_A^-(\phi)\right] \quad , \tag{6.16}$$

assuming the impurity atoms to be singly ionised.

The potential $\phi(r)$ is given by

$$\nabla^2\phi = \frac{1}{r}\,\frac{d^2(r\phi)}{dr^2} = -\frac{\rho}{\varepsilon} \quad . \tag{6.17}$$

We have assumed spherical symmetry. Now assuming that the impurity atoms have single-ionisation energies, we get

$$n(\phi) = n(0) + \frac{\partial n(0)}{\partial E_F}|e|\phi \quad ;$$

$$p(\phi) = p(0) + \frac{\partial p(0)}{\partial E_F}|e|\phi \quad ;$$

$$N_D^+(\phi) = N_D^+(0) + \frac{\partial N_D^+(0)}{\partial E_F}|e|\phi = N_D^+(0) - \frac{|e|\phi}{k_BT}\left[N_D^+(0)-\frac{N_D^+(0)^2}{N_D}\right] \quad ;$$

$$N_A^-(\phi) = N_A^-(0) + \frac{\partial N_A^-(0)}{\partial E_F}|e|\phi = N_A^-(0) + \frac{|e|\phi}{k_BT}\left[N_A^-(0)-\frac{N_A^-(0)^2}{N_A}\right] \quad .$$

Hence,

$$\rho = -\frac{e^2\phi}{k_BT}\left[\frac{\partial n(0)}{\partial\eta}-\frac{\partial p(0)}{\partial\eta}+N_D^+(0)-\frac{N_D^+(0)^2}{N_D}+N_A^-(0)-\frac{N_A^-(0)^2}{N_A}\right] \quad . \tag{6.18}$$

Since the sample is neutral,

$$p(0) - n(0) + N_D^+(0) - N_A^-(0) = 0 \quad .$$

Note that E_F is the Fermi energy and $\eta = E_F/k_BT$.

Replacing ρ by the expression (6.18), we find that the solution of (6.17) is

$$\phi(r) = (A/r)\exp(-r/\lambda) \quad , \tag{6.19}$$

where A is constant, and

$$\frac{1}{\lambda^2} = \frac{e^2}{k_BT\varepsilon}\left[\frac{\partial n(0)}{\partial\eta}-\frac{\partial p(0)}{\partial\eta}+N_D^+(0)-\frac{N_D^+(0)^2}{N_D}+N_A^-(0)-\frac{N_A^-(0)^2}{N_A}\right] \quad . \tag{6.20}$$

Since as r tends to zero, expression (6.19) must reduce to (6.15), the constant A should be taken as $Ze/(4\pi\varepsilon)$.

The effect of the free carriers is to introduce an exponential decay of the potential with the decay coefficient λ. This cofficient is often termed the Debye length.

Expression (6.20) gives λ in the most general case. It may, however, be simplified for some particular conditions. In nondegenerate materials, it simplifies to

$$\frac{1}{\lambda^2} = \frac{e^2}{\varepsilon k_B T}\left[2n(0) - \frac{N_D^+(0)^2}{N_D} + 2N_A^-(0) - \frac{N_A^-(0)^2}{N_A}\right] , \tag{6.21}$$

since $\partial n(0)/\partial\eta = n(0)$, $\partial p(0)/\partial\eta = -p(0)$, and $p(0) = n(0)+N_A^-(0)-N_D^+(0)$. If the impurity atoms are fully ionised, expression (6.21) further simplifies to

$$1/\lambda^2 = [p(0)+n(0)]e^2/\varepsilon k_B T \quad . \tag{6.22}$$

For the above condition, we get in nondegenerate materials

$$\frac{1}{\lambda^2} = \frac{1}{\varepsilon k_B T}\left[\frac{\partial n(0)}{\partial\eta} - \frac{\partial p(0)}{\partial\eta}\right] \quad . \tag{6.23}$$

On the basis of the above results, we may also obtain the effect of electron screening on other kinds of scattering potentials. We first of all note that the coefficients of a Fourier series representation of the Coulomb potential produced by a bare ionised impurity atom are

$$A(\underline{q}) = \lim_{\delta\to 0} \frac{1}{V_c}\int \frac{Ze^2}{4\pi\varepsilon r}\exp[-(i\underline{q}+\underline{\delta})\cdot\underline{r}]d\underline{r} = Ze^2/(V_c\varepsilon q^2) \quad , \tag{6.24}$$

V_c being the volume of the crystal. For the screened Coulomb potential, however, we get instead

$$A_S(\underline{q}) = Ze^2(V_c\varepsilon)^{-1}(q^2+\lambda^{-2})^{-1} \quad . \tag{6.25}$$

Electron screening thus effectively multiplies the Fourier coefficient by $q^2/(q^2+1/\lambda^2)$. The same kind of effect would be produced on the Fourier coefficients of other scattering potentials. It has been shown by EHRENREICH [6.11] from a detailed solution of the problem considering the potential due to optic vibrations in polar materials that this is indeed the case, if the thermal velocity of the electrons is larger than the velocity of lattice waves. Physically, the condition implies that the electrons may then redistribute themselves in accordance with the potential variation. However, when the phase velocity of the lattice waves is larger, the screening effects may be considered smaller and even antiscreening may occur as the electrons cannot in effect adjust to the fast-moving potential variation. In most of the cases that we deal with, the former condition applies, and we may assume that the Fourier coefficients of all scattering potentials are multiplied by $q^2/(q^2+1/\lambda^2)$ as a result of electron scattering.

Electron screening also affects the frequencies of lattice vibrations in polar materials as the interaction forces are modified by screening. We shall not derive but only quote the results obtained by EHRENREICH [6.11], who showed that the frequencies of longitudinal vibration are modified from

$$\omega_t\sqrt{\varepsilon_s/\varepsilon_\infty} \quad \text{to} \quad \omega_t\left[\left(q^2\varepsilon_s/\varepsilon_\infty+1/\lambda^2\right)/(q^2+1/\lambda^2)\right]^{\frac{1}{2}} \quad ,$$

where ω_t is the frequency of transverse vibration, and ε_s, ε_∞ are respectively the static and high-frequency permittivities of the material. It should be noted that ω_t is not affected by screening, as the transverse modes are not coupled through polar interaction.

It is of interest to note that the effects of screening discussed above are significant for ionised impurity scattering even for moderate electron concentration. It will be evident from the results presented in later sections that the effect of ionised impurity scattering becomes negligibly small for all impurity concentrations if the screening effect is neglected. It is due to the localisation of the impurity potentials by the effect of screening that the scattering contribution of ionised impurity atoms become finite. On the other hand, the effects of screening may be mostly neglected for all kinds of lattice scattering, as q is usually much larger than $1/\lambda^2$. Only for heavily doped materials does the screening effect become important, particularly at low temperatures.

6.5 Overlap Integrals

The function $|I(\underline{k},\underline{k}')|$ is often referred to as the overlap integral. Its square appears as a factor in the expression for the scattering probabilities through (6.6), and recurs in the development of transport theory quite frequently. We have mentioned earlier that for nonparabolic bands, the electron wave may be approximated by a plane wave, and we may assume that $U_{\underline{k}'}\approx U_{\underline{k}}$ and $I(\underline{k},\underline{k}')=1$. For nonparabolic bands, however, the electron wave function is more complex and we get an involved expression. We derive below the detailed expression for the case in which the electron wave functions correspond to Γ-point extrema, which is very common in compound semiconductors.

We first derive the function for the electrons. The cell periodic function for the electrons is given by (see Sect.4.5.3)

$$U_{\underline{k}_1} = ia_{\underline{k}} S\underline{\underline{\alpha}}_1 + \frac{b_{\underline{k}}}{\sqrt{2}} (X+iY)\underline{\underline{\alpha}}_2 + c_{\underline{k}} Z\underline{\underline{\alpha}}_1 \quad , \tag{6.26}$$

and

$$U_{\underline{k}_2} = ia_{\underline{k}} S\underline{\underline{\alpha}}_2 + \frac{b_{\underline{k}}}{\sqrt{2}} (X-iY)\underline{\underline{\alpha}}_1 + c_{\underline{k}} Z\underline{\underline{\alpha}}_2 \quad . \tag{6.27}$$

We remember that S is the s-type function, X,Y,Z are the components of the p-type function, and $\underline{\underline{\alpha}}_1$, $\underline{\underline{\alpha}}_2$ are the spinors. The corresponding complex conjugate functions for $\underline{k}'$ are

$$U^*_{\underline{k}'_1} = -ia_{\underline{k}'} S'\underline{\underline{\alpha}}'_1 + \frac{b_{\underline{k}'}}{\sqrt{2}} (X'-iY')\underline{\underline{\alpha}}'_2 + c_{\underline{k}'} Z'\underline{\underline{\alpha}}'_1 \quad , \tag{6.28}$$

$$U^*_{\underline{k}'_2} = -ia_{\underline{k}'} S'\underline{\underline{\alpha}}'_1 + \frac{b_{\underline{k}'}}{\sqrt{2}} (X'+iY')\underline{\underline{\alpha}}'_1 + c_{\underline{k}'} Z'\underline{\underline{\alpha}}'_2 \quad . \tag{6.29}$$

The overlap function $G(\underline{k},\underline{k}')$ may hence be expressed as

$$G(\underline{k},\underline{k}') = \frac{1}{2}\sum_{\nu,\mu} |I_{\nu\mu}(\underline{k},\underline{k}')|^2 \quad , \tag{6.30}$$

where

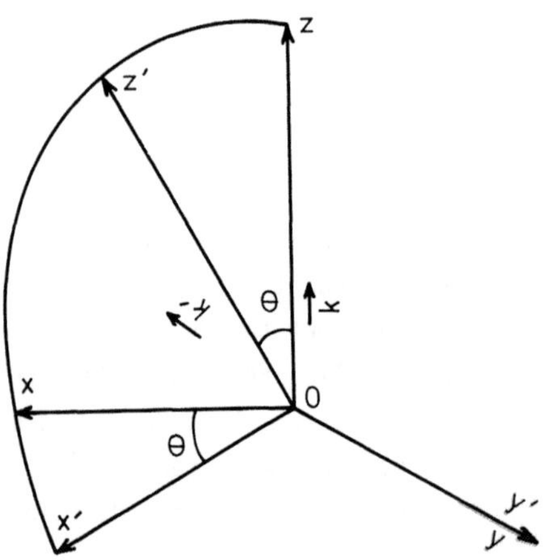

Fig. 6.3. Orientation of the wave vectors $\underline{k}$ and $\underline{k}'$

$$I_{\nu\mu}(\underline{k}',\underline{k}) = \int_{\Omega} U^{*}_{\underline{k}'\nu} U_{\underline{k}\mu} d\underline{r} \quad . \tag{6.31}$$

Here Ω is the volume of a unit cell, and $\nu\mu$ has the combination of four values.

We illustrate the detailed evaluation of $I_{\nu\mu}(\underline{k}',\underline{k})$ considering the first term $I_{11}(\underline{k}',\underline{k})$. We should remember that the direction of Z in (6.26) is chosen to be the same as that of $\underline{k}$. Similarly, the direction of Z' in (6.28) is chosen in that of $\underline{k}'$. Since the orientation of x and y directions are arbitrary, we may choose the x' and x axes to lie in the plane of $\underline{k}'$ and $\underline{k}$ (see Fig.6.3). Then, if the angle between $\underline{k}'$ and $\underline{k}$ is denoted by θ, X', Y', and Z' are related to X,Y,Z by the transformation

$$\begin{bmatrix} X' \\ Y' \\ Z' \end{bmatrix} = \begin{bmatrix} \cos\theta & 0 & -\sin\theta \\ 0 & 1 & 0 \\ \sin\theta & 0 & \cos\theta \end{bmatrix} \begin{bmatrix} X \\ Y \\ Z \end{bmatrix} \quad . \tag{6.32}$$

Similarly, the relations between $\underline{\underline{\alpha}}_1'$, $\underline{\underline{\alpha}}_2'$ and $\underline{\underline{\alpha}}_1$, $\underline{\underline{\alpha}}_2$ are [6.12]

$$\underline{\underline{\alpha}}_1' = (\cos\frac{\theta}{2})\underline{\underline{\alpha}}_1 + (\sin\frac{\theta}{2})\underline{\underline{\alpha}}_2 \quad , \tag{6.33}$$

$$\underline{\underline{\alpha}}_2' = -(\sin\frac{\theta}{2})\underline{\underline{\alpha}}_1 + (\cos\frac{\theta}{2})\underline{\underline{\alpha}}_2 \quad . \tag{6.34}$$

Using the above relations and noting that

$$\underline{\underline{\alpha}}_1 \cdot \underline{\underline{\alpha}}_1 = \underline{\underline{\alpha}}_2 \cdot \underline{\underline{\alpha}}_2 = 1 \quad ; \quad \underline{\underline{\alpha}}_1 \cdot \underline{\underline{\alpha}}_2 = \underline{\underline{\alpha}}_2 \cdot \underline{\underline{\alpha}}_1 = 0 \quad ;$$

$$\int_{\Omega} S^2 d\underline{r} = \int_{\Omega} X^2 d\underline{r} = \int_{\Omega} Y^2 d\underline{r} = \int_{\Omega} Z^2 d\underline{r} = 1 \quad ;$$

$$\int_{\Omega} SX d\underline{r} = \int_{\Omega} XY d\underline{r} = \int_{\Omega} XZ d\underline{r} = \int_{\Omega} YZ d\underline{r} = \int_{\Omega} SY d\underline{r} = \int_{\Omega} SZ d\underline{r} = 0 \quad ,$$

we get

$$\begin{aligned} \int_{\Omega} U^{*}_{\underline{k}_1'} U_{\underline{k}_1} d\underline{r} &= (\cos\frac{\theta}{2})\left[a_{\underline{k}}a_{\underline{k}'} + (b_{\underline{k}}b_{\underline{k}'} + c_{\underline{k}}c_{\underline{k}'})\cos\theta\right] \\ &\quad + \left[b_{\underline{k}}b_{\underline{k}'}/2 - (c_{\underline{k}}b_{\underline{k}'} + c_{\underline{k}'}b_{\underline{k}})/\sqrt{2}\right]\sin\theta\,\sin\frac{\theta}{2} \\ &= A(\underline{k},\underline{k}')\cos\frac{\theta}{2} + B(\underline{k},\underline{k}')\sin\frac{\theta}{2} \quad . \end{aligned} \tag{6.35}$$

Following the same procedure, we get for the other terms

$$\int U^{*}_{\underline{k}_2'} U_{\underline{k}_1} d\underline{r} = A(\underline{k},\underline{k}')\sin\frac{\theta}{2} - B(\underline{k},\underline{k}')\cos\frac{\theta}{2} \quad . \tag{6.36}$$

$$\int U^*_{\underline{k}'_1} U_{\underline{k}_2} d\underline{r} = -A(\underline{k},\underline{k}') \sin\frac{\theta}{2} + B(\underline{k},\underline{k}') \cos\frac{\theta}{2} \quad . \tag{6.37}$$

$$\int U^*_{\underline{k}'_2} U_{\underline{k}_2} d\underline{r} = A(\underline{k},\underline{k}') \cos\frac{\theta}{2} + B(\underline{k},\underline{k}') \sin\frac{\theta}{2} \quad . \tag{6.38}$$

The value of the overlap integral factor is hence

$$\begin{aligned} G(\underline{k},\underline{k}') &= \frac{1}{2}\sum_{\mu,\nu} |I_{\mu\nu}(\underline{k},\underline{k}')|^2 = A(\underline{k},\underline{k}')^2 + B(\underline{k},\underline{k}')^2 \\ &= \left[a_{\underline{k}}a_{\underline{k}'}+(b_{\underline{k}}b_{\underline{k}'}+c_{\underline{k}}c_{\underline{k}'}) \cos\theta\right]^2 + \left[b_{\underline{k}}b_{\underline{k}'}/2-(c_{\underline{k}}b_{\underline{k}'}+c_{\underline{k}'}b_{\underline{k}})/\sqrt{2}\right]^2 \sin^2\theta \\ &= a + b \cos\theta + c \cos^2\theta \quad , \end{aligned} \tag{6.39}$$

where

$$\begin{aligned} a &= (a_{\underline{k}}a_{\underline{k}'})^2 + \left[b_{\underline{k}}b_{\underline{k}'}/2-(c_{\underline{k}}b_{\underline{k}'}+c_{\underline{k}'}b_{\underline{k}})/\sqrt{2}\right]^2 \quad ; \\ b &= 2a_{\underline{k}}a_{\underline{k}'}(b_{\underline{k}}b_{\underline{k}'}+c_{\underline{k}}c_{\underline{k}'}) \quad ; \\ c &= (b_{\underline{k}}b_{\underline{k}'}+c_{\underline{k}}c_{\underline{k}'})^2 - \left[b_{\underline{k}}b_{\underline{k}}'/2-(c_{\underline{k}}b_{\underline{k}'}+c_{\underline{k}'}b_{\underline{k}})/\sqrt{2}\right]^2 \end{aligned} \tag{6.40}$$

We should note that for the two extreme values of Δ (the spin-orbit splitting), we may simplify the above expressions to the following forms.

We find from (4.76) that when $\Delta \ll E_g$, the coefficients a_k, b_k, etc., are given by

$$\begin{aligned} a_{\underline{k}} &= \left(\frac{1+\alpha E}{1+2\alpha E}\right)^{\frac{1}{2}} \quad ; \quad a_{\underline{k}'} = \left(\frac{1+\alpha E'}{1+2\alpha E'}\right)^{\frac{1}{2}} \quad ; \\ b_{\underline{k}} &= b_{\underline{k}'} = 0 \quad ; \\ c_{\underline{k}} &= \left(\frac{\alpha E}{1+2\alpha E}\right)^{\frac{1}{2}} \quad ; \quad c_{\underline{k}'} = \left(\frac{\alpha E'}{1+2\alpha E'}\right)^{\frac{1}{2}} \quad . \end{aligned} \tag{6.41}$$

Substituting in (6.40), we get for $\alpha E \ll 1$,

$$\begin{aligned} a &= \left(\frac{1+\alpha E}{1+2\alpha E}\right)\left(\frac{1+\alpha E'}{1+2\alpha E'}\right) \simeq [1-\alpha(E+E')+\alpha^2(EE'+2E^2+2E'^2)] \quad . \\ b &= 2\,\frac{(1+\alpha E)^{\frac{1}{2}}(1+\alpha E')^{\frac{1}{2}}(\alpha^2 EE')^{\frac{1}{2}}}{(1+2\alpha E)(1+2\alpha E')} \simeq 2(\alpha^2 EE')^{\frac{1}{2}} \quad . \\ c &= \frac{\alpha^2 EE'}{(1+2\alpha E)(1+2\alpha E')} \simeq \alpha^2 EE' \quad . \end{aligned} \tag{6.42}$$

Similarly, when $\Delta \gg E_g$, the coefficients are given by (see Sect.4.5.3)

$$a_{\underline{k}} = \left(\frac{1+\alpha E}{1+2\alpha E}\right)^{\frac{1}{2}} \quad , \quad a_{\underline{k}'} = \left(\frac{1+\alpha E'}{1+2\alpha E'}\right)^{\frac{1}{2}} \quad ,$$

$$b_{\underline{k}} = \left(\frac{1}{3}\cdot\frac{\alpha E}{1+2\alpha E}\right)^{\frac{1}{2}}, \quad b_{\underline{k}'} = \left(\frac{1}{3}\cdot\frac{\alpha E'}{1+2\alpha E'}\right)^{\frac{1}{2}} \quad ,$$

$$c_{\underline{k}} = \left(\frac{2}{3}\cdot\frac{\alpha E}{1+2\alpha E}\right)^{\frac{1}{2}}, \quad c_{\underline{k}'} = \left(\frac{2}{3}\cdot\frac{\alpha E'}{1+2\alpha E'}\right)^{\frac{1}{2}} \quad . \tag{6.43}$$

Putting these in (6.40),

$$a = \left(\frac{1+\alpha E}{1+2\alpha E}\right)\left(\frac{1+\alpha E'}{1+2\alpha E'}\right) + \frac{1}{4}\,\frac{\alpha^2 EE'}{(1+2\alpha E)(1+2\alpha E')}$$

$$\simeq 1 - \alpha(E+E') + \alpha^2(5EE'/4+2E^2+2E'^2) \quad ;$$

$$b = 2\left(\frac{1+\alpha E}{1+2\alpha E}\right)^{\frac{1}{2}}\left(\frac{1+\alpha E'}{1+2\alpha E'}\right)^{\frac{1}{2}}\frac{\alpha(EE')^{\frac{1}{2}}}{(1+2\alpha E)(1+2\alpha E')} \simeq 2\alpha(EE')^{\frac{1}{2}} \quad ;$$

$$c = \frac{3}{4}\,\frac{\alpha^2 EE'}{(1+2\alpha E)(1+2\alpha E')} \simeq \frac{3}{4}\,\alpha^2 EE' \quad . \tag{6.44}$$

The expressions given above for the overlap functions apply to intraband scattering in the conduction band, light-hole band, or the split-off band, as the wave functions have the same form for all these three bands (see Sect. 4.5.3). On the other hand, for the heavy-hole band, we get by using the same procedure [6.13]

$$G(\underline{k},\underline{k}') = (1+3\cos^2\theta)/4 \quad , \tag{6.45}$$

since the corresponding wave function is $(1/\sqrt{2})(X\pm iY)\underline{\underline{\alpha}}_1$ (or $\underline{\underline{\alpha}}_2$) [6.14].

In the case of holes in the valence band, we have to consider interband scattering in addition to intraband scattering. The overlap functions for the interband scattering of holes are given by [6.13]

$$G_{12}(\underline{k},\underline{k}') = (1/4)(b_{\underline{k}'}+\sqrt{2}c_{\underline{k}'})^2\sin^2\theta \quad ,$$

$$G_{21}(\underline{k},\underline{k}') = (1/4)(b_{\underline{k}}+\sqrt{2}c_{\underline{k}})^2\sin^2\theta \quad , \tag{6.46}$$

where G_{12} corresponds to scattering out of the heavy hole band, and G_{21} to scattering into the same band.

The overlap functions given above correspond to the case when the extrema are at the Γ point. In a large number of compound semiconductors, this is so, and the functions given above suffice for calculations with these. But there are also some important compound semiconductors in which the extrema are at the L or X point. Proper overlap functions for these bands are not available. However, in some analyses, the overlap functions derived above for the Γ-point extrema have been used also for these extrema [6.15].

6.6 Scattering Probability S($\underline{k}$)

We may now calculate using (6.6) the probability of an electron being scattered from the state $\underline{k}$ through the various collision processes. The probability is given by

$$S(\underline{k}) = \frac{V_c}{8\pi^3}\int \frac{2\pi}{\hbar}|M(\underline{k},\underline{k}')|^2\delta(E_{\underline{k}}-E_{\underline{k}'})d\underline{k}' \quad , \tag{6.47}$$

where V_c is the crystal volume. The material is assumed to be nondegenerate. For degenerate materials, an extra factor must be introduced to take into account the probability of the state $\underline{k}'$ being occupied. This factor is ignored in the present discussion. We obtain the above expression by integrating $T_{\underline{k}\underline{k}'}$ over all $\underline{k}'$ states to which the electron may be scattered from the $\underline{k}$ state. Note that we take the density of states in $\underline{k}'$ space to be $(1/8\pi^3)$ and not $2(1/8\pi^3)$, since the spin is not altered in the scattering process. Scattering processes in which spin is changed have been considered recently, but these processes are important only at low temperatures [6.16].

The derivation of the expressions giving S($\underline{k}$) for ionised impurity scattering, for piezoelectric and deformation potential scattering, and for polar optic phonon scattering, assuming iostropic band structure, is given in detail below. For other scattering mechanisms, however, only the final results will be given.

The evaluation of S($\underline{k}$) requires the complete expressions for M($\underline{k}$,$\underline{k}'$). As shown in the preceding sections we may write M($\underline{k}$,$\underline{k}'$) as

$$|M(\underline{k},\underline{k}')|^2 = [A(|\underline{k}-\underline{k}'|)]^2G(\underline{k},\underline{k}') \quad \text{for ionised impurity scattering} \tag{6.48a}$$

and

$$|M(\underline{k},\underline{k}')|^2 = [A(|\underline{k}-\underline{k}'|)]^2G(\underline{k},\underline{k}')\left(n_{\underline{q}}+\frac{1}{2}\pm\frac{1}{2}\right) \quad \text{for lattice scattering.} \tag{6.48b}$$

A detailed expression for G($\underline{k}$,$\underline{k}'$) has been derived earlier. Expressions for A($|\underline{k}-\underline{k}'|$) for the various scattering processes are collected in Table 6.2 (see p.114-115, this volume).

6.6.1 S(k) for Ionised Impurity Scattering

Substituting in (6.47) the expression for $A(|\underline{k}-\underline{k}'|)$ corresponding to ionised impurity scattering and that for $G(\underline{k},\underline{k}')$, we get

$$S_{imp}(\underline{k}) = \frac{V_c}{8\pi^3} N_i V_c \int \frac{2\pi}{\hbar} \frac{Z^2 e^4}{V_c^2 \varepsilon^2} [|\underline{k}-\underline{k}'|^2 + \lambda^{-2}]^{-2} G(\underline{k},\underline{k}')\delta(E_{\underline{k}}-E_{\underline{k}'})d\underline{k}' \quad . \quad (6.49)$$

We may transform the integral in (6.49) putting

$$d\underline{k}' = k'^2 \sin\theta \, d\theta d\phi dk' \quad , \quad (6.50)$$

where θ is the angle between $\underline{k}$ and $\underline{k}'$, and ϕ is the azimuthal angle. k'^2dk' may also be expressed in terms of $E_{\underline{k}'}$ as

$$k'^2dk' = \frac{1}{2}\left(\frac{2m^*}{\hbar^2}\right)^{3/2}[\gamma(E_{\underline{k}'})]^{\frac{1}{2}}\gamma'(E_{\underline{k}'})dE_{\underline{k}'} \quad , \quad (6.51)$$

where $\gamma'(E) = d\gamma/dE$. The dispersion relation can be written as

$$\hbar^2k^2/(2m^*) = \gamma(E_{\underline{k}}) \quad . \quad (6.52)$$

The function $\gamma(E)$, as noted earlier, has a complicated form in general but may be simplified to $E(1+\alpha E)$. The integral over $E_{\underline{k}'}$ may be evaluated straight away using the fact that the delta function has a nonzero value only for $E_{\underline{k}'} = E_{\underline{k}}$. The integration is thus performed by replacing k' by k in the integrand, since $E_{\underline{k}'}$ depends only on the magnitude of $\underline{k}'$. We thus get for $S_{imp}(\underline{k})$

$$S_{imp}(\underline{k}) = C_{imp}F_{imp}(\underline{k},\lambda)\gamma'(E_{\underline{k}})/k^3 \quad , \quad (6.53)$$

where

$$F_{imp}(\underline{k},\lambda) = \int_{-1}^{+1} \left[0_i(x)/(1-x+0.5k^{-2}\lambda^{-2})^2\right]dx \quad ,$$

$$C_{imp} = Z^2e^4N_im^*/8\pi\varepsilon^2\hbar^3 \quad , \quad (6.54)$$

and

$$0_i(x) = a + bx + cx^2 \quad .$$

Here a,b,c are as defined by (6.40), with E' replaced by E.

Table 6.2. Matrix elements for different kinds of scattering

Type of scattering	Scattering potential energy, ΔV	Matrix element excluding the overlap integral
1. Acoustic Phonon:		
a) Deformation Potential [Ref.6.1,Chap.2;6.17]	$E_1 \underline{\nabla} u(\underline{r}) S_c(q,\lambda)$	$E_1\left(\frac{\hbar}{2V_c\rho\omega_{\underline{q}}}\right)^{1/2} S_c(q,\lambda)(\underline{e}_{\underline{q}}\cdot\underline{q})\left(n_{\underline{q}}+\frac{1}{2}\pm\frac{1}{2}\right)^{1/2}$
b) Piezoelectric [6.18]	$\frac{eh_{pz}}{q\varepsilon}\,\underline{\nabla} u(\underline{r}) S_c(q,\lambda)$	$\frac{\lvert e\rvert h_{pz}}{\varepsilon}\left(\frac{\hbar}{2V_c\rho\omega_{\underline{q}}}\right)^{1/2} S_c(q,\lambda)\left(n_{\underline{q}}+\frac{1}{2}\pm\frac{1}{2}\right)^{1/2}$
2. Optic Phonon:		
a) Nonpolar [6.19]	$D_0 u(\underline{r}) S_c(q,\lambda)$	$D_0\left(\frac{\hbar}{2V_c\rho\omega_0}\right)^{1/2} S_c(q,\lambda)\left(n_0+\frac{1}{2}\pm\frac{1}{2}\right)^{1/2}$
b) Polar [6.20]	$\frac{u(\underline{r})e\omega_\ell}{q}\left[\frac{\rho}{V_c\varepsilon_0}\left(\frac{1}{K_\infty}-\frac{1}{K_S}\right)\right] S_c(q,\lambda)$	$\frac{\lvert e\rvert}{q}\left(\frac{1}{K_\infty}-\frac{1}{K_S}\right)^{1/2}\left(\frac{\hbar\omega_\ell}{2V_c\varepsilon_0}\right)^{1/2}\left(n_\ell+\frac{1}{2}\pm\frac{1}{2}\right)^{1/2} S_c(q,\lambda)$
3. Defect:		
a) Ionised Impurity [6.21,22]	$\frac{Ze^2}{4\pi\varepsilon r}\exp(-r/\lambda)$	$\frac{Ze^2}{V_c\varepsilon}\,\frac{1}{\lvert\underline{k}-\underline{k}'\rvert^2+\lambda^{-2}}$
b) Neutral Impurity [6.4a]	—	$\frac{2\pi\hbar^2}{m^*V_c}\left(\frac{20a_0}{4\pi k}\right)^{1/2}$
c) Dislocation:		
(i) Charged type [Ref.6.2d,p.232;6.23]	$\frac{ef}{2\pi a_D\varepsilon}K_0(r/\lambda)$	—
(ii) Strain type [6.24]	$\frac{E_1\lambda_S(1-2\nu)}{2\pi(1-\nu)}\,\frac{\sin\theta}{r}$	—
d) Alloy [6.25,Ref.6.2c]	—	$\frac{N_0^{1/2}\alpha^{1/2}(1-\alpha)^{1/2}(E_a-E_b)}{V_c^{1/2}}$
4. Intervalley Phonon [Ref.6.1,p.107]	$D_i u(\underline{r}) S_c(q,\lambda)$	$D_i\left(\frac{\hbar}{2\rho V_c\omega_i}\right)^{1/2} S_c(q,\lambda)\left(n_i+\frac{1}{2}\pm\frac{1}{2}\right)^{1/2}$

Meanings of symbols (Table 6.2):

ΔV - Potential energy of the electron due to the scattering potential
E_1 - Acoustic phonon deformation potential constant
$u(\underline{r})$ - Displacement due to a lattice wave
$\underline{q}$ - Wave vector of lattice wave
λ - Debye screening length
V_c - Volume of the crystal
ρ - Density of the material
$\omega_{\underline{q}}$ - Angular frequency of the lattice wave corresponding to the wave vector q
$\underline{e}_{\underline{q}}$ - Polarisation vector of the lattice wave
$n_{\underline{q}}$ - Occupation number of phonons with wave vector $\underline{q}$
h_{pz} - Piezoelectric constant
ε - Permittivity of the material
D_0 - Nonpolar optic phonon deformation potential constant
ω_0 - Angular frequency of the nonpolar optic wave neglecting dispersion
n_0 - Occupation number of optic phonons
K_∞ - High-frequency dielectric constant
K_S - Static dielectric constant
ω_ℓ - Longitudinal frequency of vibration for polar optic wave
ε_0 - Permittivity of free space
n_ℓ - Occupation number of polar optic phonons
Z - Ionicity of an impurity atom
$\underline{k}$ - Wave vector of the electron in the original state
$\underline{k}'$ - Wave vector of the electron in the scattered state
$S_c(q,\lambda)=q^2(q^2+\lambda^{-2})^{-1}$ - Screening factor
N_0 - Number of atoms in the crystal
a_0 - Bohr radius
r - Distance from impurity centre, or distance from dislocation axis
a_D - Distance between dislocation centres
f - Probability of occupation of a dislocation centre by an electron
K_0 - Modified Bessel function of zero order
ν - Poisson's ratio
λ_S - Unit crystallographic slip distance
θ - Angle measured from slip direction
D_i - Intervalley phonon deformation potential constant
ω_i - Intervalley phonon angular frequency
n_i - Occupation number of intervalley phonons
α - Fraction of atoms of type a
(E_a-E_b) - Difference between the band edge of atoms of type a and of type b constituting the alloy
$\pm$ - The plus sign corresponds to emission and the minus sign to the absorption process.

Also, $q=|\underline{k}-\underline{k}'|$.

The deformation potential constant E_1, piezoelectric constant h_{pz}, and the velocity of acoustic waves s are in general dependent on the direction of $\underline{q}$, the phonon wave vector relative to the crystallographic axis. Calculation of the transport coefficients taking this directional dependence into account is discussed in Sect.6.7.

It should be noted that $S_{imp}(\underline{k})$ becomes infinity if the electron screening effect is neglected, i.e., if λ is taken as infinity. This is true for all scattering mechanisms for which $A(q)$ decreases with q faster than q^{-1}. The screening effect cannot therefore be neglected for such scattering mechanisms, since it is as a result of screening that the scattering probability $S(\underline{k})$ is finite.

The function $F_{imp}(\underline{k},\lambda)$ may be evaluated analytically, but as the number of terms is rather large, we shall not do this.

6.6.2 S(k) for Piezoelectric Scattering

Substituting $A(q)$ for piezoelectric scattering from Table 6.2 into the expression (6.47), we get

$$S_{pz}(\underline{k}) = \frac{e^2\hbar_{pz}^2}{8\pi^2\varepsilon^2\rho s^2}\sum_{+,-}\int\frac{|\underline{k}-\underline{k}'|^2\omega_q}{(|\underline{k}-\underline{k}'|^2+\lambda^{-2})^2}\,\delta(E_{\underline{k}'}-E_{\underline{k}}\pm\hbar\omega_{\underline{q}})G(\underline{k},\underline{k}')\left(n_{\underline{q}}+\frac{1}{2}\pm\frac{1}{2}\right)d\underline{k}' \quad ; \tag{6.55}$$

$\sum_{+,-}$ indicates the sum of the two terms obtained by replacing the ± sign with + and - signs. The term for the + sign corresponds to a transition with the emission of a phonon and that for the - sign, to a transition with the absorption of a phonon.

The integral may be simplified noting that the maximum possible value of $|\underline{k}-\underline{k}'|$ is about 2k, which follows from the condition of conservation of energy and momentum. The value of $\hbar\omega_{\underline{q}}$ may therefore be at most $2\hbar sk$ (s is the acoustic velocity), since for low values of q, $\omega_{\underline{q}} = sq$. The average energy of the electrons at a temperature T being $(3/2)k_BT$, $2\hbar sk$ has a value of about $3\times10^{-23}\sqrt{T}$ J (s and m* may be taken approximately equal to 10^4 m/s and $0.1\times9.1\times10^{-31}$ kg). Hence the energy of the phonons is comparable to that of the electrons or to the energy corresponding to k_BT at a temperature of about 2 K. We shall assume that the lattice temperature is much higher, since the phonon scattering may be important only at high temperatures when the lattice vibrations are excited. We may then introduce the following approximations:

$$E_{\underline{k}} \pm \hbar\omega_{\underline{q}} \simeq E_{\underline{k}} \quad , \tag{6.56}$$

$$n_{\underline{q}} + \frac{1}{2} \pm \frac{1}{2} \simeq n_{\underline{q}} \simeq k_BT/(\hbar\omega_{\underline{q}}) \quad . \tag{6.57}$$

The first approximation implies that the collisions are elastic, since $\hbar\omega_{\underline{q}}$, the energy gained or lost in a collision, is neglected. The second approximation is often called the equipartition approximation, as the energy for each mode of vibration is the same and equal to k_BT when this approximation is valid.

We may thus simplify (6.55) to

$$S_{pz}(\underline{k}) = \frac{e^2 h_{pz}^2 \, k_B T}{4\pi^2 \hbar \rho s^2 \varepsilon^2} \int \frac{|\underline{k}-\underline{k}'|^2}{(|\underline{k}-\underline{k}'|^2+\lambda^{-2})^2} \delta(E_{\underline{k}'}-E_{\underline{k}}) G(\underline{k},\underline{k}') d\underline{k}' \quad . \qquad (6.58)$$

It may be noted that when $\hbar\omega_q$ is neglected in comparison to E_k, the delta function is independent of q, i.e., of the angle between $\underline{k}$ and $\underline{k}'$, and depends only on the magnitudes of $\underline{k}$ and $\underline{k}'$. Instead of applying the usual procedure of converting the integral over $\underline{k}'$ to one over $\underline{q}$, as is often done [Ref.6.1, p.107], we may therefore apply the procedure of the preceding section. This procedure simplifies the integration considerably, particularly when nonparabolic bands are involved. We may apply this procedure for the evaluation of scattering probabilities of all kinds of lattice scattering, since for acoustic phonon scattering (assumed elastic) and for optic phonon scattering (when the dispersion of ω_q with q is neglected), the delta function is independent of the angle between $\underline{k}$ and $\underline{k}'$.

Integrating over $E_{k'}$ using the property of the delta function and over θ (the angle between $\underline{k}$ and $\underline{k}'$) as in the case of ionised impurity scattering, we obtain

$$S_{pz}(\underline{k}) = C_{pz} F_{pz}(\underline{k},\lambda) \gamma'(E_{\underline{k}})/k \quad , \qquad (6.59)$$

where

$$F_{pz}(\underline{k}, \ ) = \int_{-1}^{+1} \left[0_i(x)(1-x)(1-x+0.5k^{-2}\lambda^{-2})^{-2}/2 \right] dx \quad ,$$

and

$$C_{pz} = e^2 h_{pz}^2 \varepsilon^{-2} k_B T m^*/2\pi\hbar^3 \rho s^2 \quad .$$

We find that $F_{pz}(\underline{k},\lambda)$ is finite only if λ is finite. It is infinity if the screening effect is neglected.

6.6.3 S(k) for Deformation-Potential Acoustic Phonon Scattering

The expression for $S(\underline{k})$ corresponding to deformation-potential acoustic phonon scattering is derived following the same procedure as outlined above for piezoelectric scattering. Using the matrix element given in Table 6.2 and assuming that $\hbar\omega_{\underline{q}} << E$ and $n_{\underline{q}} + \frac{1}{2} \pm \frac{1}{2} \simeq k_B T/\hbar\omega_{\underline{q}}$, we get

$$S_{ac}(\underline{k}) = \frac{E_1^2 k_B T}{4\pi^2 \rho \hbar s^2} \int \frac{|\underline{k}-\underline{k}'|^4}{(|\underline{k}-\underline{k}'|^2+\lambda^{-2})^2} G(\underline{k},\underline{k}')\delta(E_{\underline{k}}-E_{\underline{k}'})d\underline{k}' \quad . \tag{6.61}$$

Evaluating the integral using the property of the delta function, we may reduce S_{ac} to the form

$$S_{ac}(\underline{k}) = C_{ac}F_{ac}(\underline{k},\lambda)\gamma'(E_{\underline{k}})k \quad , \tag{6.62}$$

where

$$F_{ac}(\underline{k},\lambda) = \int_{-1}^{+1} \left[0_i(x)(1-x)^2(1-x+0.5k^{-2}\lambda^{-2})^{-2}/2\right]dx \quad ,$$

and

$$C_{ac} = E_1^2 k_B T m^*/\pi\rho\hbar^3 s^2 \quad .$$

In contrast to the results for the two kinds of scattering considered earlier, $S_{ac}(\underline{k})$ remains finite even when the screening effect is neglected. The effects of screening may be neglected for moderate electron concentration at not-too-low temperature. At very low temperatures the above formula does not apply in any case, since we cannot assume that $\hbar\omega_{\underline{q}} << E$ or $n_{\underline{q}} + \frac{1}{2} \pm \frac{1}{2} \simeq k_B T/\hbar\omega_{\underline{q}}$. When screening effect is neglected, we get

$$F_{ac}(\underline{k}) = a + c/3 \quad , \tag{6.64}$$

where a and c are given by (4.76). The above form is commonly employed in the analysis of the effects of deformation-potential acoustic phonon scattering.

6.6.4 $S(\underline{k})$ for Polar Optic Phonon Scattering

The energy of optic phonons is usually comparable to that of the electrons at ordinary temperatures since the equivalent temperature of the optic phonons of almost all semiconductors lies between 150 K and 500 K. We cannot therefore neglect the energy of the phonons in comparison to that of the electrons, and the collisions should be considered to be inelastic. The equipartition approximation is also not applicable, and the full phonon distribution function for n_q sould be used. However, we find from the dispersion curves (see Fig.6.1) that the optic phonon frequency varies slowly with the wave number. Hence we may assume that ω_ℓ is independent of q, for the small change in q caused by collisions. We may treat ω_ℓ and n_q as constants and apply the same procedure for the evaluation of the integral in $S(\underline{k})$ used in the earlier cases.

Substituting A(q) from Table 6.2 in (6.47), we get

$$S_{pop}(\underline{k},\lambda) = \frac{e^2\omega_\ell\left(K_\infty^{-1}-K_S^{-1}\right)}{8\pi^2\varepsilon_0}\sum_{+,-}\int\frac{|\underline{k}-\underline{k}'|^2G(\underline{k},\underline{k}')}{(|\underline{k}-\underline{k}'|^2+\lambda^{-2})^2}\,\delta(E_{\underline{k}}-E_{\underline{k}'}\pm\hbar\omega_\ell)$$

$$\times\left(n_\ell+\frac{1}{2}\pm\frac{1}{2}\right)d\underline{k}' \quad ; \qquad (6.65)$$

where $n_\ell = [\exp(\hbar\omega_\ell/k_BT)-1]^{-1}$. The summation is to be carried out over the plus and minus signs, to take into account the absorption and emission processes. The integration may be carried out replacing $d\underline{k}'$ by (6.50) and using the property of the delta function. In this case, the integration over $E_{\underline{k}'}$ results in the replacement of $E_{\underline{k}'}$ by $E_{\underline{k}}\pm\hbar\omega_\ell$ corresponding to the two scattering processes, and k' takes the corresponding values. We get

$$S_{pop}(\underline{k}) = C_{pop}k^{-2}\sum_{+,-}k_\pm\gamma'(E_k\pm\hbar\omega_\ell)F_{pop\pm}(\underline{k},\lambda)\left(n_\ell+\frac{1}{2}\pm\frac{1}{2}\right) \quad , \qquad (6.66)$$

where

$$F_{pop\pm}(\underline{k},\lambda) = \int_{-1}^{+1} 0_i(x)\left[1-2(k_\pm/k)x+(k_\pm/k)^2\right]\left[1-2(k_\pm/k)x+(k_\pm/k)^2+1/(k\lambda)^2\right]^{-2}dx \; ,$$

$$C_{pop} = \frac{e^2\omega_\ell m^*}{4\pi\varepsilon_0\hbar^2}\left(K_\infty^{-1}-K_S^{-1}\right) \quad ,$$

and $k_\pm$ are the wave vectors corresponding to $E\pm\hbar\omega_\ell$.

6.6.5 S(k) for Intervalley and Nonpolar Optic Phonon Scattering

The scattering mechanisms discussed in the preceding sections must be considered in the evaluation of the mobility of electrons occupying the Γ-point minimum. In some compound semiconductors, electrons occupy L-point or X-point minima. For these minima, two other scattering mechanisms must be considered in addition: intervalley, and nonpolar optic phonon scattering. The matrix elements for these two kinds of scattering are similar. Using the expressions for the matrix elements given in Table 6.2 and following the procedure outlined for polar optic phonon scattering, we obtain the following expression for $S(\underline{k})$ for nonpolar optic phonon scattering:

$$S_{op}(\underline{k}) = \frac{D_0^2}{8\pi^2\rho\omega_0}\sum_{+,-}\int\frac{|\underline{k}-\underline{k}'|^4}{(|\underline{k}-\underline{k}'|^2+\lambda^{-2})^2}\,G(\underline{k},\underline{k}')\delta(E_{\underline{k}}-E_{\underline{k}'}\pm\hbar\omega_0)\left(n_0+\frac{1}{2}\pm\frac{1}{2}\right)d\underline{k}'$$

$$= C_{op}\sum_{+,-}k_{\pm}\gamma'(E_{\underline{k}}\pm\hbar\omega_0)\left(n_0+\frac{1}{2}\pm\frac{1}{2}\right)F_{op\pm}(\underline{k},\lambda) \quad , \qquad (6.67)$$

where

$$F_{op\pm}(\underline{k},\lambda) = \int_{-1}^{+1}\left(1-2\frac{k_{\pm}}{k}x+\frac{k_{\pm}^2}{k^2}\right)^2 O_i(x)\left(1-2\frac{k_{\pm}}{k}x+\frac{k_{\pm}^2}{k^2}+\frac{1}{\lambda^2k^2}\right)^{-2}dx \quad ,$$

$$C_{op} = D_0^2m^*\Big/\left(4\pi\rho\hbar^2\omega_0\right) \quad .$$

Here $k_{\pm}$ are the wave vectors corresponding to $E_{\underline{k}}\pm\hbar\omega_0$, and are given by

$$k_{\pm}^2 = (2m^*/\hbar^2)\gamma(E_{\underline{k}}\pm\hbar\omega_0) \quad .$$

Also, $n_0 = [\exp(\hbar\omega_0/k_BT)-1]^{-1}$. The expression for intervalley scattering is obtained by replacing ω_0 and D_0 by ω_i and D_i respectively.

The expression is also applicable to nonequivalent intervalley scattering, in which an electron is scattered to another valley with its energy minimum separated by ΔE from the original valley. In that case, ΔE must be subtracted from the argument of the function γ in the expression for $k_{\pm}$.

6.7 Scattering Probabilities for Anisotropic Bands

All the formulae derived in the preceding sections are based on the assumption that the effective mass is isotropic, which applies to electrons occupying the Γ-point minimum of the Brillouin zone. In most compound semiconductors this condition applies, but in some compounds, electrons occupy X- or L-point minima, for which the effective mass is anisotropic. Complete expressions for the nonparabolic E-$\underline{k}$ relation, overlap integrals, and electron screening are not yet well established. Work is in progress, particularly for lead chalcogenides, which are important materials having minima at the L point of the Brillouin zone. The exact form of the E-$\underline{k}$ relation and the overlap integrals not being known, the expressions derived for the case of isotropic effective mass are often used. Even with this simplification, the expressions for the scattering probabilities are fairly complex for materials with anisotropic effective masses. We derive in this section the formulae obtainable with the assumptions mentioned above.

6.7.1 Herring-Vogt Transformation

The E-$\underline{k}$ relation for X- or L-point minima may be expressed as

$$(\hbar^2/2)\underline{k}\cdot\underline{\underline{M}}\cdot\underline{k} = E(1+\alpha E) = \gamma(E) \quad , \tag{6.68}$$

where

$$\underline{\underline{M}} = \begin{bmatrix} 1/m_1 & 0 & 0 \\ 0 & 1/m_2 & 0 \\ 0 & 0 & 1/m_3 \end{bmatrix} \quad .$$

As discussed earlier, the constant-energy surfaces for the minima at X or L points are ellipsoidal. In the above representation, the reference axes are chosen along the directions of the principal axes of the ellipsoid. Since, in general, the constant-energy surfaces are spheroidal, two of the effective mass components will be equal. The integrals involved in the various scattering probabilities may be simplified for the above E-$\underline{k}$ relation by working in a transformed $\underline{k}$ space instead of in the direct $\underline{k}$ space. The transformed $\underline{k}$ vector, $\underline{k}^*$, is related to $\underline{k}$ by the relation

$$\underline{k}^* = \underline{\underline{\alpha}} \cdot \underline{k} \quad , \tag{6.69}$$

where

$$\underline{\underline{\alpha}}^{\frac{1}{2}} \equiv \begin{bmatrix} \alpha_1^{\frac{1}{2}} & 0 & 0 \\ 0 & \alpha_2^{\frac{1}{2}} & 0 \\ 0 & 0 & \alpha_3^{\frac{1}{2}} \end{bmatrix} \quad ;$$

$\alpha_1 = m_0/m_1$, $\alpha_2 = m_0/m_2$, $\alpha_3 = m_0/m_3$; m_0 is the free electron mass. This transformation was first used by HERRING and VOGT [6.26] and is therefore often referred to by their name.

The E-$\underline{k}^*$ relation in the transformed space is

$$\hbar^2 k^{*2}/(2m_0) = \gamma(E) \quad , \tag{6.70}$$

and the constant-energy surfaces are spherical in $\underline{k}^*$ space. As a result, the integrals may be conveniently evaluated using polar coordinates.

6.7.2 Scattering Integrals

The volume element $d\underline{k}'$ in the transformed space may be written as

$$d\underline{k}' = (m_1 m_2 m_3)^{\frac{1}{2}} m_0^{-3/2} d\underline{k}'^* = (m_d/m_0)^{3/2} k'^{*2} \sin\theta^* \, d\theta^* d\phi^* dk'^* \quad , \tag{6.71}$$

where θ^* and ϕ^* are the polar and azimuthal angles of $\underline{k}'^*$. If the integrand in the scattering integral is independent of $\underline{k}$ and $\underline{k}'$, the only modification that occurs in the scattering formulae is the introduction of a factor $(m_d/m_0)^{3/2}$. When, however, $\underline{k}$ and $\underline{k}'$ are involved, we get complex modified expressions, which we now derive.

The first step in the derivation is writing the term $|\underline{k}-\underline{k}'|$ in terms of the transformed vectors. As shown in Fig.6.4, we choose the z axis along the direction of a principal axis of the constant-energy ellipsoid, and the y axis in the plane containing the z axis and $\underline{k}^*$. Let $\theta_{\underline{k}}$ be the polar angle of $\underline{k}^*$ with respect to the z axis. With the chosen directions of the coordinate axes, the azimuthal angle of $\underline{k}^*$ is zero. Also we note that the effective mass for the z direction is m_ℓ, and for both the x and y directions, m_t, the constant-energy surface being spheroidal. Let θ and ϕ denote the polar and azimuthal angles of k'^* measured with respect to the direction of $\underline{k}^*$ and y-z

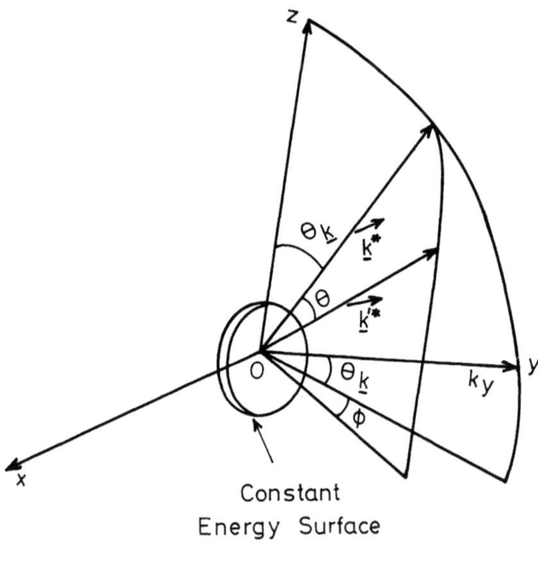

Fig. 6.4. Coordinate axes for ellipsoidal constant-energy surfaces

plane, which contains $\underline{k}^*$. The components of $\underline{k}$ and $\underline{k}'$ corresponding to $\underline{k}^*$ and $\underline{k}'^*$ along the x, y, z directions are given respectively by 0, $\alpha_t^{-\frac{1}{2}}k^* \sin\theta_{\underline{k}}$, $\alpha_\ell^{-\frac{1}{2}}k^* \cos\theta_{\underline{k}}$; and $\alpha_t^{-\frac{1}{2}}k'^* \sin\theta \sin\phi$, $\alpha_t^{-\frac{1}{2}}k'^*(\sin\theta_{\underline{k}} \cos\theta + \cos\theta_{\underline{k}} \sin\theta \cos\phi)$, $\alpha_\ell^{-\frac{1}{2}}k'^*(\cos\theta_{\underline{k}} \cos\theta - \sin\theta_{\underline{k}} \sin\theta \cos\phi)$. The term $|\underline{k}-\underline{k}'|^2$ may hence be expressed in terms of $\underline{k}^*$ and $\underline{k}'^*$ as

$$\begin{aligned}|\underline{k}-\underline{k}'|^2 &= \alpha_t^{-1}(k^{*2}-2k^*k'^* \cos\theta+k'^{*2})\\ &\quad + \left(\alpha_\ell^{-1}-\alpha_t^{-1}\right)\left[k^* \cos\theta_{\underline{k}}-k'^*(\cos\theta \cos\theta_{\underline{k}} - \sin\theta_{\underline{k}} \sin\theta \cos\phi)\right]^2\\ &= \psi(k^*,k'^*,\theta_{\underline{k}},\cos\theta,\phi) \quad , \end{aligned} \tag{6.72}$$

where $\alpha_\ell = m_0/m_\ell$, $\alpha_t = m_0/m_t$. Using the above expression for $|\underline{k}-\underline{k}'|^2$, we may now generalise the expressions derived earlier for the various scattering processes. The modified expressions are

$$\begin{aligned}S_{imp}(\underline{k}^*,\theta_{\underline{k}},\lambda) &= \frac{2}{\pi} C_{imp} \frac{m_d^{3/2}}{m_0^{\frac{1}{2}}m^*} k^*\gamma'(E_{\underline{k}},\theta_{\underline{k}})\\ &\quad \times \int_{-1}^{+1}\int_0^{2\pi} O_i(x)\left[\psi(k^*,k'^*,\theta_{\underline{k}},x,\phi)+\lambda^{-2}\right]^{-2}dxd\phi \quad , \end{aligned} \tag{6.73}$$

where $O_{i(x)} = a + bx + cx^2$, $k'^* = k^*$,

$$\begin{aligned}S_{pz}(\underline{k}^*,\theta_{\underline{k}},\lambda) &= \frac{C_{pz}}{2\pi m^*} \frac{m_d^{3/2}}{m_0^{\frac{1}{2}}} \gamma'(E_{\underline{k}},\theta_{\underline{k}})k^*\\ &\quad \times \int_{-1}^{+1}\int_0^{2\pi} O_i(x)\psi(k^*,k'^*,\theta_{\underline{k}},x,\phi)\left[\psi(k^*,k'^*,\theta_{\underline{k}},x,\phi)+\lambda^{-2}\right]^{-2}dxd\phi \quad , \end{aligned} \tag{6.74}$$

$$\begin{aligned}S_{ac}(\underline{k}^*,\theta_{\underline{k}},\lambda) &= \frac{C_{ac}}{4\pi} \frac{m_d^{3/2}}{m^*m_0^{\frac{1}{2}}} \gamma'(E_{\underline{k}},\theta_{\underline{k}})k^*\\ &\quad \times \int_{-1}^{+1}\int_0^{2\pi} \left[\psi(k^*,k'^*,\theta_{\underline{k}},x,\phi)\right]^2 O_i(x)\left[\psi(k^*,k'^*,\theta_{\underline{k}},x,\phi)+\lambda^{-2}\right]^{-2}dxd\phi \quad , \end{aligned} \tag{6.75}$$

$$\begin{aligned}S_{pop}(\underline{k}^*,\theta_{\underline{k}},\lambda) &= \left[C_{pop}m_d^{3/2}/\left(2\pi m^* m_0^{\frac{1}{2}}\right)\right]\\ &\quad \times \sum_{+,-} k_\pm^*\gamma'(E_{k^*}\pm\hbar\omega_\ell)\left(n_\ell+\frac{1}{2}\pm\frac{1}{2}\right)\int_{-1}^{+1}\int_0^{2\pi} O_i(x)\psi(k^*,k_\pm^*,\theta_{\underline{k}},x,\phi)\\ &\quad \times \left[\psi(k^*,k_\pm^*,\theta_{\underline{k}},x,\phi)+\lambda^{-2}\right]^{-2}dxd\phi \quad , \end{aligned} \tag{6.76}$$

where $k_{\pm}^{*} = \left[(2m_0/\hbar^2)\gamma(E_{\underline{k}*} \pm \hbar\omega_{\ell})\right]^{\frac{1}{2}}$.

$$S_{\substack{op\\int}}(\underline{k}^{*},\theta_{\underline{k}},\lambda) = C_{\substack{op\\int}} \frac{m_d^{3/2}}{\pi m^{*} m_0^{\frac{1}{2}}} \sum_{+,-} k_{\pm}^{*}\gamma'\left(E_{k*} \pm \hbar\omega_{0_i}\right)\left(n_{0_i} + \frac{1}{2} \pm \frac{1}{2}\right)$$

$$\times \int_{-1}^{+1}\int_{0}^{2\pi} \left[\psi(k^{*},k_{\pm}^{*},\theta_{\underline{k}},x,\phi)\right]^2 0_i(x)\left[\psi(k^{*},k_{\pm}^{*},\theta_{\underline{k}},x,\phi)+\lambda^{-2}\right]^{-2} dx d\phi \ . \qquad (6.77)$$

We find that the scattering probability in all cases now depends not only on the magnitude of the wave vector, but also on its direction relative to the major axis of the constant-energy ellipsoids. The integrals cannot, in general, be evaluated analytically, and we are required to evaluate numerically a double integral in each case. Because of this complexity in the computations, the transport properties of materials with ellipsoidal band structure have not been much studied by treating the scattering probabilities in the manner discussed above.

The equations above may be simplified in the case of deformation-potential acoustic phonons and nonpolar optic or intervalley phonons when the screening effect of the electrons is neglected. The formulae are then the same as those for the case of isotropic band structure, but multiplied by $\left(m_d^{3/2} m^{*-3/2}\right)$. Since the scattering mechanism mentioned above predominates in elemental semiconductors, these materials could be studied extensively without undertaking extensive computations even though the band structures are anisotropic. In compound semiconductors, on the other hand, other scattering mechanisms (piezoelectric, impurity, and polar optic phonon scattering) being equally important, detailed calculations of the transport parameters becomes extremely laborious when the band structure is anisotropic. Properly calculated theoretical values of the transport coefficients of these materials are therefore not yet available even though the basic formulae have been worked out. We shall discuss in Sect.7.7 the formulae that may be used to evaluate the transport coefficients when impurity-atom scattering or polar optic phonon scattering predominate in a material with anisotropic band structure.

6.8 S($\underline{k}$) for Neutral Impurity, Alloy, and Crystal-Defect Scattering

Neutral impurity, alloy, and crystal-defect scattering are additional scattering mechanisms which are also important in determining the transport coefficients of compound semiconductors. Detailed formulae giving the scattering probability for these mechanisms that take account of all the complexities of band structure are not yet available. The perturbing potentials produced by the defects caused by neutral impurities, alloying, and dislocations are not of the same nature as those due to the ionised impurities or lattice vibrations, and the expressions for the scattering probability for these scattering mechanisms is based on a different formalism. We shall only quote the available results with references, and shall not discuss the derivation.

6.8.1 Neutral-Impurity Scattering

Under degenerate conditions and at very low temperatures, a large fraction of the impurity atoms are un-ionised. These neutral impurities do not produce a perturbing potential of the kind discussed before. However, an electron that comes within a short distance of the neutral atom may interact with it, and its momentum may change through a process in which the bound and free electrons exchange places. Expressions for the collision cross section in collisions with neutral atoms have been derived by ERGINSOY and by SCLAR [6.4] for parabolic bands. These results have received wide attention, although there is an important difference between the two. ERGINSOY's formula is

$$S_N(\underline{k}) = \frac{N_n \hbar 20 a_0}{m^*} \quad , \tag{6.78}$$

while SCLAR's is

$$S_N(\underline{k}) = \frac{\pi N_n e^4}{\sqrt{2m^*}\, E^{3/2}} \left[\ln(1+k\lambda)^2 - 1/(1+k^{-2}\lambda^{-2})\right] \quad , \tag{6.79}$$

with N_n the concentration of neutral impurity atoms.

It should be noted that ERGINSOY's formula predicts an energy-independent scattering probability, whereas according to SCLAR's formula, the scattering probability should depend on the electron energy. The nature of neutral-impurity scattering has not yet been analysed in sufficient detail to allow us to decide which formula should be taken. The conditions under which this type of scattering is important introduce other factors such as degeneracy, the effects of which are difficult to assess accurately. The exact form of neutral-impurity scattering is also expected to vary with the atomic nature of the impurity. Although a definite conclusion cannot be drawn at this stage, some recent work appears to support the SCLAR expression [6.27].

6.8.2 Alloy Scattering

In mixed semiconducting compounds, the composition is not strictly homogeneous, even in carefully grown crystals. Due to random fluctuations in the composition, randomly distributed potential discontinuities occur. These act as scattering centres, just like the deformation potentials associated with lattice waves. An expression for the scattering probability for these discontinuities for parabolic bands has been quoted by MAKOWSKI and GLICKSMAN [6.26]:

$$S_{all}(\underline{k}) = \sqrt{2}\,\pi^{-1}\hbar^{-4}N_0(1-\alpha)\alpha(E_a-E_b)^2 m^{*3/2}E^{1/2} \quad . \qquad (6.80)$$

Recent analyses [6.1,2c,25,28] using this expression apparently indicate that the difference between the observed mobilities in mixed crystals and the theoretical values may be explained on the basis of this expression. It also appears that this kind of scattering is of a significance comparable to that of lattice scattering. A more detailed study of alloy scattering may therefore be important for the understanding of electron transport in compound semiconductors. The formula given above may be taken as the starting point of such a study, but modifications may be needed.

6.8.3 Defect Scattering

All real semiconductor crystals contain imperfections in the arrangement of the atoms. The dislocations of the atoms from the perfect-crystal positions may affect the transport of electrons in two ways. The strains produced at the dislocations may cause potentials to appear which scatter electrons. Charges also collect at the dislocations which often act as acceptors, and the electrons may be scattered by these charges. Analyses reported in the literature [6.23] suggest that the effect of the former is negligible, but the latter effect may be significant. An expression for the scattering probability for charged dislocations valid in parabolic bands is [6.23]

$$S_{dis}(\underline{k}) = \frac{Ne^4f^2\lambda}{8\varepsilon^2a_D^2m^{*2}}\left[V_t^2+\hbar^2/(4m^{*2}\tau_D^2)\right]^{-3/2} \quad , \qquad (6.81)$$

where V_t is the component of electron velocity perpendicular to the dislocation lines, N is the density of dislocation centres, and τ_D is dielectric relaxation time.

We should, however, mention that in most crystals, the contribution of defect scattering turns out to be negligible in comparison to that of other scattering mechanisms. Although such scattering can take place and may have to be taken into account to explain results in some crystals, most of the available results on electron transport in compound semiconductors are usually explained neglecting it.

6.9 Conclusions

In this chapter, we derived a number of expressions for scattering probabilities due to the various possible scattering mechanisms. The expressions are complete in the sense that the effects of band nonparabolicity, overlap integrals, and electron screening are all included. In some materials, these effects may be negligible, and the corresponding expressions may be obtained simply by putting α in the expression for $\gamma(E)$ equal to zero, and λ equal to infinity.

The expressions for the scattering probabilities form the basis of the theory of electron transport presented in the following chapters. It should be evident that the expressions involve constants of the materials which can usually be measured by independent methods, some of which are described in [6.2]. The object of the theory of transport is to obtain formulae for the various transport coefficients which may be used to evaluate the coefficients with values of physical constants determined by other methods. Some of these formulae will be derived in the following chapters using the expressions for scattering probabilities derived here. We shall also compare the experimental results with theoretical values for important compounds in a later chapter. Suffice it to say at this stage that great success has been achieved in recent years in computing theoretically the values of transport coefficients which agree well with experiments in many materials. The agreement has progressed in the course of the last decade from an agreement in order to an agreement within a few per cent. Interestingly, all these computations are performed using the expressions given in this chapter, which, although based on some approximations, are found to yield good results.

We have mainly considered scattering probabilities for electrons with isotropic or anisotropic effective masses. We have not considered scattering probabilities for holes. The constant-energy surfaces for holes are ellipsoidal in IV-VI compounds, but in all other compounds, they are warped, and the corresponding E-$\underline{k}$ relation is much more involved. The formulae derived in this chapter may thus be applied to the case of holes in IV-VI compounds, but they are not strictly applicable to holes in other compounds. The task of deriving expressions for the scattering probability of holes taking into consideration the exact E-$\underline{k}$ relation is so difficult that very few attempts have so far been made [6.13,29]. It is, however, quite common to study the transport properties of holes by assuming an equivalent spherical constant-energy surface for them, so that the formulae derived in this chapter can be used in such studies of hole transport in compound semiconductors.

7. The Boltzmann Transport Equation and Its Solution

The theory of electron transport is concerned with the evaluation of macroscopic transport coefficients such as electrical and thermal conductivity, Hall coefficient, diffusivity, etc., from a knowledge of other independently measurable physical constants of the medium. The theory is also concerned with the understanding of the variation of the transport coefficients with electric and magnetic fields, temperature, and composition of the materials. The macroscopic average quantities are basically determined by the microscopic behaviour of the particles constituting the systems. Since the microscopic motion is intractable, the statistical behaviour of the particles is studied in the evaluation of the transport coefficients. Such an analysis starts from the transport equation, the solution of which gives a function related to the position and velocity of the particles at different times.

7.1 The Liouville Equation and the Boltzmann Equation

The general equation of motion of the particles yields an equation known as the Liouville equation, which defines how the distribution of the particles changes with time in the phase space as a result of the action of the forces acting on the particles. The equation may be stated as [7.1]

$$\frac{\partial D_N}{\partial t} = \left\{H_N, D_N\right\}' = \sum_{i=1}^{N}\left(\frac{\partial H_N}{\partial \underline{r}_i}\frac{\partial D_N}{\partial \underline{p}_i} - \frac{\partial H_N}{\partial \underline{p}_i}\frac{\partial D_N}{\partial \underline{r}_i}\right) , \tag{7.1}$$

where D_N may be interpreted as giving the probability of particle 1 being located in the interval $d\underline{r}_1 d\underline{p}_1$ around $\underline{r}_1, \underline{p}_1$, particle 2 being located in $d\underline{r}_2 d\underline{p}_2$ around $\underline{r}_2\underline{p}_2$, and so on ($\underline{r}$ and $\underline{p}$ are the position and momentum vectors); H_N is the Hamiltonian of the N particles constituting the system.

Although the Liouville equation gives a complete description of the evolution of the system in time under the action of the forces, it is almost impossible to evaluate transport coefficients using this equation. The equation which has proved much more successful in the study of transport processes in gases is the Boltzmann equation, which may be derived intuitively by considering the flow of particles under the action of external and diffusive forces in the six-dimensional $\underline{p}$-$\underline{r}$ space, and assuming the collision events to be spaced in time and space at discrete intervales, and to be of infinitesimally small duration and extent in the scale of observations. The Boltzmann equation gives the probability distribution of a single particle in the phase space as a function of time. This one-particle distribution is assumed to be representative of the distribution of all the particles. In deriving the collision term it is also assumed that the time development is determined by the distribution at any particular observation time. In other words, information relating to the initial condition is neglected and any correlations present at the time of observation are completely ignored.

The equation may also be derived from the Liouville equation, as the one-particle distribution is related to the integral of D_N over the coordinates of all other particles except one. This derivation requires the assumptions mentioned above and also shows that the Boltzmann equation describes the dynamics of the multiparticle system in the zeroth-order approximation in density, so that it should not be applied to liquids or solids.

The transport of electrons in ionised gases as well as of electrons in solids is also dealt with in most of the analyses starting from the Boltzmann equation. We shall not discuss the approximations implied by the application of Boltzmann equation for the analysis of transport of electrons in semiconductors either with or without a magnetic field [7.2]. However, in spite of all the doubts expressed about the validity of the Boltzmann equation or of the scrutiny to which the equation has been subjected, almost all the results available on electron transport in semiconductors are derived from it. The best justification for the use of the equation lies in the good agreement that is observed between these results and the experiments. The equation may be taken, in any case, as empirically well justified. Since our purpose is to present the theory of electron transport in current use, we shall take the Boltzmann equation as our starting point. In Sect.7.2 we give its intuitive derivation, which may not be very elegant but produces the equation with very simple arguments. The rest of this chapter is concerned with solutions of this equation for the various kinds of collision mechanisms discussed earlier.

7.2 The Boltzmann Transport Equation

Let $f(\underline{r},\underline{k},t)$ be a function, often referred to as the distribution function, giving the probability of occupation of a state by an electron at $(\underline{r},\underline{k})$ in phase space at time t. This statement really violates the uncertainty principle [7.3] according to which both the position and the momentum of the electron cannot be specified at the same time. However, in most transport problems of interest, the location of the electrons need only be fixed on a macroscopic scale. We may assume that $\underline{r}$ is fixed within an interval $\Delta\underline{r}$, and that the electron is represented by a wave packet with a spread of $\Delta\underline{k}$ in $\underline{k}$ such that $\Delta\underline{r}\Delta\underline{k} \geq 1$. The specification of $\underline{r}$ and $\underline{k}$ in $f(\underline{r},\underline{k},t)$ is thus correct to within $\Delta\underline{r}$ and $\Delta\underline{k}$.

We also assume that the probability of occupation of other states by other electrons at other positions of the phase space is also given by the same function. The number of electrons occupying a volume element $d\underline{k}$ in $\underline{k}$ space per unit volume of the crystal centred round $\underline{r}$ is hence given by

$$dn = (1/4\pi^3) f(\underline{r},\underline{k},t) d\underline{k} \quad , \tag{7.2}$$

since the number of states in the same volume is $(1/4\pi^3)d\underline{k}$. The distinct number of states is $(1/8\pi^3)d\underline{k}$, but each state can accommodate two electrons of opposite spin.

The equation giving f is obtained using the following intuitive arguments.

The electrons occupying points around $\underline{r}'$, $\underline{k}'$ at time t_0-dt will occupy points around $\underline{r}_0$, $\underline{k}_0$ at time t_0. If the motion of the electrons is not affected by any force other than the external accelerating forces, then $\underline{r}'$ and $\underline{k}'$ are related to $\underline{r}_0$ and $\underline{k}_0$ by the following relations:

$$\underline{r}' = \underline{r}_0 - (1/\hbar)\underline{\nabla}_{\underline{k}} E dt \quad , \tag{7.3}$$

$$\underline{k}' = \underline{k}_0 - (\underline{F}/\hbar) dt \quad , \tag{7.4}$$

where $(1/\hbar)\underline{\nabla}_{\underline{k}}E$ is the velocity of the wave packet representing an electron of energy E, and $\underline{F}$ is the external force.

The volume element occupied at time t_0-dt by the electrons which occupy a volume $d\underline{r}\,d\underline{k}$ at time t, is $d\underline{r}'d\underline{k}' = d\underline{r}d\underline{k} + O(dt)$. In the limit of infinitesimal time interval, we then have

$$\left(V_c/4\pi^3\right) f\left(\underline{r}_0 - \hbar^{-1}\underline{\nabla}_{\underline{k}} E dt, \underline{k}_0 - \underline{F}\hbar^{-1}dt, t_0 - dt\right) d\underline{r}'d\underline{k}' = \left(V_c/4\pi^3\right) f(\underline{r}_0,\underline{k}_0,t) d\underline{r}d\underline{k} \,. \tag{7.5}$$

We should, however, note that there are intrinsic physical processes which would affect the motion of the electrons. Firstly, electrons may come across other electrons in their path with which they may interact through the Coulomb field. This interaction would depend on the density of the electrons, and in nondegenerate materials we may ignore such possible interactions. However, in degenerate materials, the electron-electron interaction would be important enough to merit careful consideration.

Even if we neglect the electron-electron interaction, we cannot omit from consideration the possibility of electrons colliding with impurity atoms, or exchanging energy and momentum with the lattice atoms through the potentials arising from vibrations. The effect of such collisions would be to remove some of the electrons occupying a volume $d\underline{k}$ around $\underline{k}$ to other points in $\underline{k}$ space, and to transfer other electrons from other points in $\underline{k}$ space to the volume $d\underline{k}$ around $\underline{k}$. As a result, the distribution function f would change. Indicating the rate of change due to the collision process as $\partial_c f$, and equating the numbers of electrons, we get

$$f(\underline{r}_0,\underline{k}_0,t)d\underline{k} = f\left(\underline{r}_0-\hbar^{-1}\underline{\nabla}_{\underline{k}}E\,dt,\underline{k}_0-\hbar^{-1}\underline{F}\,dt,\ t_0-dt\right)d\underline{k} + \partial_c f\,d\underline{k}\,dt \quad . \tag{7.6}$$

The left-hand side gives the number of electrons in $d\underline{k}$ at time. The first term on the right-hand side gives the number of electrons arriving in the time interval dt, and the second term, the increase in the time interval dt of the number of electrons at $(\underline{r}_0,\underline{k}_0)$ in volume element $d\underline{k}$ due to collisions.

Expanding $f\left(\underline{r}_0-\hbar^{-1}\underline{\nabla}_{\underline{k}}E\ dt,\underline{k}_0-\hbar^{-1}\underline{F}\ dt,t_0-dt\right)$ in a Taylor's series and retaining only the first-order term, we get

$$\left.\frac{\partial f}{\partial t}\right|_{t_0,r_0,k_0} = -\frac{1}{\hbar}\underline{\nabla}_{\underline{k}}E\cdot\underline{\nabla}_{\underline{r}}f\Big|_{t_0,r_0,k_0}-\frac{\underline{F}}{\hbar}\cdot\underline{\nabla}_{\underline{k}}f\Big|_{t_0,r_0,k_0}+\partial_c f\Big|_{t_0,r_0,k_0} \quad . \tag{7.7}$$

At this stage a subtle transformation is made to derive the Boltzmann equation [7.4]. Équation (7.7) is really a relation between the various differential coefficients of f at any particular time of observation. However, we get a differential equation if we omit the subscripts, implying that the relation holds for any time and position in phase space. The resulting equation is

$$\frac{\partial f}{\partial t} = -(1/\hbar)\underline{\nabla}_{\underline{k}}E\cdot\nabla_{\underline{r}}f - (\underline{F}/\hbar)\cdot\underline{\nabla}_{\underline{k}}f + \partial_c f \quad . \tag{7.8}$$

It is further assumed that the collision term (the detailed expression for which is derived in the following section) is determined only by the value of f at the time t at which observation starts.

Equation (7.8) is the Boltzmann equation. A strong criticism often advanced against this equation is that, being a first-order equation in t, it implies that the distribution function may be worked out at all future times, if it is known at any particular time. Therefore, we do not need to know the value of the function at earlier times. This in a way results in the discarding of information at every instant. It is not readily possible to justify this discarding. However, if we assume that any correlation that may have existed between the motion of electrons occupying different positions dies out quickly and that a disorder exists at all times, we may compute the collision effects statistically. This hypothesis is known as the Stoßzahlansatz (from the German word, meaning "collision-number assumption"). We shall not consider this point further [7.1].

As mentioned earlier, the Boltzmann equation is empirically well justified, and we shall use it in that context forgetting about the lack of rigour in its derivation. We should, however, keep in mind that the assumption of disorder restricts the application of the equation to systems which are known to be disordered physically. We should not apply the equation to study the streamy motion of electrons as the assumption of disorder is not applicable for such motion. However, transport phenomena are quite often analysed using the Boltzmann equation even in this situation.

In the study of photoconduction phenomena, a modified form of the equation has been used recently which is of some interest [7.5]. In these phenomena, electrons are excited from the valence band or impurity levels into the conduction band by the absorption of incident light. These electrons are available for conduction for a time before they recombine or fall back to the levels from which they were excited. Let $G(\underline{r},\underline{k},t)$ and $R(\underline{r},\underline{k},t)$ be two functions giving the rates of generation and recombination of electrons of wave vector $\underline{k}$ at time t at position $\underline{r}$. These two processes would also alter the number of electrons at any position and may be treated as extra terms to be added to the collision term. We may hence generalise the Boltzmann equation to apply to photoconduction phenomena by including these terms. The modified equation is

$$\partial_t f = -(1/\hbar)\underline{\nabla}_k E \cdot \underline{\nabla}_r f - (\underline{F}/\hbar) \cdot \underline{\nabla}_k f + \partial_c f + G - R \quad , \qquad (7.9)$$

where f, G, and R are all functions of $\underline{r}$, $\underline{k}$, and t.

It should be pointed out that in the analysis of electron transport in semiconductors, the last two terms should in principle always be added. But in the absence of external excitation, we may take the sum of these two terms to be zero, unless we are considering the phenomena on a very microscopic time scale.

7.3 The Collision Integral

As mentioned earlier, the detailed form of the collision term is derived in the Boltzmann equation by assuming that the number of electrons transferred at any time from one point of the $\underline{k}$ space to another by collisions is determined by the instantaneous value of the distribution function. It is also assumed that the duration of the collision is negligibly small, that it is a localised event, and that it does not depend on the position of the particles. The probability of a transition from a point around $\underline{k}$ to a point around $\underline{k}'$ therefore depends only on $\underline{k}$ and $\underline{k}'$, and is given by $S(\underline{k},\underline{k}')$, expressions for which have been derived in Sect.6.6 for various scattering mechanisms.

In deriving the expression for the collision term we should also distinguish between electron-electron collisions and collision of the electrons with impurity atoms or lattice atoms. We consider first the latter kind. One of the important assumptions that we shall make is that the collisions do not affect the condition of the impurity atoms or the distribution of the phonons characterising the lattice vibrations. The latter assumption is not, however, valid under certain experimental conditions. One example is the transport of electrons at low temperatures under the action of a large electric field [7.6]. As the momentum and energy of the electrons may be very high, the momentum and energy transferred from the electrons to the lattice atoms may also be sufficiently high to affect the phonon population. We shall, however, exclude from consideration such experimental conditions. We may hence assume that the number of electrons transferred from one point to another is determined by the distribution function of the electrons alone.

The number of electrons scattered per unit time from $d\underline{k}$ around $\underline{k}$ to $d\underline{k}'$ around $\underline{k}'$ is given by

$$\left(V_c/4\pi^3\right)\left(V_c/8\pi^3\right)f(\underline{r},\underline{k},t)[1-f(\underline{r},\underline{k}',t)]S(\underline{k},\underline{k}')d\underline{k}d\underline{k}' \quad , \tag{7.10}$$

since the number of electrons which may be scattered is $\left(V_c/4\pi^3\right)f(\underline{r},\underline{k},t)d\underline{k}$, and the number of vacancies which these electrons may occupy after scattering is $V_c/(2\pi)^3[1-f(\underline{r},\underline{k}',t)]d\underline{k}'$, V_c being the volume of the crystal. It should be noted that the spin of the electron is assumed to be conserved in the collision process [7.7].

Similarly, the number of electrons which are scattered per unit time from $d\underline{k}'$ around $\underline{k}'$ to $d\underline{k}$ around $\underline{k}$ is

$$\left(V_c/4\pi^3\right)\left(V_c/8\pi^3\right)f(\underline{r},\underline{k}',t)[1-f(\underline{r},\underline{k},t)]S(\underline{k}',\underline{k})d\underline{k}d\underline{k}' \quad . \tag{7.11}$$

The rate of increase of the number of electrons in $d\underline{k}$ around $\underline{k}$ per unit volume of the crystal is hence

$$-(1/4\pi^3)(1/8\pi^3)\int_{\underline{k}'}[f(1-f_1)S-f_1(1-f)S_1]d\underline{k}d\underline{k}' \quad , \tag{7.12}$$

where we have used the following abbreviations:

$$f(\underline{r},\underline{k},t) = f \quad ; \quad f(\underline{r},\underline{k}',t) = f_1 \quad ;$$

$$S(\underline{k},\underline{k}') = S \quad ; \quad S(\underline{k}',\underline{k}) = S_1 \quad .$$

The integration is to be carried out over all values of $\underline{k}'$, as an electron may be scattered from the interval $d\underline{k}$ around $\underline{k}$ to any other point of $\underline{k}$ space, and an electron from any other point may be scattered into $d\underline{k}$.

The collision term is thus given by

$$\partial_c f = -\left(V_c/8\pi^3\right)\int_{\underline{k}'}[f(1-f_1)S-f_1(1-f)S_1]d\underline{k}' \quad . \tag{7.13}$$

Note that the term is an integral, which is often referred to as the collision integral. The presence of this nonlinear integral term makes the Boltzmann equation an integro-differential equation, which cannot be solved by analytic methods. We shall discuss in the following sections the methods which are used to obtain approximate solutions.

7.4 Linearised Boltzmann Equation

The first step for obtaining an analytic solution of the Boltzmann equation is the linearisation of the equation. We derive in this section the linearised equation which is applicable for low accelerating forces. We also assume, for convenience, that the semiconductor sample is homogeneous, and that the accelerating force is due to a uniform electric field alone.

In the absence of an electric field, the distribution function of the electrons must be the Fermi-Dirac function for degenerate materials, which sim-

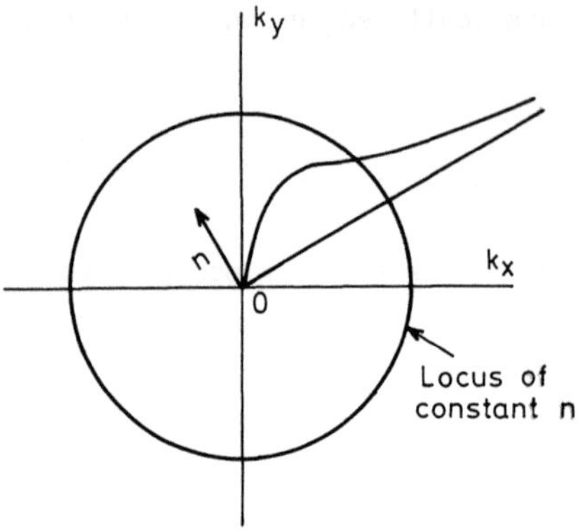

Fig. 7.1. Electron distribution under equilibrium condition in k space for scalar effective mass. a) Locus of constant electron concentration. b) Relative electron concentration per unit energy interval

plifies to the Maxwellian function for nondegenerate materials. This function may be derived from statistical considerations, and gives the most probable distribution of the electrons in the energy space. A convenient way of visualising it is to draw the locus of points in $\underline{k}$ space for which the distribution function has the same value. Such a locus in two dimensions together with a curve showing the number of electrons per unit energy interval is shown in Fig.7.1 for electrons with a scalar effective mass.

In the absence of accelerating and diffusive forces, the Boltzmann equation reduces in the steady state to

$$\partial_c f = 0 \quad . \tag{7.14}$$

The Fermi function thus makes the collision integral zero. In fact, it makes the integrand zero for all values of the electron energy. The transfer of electrons under the equilibrium conditions is actually balanced in detail. The number of electrons transferred from $d\underline{k}$ around $\underline{k}$ to $d\underline{k}'$ around $\underline{k}'$ is exactly balanced by the inverse transfer, i.e. the number of electrons transferred from $d\underline{k}'$ to $d\underline{k}$. This is the so-called "principle of detailed balance", which may be shown to follow from statistical considerations [7.8].

Denoting the Fermi function by f_0, we thus find that

$$f_0(E)[1-f_0(E')]S(\underline{k},\underline{k}') - f_0(E')[1-f_0(E)]S(\underline{k}',\underline{k}) = 0 \quad , \tag{7.15}$$

where E and E' denote respectively the energies corresponding to the wave vectors $\underline{k}$ and $\underline{k}'$.

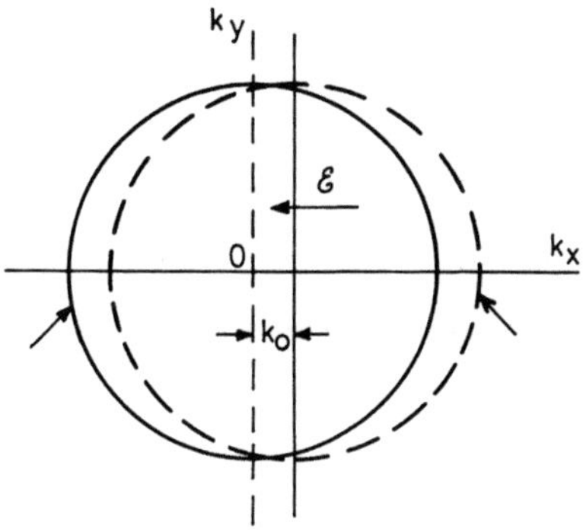

Fig. 7.2. Electron distribution in an electric field. Solid line – locus of constant electron concentration in equilibrium. Dotted line – locus of constant electron concentration in an electric field

The distribution function of the electrons is altered when the electric field is applied. This establishes some order in the motion of the electrons, which is otherwise fully disordered. Let us first assume that the electrons have an isotropic effective mass. The electric field accelerates the electrons in a direction opposite to it, and if we assume that all the electrons attain the same drift velocity, the distribution function also shifts in the same direction (as shown in Fig.7.2) by an amount equal to the drift component of the wave vector. The actual velocity of an electron, however, depends on its energy and the direction of its wave vector. The change in the distribution function thus also depends on the energy and the direction of the electron wave vector. The perturbation in the distribution function, however, is biased towards the direction of the field. We can express the distribution function in a spherical harmonic series as

$$f(\underline{k}) = \sum_n f_n(E)P_n(\cos\theta) \quad , \tag{7.16}$$

where the $f_n(E)$ are unknown functions to be determined by solving the Boltzmann equation, $P_n(\cos\theta)$ is the Legendre polynomial of order n, and θ is the angle between the electron wave vector and the applied field $\underline{\mathscr{E}}$.

Substituting the series in the Boltzmann equation and using the orthogonal property of the Legendre polynomials, we get a set of equations relating the $f_n(E)$. In principle, the set includes an infinite number of equations. It is, however, found that the low-field transport coefficients may be fairly accurately derived by retaining only the first two terms. The equations relating these two terms are

$$(1/3)(e\mathscr{E}/\hbar)(\partial_k f_1+2f_1/k) = -\left(V_c/8\pi^3\right)\int\Big\{f_0(E)[1-f_0(E')]S(\underline{k},\underline{k}') - f_0(E')[1-f_0(E)]S(\underline{k}',\underline{k})\Big\}d\underline{k}' \quad , \tag{7.17}$$

$$(e\mathscr{E}/\hbar)\partial_k f_0 = -\left(V_c/8\pi^3\right)\int\Big[f_1(E)\Big\{[1-f_0(E')]S(\underline{k},\underline{k}')+f_0(E')S(\underline{k}',\underline{k})\Big\} - (\cos\theta'/\cos\theta)f_1(E')\Big\{f_0(E)S(\underline{k},\underline{k}')+[1-f_0(E)]S(\underline{k}',\underline{k})\Big\}\Big]d\underline{k}' \quad , \tag{7.18}$$

where θ' is the angle between $\underline{k}'$ and $\underline{\mathscr{E}}$. In deriving the above equations we have assumed that f_1 is much smaller than f_0, and we have neglected $f_1(E)f_1(E')$ in comparison to other terms. We have also assumed that $S(\underline{k},\underline{k}')$ depends only on the magnitude of $|\underline{k}-\underline{k}'|$, and not on the individual orientation of $\underline{k}$ or $\underline{k}'$. This implies that $S(\underline{k},\underline{k}')$ involves only the angle between $\underline{k}$ and $\underline{k}'$. It should be evident from the expressions for $S(\underline{k},\underline{k}')$ given in Sect.6.6 that this assumption is valid for all scattering mechanisms if the effective mass of the electrons is isotropic. On the other hand, for an anisotropic electron effective mass, the assumption would be valid only for those scattering mechanisms for which the matrix element is independent of $|\underline{k}-\underline{k}'|$. It is not, therefore, valid for scattering mechanisms such as ionised impurity scattering, polar optic phonon scattering, etc. in materials with anisotropic effective mass. In dealing with the transport coefficients of materials for which the electron effective mass is anisotropic and in which these scattering mechanisms predominate, the collision terms containing different f_n cannot be separated as in the above equations, and the equations are coupled on the right-hand side as well. We shall discuss the methods to be used for solving the Boltzmann equation for such cases in Sect.7.6.

We find from (7.18) that the odd part of the distribution function $f_1(E)$ is proportional to the electric field $\mathscr{E}$. The left-hand side of (7.17) is therefore proportional to $\mathscr{E}^2$. For low values of the electric field, we may neglect this term. We shall see later that at high electric fields, this term being significant, the characteristics of high-field transport are radically different from those of low-field transport. However as we are considering low-field transport at present, we may take the left-hand side of (7.17) as zero. We find on comparing (7.17) with (7.15) that f_0, the even part of the distribution function, is the same as the Fermi-Dirac function.

The Boltzmann equation may thus be transformed for low fields to the following linear form:

$$(e\mathscr{E}/\hbar)\partial_k f_0 = -\left(V_c/8\pi^3\right)\int\Big\{f_1(E)\Big[(1-f_0(E'))S(\underline{k},\underline{k}')+f_0(E')S(\underline{k}',\underline{k})\Big]$$
$$-(\cos\theta'/\cos\theta)f_1(E')\Big[f_0(E)S(\underline{k},\underline{k}')+(1-f_0(E))S(\underline{k}',\underline{k})\Big]\Big\}d\underline{k}' \quad , \tag{7.19}$$

where f_0 is the Fermi-Dirac function.

7.5 Simplified Form of the Collision Terms

The collision terms in the linearised Boltzmann equation may be further simplified taking advantage of the property of the delta function involved in $S(\underline{k},\underline{k}')$.

Before deriving the simplified expression, we introduce for the sake of algebraic simplicity a function $\phi(E)$, in place of $f_1(E)$, obeying the relation

$$f_1(E) = -(e\hbar\mathscr{E}/m^*)k\partial_E f_0\phi(E) \quad . \tag{7.20}$$

Replacing $f_1(E)$, using (7.15), and noting that

$$\partial_E f_0 = -f_0(1-f_0)/(k_BT) \quad , \tag{7.21}$$

we get for the collision term

$$\partial_c f = -(e\hbar\mathscr{E}/m^*k_BT)k\cos\theta\left(V_c/8\pi^3\right)$$
$$\times\int f_0(E)[1-f_0(E')][\phi(E)-(k'\cos\theta'/k\cos\theta)\phi(E')]S(\underline{k},\underline{k}')d\underline{k}' \quad . \tag{7.22}$$

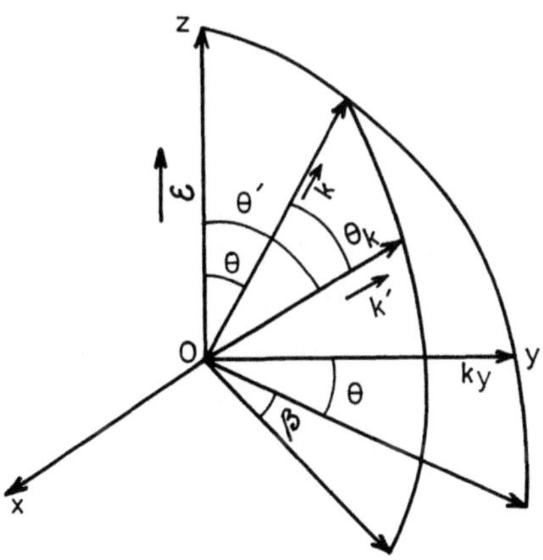

Fig. 7.3. Reference coordinate axes and the vectors $\underline{\mathscr{E}}$, $\underline{k}$, and $\underline{k}'$

We may also express $\cos\theta'$ in terms of the angle between $\underline{k}$ and $\underline{k}'$. Figure 7.3 shows the geometry of the situation, from which we find that

$$\cos\theta' = \cos\theta \cos\theta_k - \sin\theta \sin\theta_k \cos\beta \quad , \tag{7.23}$$

where θ_k is the angle between $\underline{k}$ and $\underline{k}'$, and β is the angle between the planes containing $\underline{k}$ and $\underline{k}'$, and $\underline{k}$ and $\underline{\mathscr{E}}$.

Since none of the terms other than $\cos\theta'$ in (7.19) contains β, the contribution of the term multiplied by $\cos\beta$ becomes zero when we integrate over β from 0 to 2π. We thus get

$$\partial_c f = -(e\hbar\mathscr{E}/m^*k_BT)k \cos\theta\left(V_c/8\pi^3\right)$$
$$\int f_0(E)[1-f_0(E')][\phi(E)-(k'\cos\theta_k/k)\phi(E')]S(\underline{k},\underline{k}')d\underline{k}' \quad . \tag{7.24}$$

To simplify this term further, we consider elastic and inelastic scattering processes separately.

7.5.1 Collision Terms for Elastic Scattering

As explained in Chap.6, acoustic phonon scattering of both kinds (deformation potential and piezoelectric) and defect scattering of different kinds (ionised and neutral impurity-atom scattering, alloy scattering, and crystal-defect scattering) do not alter the energy of the electron. For all these elastic scattering processes, $E' \approx E$. Equation (7.24) may thus be simplified to

$$\partial_c f = -(e\hbar\mathscr{E}/m^*k_BT)k \cos\theta\; f_0(E)[1-f_0(E)]S(\text{out+in};k)\phi(E) \quad , \tag{7.25}$$

where

$$S(\text{out+in};k) = (V_c/8\pi^3)\int S(\underline{k},\underline{k}')(1-\cos\theta_k)d\underline{k}' \quad . \tag{7.26}$$

It includes the effects of both out-scattering from and in-scattering to, $\underline{k}$.

Comparing S(out+in;k) with S(k) defined in Chap.6, we find that S(out+in;k) involves only an additional factor $(1-\cos\theta_k)$ in the integrand. We may therefore write expressions for S(out+in;k) by inspection of those for S(k). We give below the expressions for the important elastic scattering processes.

Ionised impurity scattering:

$$S(\text{out+in};k;\text{imp}) = \left(C_{imp}/k^3\right)\gamma'(E)\int_{-1}^{+1}(1-x)Q_i(x)(1-x+0.5k^{-2}\lambda^{-2})^{-2}dx \quad . \tag{7.27}$$

Piezoelectric scattering:

$$S(out+in;k;piezo) = (C_{pz}/2k)\gamma'(E)\int_{-1}^{+1}(1-x)^2 O_i(x)(1-x+0.5k^{-2}\lambda^{-2})^{-2}dx . \quad (7.28)$$

Deformation-potential acoustic scattering:

$$S(out+in;k;ac)\ (C_{ac}k\gamma'(E)/2)\int_{-1}^{+1}(1-x)^3 O_i(x)(1-x+0.5k^{-2}\lambda^{-2})^{-2}dx \quad . \quad (7.29)$$

The symbols used in the above expressions have the same significance as those defined for S(k) (see Sect.6.6).

7.5.2 Collision Terms for Inelastic Scattering

Optic phonon scattering of both kinds (nonpolar and polar) and intervalley scattering are such that, at ordinary temperatures, the energy gained or lost by the electrons is comparable with its own energy. We cannot therefore assume that $E \approx E'$, and a simplification of the kind discussed above is not possible. Following the procedure outlined for the derivation of S(k), we get the following expressions for these scattering processes.

Polar optic phonon scattering:

$$\left.\frac{\partial f}{\partial t}\right|_{pop} = -(e\hbar\mathcal{E}/m^*k_BT)k\cos\theta\ f_0(E)[1-f_0(E)]$$
$$\left[\phi(E)S(out,k,pop)-\phi(E+\hbar\omega_\ell)S(in_+,k,pop)-\phi(E-\hbar\omega_\ell)S(in_-,k,pop)\right] \quad , \quad (7.30)$$

where

$$S(out,k,pop) = k_+(n_\ell+1)f_0(E+\hbar\omega_\ell)[f_0(E)]^{-1}F(out_+,k,pop,\lambda)$$
$$+ k_-n_\ell H(E-\hbar\omega_\ell)f_0(E-\hbar\omega_\ell)[f_0(E)]^{-1}F(out_-,k,pop,\lambda) \quad ,$$

$$S(in_+,k,pop) = k_+(n_\ell+1)f_0(E+\hbar\omega_\ell)[f_0(E)]^{-1}F(in_+,k,pop,\lambda) \quad ,$$

$$S(in_-,k,pop) = k_-n_\ell f_0(E-\hbar\omega_\ell)[f_0(E)]^{-1}F(in_-,k,pop,\lambda)H(E-\hbar\omega_\ell) \quad ,$$

n_ℓ is the phonon occupation number; $H(E-\hbar\omega_\ell)$ is the Heaviside unit function; k_+ and k_- are respectively the wave vectors corresponding to the electron energies $E+\hbar\omega_\ell$ and $E-\hbar\omega_\ell$.

$$F(out_\pm;k,\lambda,pop) = \left(C_{pop}/k^2\right)\gamma'_\pm \int_{-1}^{+1} O_i(x)\left[1-2(k_\pm/k)x+(k_\pm/k)^2\right] \times \left[1-2(k_\pm/k)x+(k_\pm/k)^2+1/(k^2\lambda^2)\right]^{-2}dx \quad . \tag{7.31}$$

$$F(in_\pm;k,\lambda,pop) = \left(C_{pop}/k^2\right)\gamma'_\pm \int_{-1}^{+1} xO_i(x)\left[1-2(k_\pm/k)x+(k_\pm/k)^2\right] \times \left[1-2(k_\pm/k)x+(k_\pm/k)^2+1/(k^2\lambda^2)\right]^{-2}dx \quad . \tag{7.32}$$

Nonpolar optic and intervalley phonon scattering:

$$(\partial f/\partial t)_{op} = -(e\hbar\mathscr{E}/m^*k_BT)k\cos\theta\, f_0(E)[1-f_0(E)] \left[\phi(E)S(out;k;op)-\phi(E+\hbar\omega_0)S(in_+;k;op)-\phi(E-\hbar\omega_0)S(in_-;k;op)\right] \quad , \tag{7.33}$$

where

$$S(out;k;op) = k_+(n_0+1)\left[f_0(E+\hbar\omega_0)/f_0(E)\right]F(out_+;k;op;\lambda) + k_-H(E-\hbar\omega_0)n_0\left[f_0(E-\hbar\omega_0)/f_0(E)\right]F(out_-;k;op;\lambda) \quad ,$$

$$F(out_\pm;k;op;\lambda) = C_{op}\gamma'_\pm \int_{-1}^{+1} O_i(x)\left[1-2(k_\pm/k)x+(k_\pm/k)^2\right]^2 \times \left[1-2(k_\pm/k)x+(k_\pm/k)^2+k^{-2}\lambda^{-2}\right]^{-2}dx \quad ,$$

$$S(in_+;k;op) = k_+(n_0+1)\left[f_0(E+\hbar\omega_0)/f_0(E)\right]F(in_+;k;op;\lambda) \quad ,$$

$$S(in_-;k;op) = k_-n_0\left[f_0(E-\hbar\omega_0)/f_0(E)\right]F(in_-;k;op;\lambda)H(E-\hbar\omega_0) \quad ,$$

$$F(in_\pm;k;op;\lambda) = C_{op}\gamma' \int_{-1}^{+1} xO_i(x)\left[1-2(k_\pm/k)x+(k_\pm/k)^2\right]^2 \times \left[1-2(k_\pm/k)x+(k_\pm/k)^2+k^{-2}\lambda^{-2}\right]^{-2}dx \quad .$$

The expression for $(\partial f/\partial t)_{int}$ is obtained by replacing C_{op} by C_{int}, $\hbar\omega_0$ by $\hbar\omega_i$, and n_0 by n_i.

It should be noted that the collision term for inelastic scattering processes, in general, involves $\phi(E)$, $\phi(E+\hbar\omega_j)$, and $\phi(E-\hbar\omega_j)$. However, when the temperature of the sample is much higher than the phonon temperature $\hbar\omega_j/k_B$, we may approximate $\phi(E\pm\hbar\omega_j)$ by $\phi(E)$ and reduce the collision term to the form

$$\partial_c f = -(e\hbar\mathscr{E}/m^*k_BT)k\cos\theta\, f_0(E)[1-f_0(E)]\phi(E)S(out+in;k;\lambda) \quad ,$$

where

$$S(out+in;k;\lambda) = S(out;k;\lambda) - S(in_+;k;\lambda) - S(in_-;k;\lambda) \quad .$$

Similar simplifications may be made for nonpolar optic and intervalley scattering even at ordinary temperatures if the effects of electron screening and overlap integrals are negligible. The matrix elements for these scattering mechanisms are independent of $|\underline{k}-\underline{k}'|$ and hence of $\cos\theta_k$, when the effects of electron screening and overlap integrals are negligible. For these conditions $S(in;k;\lambda)$ becomes zero, since the integrand in the function $F(in_{\pm};k;\lambda)$ is proportional to $\cos\theta_k$. The collision term may then be simplified to

$$(\partial f/\partial t)_c = -(e\hbar\mathscr{E}/m^*k_BT)k \cos\theta \; f_0(E)[1-f_0(E)]\phi(E)S(out;k) \quad . \tag{7.34}$$

7.6 Solution of the Boltzmann Equation

Using the results of the preceding section, the Boltzmann equation may be written as

$$L_c\phi = 1/\gamma'(E) \quad , \tag{7.35}$$

where

$$L_c\phi = S_0(E)\phi(E) - \sum_j \sum_{+,-} S_{j\pm}(k)\phi(E\pm\hbar\omega_j) \quad .$$

$S_0(E)$ is the sum of both the out-scattering and in-scattering contributions of all the elastic scattering processes and the out-scattering contributions of the inelastic scattering processes. The second term represents the in-scattering contribution of the inelastic scattering processes. We discuss below methods of solution of this equation.

7.6.1 Relaxation-Time Approximation

In the absence of inelastic scattering processes, the equation assumes the simple form

$$S_0(E)\phi(E) = 1/\gamma'(E) \quad . \tag{7.36}$$

The solution of (7.36) is

$$\phi(E) = [\gamma'(E)S_0(E)]^{-1} \quad , \tag{7.37}$$

and the complete distribution function is then given by

$$f(\underline{k}) = f_0(E) - (e\hbar\mathscr{E}/m^*)k\cos\theta\ \partial_E f_0[\gamma'(E)S_0(E)]^{-1} \quad . \tag{7.38}$$

This solution is often referred to as being based on the relaxation-time approximation. The corresponding collision term may be written as

$$(\partial f/\partial t)_{\mathrm{coll}} = -(f-f_0)/\tau(E) \quad , \tag{7.39}$$

where

$$\tau(E) = [S_0(E)]^{-1} \quad . \tag{7.40}$$

This equation implies that the perturbed distribution function will relax to the equilibrium function when the perturbing field is removed, with a time constant $\tau(E)$, through the collision processes. Hence the name relaxation-time approximation.

We note that the method may be applied exactly when the important collision processes are all elastic. It is also applicable when the inelastic processes include only the nonpolar optic phonon and intervalley phonon scattering, and the effects of electron screening and overlap integrals are negligible. If polar optic phonon scattering is also important, the method is applicable only at very high temperatures. In compound semiconductors, as we have repeatedly stressed, polar optic phonon scattering is the dominant scattering process even at temperatures as low as that of liquid nitrogen, so that this method (although widely used a decade ago) gives inaccurate results.

Accurate solutions of the equation when polar optic phonon scattering is important may be obtained only by applying variational or numerical methods, which are discussed in the following sections. For the development of these methods we shall assume that polar optic phonon scattering is the only inelastic scattering process and also that it is characterised by a single phonon frequency. Extension of the methods to deal with more than one inelastic scattering process characterised by different phonon frequencies will be discussed in a later section.

7.6.2 Variational Method

Variational principle: The variational method for the solution of the Boltzmann equation relies on two properties of the collision term, discussed below [7.9]. For this discussion it is convenient to express the distribution function in the form

$$f(\underline{k}) = f_0(E) - \psi(\underline{k})\partial_E f_0 \quad , \qquad (7.41)$$

instead of in the form used earlier. We note that $\psi(\underline{k})$ is related to $\phi(E)$ by the relation

$$\psi(\underline{k}) = (e\hbar\mathscr{E}/m^*)k\cos\theta \cdot \phi(E) \quad . \qquad (7.42)$$

Using this distribution function we get for the collision term

$$\partial_c f = -\left[V_c/(8\pi^3 k_B T)\right]\int f_0(E)[1-f_0(E')][\psi(\underline{k})-\psi(\underline{k}')]S(\underline{k},\underline{k}')d\underline{k}' \quad . \qquad (7.43)$$

Introducing an operator L given by

$$L\psi(\underline{k}) = \left[V_c/(8\pi^3 k_B T)\right]\int f_0(E)[1-f_0(E')]S(\underline{k},\underline{k}')[\psi(\underline{k})-\psi(\underline{k}')]d\underline{k}' \quad , \qquad (7.44)$$

the collision term may be written as

$$\partial_c f = -L\psi(\underline{k}) \quad .$$

The operator L is self-adjoint and symmetric, as is evident from the following expressions:

$$\begin{aligned}\left(8\pi^3 k_B T/V_c\right)\int F(\underline{k})LG(\underline{k})d\underline{k} &= \iint F(\underline{k})f_0(E)[1-f_0(E')][G(\underline{k})-G(\underline{k}')]S(\underline{k},\underline{k}')d\underline{k}'d\underline{k} \\ &= \iint F(\underline{k}')f_0(E)[1-f_0(E')][G(\underline{k}')-G(\underline{k})]S(\underline{k},\underline{k}')d\underline{k}d\underline{k}' \\ &= \frac{1}{2}\iint [F(\underline{k})-F(\underline{k}')][G(\underline{k})-G(\underline{k}')]f_0(E)[1-f_0(E')]S(\underline{k},\underline{k}')d\underline{k}'d\underline{k} \quad , \end{aligned} \qquad (7.45)$$

F and G being arbitrary functions. In the second step $\underline{k}$ and $\underline{k}'$ have been interchanged and $f_0(E')[1-f_0(E)]S(\underline{k}',\underline{k})$ has been replaced by $f_0(E)[1-f_0(E')]S(\underline{k},\underline{k}')$ [see (7.15)]. In the third step, the two equivalent expressions have been added together and divided by 2. It follows from the symmetry in the expression that

$$\int F(\underline{k})LG(\underline{k})d\underline{k} = \int G(\underline{k})LF(\underline{k})d\underline{k} \quad , \tag{7.46}$$

i.e. the operator is self-adjoint. Also,

$$\int G(\underline{k})LG(\underline{k})d\underline{k} = (1/2)\int [G(\underline{k})-G(\underline{k}')]^2 f_0(E)[1-f_0(E')]S(\underline{k},\underline{k}')d\underline{k}d\underline{k}' \quad .$$

In the integrand on the right-hand side, $[G(\underline{k})-G(\underline{k}')]^2$ being a square term is positive, the product $f_0(E)[1-f_0(E')]$ is positive and $S(\underline{k},\underline{k}')$, being a transition probability is also positive. The integrand is, therefore, always positive and

$$\int G(\underline{k})LG(\underline{k})d\underline{k} \geq 0 \quad . \tag{7.47}$$

On the basis of the above two properties of the collision operator we may derive a variational principle. Let $f_0(E)-F(\underline{k})\partial_E f_0$ be a function satisfying the Boltzmann equation. Then

$$LF(\underline{k}) = -(e/\hbar)(\underline{\mathscr{E}}\cdot\underline{\nabla}_k E)\partial_E f_0 \quad . \tag{7.48}$$

Multiplying by $F(\underline{k})$ and integrating over $\underline{k}$, we get

$$\int F(\underline{k})LF(\underline{k})d\underline{k} = -(e/\hbar)\int(\underline{\mathscr{E}}\cdot\underline{\nabla}_k E)F(\underline{k})\partial_E f_0 d\underline{k} \quad . \tag{7.49}$$

Let $G(\underline{k})$ be another function which does not satisfy the Boltzmann equation (7.48), but does satisfy the above integral relation, i.e.

$$\int G(\underline{k})LG(\underline{k})d\underline{k} = -(e/\hbar)\int(\underline{\mathscr{E}}\cdot\underline{\nabla}_k E)G(\underline{k})\partial_E f_0 d\underline{k} \quad , \tag{7.50}$$

but

$$LG(\underline{k}) \neq -(e/\hbar)(\underline{\mathscr{E}}\cdot\underline{\nabla}_k E)\partial_E f_0 \quad . \tag{7.51}$$

We find that

$$\int (F-G)L(F-G)d\underline{k} = \int FL(F)d\underline{k} - \int GL(G)d\underline{k} \quad , \tag{7.52}$$

since

$$\int FL(G)d\underline{k} = \int GL(F)d\underline{k} = -(e/\hbar)\int G(\underline{\mathscr{E}}\cdot\underline{\nabla}_k E)\partial_E f_0 d\underline{k} = \int GL(G)d\underline{k} \quad . \tag{7.53}$$

Since the left-hand side of (7.52) is non-negative,

$$\int GL(G)d\underline{k} \le \int FL(F)d\underline{k} \quad . \tag{7.54}$$

We may conclude, in other words, that the integral $\int FL(F)d\underline{k}$ will be maximised if F satisfies the Boltzmann equation. In the variational method, advantage is taken of this result to obtain values of the transport coefficients with maximum possible accuracy using an arbitrary function for the distribution function. The basic idea is to choose for $\psi(\underline{k})$ a convenient function satisfying (7.48) and then adjust it so as to make $\int \psi(\underline{k})L\psi(\underline{k})d\underline{k}$ a maximum [Ref.7.8, p.304].

We shall illustrate this method with an example. We shall first express, for the sake of convenience, the variational principle in terms of the function $\phi(E)$ used earlier.

Replacing $\psi(\underline{k})$, we get for the integral

$$\int \psi(\underline{k})L\psi(\underline{k})d\underline{k} = (e\hbar\mathscr{E}/m^*)^2 \int k\cos\theta\ \phi(E)\Big\{\Big(V_c/8\pi^3k_BT\Big)\int f_0(E)[1-f_0(E')]$$

$$\times[k\cos\theta\,\phi(E)-k'\cos\theta'\,\phi(E')]S(\underline{k},\underline{k}')d\underline{k}'\Big\}d\underline{k} \quad . \tag{7.55}$$

Simplifying, we get for the variation principle that

$$-\int_0^\infty \phi(E)L_{c1}[\phi(E)]\partial_E f_0 dE \tag{7.56}$$

should be maximum, where

$$-L_{c1}[\phi(E)] = 2(\hbar^2/2m^*)^{5/2}k^4(\partial E/\partial k)^{-1}\Big(V_c/8\pi^3\Big)\int [1-f_0(E')][1-f_0(E)]^{-1}$$

$$\times[(k'\cos\theta'/k\cos\theta)\phi(E')-\phi(E)]S(\underline{k},\underline{k}')d\underline{k}'$$

$$= 2(\hbar^2/2m^*)^{5/2}k^4(\partial E/\partial k)^{-1}\Big[-S_0(E)\phi(E)+\sum_j\sum_{+,-}S_{j,\pm}(E)\phi(E\pm\hbar\omega_j)\Big] \quad . \tag{7.57}$$

S_0, S_j are as defined earlier. Equation (7.48) also assumes the form

$$\int_0^\infty \gamma^{3/2}\phi(E)\partial_E f_0 dE = \int_0^\infty \phi(E)L_{c1}[\phi(E)]\partial_E f_0 dE \quad . \tag{7.58}$$

We have assumed that the dispersion relation is given by

$$\hbar^2k^2/(2m^*) = \gamma(E) \quad . \tag{7.59}$$

For a parabolic dispersion relation, the above equations reduce to

$$\int_0^\infty E^{3/2}\phi(E)\partial_E f_0 dE = \int_0^\infty \phi(E)L_{c1}[\phi(E)]\partial_E f_0 dE \quad , \tag{7.60}$$

where

$$-L_{c1}[\phi(E)] = E^{3/2}\left[-S_0(E)\phi(E)+\sum_j \sum_{+,-} S_{j,\pm}\,(E\pm\hbar\omega_j)\right] = E^{3/2}L_c\phi \quad . \tag{7.61}$$

Solution by the variational method: We illustrate the method using a power series for ϕ, which is also the form of the function commonly used when this method is applied.

Replacing ϕ by $\sum_{r=0}^{n} c_r E^r$ in (7.56) and (7.58), the principle becomes that $\sum_{r,s} c_r c_s d_{rs}$ should be maximum, and we have

$$\sum_{r,s} c_r c_s d_{rs} - \sum c_r \alpha_r = 0 \quad , \tag{7.62}$$

where the c_r are the unknown constants in the chosen arbitrary function for ϕ. The constants α_r and d_{rs} are given by the expressions

$$\alpha_r = \int_0^\infty E^r \gamma^{3/2} \partial_E f_0 dE \quad , \tag{7.63}$$

$$d_{rs} = \int_0^\infty E^r L_{c1}(E^s)\partial_E f_0 dE \quad . \tag{7.64}$$

We note that if the c_r were chosen to satisfy (7.62), the arbitrary function would satisfy (7.58) as required. The variation principle should therefore be satisfied by chosing c_r subject to the restriction of (7.62). This may easily be done using Lagrange's method of unknown multipliers. The equations giving the c_r are then found to be [7.10]

$$\sum_s d_{rs} c_s - \alpha_r = 0 \quad . \tag{7.65}$$

We may now evaluate the coefficients in the power series for ϕ by solving the set of linear algebraic equations obtained by giving r and s different values in accordance with the chosen order of the polynomial.

The transport coefficients may now be evaluated using for the distribution function

$$f = f_0 - (e\mathscr{E}\hbar/m^*)k\cos\theta\,\partial_E f_0 \sum_r c_r E^r \quad , \tag{7.66}$$

with the values of the c_r determined above. Though the distribution function used above does not exactly satisfy the Boltzmann equation for all electron energies, the results are found to be fairly accurate when a few terms are used in the series. This is because the coefficients are adjusted to make the chosen function as close as possible to the exact solution through the use of the variation principle. One may expect that the accuracy should improve as more and more terms are included in the series. This is found to be true, particularly when small number of terms are used in the series.

Comments on the method: Some years ago the variational method was the only method commonly used for the evaluation of the transport coefficients when the relaxation-time approximation was inapplicable [7.11]. It was a convenient method as the integrals, the α_r, and the d_{rs} may be analytically evaluated in terms of tabulated functions for nondegenerate or extremely degenerate materials with a parabolic band structure [7.10]. The problem was thus reduced only to a solution of a set of linear algebraic equations. For materials with nonparabolic bands the method was also often used, but then the integrals have to be evaluated numerically [7.11]. The method for such bands becomes less advantageous as by applying other numerical methods developed in course of the last few years one may obtain accurate results with less computation.

The method has two other disadvantages. Firstly, it is difficult to determine beforehand the number of terms required in the series to ensure a desired accuracy. It is even difficult to determine the correct value and estimate the accuracy of the values obtained by using a finite number of terms. In actual calculations, therefore, series involving as many as sixteen terms have been used [7.12]. The inclusion of such a large number of terms makes the computation an enormous task, particularly for materials with nonparabolic bands.

The variational method has, therefore, been little used in recent years and most of the calculations reported in the literature in the last few years have been done with other numerical methods, which are discussed in the following sections. It is, however, worthwhile to note that if we are interested only in the drift mobility, for which the variational principle is theoretically derived, this method may still be used with computational advantage, provided we are able to make a proper choice of the arbitrary function. Taking advantage of our knowledge of the exact form of the distribution functions, which has emerged through the application of other numerical methods, we may make such a choice and approximate the distribution function with two or three terms in the series, so that we can still obtain a good accuracy evaluating very few integrals numerically.

7.6.3 Matrix Method

Referring to the reduced form of the Boltzmann equation, (7.35), we find that we are required to solve a set of simultaneous linear algebraic equations. Let us assume that there is one kind of optic phonon of frequency ω_ℓ and let us consider a particular value of electron energy E_i, such that $0 < E_i < \hbar\omega_\ell$. The value of $\phi(E_i)$ is then determined by the following set of equations:

$$S_0(E_i)\phi(E_i) - S_+(E_i)\phi(E_i+\hbar\omega_\ell) = [\gamma'(E_i)]^{-1} \quad ,$$

$$S_0(E_i+\hbar\omega_\ell)\phi(E_i+\hbar\omega_\ell) - S_+(E_i+\hbar\omega_\ell)\phi(E_i+2\hbar\omega_\ell) - S_-(E_i+\hbar\omega_\ell)\phi(E_i)$$
$$= [\gamma'(E_i+\hbar\omega_\ell)]^{-1} \quad ,$$

$$\text{---} = \text{--------} \quad ,$$

$$S_0(E_i+\ell\hbar\omega_\ell)\phi(E_i+\ell\hbar\omega_\ell) - S_+(E_i+\ell\hbar\omega_\ell)\phi(E_i+(\ell+1)\hbar\omega_\ell)$$
$$- S_-(E_i+\ell\hbar\omega_\ell)\phi(E_i+(\ell-1)\hbar\omega_\ell) = [\gamma'(E_i+\ell\hbar\omega_\ell)]^{-1} \quad ,$$

$$\text{---} = \text{--------} \quad . \tag{7.67}$$

The set of equations is infinite in extent. The general form of the equations is given in the third above, and we get successive equations by giving positive integral values to ℓ, starting from 0 and putting $\phi(E_i-\hbar\omega_\ell)$ equal to zero. The equations imply that the value of $\phi(E_i)$ is determined by the in-scattering from the state $E_i+\hbar\omega_\ell$, whereas $\phi(E_i+\ell\hbar\omega_\ell)$ is determined by the in-scattering from the states $E_i+(\ell\pm1)\hbar\omega_\ell$. In principle, therefore, $\phi(E_i)$ depends on its values for electron energies of $E_i+\hbar\omega_\ell$, $E_i+2\hbar\omega_\ell$, ..., $E_i+\ell\hbar\omega_\ell$, ..., and so on through the in-scattering terms.

In the matrix method developed by GRIGOR'EV et al. [7.13], the set of equations is truncated at the nth equation. We then find that we have n equations but n+1 unknowns $\phi(E_i)$, $\phi(E_i+\hbar\omega_\ell)$, ..., $\phi(E_i+n\hbar\omega_\ell)$. The number of unknowns is reduced by one to obtain a solution by assuming that

$$\phi(E_i+(n-1)\hbar\omega_\ell) \simeq \phi(E_i+n\hbar\omega_\ell) \simeq \phi(E_i+(n+1)\hbar\omega_\ell) \quad . \tag{7.68}$$

The quantity $\phi(E_i+n\hbar\omega_\ell)$ is then given by

$$\phi(E_i+n\hbar\omega_\ell) = [\gamma'(E_i+n\hbar\omega_\ell)]^{-1}[S_0(E_i+n\hbar\omega_\ell)-S_+(E_i+n\hbar\omega_\ell)-S_-(E_i+n\hbar\omega_\ell)]^{-1} \quad . \tag{7.69}$$

In other words, $\phi(E_i+n\hbar\omega_\ell)$ is assumed to be determined by the relaxation-time approximation. In some analyses, the contributions of S_+ and S_- terms have also been neglected [7.14]. It should be mentioned that $\phi(E_i+n\hbar\omega_\ell)$ determined in the above manner is very approximate, but its contribution to $\phi(E_i)$ being negligible, the error arising from this approximation is also negligible.

On using the value of $\phi(E_i+n\hbar\omega_\ell)$ so determined, we get n equations with n unknowns. The unknowns are then determined by applying a numerical method for the inversion of the matrix formed by $S_0(E_i)$, $S_0(E_i+\hbar\omega_\ell)$, etc.

Having determined $\phi(E_i)$, $\phi(E_i+\hbar\omega_\ell)$, ..., $\phi(E_i+(n-1)\hbar\omega_\ell)$, the process is repeated for other chosen values of E_i in the interval $0 < E_i < \hbar\omega_\ell$. In the end, therefore, the complete solution for $\phi(E)$ is obtained for selected values of the electron energy. The transport coefficients may now be obtained by using these values of $\phi(E)$ with a numerical method of integration.

We note that due to the termination procedure used, the solution for lower values of E, as mentioned earlier, should not be subject to a large error, as the contribution of the approximated term is small for $\phi(E)$ corresponding to low values of E. The error in the values of $\phi(E)$ for higher values of E may, however, be large; but this error is not usually reflected in the values of the transport coefficients as the contribution of the higher-energy electrons to the average value of the transport coefficients is small, due to their small number. In an actual computation, the value of n is selected by trial so that the values of the transport coefficients do not change significantly as n is further increased.

In a variation of the method [7.15], the solution of the equations is obtained using the following procedure. Let us put $E_i = x\hbar\omega_\ell$. We may then by the process of elimination express $\phi(E_i+n\hbar\omega_\ell)$, i.e. $\phi((x+n)\hbar\omega_\ell)$, as

$$\phi((x+n)\hbar\omega_\ell) = \chi(x,n) + \xi(x,n)\phi(x\hbar\omega_\ell) \quad , \tag{7.70}$$

where $\chi(x,n)$ and $\xi(x,n)$ are determined by the values of S_0 and $S_\pm$ and hence by the values of x and n. The values of these functions are obtained using the same process of elimination putting $\phi(x\hbar\omega_\ell)$ first equal to zero and next to one. The resultant values of $\phi((n+x)\hbar\omega_\ell)$ are related to $\chi(x,n)$ and $\xi(x,n)$ by the relations

$$\phi^{(0)}((n+x)\hbar\omega_\ell) = \chi(x,n) \quad , \tag{7.71}$$

$$\phi^{(1)}((n+x)\hbar\omega_\ell) = \chi(x,n) + \xi(x,n) \quad , \tag{7.72}$$

where $\phi^{(0)}((n+x)\hbar\omega_\ell)$ and $\phi^{(1)}((n+x)\hbar\omega_\ell)$ are the values of $\phi((x+n)\hbar\omega_\ell)$ corresponding to $\phi(x\hbar\omega_\ell)$ equal to zero and unity respectively.

The value of $\phi(x\hbar\omega_\ell)$ is then given by

$$\phi((x+n)\hbar\omega_\ell) = \phi^{(0)}((n+x)\hbar\omega_\ell) + \left[\phi^{(1)}((x+n)\hbar\omega_\ell)-\phi^{(0)}((x+n)\hbar\omega_\ell)\right]\phi(x\hbar\omega_\ell) \quad , \tag{7.73}$$

the value of $\phi((n+x)\hbar\omega_\ell)$ being determined by the relaxation-time-approximation method described earlier.

Knowing $\phi(x\hbar\omega_\ell)$, we may in principle evaluate $\phi((x+1)\hbar\omega_\ell)$, $\phi((x+2)\hbar\omega_\ell)$, and so on, using successive equations. However, as a small error in $\phi(x\hbar\omega_\ell)$ propagates in the course of computations, values of $\phi((x+\ell)\hbar\omega_\ell)$ for different values of ℓ are also obtained using a similar procedure. We may now express $\phi((n+x)\hbar\omega_\ell)$ as

$$\begin{aligned}\phi((n+x)\hbar\omega_\ell) &= \chi_\ell(x,n,\ell) + \xi_\ell(x,n,\ell)\phi((\ell+x)\hbar\omega_\ell)\\ &\quad + \xi_{\ell-1}(x,n,\ell)\phi((\ell-1+x)\hbar\omega_\ell) \quad ,\end{aligned} \tag{7.74}$$

where χ_ℓ, ξ_ℓ, and $\xi_{\ell-1}$ are functions similar to χ and ξ, determined by $S_0(E_i+\hbar\omega_\ell)$, $S_\pm(E_i+\hbar\omega_\ell)$, $S_0(E_i+2\hbar\omega_\ell)$, $S_\pm(E_i+2\hbar\omega_\ell)$, and so on. These three constants are determined, as in the earlier case, by evaluating the values of $\phi(\hbar\omega_\ell(n+x))$ for three chosen combination of values of $\phi(\hbar\omega_\ell(\ell+x))$ and $\phi(\hbar\omega_\ell(\ell-1+n))$, namely (0,0), (0,1) and (1,0). Indicating these values by the superscripts 00, 01, 10, we may express the constants as

$$\chi(x,n,\ell) = \phi^{(00)}((x+n)\hbar\omega_\ell) \quad , \tag{7.75}$$

$$\xi_{\ell-1}(x,n,\ell) = \phi^{(01)}((x+n)\hbar\omega_\ell) - \phi^{(00)}((x+n)\hbar\omega_\ell) \quad , \tag{7.76}$$

$$\xi_\ell(x,n,\ell) = \phi^{(10)}((x+n)\hbar\omega_\ell) - \phi^{(00)}((x+n)\hbar\omega_\ell) \quad . \tag{7.77}$$

We may then evaluate $\phi((x+\ell)\hbar\omega_\ell)$ using these values of the constants, replacing $\phi((x+n)\hbar\omega_\ell)$ by the relaxation-time-approximation value, and $\phi((\ell-1+x)\hbar\omega_\ell)$ by its value determined in the earlier step.

The variation of the matrix method discussed above was first developed by DELVES [7.15]. It has been applied recently to the calculation of the transport coefficients in GaAs and GaP [7.16]. It should be evident that the solution of the equations is so mechanised in this method that the propagation of errors is minimised. This particular feature may be an advantage when the matrix of the S_0 and $S_\pm$ quantities is ill behaved.

7.6.4 Iteration Method

In the iteration method, first reported by RODE [7.17], the values of $\phi(E)$ are determined by applying the Ritz iteration method [7.18] for the solution of linear algebraic equations. In the first iteration step, the values of $\phi(E_i+\ell\hbar\omega_\ell)$ are obtained from the successive equations putting the two other terms, i.e. $\phi(E_i+(\ell\pm1)\hbar\omega_\ell)$, equal to zero. Specifically, these values, indicated by the superscript (1), are given by

$$\phi^{(1)}(E_i+\ell\hbar\omega_\ell) = [\gamma'(E_i+\ell\hbar\omega_\ell)S_0(E_i+\ell\hbar\omega_\ell)]^{-1} \quad . \tag{7.78}$$

In the second iteration step, $\phi(E_i+\ell\hbar\omega_\ell)$ are evaluated using for $\phi(E_i+((\ell\pm1)\hbar\omega_\ell)$ the values obtained in the previous iteration step. Thus, the values of $\phi(E_i+\ell\hbar\omega_\ell)$ in the nth iteration step, indicated by the superscript (n), are given by

$$\phi^{(n)}(E_i+\ell\hbar\omega_\ell) = \Big[1/\gamma'(E_i+\ell\hbar\omega_\ell)+S_+(E_i+\ell\hbar\omega_\ell)\phi^{(n-1)}(E_i+(\ell+1)\hbar\omega_\ell)$$
$$+S_-(E_i+\ell\hbar\omega_\ell)\phi^{(n-1)}(E_i+(\ell-1)\hbar\omega_\ell)\Big]\Big/S_0(E_i+\ell\hbar\omega_\ell) \quad . \tag{7.79}$$

This iteration procedure is repeated till the values determined in successive steps of iteration agree to within a specified accuracy.

This method of iteration has been found to be convergent and, in most of the cases to which it has been applied, the results converge to within 1 in 10^3 in between six to ten (see [Ref.7.18, p.1152] for a discussion of the rate of convergence of this method). Convergence is assured if the coefficients of the diagonal terms are the largest of all the coefficients. In other words, we may expect a rapid convergence when $\phi(E)$ is mostly determined by $S_0(E)$ and the contribution of the in-scattering term is small. This condition is generally satisfied in most transport calculations. In fact, it is found that the iteration procedure converges almost exponentially and the difference between the values of $\phi(E)$ in successive iterations is larger than the error in the iterated values. We may therefore use this difference as an indication of the error and terminate the iteration when the desired accuracy is obtained.

Although RODE used the Ritz iteration method in his formulation, the Gauss-Seidel method [Ref.7.18, p.1155] may also be used to hasten the convergence. In this method, the value of $\phi(E_i)$ is determined in the first iteration step from the first equation putting $\phi(E_i+\hbar\omega_\ell)$ equal to zero. On the other hand, $\phi(E_i+\ell\hbar\omega_\ell)$ is obtained by putting $\phi(E_i+(\ell+1)\hbar\omega_\ell)$ equal to zero, but

$\phi(E_i+(\ell-1)\hbar\omega_\ell)$ is replaced by its value obtained by solving the preceding equation. In the second iteration step, $\phi(E_i)$ is determined from the first equation by replacing $\phi(E_i+\hbar\omega_\ell)$ by its value determined in the previous step. $\phi(E_i+\ell\hbar\omega_\ell)$ is, however, determined from the $(\ell+1)^{th}$ equation replacing $\phi(E_i+(\ell-1)\hbar\omega_\ell)$ by its value determined from the ℓth equation and $\phi(E_i+(\ell+1)\hbar\omega_\ell)$ by its value determined in the previous iteration step. In other words, the following equation is used to obtain $\phi(E_i+\ell\hbar\omega_\ell)$ in the nth iteration step:

$$\phi^{(n)}(E_i+\ell\hbar\omega_\ell) = \Big[1/\gamma'(E_i+\ell\hbar\omega_\ell)+S_+(E_i+\ell\hbar\omega_\ell)\phi^{(n-1)}(E_i+(\ell+1)\hbar\omega_\ell)$$
$$+S_-(E_i+\ell\hbar\omega_\ell)\phi^{(n)}(E_i+(\ell-1)\hbar\omega_\ell)\Big]/S_0(E_i+\ell\hbar\omega_\ell) \quad . \qquad (7.80)$$

Successive iteration steps are carried out by repeating this procedure. Some trials by the author indicate that convergence may be hastened if this method is used for iteration instead of that used by RODE.

We note that in the iteration method, the Boltzmann equation is reduced to a set of linear algebraic equations as in the matrix method. The solutions are, however, obtained by applying an iteration method. We may therefore expect that this method will have the advantages of an iteration method. Some of these advantages are: self-correction of the errors, smaller memory requirement, and often smaller number of computations. Of these advantages, the most important advantage is that the method is self-correcting, and is therefore preferable to the matrix method.

7.6.5 Monte Carlo Method

In the methods discussed in the previous sections, the Boltzmann equation is reduced to a linear difference equation by expanding the distribution function in two terms, one of which is the same function as in the field-free case, while the second one is a perturbation term proportional to the cosine of the angle between the wave vector and the direction of the field.

We describe in this section a method in which the Boltzmann equation is not directly solved, but the distribution function and the transport coefficients are evaluated by the simulation of electron trajectories using random numbers. The method is known as the Monte Carlo method. It has been applied in different forms to solve transport problems in various kinds of physical phenomena [7.19]. Historically, the method was first employed in the study of electron transport in semiconductors in connection with investigations in high fields or the so-called hot-electron transport [7.20]. But, in some recent studies, the method has also been used to obtain low-field transport parameters [7.21]. It also promises to provide an exact method for the evaluation of transport coefficients under quantum conditions and for thin samples. We shall therefore merely outline the method here, postponing until Chap.11 a detailed discussion in connection with hot-electron transport.

As mentioned earlier, the basic principle of the method is to work out the trajectory of an electron. The trajectory is determined by the collision processes and external forces. The effect of external forces is deterministic but the collision processes affect the trajectories in a probabilistic manner. The former effect may be calculated by applying the classical laws of motion, but the latter effect is required to be evaluated by applying the probability theory. This is done in the Monte Carlo method by generating uniformly distributed random numbers [7.22] and by using them to determine the time between collisions and the nature of the collisions using the following procedure.

Let us assume that the wave vector of the electron at the starting time t_0 is $\underline{k}_0$. The wave vector of the electron at a subsequent time t is

$$\underline{k}(t) = \underline{k}_0 + (\underline{F}/\hbar)(t-t_0) \quad , \quad \text{since} \quad \underline{F} = \frac{d}{dt}(\hbar\underline{k}) \quad , \tag{7.81}$$

where $\underline{F}$ is the external force on the electron. However, as the electron moves it may suffer a collision at any time. The collision is a random event and for any time t there is a finite probability of a collision. Let the probability of collision at time t be P(t), and let r_1 be a random number uniformly distributed between 0 and 1. Then, according to probability theory, if the collisions are random events, we may relate r_1 to the collision time t_c by

$$r_1 = \int_0^{t_c} P(t)dt \Big/ \int_0^{\infty} P(t)dt \quad . \tag{7.82}$$

However, if $S(\underline{k}(t))$ be the scattering probability for an electron with wave vector $\underline{k}(t)$, then the number of electrons suffering collisions in the interval between t and t + dt is $n(t)S(\underline{k}(t))dt$, where n(t) represents the number of electrons which have not suffered a prior collision. The number of such electrons is reduced by the above number due to collisions in the interval between t and t + dt. Hence,

$$-dn(t) = n(t)S(\underline{k}(t))dt \quad . \tag{7.83}$$

Integrating, we get

$$n(t) = n(0) \exp\left[-\int_0^t S(\underline{k}(t'))dt'\right] . \tag{7.84}$$

The probability of collision for an electron is therefore given by

$$P(t) = n(t)S(\underline{k}(t))/n(0) = S(\underline{k}(t)) \exp\left[-\int_0^t S(\underline{k}(t'))dt'\right] . \tag{7.85}$$

Relation (7.82) may now be written as

$$r_1 = \int_0^{t_c} S(\underline{k}(t)) \exp\left[-\int_0^t S(\underline{k}(t'))dt'\right]dt$$
$$\times \left\{\int_0^\infty S(\underline{k}(t)) \exp\left[-\int_0^t S(\underline{k}(t'))dt'\right]dt\right\}^{-1} .$$

In principle we may therefore obtain t_c, the time of collision, using the random number r_1 and the expressions for $S(\underline{k})$ derived earlier.

Once t_c is determined, we may plot the trajectory of the electron from 0 to t_c using the following expressions:

$$x = (\hbar/m^*)k_{x0}(t_c-t_0) , \tag{7.86}$$

$$y = (\hbar/m^*)k_{y0}(t_c-t_0) , \tag{7.87}$$

$$z = (\hbar/m^*)\left[k_{z0}(t_c-t_0)+(F/2\hbar)(t_c-t_0)^2\right] , \tag{7.88}$$

where k_{x0}, k_{y0}, k_{z0} are the x, y, and z components of $\underline{k}_0$ at $t = t_0$. The z direction has been chosen in the direction of the external force. The E-$\underline{k}$ relation has been assumed to be parabolic and isotropic for the sake of simplicity. Modification of the expressions for nonparabolic or anisotropic band structure is, however, a simple task.

In order to plot the trajectory, we are required to know the wave vector of the electron after the collision, which occurs at t_c. For this purpose, a second random number r_2, distributed uniformly between 0 and 1, is generated. If now S_1, S_2, S_3, etc., denote the scattering probability for the different kinds of scattering, and if r_2 satisfies the relation

$$\sum_{i=1}^{n-1} S_i / \sum_{i=1}^{m} S_i < r_2 < \sum_{i=1}^{n} S_i / \sum_{i=1}^{m} S_i , \tag{7.89}$$

then we may conclude that the nth kind of scattering will take place. The total number of kinds of scattering is m, and n is less than or equal to m. The above relation is tested increasing n in steps of 1. We should note that the absorption and emission processes should be considered as two different kinds of scattering. It may also be mentioned that the inequality test may be performed by ordering the $S(\underline{k})$ in any order. But, computation time is saved if the scattering mechanisms with larger scattering probabilities are considered earlier.

Having determined the kind of scattering, the wave vector of the electron after the collision is obtained by generating two more random numbers r_3 and r_4, distributed uniformly between 0 and 1. The energy of the electron after collision is given by $E' + \Delta E$, where ΔE is the change in energy induced by collision and E' is the energy of the electron immediately before collision. It is given by

$$E' = \hbar^2 k^2(t_c)/(2m^*) \quad . \tag{7.90}$$

As discussed earlier for acoustic phonon and defect scattering, we may take ΔE equal to zero. For optic phonon scattering, it is equal to $\pm\hbar\omega_0$.

The magnitude of the electron wave vector after collision is thus given by

$$k_0 = [2m^*(E'+\Delta E)]^{\frac{1}{2}}/\hbar \quad . \tag{7.91}$$

For the determination of the orientation of the wave vector we note that the expressions for the scattering probability (given in Sect.6.6) involve only the angle between $\underline{k}$ and $\underline{k}'$ (and not their individual orientations) for isotropic band structure. It is therefore convenient to choose the direction of $\underline{k}$ as the polar axis of the coordinate system, as shown in Fig.7.4. For this coordinate system $S(\underline{k},\underline{k}')$ is independent of the azimuthal angle ϕ.

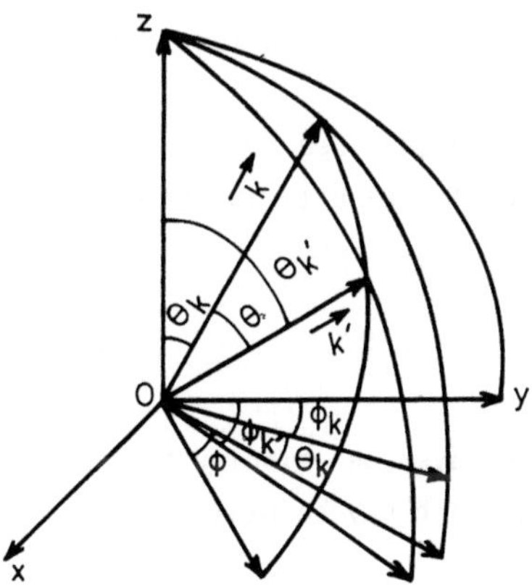

Fig. 7.4. Vectors $\underline{k}$ and $\underline{k}'$ in reference to the fixed cartesian coordinate axes

The angle ϕ is therefore obtainable by using the random number r_3 from the relation

$$r_3 = \int_0^\phi d\phi'/\int_0^{2\pi} d\phi = \phi(2\pi)^{-1} \quad . \tag{7.92}$$

The scattering probability $S(\underline{k},\underline{k}')$, on the other hand, may depend on the polar angle θ. This angle is therefore obtained in general from the following relation using the other random number r_4:

$$r_4 = \int_0^\theta S(\underline{k},\underline{k}')\sin\theta'\,d\theta'/\int_0^\pi S(\underline{k},\underline{k}')\sin\theta'\,d\theta' \quad . \tag{7.93}$$

The electron wave vector after the collision occurring at t_c is now completely known, but it is stated with reference to the orientation of $\underline{k}$. We may, however, obtain the components of $\underline{k}$ in a fixed coordinate system using the relations

$$\cos\theta_{\underline{k}'} = \cos\theta\,\cos\theta_{\underline{k}} - \sin\theta\,\sin\theta_{\underline{k}}\cos\phi \quad , \tag{7.94}$$

$$\sin\theta_{\underline{k}'}\,\cos\phi_{\underline{k}'} = \sin\theta\,\cos\phi\,\cos\theta_{\underline{k}}\,\cos\phi_{\underline{k}} - \sin\theta\,\sin\phi\,\sin\phi_{\underline{k}} + \cos\theta\,\sin\theta_{\underline{k}}\,\cos\phi_{\underline{k}} \quad , \tag{7.95}$$

$$\sin\theta_{\underline{k}'}\,\sin\phi_{\underline{k}'} = \sin\theta\,\cos\phi\,\cos\theta_{\underline{k}}\,\sin\phi_{\underline{k}} + \sin\theta\,\sin\phi\,\cos\phi_{\underline{k}} + \cos\theta\,\sin\theta_{\underline{k}}\,\sin\phi_{\underline{k}} \quad , \tag{7.96}$$

where $\theta_{\underline{k}}$ and $\phi_{\underline{k}}$ are respectively the polar and azimuthal angles of $\underline{k}$ for the fixed coordinate system, and $\theta_{\underline{k}'}$ and $\phi_{\underline{k}'}$, the corresponding angles for $\underline{k}'$.

The components of the wave vector k_{x0}, k_{y0}, and k_{z0} being now completely known, we may repeat the procedure outlined above to plot the next electron trajectory, and so on. Repeating the calculations a large number of times we get representative trajectories of an electron moving in an actual semiconductor sample.

The transport coefficients of interest are then obtained from the knowledge of these trajectories. For example, the drift velocity in the direction of the field is given by

$$v_z = (\hbar/m^*)\sum_i\left(k_{z0}t_{ci} + F\hbar^{-1}t_{ci}^2/2\right)\left(\sum_i t_{ci}\right)^{-1} \quad . \tag{7.97}$$

It should be noted that the evaluation of the average velocity by the above formula is based on the ergodic theorem, which may be stated as [7.23]

$$\int \phi(\underline{k}) f'(\underline{k}) d\underline{k} = \tau^{-1} \int_0^{\tau} \phi(\underline{k}(t)) dt + O(\tau^{-\frac{1}{2}}) \quad . \tag{7.98}$$

Stated in words, the theorem states that the ensemble average for a particular variable is the same as the time average of the same variable for a single particle, provided it is observed for a sufficiently long time. Application of this theorem is justified if the motion of the particle is observed in the presence of other particles, i.e. collisions with other particles are taken into account. It may then be assumed that the particle passes through all possible states and stays in each state for a time in accordance with the distribution function for the particles. However, we note that in the procedure outlined above, the motion of a single particle is observed and collisions with other particles are not taken into account. The validity of the assumption of the equivalence of the time average and the ensemble average becomes therefore difficult to ensure, particularly at low fields. A simple example illustrates this point. Let us assume that the collision processes include elastic processes and an inelastic optic phonon collision process, which may alter the energy of the electron by $\pm\hbar\omega_0$. Let the trajectory of the electron be observed in the absence of any external field. It should be evident that if the electron is assumed to have an initial energy E such that $0 < E < \hbar\omega_0$, then it can only be in the states E, $E + \hbar\omega_0$, $E + 2\hbar\omega_0$, ..., $E + n\hbar\omega_0$, and so on. The data obtainable by observing the electron would be related to these energy states, and we cannot extract information related to the ensemble by observing one electron, however long we may observe it. If, however, electron collisions could be taken into account, the energy of the electron would have changed through such collisions, and it would have moved through all possible energies. In that case, the time average would really have represented the ensemble average, provided data were taken for a long enough time.

The above picture is changed to some extent when an external field is present. In the presence of the field the electron gains energy and is made to move through different points of the energy space. We may expect that if the electron is observed for a sufficiently long time, the trajectories of the single electron will yield data related to the characteristics of the ensemble determined by the external field and by collisions excluding the electron-electron collisions. The truth of the above statement may be accepted intuitively without a mathematical proof. However, it should also be evident that the electron may have to be observed for a very very long time, so that it traverses whole of the energy space. The required time of observation would decrease with increase in the strength of the external field as the electron will be rapidly moved in energy space by a large field. Hence, at high fields, the method may be applied with confidence to obtain transport coefficients and distribution function in a straightforward manner. At low fields, however, because of the time limitation, the simple method cannot be conveniently applied. It is, however, possible to evaluate transport coefficients by modifying the method, taking advantage of the fact that, at low fields, the distribution of the electron in the energy space may be taken as known. It is either Maxwellian or Fermi-Dirac, and by weighting the trajectories with these functions we may devise suitable formulae for evaluating transport coefficients from the observed trajectories of the electron for various energies. We shall discuss these modifications of the straightforward Monte Carlo method in the next chapter. It should, however, be evident from this discussion that the Monte Carlo method is not a convenient method for solving problems for which alternative methods discussed earlier can be used.

7.7 Method of Solution for Anisotropic Coupling Constants and Anisotropic Electron Effective Mass

Solutions of the Boltzmann equation have been discussed above assuming the electron effective mass to be isotropic. The scattering parameters (e.g. deformation potential constant, piezoelectric coupling constant, acoustic velocity) were also taken as isotropic. In most semiconductors, these parameters are not isotropic, and the electron effective mass is also anisotropic in many compound semiconductors. The methods of solution discussed earlier may be used in principle to obtain solutions taking these complexities into account. However, the mathematical steps become rather involved. We shall consider in this section the case of electrons with spheroidal constant-energy surfaces. While developing the solutions we shall also indicate how the direction-dependent scattering parameters are to be averaged.

We assume that the E-$\underline{k}$ relation for the electron is

$$E = (\hbar^2/2)\left[\left(k_x^2/m_x\right)+\left(k_y^2/m_y\right)+\left(k_z^2/m_z\right)\right] \quad . \tag{7.99}$$

The parabolic relation is assumed for the sake of mathematical simplicity. Effects of nonparabolicity may be easily incorporated in the final results. We note that in the above representation, the x,y,z directions are chosen along the directions of the principal axes of the constant-energy spheroids, which are usually either in the <100> or <111> crystallographic directions. Electrons with the above dispersion relation are accelerated, in general, in a direction different from that of the applied electric field, unless the field is in x, y, or z direction. The formulation of the method for an arbitrary direction of the electric field becomes, therefore, algebraically complex. The transport parameters are, however, independent of the magnitude of the electric field for low values of the field. To simplify the mathematics, we may first evaluate the coefficients for the directions of the principal axes and then obtain the coefficients for arbitrary direction of the field by adding together the contributions from the components of the field in the directions of the principal axes, as clarified below. Let the electric field be given by

$$\underline{\mathscr{E}} = \hat{\underline{i}}\,\mathscr{E}_x + \hat{\underline{j}}\,\mathscr{E}_y + \hat{\underline{k}}\,\mathscr{E}_z \quad . \tag{7.100}$$

We may first evaluate the current density components J_x, J_y, J_z. These components are individually determined by $\mathscr{E}_x$, $\mathscr{E}_y$, and $\mathscr{E}_z$, since for the x, y, z directions, electrons are accelerated in the direction of the electric field. The total current density may then be obtained by adding J_x, J_y, J_z vectorially. It should be noted that due to the anisotropy of the scattering parameters and of the electron effective mass, the ratio of the current density and electric field will be different for the three directions, and we are required to obtain solutions of the Boltzmann equation for the three directions of the electric field separately. We discuss below the solution procedure, considering the x component of the electric field.

The Boltzmann equation is

$$(e\,\mathscr{E}_x/\hbar)(\partial f/\partial k_x) = \partial_c f \quad . \tag{7.101}$$

As in the isotropic case, we assume that the solution may be written as

$$f = f_0 - \phi_1(\underline{k})\partial_E f_0 \quad , \tag{7.102}$$

where f_0 is the equilibrium Fermi-Dirac function.

Substituting (7.102) in the Boltzmann equation, and rearranging the collision term (see Sect.7.4), we get

$$(e\mathscr{E}_x/\hbar)(\partial f_0/\partial k_x) = V_c(8\pi^3)^{-1}\partial_E f_0 \int [1-f_0(E_{k'})][1-f_0(E_{\underline{k}})]^{-1} \times S(\underline{k},\underline{k}')[\phi_1(\underline{k})-\phi_1(\underline{k}')]d\underline{k}' \quad . \tag{7.103}$$

Introducing the transformation [7.24]

$$\underline{k}^* = \underline{\underline{\alpha}}^{\frac{1}{2}}\cdot\underline{k} \quad , \tag{7.104}$$

where

$$\underline{\underline{\alpha}}^{\frac{1}{2}} \equiv \begin{bmatrix} (m_0/m_x)^{\frac{1}{2}} & 0 & 0 \\ 0 & (m_0/m_y)^{\frac{1}{2}} & 0 \\ 0 & 0 & (m_0/m_z)^{\frac{1}{2}} \end{bmatrix} , \tag{7.105}$$

and putting

$$\mathscr{E}_x/m_x^{\frac{1}{2}} = \mathscr{E}_x^*/m_0^{\frac{1}{2}} \quad ; \quad (e\hbar\mathscr{E}_x^* k^*/m_0)\phi(\underline{k}^*) = \phi_1(\underline{k}) \quad , \tag{7.106}$$

we obtain

$$\cos\theta_{\underline{k}^*} = m_d^{3/2}\left(8\pi^3 m_0^{3/2}\right)^{-1} \int [1-f_0(\underline{k}'^*)][1-f_0(\underline{k}^*)]^{-1} \times S(\underline{k}^*,\underline{k}'^*)[\phi(\underline{k}^*)-\phi(\underline{k}'^*)]d\underline{k}'^* \quad , \tag{7.107}$$

where θ_{k^*} is the polar angle of k^* with x as the polar axis.

We note that $S(\underline{k}^*,\underline{k}'^*)$ may, in general, be written as

$$S(\underline{k}^*,\underline{k}'^*) = C(\theta_{\underline{k}^*},\theta_{\underline{k}'^*},\phi_{\underline{k}^*},\phi_{\underline{k}'^*})A(|\underline{k}-\underline{k}'|)\frac{2\pi}{\hbar}\delta(E_{\underline{k}^*}-E_{\underline{k}'^*}+\Delta E) \quad , \tag{7.108}$$

where ΔE represents the change in the electron energy on scattering, the scattering parameters are included in the function C, and these will depend on the orientation of the phonon wave vector $\underline{q}$ ($=\underline{k}'\pm\underline{k}$) in relation to the principal axes of the constant-energy ellipsoids. The function A is the wave vector dependent component of $S(\underline{k},\underline{k}')$ and varies with the kind of scattering. Since, the principal axes of the constant energy ellipsoids are in either <100> or <111> directions, the coupling constant C has azimuthal symmetry (i.e. in φ). The perturbation function ϕ may be expanded, in general, in terms of the spherical harmonics $P_\ell^m(\cos\theta_{k^*})$. However, using the expected symmetry, we shall take $m=0$ and expand $\theta(k^*)$ as

$$\phi(\underline{k}^*) = \sum_{\ell=0}^{\infty} a_{\ell x}(E_{\underline{k}})P_\ell(\cos\theta_{\underline{k}*}) \quad , \tag{7.109}$$

where $P_\ell(\cos\theta)$ is the Legendre polynomial of order ℓ and argument $\cos\theta$. The subscript x is introduced to indicate the coefficients of the solution when the electric field is applied in the x direction. We also note that, since $S(\underline{k}^*,\underline{k}'^*)$ is an odd function of $\cos\theta_{\underline{k}*}$, only the odd terms in the series need to be considered.

Equation (7.107) may now be written as

$$\begin{aligned}\cos\theta_{\underline{k}*} = &\left[\sqrt{2}\, m_d^{3/2}/(4\pi^2\hbar^4)\right]\int [1-f_0(E_{\underline{k}'*})][1-f_0(E_{\underline{k}*})]^{-1} \\ &\times C(\theta_{\underline{k}*},\phi_{\underline{k}*},\theta_{\underline{k}'*},\phi_{\underline{k}'*})A\Big\{\Big[(m_x/m_0)(k^*\cos\theta_{\underline{k}*}-k'^*\cos\theta_{\underline{k}'*})^2 \\ &+(m_y/m_0)(k^*\sin\theta_{\underline{k}*}\cos\phi_{\underline{k}*}-k'^*\sin\theta_{\underline{k}'*}\cos\phi_{\underline{k}'*})^2 \\ &+(m_z/m_0)(k^*\sin\theta_{\underline{k}*}\sin\phi_{\underline{k}*}-k'^*\sin\theta_{\underline{k}'*}\sin\phi_{\underline{k}'*})^2\Big]^{\frac{1}{2}}\Big\} \\ &\times \sum\left[a_{\ell x}(E_{\underline{k}*})P_\ell(\cos\theta_{\underline{k}*})-a_{\ell x}(E_{\underline{k}'*})P_\ell(\cos\theta_{\underline{k}'*})\right] \\ &\times \delta(E_{\underline{k}'*}-E_{\underline{k}*}-\Delta E)E_{\underline{k}'*}^{\frac{1}{2}}\sin\theta_{\underline{k}'*}\,d\theta_{\underline{k}'*}d\phi_{\underline{k}'*}dE_{\underline{k}'*} \quad .\end{aligned} \tag{7.110}$$

The integral on the right-hand side of the above equation may be simplified using the property of the delta function: integration over $E_{\underline{k}'*}$ simply causes $E_{\underline{k}'*}$ to be replaced by $E_{\underline{k}'*}+\Delta E$ and k'^* by $(2m_0/\hbar^2)^{\frac{1}{2}}(E_{\underline{k}*}+\Delta E)^{\frac{1}{2}}$. We get accordingly

$$\begin{aligned}\cos\theta_{\underline{k}*} = &\left[m_d^{3/2}/(2^{3/2}\pi^2\hbar^4)\right]\int [1-f_0(E_{\underline{k}*}+\Delta E)][1-f_0(E_{\underline{k}*})]^{-1} \\ &\times C(\theta_{\underline{k}*},\phi_{\underline{k}*},\theta_{\underline{k}'*},\phi_{\underline{k}'*})A\Big((2/\hbar^2)\Big\{m_x\Big[E_{\underline{k}*}^{\frac{1}{2}}\cos\theta_{\underline{k}*}-(E_{\underline{k}*}+\Delta E)^{\frac{1}{2}}\cos\theta_{\underline{k}'*}\Big]^2 \\ &+m_y\Big[E_{\underline{k}*}^{\frac{1}{2}}\sin\theta_{\underline{k}*}\cos\phi_{\underline{k}*}-(E_{\underline{k}*}+\Delta E)^{\frac{1}{2}}\sin\theta_{\underline{k}'*}\cos\phi_{\underline{k}'*}\Big]^2 \\ &+m_z\Big[E_{\underline{k}*}^{\frac{1}{2}}\sin\theta_{\underline{k}*}\sin\phi_{\underline{k}*}-(E_{\underline{k}*}+\Delta E)^{\frac{1}{2}}\sin\theta_{\underline{k}'*}\sin\phi_{\underline{k}'*}\Big]^2\Big\}^{\frac{1}{2}}\Big) \\ &\times \sum\left[a_{\ell x}(E_{\underline{k}*})P_\ell(\cos\theta_{\underline{k}*})-a_{\ell x}(E_{\underline{k}*}+\Delta E)P_\ell(\cos\theta_{\underline{k}'*})\right] \\ &\times (E_{\underline{k}*}+\Delta E)^{\frac{1}{2}}\sin\theta_{\underline{k}'*}\,d\theta_{\underline{k}'*}d\phi_{\underline{k}'*} \quad .\end{aligned} \tag{7.111}$$

Before proceeding further we distinguish three cases: elastic, randomising, and nonrandomising inelastic scattering.

7.7.1 Solution for Elastic Collisions

For elastic collisions, we may put $\Delta E = 0$ and therefore

$$E_{\underline{k}'*} = E_{\underline{k}*} \quad .$$

Equation (7.111) takes the following form for elastic collisions:

$$\begin{aligned}\cos\theta_{\underline{k}*} = m_d^{3/2}(2\sqrt{2}\,\pi^2\hbar^4)^{-1}\int\sum\Big[a_{\ell x}(E_{\underline{k}*})\Big(P_\ell(\cos\theta_{\underline{k}*})-P_\ell(\cos\theta_{\underline{k}'*})\Big)\\ \times C(\theta_{\underline{k}*},\theta_{\underline{k}'*},\phi_{\underline{k}*},\phi_{\underline{k}'*})A(E_{\underline{k}*},\theta_{\underline{k}*},\theta_{\underline{k}'*},\phi_{\underline{k}*},\phi_{\underline{k}'*})\\ \times E_{\underline{k}*}^{\frac{1}{2}}\sin\theta_{\underline{k}'*}\,d\theta_{\underline{k}'*}d\phi_{\underline{k}'*} \quad .\end{aligned} \tag{7.112}$$

Note that C includes in the above representation both the emission and absorption terms. We may reduce the above equation to a linear simultaneous equation using the orthogonal properties of the Legendre polynomials. It is found that the magnitudes of the a_ℓ decrease with increase in ℓ. The current density is affected by less than 10% [7.16] if all the terms higher than $a_1(E_k)$ are neglected. If the first term alone is retained, $a_1(E_k)$ is given by

$$\begin{aligned}a_{1x}^{-1}(E_{\underline{k}}) = -3m_d^{3/2}(8\sqrt{2}\,\pi^3\hbar^4)^{-1}\int P_1(\cos\theta_{\underline{k}*})\Big[\int\Big(P_1(\cos\theta_{\underline{k}*})-P_1(\cos\theta_{\underline{k}'*})\Big)\\ \times C(\theta_{\underline{k}*},\theta_{\underline{k}'*},\phi_{\underline{k}*},\phi_{\underline{k}'*})A(E_{\underline{k}*},\phi_{\underline{k}*},\phi_{\underline{k}'*},\theta_{\underline{k}*},\theta_{\underline{k}'*})E_{\underline{k}*}^{\frac{1}{2}}\\ \times \sin\theta_{\underline{k}'*}\,d\theta_{\underline{k}'*}d\phi_{\underline{k}'*}\Big]d(\cos\theta_{\underline{k}*})d\phi_{\underline{k}*} \quad .\end{aligned} \tag{7.113}$$

Integrals in the expression for a_{1x} and in the similar expression for a_{1y} have been evaluated for deformation-potential acoustic phonon scattering in elemental semiconductors [7.24]. The detailed steps are rather involved and the interested reader should consult [7.24]. It is found that a_{1x} and a_{1y} may be expressed as

$$a_{1x}^{-1} = C_{ac}E^{\frac{1}{2}}\left(\xi_\parallel\Xi_d^2+\eta_\parallel\Xi_d\Xi_u+\zeta_\parallel\Xi_u^2\right) \quad , \tag{7.114}$$

$$a_{1y}^{-1} = C_{ac}E^{\frac{1}{2}}\left(\xi_\perp\Xi_d^2+\eta_\perp\Xi_d\Xi_u+\zeta_\perp\Xi_u^2\right) \quad , \tag{7.115}$$

where

$$C_{ac} = (3m_d^{3/2}k_BT)(2\sqrt{2}\,\pi\hbar^4c_\ell)^{-1} \quad ,$$

$$c_\ell = C_{12} + 2C_{44} + (3/5)(C_{11}-C_{12}-2C_{44}) \quad ,$$

C_{11}, C_{12}, C_{44} are the elastic constants, defined as usual [7.25]; Ξ_u is the deformation potential for uniaxial shear and Ξ_d, that for dilation. Values of $\xi_{\parallel}$, $\eta_{\parallel}$, $\zeta_{\parallel}$, $\xi_{\perp}$, $\eta_{\perp}$, $\zeta_{\perp}$ are determined by the mass and elastic anisotropy ratios. Exact values vary from material to material. Values have been worked out for silicon and germanium. The values are given below:

	$\xi_{\parallel}$	$\eta_{\parallel}$	$\zeta_{\parallel}$	$\xi_{\perp}$	$\eta_{\perp}$	$\zeta_{\perp}$
Germanium	1.33	2.46	1.29	1.33	1.66	1.09
Silicon	1.40	2.40	1.62	1.33	1.15	1.07

Such values, however, have not yet been worked out for all compound semiconductors [7.29].

It may be mentioned in this connection that the theory of anisotropic scattering has been developed above assuming the electron effective mass to be anisotropic. However, even for an isotropic electron effective mass, the perturbation in the distribution function may be anisotropic due to anisotropy in the coupling parameters, which has been so far neglected. For example, when we consider deformation-potential acoustic phonon scattering for electrons at the Γ point, the scattering may be anisotropic due to anisotropy in the elastic constants. It is usual to use for such cases average values of the elastic constants, which have to be evaluated following the procedure outlined above. We quote the result obtained by ZOOK [7.26]. The average value of elastic modulus, which should be used in the formulae given earlier, is the same as given in (7.114-5) for cubic crystals. But, for hexagonal crystals, the expression is

$$c_{\ell} = (2C_{11}+C_{33})/3 - 2(C_{11}+C_{33}-2C_{13}-4C_{44})/15 \quad . \tag{7.116}$$

Similar expressions for piezoelectric acoustic phonon scattering have not been worked out for the general case of anisotropic effective mass. ZOOK [7.26], however, shows that for isotropic effective mass, the average values of piezoelectric coupling constants to be used in the expressions are as given below.

For cubic crystal structure, the coupling constant for piezoelectric scattering was written as

$$h_{pz}^2/\varepsilon_0 \rho s^2 \quad ,$$

where h_{pz} was defined as the piezoelectric coefficient, and s as the acoustic velocity. When the average is properly worked out, we find that this term is really given by [7.27]

$$h_{pz}^2/\varepsilon_0\rho s^2 \rightarrow \left(h_{14}^2/\varepsilon_0\right)\left(12c_\ell^{-1}+16c_t^{-1}\right)(1/35) \quad , \tag{7.117}$$

where

$$c_\ell = (1/5)(3C_{11}+2C_{12}+4C_{44}) \quad , \tag{7.118}$$

and

$$c_t = (1/5)(C_{11}-C_{12}+2C_{44}) \quad , \tag{7.119}$$

and h_{14} is the piezoelectric coefficient; C_{11}, C_{12}, C_{44} are the components of the elasticity tensor.

For hexagonal crystal structure, the corresponding term is a tensor, the components of which are [7.27]

$$h_\perp^2/\rho u_p^2 = 4h_{15}^2\left(21+6h_x/h_{15}+h_x^2/h_{15}^2\right)/\,105c_t + h_{33}^2\left(21-24h_x/h_{33}+8h_x^2/h_{33}^2\right)/\,105\,c_\ell \quad , \tag{7.120}$$

$$h_\parallel^2/\rho u_p^2 = 2h_{15}^2\left(21+18h_x/h_{15}+5h_x^2/h_{15}^2\right)/\,105\,c_t + h_{33}^2\left(63-36h_x/h_{33}+8h_x^2/h_{33}^2\right)/\,105\,c_\ell \quad , \tag{7.121}$$

where

$$c_\ell = (2C_{11}+C_{33})/3 - 2C_x/15 \quad ; \quad c_t = C_{44} + 2C_x/15 \quad ;$$

$$C_x = C_{11} + C_{33} - 2C_{13} - 4C_{44} \quad ; \quad h_x = h_{33} - h_{31} - 2h_{15} \quad .$$

The h_{ij} and c_{ij} are the components of the piezoelectric tensor and the elasticity tensor respectively.

The same method has been used to derive expressions for the perturbation in the distribution function due to ionised impurity scattering. In this case, the anisotropy is introduced mostly by the anisotropy of the effective mass. We shall only quote the result for silicon, and refer the reader to the paper of SAMOILOVICH [7.28a] for the details of the derivation. The formulae obtained are

$$a_{1x}^{-1} = N_i\pi e^4[\ell n b_\parallel - 2.34+7.88/b_\parallel]\,(4\pi\varepsilon_0 K_s)^{-2}(2m_\parallel)^{-\frac{1}{2}}\,E^{-3/2}(1+g_\parallel)^{-1} \quad , (7.122)$$

$$a_{1y}^{-1} = N_i\pi e^4[\ell n b_\parallel - 1.75+4.64/b_\parallel]\,(4\pi\varepsilon_0 K_s)^{-2}(2m_\perp)^{-\frac{1}{2}}E^{-3/2}(1+g_\perp)^{-1} \quad , (7.123)$$

where

$$g_\parallel = 0.3\,[\log_{10}(b_\parallel)-1] \quad , \quad g_\perp = 0.192 - 0.067[5/\log_{10}(b_\parallel)] \quad ,$$

$$b_\parallel = 8\varepsilon_0 K_s m_\parallel k_B T E\left(h_{pz}^2 e^2 n'\right)^{-1} \quad , \quad n' = n + (n+N_a)[1-(n+N_a)/N_d] \quad ;$$

the other symbols being as before.

In the solution discussed above we retained only one term in the series for θ. We may easily extend the method to include higher-order terms. For example, if we retain two terms, we get from (7.111) the following two equations:

$$A_{11}a_{1x}(E_{\underline{k}*}) + A_{13}a_{3x}(E_{\underline{k}*}) = 1 \quad , \qquad (7.124)$$

$$A_{31}a_{1x}(E_{\underline{k}*}) + A_{33}a_{3x}(E_{\underline{k}*}) = 0 \quad , \qquad (7.125)$$

where

$$A_{11} = B\int P_1(\cos\theta_{\underline{k}*})d(\cos\theta_{\underline{k}*})d\phi_{\underline{k}*}\int\left[P_1(\cos\theta_{\underline{k}*})-P_1(\cos\theta_{\underline{k}'*})\right]$$
$$\times\, C(\theta_{\underline{k}*},\theta_{\underline{k}'*},\phi_{\underline{k}*},\phi_{\underline{k}'*})A(E_{\underline{k}*},\theta_{\underline{k}*},\theta_{\underline{k}'*},\phi_{\underline{k}*},\phi_{\underline{k}'*})E_{\underline{k}*}^{\frac{1}{2}}d(\cos\theta_{\underline{k}'*})d\phi_{\underline{k}'*} \quad ,$$

$$A_{13} = B\int P_1(\cos\theta_{\underline{k}*})d(\cos\theta_{\underline{k}*})d\phi_{\underline{k}*}\int\left[P_3(\cos\theta_{\underline{k}*})-P_3(\cos\theta_{\underline{k}'*})\right]$$
$$\times\, CAE_{\underline{k}*}^{\frac{1}{2}}d(\cos\theta_{\underline{k}'*})d\phi_{\underline{k}'*} \quad ,$$

$$A_{31} = B\int P_3(\cos\theta_{\underline{k}*})d(\cos\theta_{\underline{k}*})d\phi_{\underline{k}*}\int\left[P_1(\cos\theta_{\underline{k}*})-P_1(\cos\theta_{\underline{k}'*})\right]$$
$$\times\, CAE_{\underline{k}*}^{\frac{1}{2}}d(\cos\theta_{\underline{k}'*})d\phi_{\underline{k}'*} \quad ,$$

$$A_{33} = B\int P_3(\cos\theta_{\underline{k}*})d(\cos\theta_{\underline{k}*})d\phi_{\underline{k}*}\int\left[P_3(\cos\theta_{\underline{k}*})-P_3(\cos\theta_{\underline{k}'*})\right]$$
$$\times\, CAE_{\underline{k}*}^{\frac{1}{2}}d(\cos\theta_{\underline{k}'*})d\phi_{\underline{k}'*} \quad ,$$

$$B = 3m_d^{3/2}(8\sqrt{2}\,\pi^3\hbar^4)^{-1} \quad .$$

The A_{ij} may be evaluated by using the detailed expressions for the functions A and C. Usually the integration is so complex that numerical methods [7.29] have to be employed. We find that now a_{12} is given by

$$a_{1x}^{-1} = A_{11} - (A_{13}A_{31}/A_{33}) \quad . \tag{7.126}$$

7.7.2 Solution for Randomising Collisions

The solution for randomising collisions has a much simpler form than for the other two kinds of scattering. In randomising collisions, both the coupling constant C and the scattering function A are independent of the angles of $\underline{k}$ and $\underline{k}'$ as well as of the magnitude of $(\underline{k}-\underline{k}')$. So, (7.111) reduces to

$$\cos\theta_{\underline{k}*} = \sqrt{2}\, m_d^{3/2}(\pi\hbar^4)^{-1}\left[1-f_0(E_{\underline{k}*}+\Delta E)\right]\left[1-f_0(E_{\underline{k}*})\right]^{-1}AC \times\left[\sum a_{\ell x}(E_{\underline{k}*})P_\ell(\cos\theta_{\underline{k}*})\right](E_{\underline{k}}+\Delta E)^{\frac{1}{2}} \quad . \tag{7.127}$$

Evidently, only a_{1x} has a nonzero value, given by

$$a_{1x}^{-1} = \sqrt{2}\, m_d^{3/2}(\pi\hbar^4)^{-1}\left[1-f_0(E_{\underline{k}*}+\Delta E)\right]\left[1-f_0(E_{\underline{k}*})\right]^{-1}AC(E_{\underline{k}}+\Delta E)^{\frac{1}{2}} \quad . \tag{7.128}$$

Note that the expression on the right-hand side should represent the sum of the emission and absorption terms corresponding to which ΔE has the values $-\hbar\omega_i$ and $+\hbar\omega_i$, ω_i being the phonon frequency involved in the randomising collision.

The above expression differs from that given earlier in only one term. The effective mass m* is now replaced by the density-of-states effective mass. We recall that intervalley phonon and nonpolar optic phonon scattering are the only two scattering mechanisms for which the scattering is randomising. Therefore, for these cases the anisotropy in the effective mass causes a very small alteration. We also note that a_{1y} and a_{1z} would be identical to a_{1x}, since the expression for the latter involves no directional term. This kind of scattering therefore remains isotropic even when the mass is anisotropic.

7.7.3 Solution for Nonrandomising Inelastic Collisions

Equation (7.111) assumes the most general form when the collisions are nonrandomising and inelastic. If we retain only one term and reduce the equation to an algebraic equation, we get

$$A_{11}(E_{\underline{k}})a_{1x}(E_{\underline{k}}) - B_+(E_{\underline{k}})a_{1x}(E_{\underline{k}}+\Delta E) - B_-(E_{\underline{k}})a_{1x}(E_{\underline{k}}-\Delta E) = 1 \quad , \tag{7.129}$$

where

$$A_{11}(E_{\underline{k}}) = B\int P_1(\cos\theta_{\underline{k}*})d(\cos\theta_{\underline{k}*})d\phi_{\underline{k}*}$$
$$\int P_1(\cos\theta_{\underline{k}'*})\sum_{+,-} C_{\pm}A_{\pm}(E_{\underline{k}}\pm\Delta E)^{\frac{1}{2}}d(\cos\theta_{\underline{k}'*})d\phi_{\underline{k}'*} \quad ,$$

$$B_+(E_{\underline{k}}) = B\int P_1(\cos\theta_{\underline{k}*})d(\cos\theta_{\underline{k}*})d\phi_{\underline{k}*}$$
$$\int P_1(\cos\theta_{\underline{k}'*})C_+A_+(E_{\underline{k}}+\Delta E)^{\frac{1}{2}}d(\cos\theta_{\underline{k}'*})d\phi_{\underline{k}'*} \quad ,$$

$$B_-(E_{\underline{k}}) = B\int P_1(\cos\theta_{\underline{k}*})d(\cos\theta_{\underline{k}*})d\phi_{\underline{k}*}$$
$$\int P_1(\cos\theta_{\underline{k}'*})C_-A_-(E_{\underline{k}}-\Delta E)^{\frac{1}{2}}d(\cos\theta_{\underline{k}'*})d\phi_{\underline{k}'*} \quad .$$

It may be recalled that only polar optic phonon scattering is nonrandomising and inelastic. The detailed forms of $C_{\pm}$ and $A_{\pm}$ for this scattering are

$$C_{\pm} = e^2\left(K_{\infty}^{-1}-K_S^{-1}\right)(\hbar\omega_{\ell}/2\varepsilon_0)\left(n_{\ell}+\frac{1}{2}\pm\frac{1}{2}\right) = 2\pi\hbar^3 m^{*-1} C_{pop}\left(n_{\ell}+\frac{1}{2}\pm\frac{1}{2}\right) \quad ,$$

$$A_{\pm} = \left(2m_x/\hbar^2\right)^{-1}\left\{\left[E_{\underline{k}}^{\frac{1}{2}}\cos\theta_{\underline{k}*}-(E_{\underline{k}}\pm\Delta E)^{\frac{1}{2}}\cos\theta_{\underline{k}'*}\right]^2\right.$$
$$+(m_y/m_x)\left[E_{\underline{k}}^{\frac{1}{2}}\sin\theta_{\underline{k}*}\cos\phi_{\underline{k}*}-(E_{\underline{k}}\pm\Delta E)^{\frac{1}{2}}\sin\theta_{\underline{k}'*}\cos\phi_{\underline{k}'*}\right]^2$$
$$\left.+(m_z/m_x)\left[E_{\underline{k}}^{\frac{1}{2}}\sin\theta_{\underline{k}*}\sin\phi_{\underline{k}*}-(E_{\underline{k}}\pm\Delta E)^{\frac{1}{2}}\sin\theta_{\underline{k}'*}\sin\phi_{\underline{k}'*}\right]^2\right\}^{-1} \quad , \quad \Delta E = \hbar\omega_{\ell} \quad .$$

The coefficients A_{11}, B_+, and B_- may be evaluated numerically. We now find that the equation to be solved has the same form as that of (7.67), and therefore the methods discussed for the solution of the latter may also be applied to this case. However, to evaluate the mobility, we are required to solve another set of similar equations for a_{1y} and a_{1z}.

The method of solution may be extended in principle, as in the earlier case, to include higher-order terms. However, inclusion of terms higher than $a_3(E_{\underline{k}})$ makes the algebra exceedingly complex. Even when terms up to $a_3(E_{\underline{k}})$ are included, we get a set of coupled difference equations. We shall discuss the solution of these equations in connection with the galvanomagnetic coefficients for the isotropic case, where a similar set also arises.

7.8 Conclusions

We have presented in this chapter the various methods available for the solution of the Boltzmann equation. We have also discussed the Monte Carlo method, which provides an alternative approach for the evaluation of transport coefficients. It has been shown that for low electric fields, we may reduce the Boltzmann equation to a linear equation when electron-electron collisions are unimportant. This equation may be solved directly if the collision processes are elastic or randomising. On the other hand, when the collision processes include polar optic phonon scattering, which is inelastic as well as nonrandomising, The Boltzmann equation may be solved only by applying numerical methods. Since electron scattering in compound semiconductors is dominated by this kind of scattering, accurate theoretical values of transport coefficients for these materials may be obtained only numerically, by means such as the variational method, the matrix method, and the iteration method. The variational method, which was widely used a few years ago, has been superceded by the other two methods, mainly because of their mathematical simplicity. These latter methods are in principle equivalent, and the choice of one or the other depends on individual preference. In the experience of the author, however, the iteration method appears to be more convenient, for the following reasons. The programming on the computer is easier. The accuracy of the method may be estimated at each stage of iteration. It may easily be extended to compute all the low-field transport coefficients. Lastly, the method is self-correcting. A computer program for this method is given in Appendix B.

It is also evident from the outline of the Monte Carlo method presented in this chapter that the direct evaluation of low-field transport coefficients by this method is beset with difficulties. In fact, the Monte Carlo method does not have any special advantages, and indeed it requires more computation time. It should not therefore be considered as an alternative to numerical methods. However, for problems such as transport in thin films and transport in quantising magnetic fields, the Monte Carlo method is the only available method for accurate computations.

8. Low-Field DC Transport Coefficients

Electron transport in semiconductors is usually studied in the presence of various external forces such as electric fields with or without magnetic fields, high-frequency electromagnetic fields with or without a DC magnetic field, temperature gradients, or electron density gradients. Some of these studies are aimed at the discovery of new phenomena with possible practical applications, but a majority of them are used for characterising the materials. We shall discuss in this chapter the transport coefficients related to electron transport caused by a small DC electric field. We shall also consider the effect of a DC magnetic field. In addition we shall deal with the transport of electrons due to a density gradient and also due to a temperature gradient. The coefficients related to electron transport under these conditions are drift mobility, Hall mobility, magnetoresistance, diffusion coefficient, and thermoelectric power. By studying these various coefficients, it is possible to determine the suitability of a material for applications in a particular device. The basic electron parameters such as the effective mass and number density may also be determined from such studies. In addition the importance of various scattering mechanisms may also be estimated. To extract all this information, the theoretical values of the coefficients have to be derived accurately. The principles of the methods used for this purpose have been discussed in the preceding chapter. In this chapter, we derive formulae giving the various coefficients, and give some examples to illustrate the methods.

A vast wealth of experimental data has been collected on transport coefficients of compound semiconductors in course of the last decade. We shall review in a later chapter the information that has been gleaned from these data. However, a large number of these results await further analysis and we shall indicate areas where theory may be fruitfully employed to elicit new results. The formulae given in this chapter would be useful for such an analysis.

8.1 Evaluation of Drift Mobility

Electrons in a semiconductor attain a nonzero average velocity in the presence of an electric field. The magnitude of the average velocity per unit applied electric field is defined as the drift mobility. Experimental values of drift mobility are mostly obtained by taking a rectangular sample and measuring the current produced by a DC field. The current density is related to drift mobility by the equation

$$\underline{J} = n|e|\mu_d \underline{\mathscr{E}} \quad , \tag{8.1}$$

where n is the electron concentration, $|e|$ is the magnitude of its change, μ_d is its drift mobility, $\underline{J}$ is the current density, and $\underline{\mathscr{E}}$ is the electric field intensity. We may thus obtain μ_d from $J/\mathscr{E}$ provided n is known. The electron concentration n is usually determined from the Hall coefficient, as discussed in the following section.

It should be evident that for the analysis of drift mobility the Boltzmann equation has the form

$$-(e\,\mathscr{E}_x/\hbar)(\partial f/\partial k_x) + \partial_c f = 0 \quad , \tag{8.2}$$

where the DC electric field $\mathscr{E}_x$ is assumed to be applied in the x direction.

The distribution function $f(\underline{k})$ may be expanded as

$$f(\underline{k}) = f_0(E) - (e\hbar\,\mathscr{E}_x/m^*)\partial_E f_0 \phi(E) k_x \quad , \tag{8.3}$$

where $\phi(E)$ is the perturbation function, and m* is the electron effective mass. The band structure is assumed to be isotropic. The current density is given by

$$\begin{aligned} J &= (4\pi^3)^{-1}(e/\hbar)\int(\partial E/\partial k_x) f(\underline{k}) d\underline{k} \\ &= -(2/3I_n)\left(ne^2\,\mathscr{E}_x/m^*\right)\int_0^\infty [\gamma(E)]^{3/2}\partial_E f_0 \phi(E) dE \quad , \end{aligned} \tag{8.4}$$

since $\gamma(E) = \hbar^2 k^2/(2m^*)$, $n = (4\pi^3)^{-1}\int f_0(E) d\underline{k}$ and $I_n = \int_0^\infty [\gamma(E)]^{\frac{1}{2}}\gamma'(E) f_0 dE$.
The drift mobility is hence given by

$$\mu_d = -2|e|(3I_n m^*)^{-1}\int_0^\infty [\gamma(E)]^{3/2}\phi(E)\partial_E f_0 dE \quad . \tag{8.5}$$

We may evaluate ϕ by applying the methods discussed in the previous chapter, and hence obtain μ_d. An analytic expression for ϕ may be obtained when the relaxation-time approximation is applicable. It is also possible to bring out a few characteristic features of the drift mobility limited by different scattering mechanisms from these analytic expressions. We discuss these analytic formulae and the characteristics below.

8.1.1 Formulae for Relaxation-Time Approximation

It was shown in Chap.7 that for the conditions under which the relaxation-time approximation applies, the perturbation function $\phi(E)$ is the same as $\tau(E)$, where $\tau(E)$ is given by

$$[\tau(E)]^{-1} = V_c(8\pi^3)^{-1}\int(1-\cos\theta)S(\underline{k},\underline{k}')d\underline{k}' \quad . \tag{8.6}$$

The drift mobility is hence given by

$$\mu_d = -2|e|(3m^*I_n)^{-1}\int_0^\infty \gamma^{3/2}\tau\partial_E f_0 dE = |e|\langle\tau\rangle/m^* \quad , \tag{8.7}$$

where

$$\langle\tau\rangle = -2(3I_n)^{-1}\int_0^\infty \gamma^{3/2}\tau\partial_E f_0 dE \quad . \tag{8.8}$$

The above expression for the drift mobility may be simplified if the sample is extremely degenerate, or if the sample is nondegenerate and the electron energy-band structure is parabolic. Under conditions of extreme degeneracy, $\partial_E f_0$ behaves as a delta function with argument $E-E_F$. The integral may thus be obtained straightaway by replacing E in the integral by the Fermi energy E_F. The drift mobility for extremely degenerate materials is thus

$$\mu_d = 2|e|(3I_n m^*)^{-1}[\gamma(E_F)]^{3/2}\tau(E_F) \quad . \tag{8.9}$$

The evaluation of the drift mobility for degenerate materials thus essentially reduces to the evaluation of $\tau(E_F)$. We give in Table 8.1 expressions for $\tau(E)$ for the different kinds of scattering to which the relaxation-time approximation applies. We have assumed a parabolic energy-wave-vector relation. These expressions are obtained by evaluating the integrals in (8.6). It should be mentioned that in any material, more than one kind of scattering usually limits the mobility. Since the combined scattering probability is

Table 8.1. Relaxation times for materials with isotropic parabolic energy bands

Scattering mechanism	Relaxation time $\tau(E)$	Index of energy dependence	Index of mass dependence
1. Acoustic phonon:			
a) Deformation potential	$\frac{\pi\rho\hbar^4 s^2}{\sqrt{2}\,E_1^2 m^{*3/2} k_B T} E^{-1/2}$	$-\frac{1}{2}$	$-\frac{3}{2}$
b) Piezoelectric	$\frac{2\sqrt{2}\,\pi\hbar^2\rho s^2}{(eh_{pz}/\varepsilon)^2 m^{*1/2} k_B T} E^{1/2}$	$\frac{1}{2}$	$-\frac{1}{2}$
2. Nonpolar optic phonon	$\frac{\sqrt{2}\,\pi\rho\hbar^3\omega_0 F_{op}(E)}{D_0^2 m^{*3/2} n_0}$;	$\sim -\frac{1}{2}$	$-\frac{3}{2}$
	$F_{op}(E) = \left[(E+\hbar\omega_0)^{1/2} + H(E-\hbar\omega_0)(n_0+1)n_0^{-1}(E-\hbar\omega_0)^{1/2}\right]^{-1}$.		
3. Impurity atom:			
a) Ionised	$\frac{16\sqrt{2}\,\pi\varepsilon^2 m^{*1/2}}{N_i Z^2 e^4} F_{imp}(E)$;	$\frac{3}{2}$	$\frac{1}{2}$
	$F_{imp}(E) = E^{3/2}\left[\ln(1+4k^2\lambda^2) - \frac{4k^2\lambda^2}{1+4k^2\lambda^2}\right]^{-1}$.		
b) Neutral	$(N_n\ 20\hbar a_0/m^*)^{-1}$	0	+ 1
4. Alloy	$\frac{\pi\hbar^4}{\sqrt{2}\,N_0\,\alpha(1-\alpha)(E_a-E_b)^2 m^{*3/2}} E^{-1/2}$	$-\frac{1}{2}$	$-\frac{3}{2}$

Table 8.2. Expressions for the drift mobility

Scattering mechanism	Drift mobility	Index of mass dependence	Index of temperature dependence
1. Acoustic phonon:			
a) Deformation potential	$\frac{2(2\pi)^{1/2}\rho\hbar^4 s^2 \lvert e\rvert}{3E_1^2 m^{*5/2}(k_BT)^{3/2}}$	$-\frac{5}{2}$	$-\frac{3}{2}$
b) Piezoelectric	$\frac{16\sqrt{2\pi}\,\rho\hbar^2 s^2 \lvert e\rvert}{3(eh_{pz}/\varepsilon)^2 m^{*3/2}(k_BT)^{1/2}}$	$-\frac{3}{2}$	$-\frac{1}{2}$
2. Nonpolar optic phonon	$\frac{2\sqrt{2\pi}\,\rho\hbar^4 \lvert e\rvert \omega_0^2}{3D_0^2 m^{*5/2}(k_BT)^{3/2}}\ , \quad k_BT \gg \hbar\omega_0\ .$	$-\frac{5}{2}$	$-\frac{3}{2}$
	$\frac{\sqrt{2}\,\pi\rho\hbar^2(\hbar\omega_0)^{1/2}\lvert e\rvert}{D_0^2 m^{*5/2} n_0}\ , \quad k_BT \ll \hbar\omega_0\ .$	$-\frac{5}{2}$	-
3. Impurity atom:			
a) Ionised [8.2]	$\frac{2^7(2\pi)^{1/2}\varepsilon^2(k_BT)^{3/2}}{N_i Z^2 \lvert e\rvert^3 m^{*1/2}}\, c_i\ ;$ $c_i = \ln\left[1+(28\pi\varepsilon k_BT/2Ze^2N_c^{1/3})^2\right]^{-1}\ .$	$-\frac{1}{2}$	$\frac{3}{2}$
b) Neutral	$\lvert e\rvert(20N_n\hbar a_0)^{-1}$	0	0
4. Alloy	$\frac{2(2\pi)^{1/2}\hbar^4\lvert e\rvert}{3N_0\alpha(1-\alpha)(E_a-E_b)^2 m^{*5/2}(k_BT)^{1/2}}$	$-\frac{5}{2}$	$-\frac{1}{2}$

given by $\sum_i S_i(\underline{k},\underline{k}')$, the resultant mobility is $1/\sum_i(1/\mu_i)$. This method of combination of the mobilities is the MATTHIESSEN rule [8.1]. We find that the rule applies exactly in the case of extremely degenerate materials, but is generally inexact in other cases.

Simple expressions for the drift mobility may also be obtained, as mentioned earlier, for nondegenerate samples with parabolic energy bands. Exact analytic expressions may be derived, however, only for scattering mechanisms characterised by a $\tau(E)$ varying with E as E^r. Hence, we may obtain exact expressions for acoustic phonon, neutral impurity, and alloy scattering only if the electron screening effect is neglected. The drift mobility for ionised impurity scattering, however, turns out to be infinity when the electron screening effect is neglected. The expression for the ionised-impurity-scattering-limited mobility is therefore obtained by replacing the electron screening term by its average values [Ref.8.2, p.316] taking into account the slow-variation character of the logarithmic term with energy. We have collected in Table 8.2 the expressions for the drift mobility in nondegenerate samples limited by the various scattering mechanisms under the simplifying conditions mentioned above.

The expressions given in the table have been used extensively in the analysis of scattering mechanisms in semiconductors. We find that the mobilities determined by the various kinds of scattering vary with the lattice temperature and the effective mass, obeying different power laws. The characteristic power laws are also listed in Table 8.2. The dependences of mobility on lattice temperature and effective mass listed above are often utilised to determine the nature of the predominant scattering mechanisms in a material. Of particular interest is the temperature dependence of deformation potential acoustic phonon scattering. The -3/2 power law for its temperature dependence has been used to conclude that at ordinary temperatures it is the predominant scattering mechanisms.

As for degenerate materials, it has been customary to use MATTHIESSEN's rule to obtain the resultant mobility also of nondegenerate materials when more than one scattering mechanism predominate. It should, however, be evident from the expressions for $\tau(E)$ that the rule is a gross simplification in the case of nondegenerate materials and may introduce significant errors [8.3].

The drift mobility limited by optic phonon scattering is often evaluated analytically assuming that E is much larger than $\hbar\omega_0$, so that $(E+\hbar\omega_0)^{\frac{1}{2}}$ may be replaced by $E^{\frac{1}{2}}$. With this assumption, polar optic phonon scattering may also be analysed in the relaxation-time formalism. The results so obtained would be mostly incorrect as E is really comparable with or smaller than $\hbar\omega_0$, for most experimental conditions. Such expressions have, however, lost whatever significance they may have had with the emergence of the accurate numerical methods.

We have discussed above analytic expressions for the drift mobility, commonly used in the past in the analysis of transport results, and we have also indicated the conditions required for their validity. At present, however, these formulae are rarely used for comparison with experiments, as we may evaluate the exact mobility numerically without introducing any simplifying assumption.

Finally, we should mention that our discussion has been confined to the case of an isotropic E-$\underline{k}$ relation. The generalisation of the formulae to the case of ellipsoidal band structure may be easily carried out for isotropic scattering following the procedure outlined above. A detailed discussion is given in Sect.8.2.

As mentioned earlier, the relaxation-time formalism is not applicable to polar optic phonon scattering. Since the drift mobility in compound semiconductors is mostly determined by this scattering mechanism, the formulae given in the preceding tables are of little use. We have to employ numerical methods to obtain exact theoretical values of the drift mobility. The formulae relevant to the numerical methods are given below.

8.1.2 Evaluation by the Variational Method

The perturbation function $\phi(E)$ is usually taken as a power series in the variational method. Expressing $\phi(E)$ as

$$\phi(E) = \sum c_r E^r \quad , \tag{8.10}$$

we find that the drift mobility is given by

$$\mu_d = -2|e|(3m^*I_n)^{-1} \sum c_r \int_0^\infty [\gamma(E)]^{3/2} E^r \partial_E f_0 dE = 2|e|(3m^*I_n)^{-1} \sum c_r \alpha_r \quad , \tag{8.11}$$

where α_r is as defined earlier, i.e.

$$\alpha_r = -\int_0^\infty E^r [\gamma(E)]^{3/2} \partial_E f_0 dE \quad . \tag{8.12}$$

It may be recalled that the c_r are the solutions of the equations

$$\sum c_r d_{rs} - \alpha_r = 0 \quad , \tag{8.13}$$

where

$$d_{rs} = \int_0^\infty E^r L_c(E^s) \partial_E f_0 dE \quad .$$

Solving the equations, we find that μ_d may be written as

$$\mu_d = |e|(m^* I_n)^{-1} \sum D_{mn}/D \quad , \tag{8.14}$$

where the D_{mn} are the elements of the bordered determinant

$$\begin{vmatrix} 0 & \alpha_1 & \cdot\cdot & \cdot\cdot & \alpha_n \\ \alpha_1 & d_{11} & \cdot\cdot & \cdot\cdot & d_{1n} \\ \cdot\cdot & \cdot\cdot & \cdot\cdot & \cdot\cdot & \cdot\cdot \\ \cdot\cdot & \cdot\cdot & \cdot\cdot & \cdot\cdot & \cdot\cdot \\ \alpha_n & d_{n1} & \cdot\cdot & \cdot\cdot & d_{nn} \end{vmatrix} \quad , \tag{8.15}$$

and D is the value of this determinant.

The integrals involved in α_r and d_{rs} may be analytically evaluated if the band structure is isotropic and parabolic and if the sample is either extremely degenerate or nondegenerate. The expressions for these constants for the two cases for polar optic phonon scattering are given in [8.4].

It should be remembered that the value of the drift mobility obtained by using the variational method depends on the number of terms included in the series. It is found that the value reaches a limit when more than five or six terms are included. Even then the value shows some alterations with the number of terms included in the power series, so that a definite conclusion about the correct value is difficult. However, the value obtained by a small number of terms, even only one term, may be correct to within 10-20%.

The analytic expressions are, however, of limited use as for almost all the materials the effects of band nonparabolicity are required to be included. The integrals for nonparabolic bands may be evaluated only numerically and the advantages of the variational method then disappear. However, for preliminary estimates the method is often used even for nonparabolic bands. Some computational advantages may be obtained by using the method with functions more appropriate than a power series. With suitably chosen functions, the method competes well with other methods discussed in the following sections.

8.1.3 Evaluation by Matrix and Iteration Methods

The matrix and iteration methods give values of the perturbation function ϕ at some chosen values of energy E and $E + n\hbar\omega_\ell$ (E lies between 0 and $\hbar\omega_\ell$, n is an integer ranging between 1 and ∞, and ω_ℓ is the optic phonon frequency). Since the drift mobility is evaluated numerically at the end, it is convenient to choose the energy values E by the Gaussian quadrature method. We may normalise the energy values by $\hbar\omega_\ell$ for this purpose and choose a suitable number of normalised energy values W_i in the range 0 to 1, using the formula

$$W_i = a_i \quad ,$$

where the values of a_i for different numbers of chosen points in the interval 0 to 1 for the Gaussian quadrature method are listed in Appendix C.

Applying the matrix or the iteration method we then obtain values of ϕ at these chosen values of W_i. The value of mobility is then obtained from

$$\mu_d = -2|e|(3m^*)^{-1} \sum_{j=1}^{m} \sum_{i=1}^{n} b_i(\gamma_{ij}(E_i)/\hbar\omega_\ell)^{3/2}(\partial f_0/\partial W_i)_{ij}\phi(W_{ij})$$
$$\times \left[\sum_{j=1}^{m} \sum_{i=1}^{n} b_i(\gamma_{ij}(E_i)/\hbar\omega_\ell)^{\frac{1}{2}}(f_0)_{ij}\gamma'_{ij} \right]^{-1} \quad , \tag{8.16}$$

where the subscript ij denotes the value of the variable for the ith point in the jth interval of energy. For example, the value of W_{ij} is $(j-1) + a_i$. The above sums really represent the integrals in the formula for μ_d. The b_i are the weighting factors of the Gaussian quadrature method. These are also listed in Appendix C. A computer program for the evaluation of mobility using this method is given in Appendix B.

8.1.4 Evaluation by the Monte Carlo Method

It has been mentioned earlier that the low-field drift mobility cannot be evaluated conveniently by the Monte Carlo method. The main difficulty is evident when we consider the expression for the drift velocity

$$v_d = (\hbar/m^*)\left[\sum k_{xi}t_i + (e\mathscr{E}/\hbar) \sum t_i^2/2 \right]\left(\sum t_i \right)^{-1} \quad , \tag{8.17}$$

where t_i is a particular time of flight, and k_{xi} is the initial value of the component of the wave vector in the direction of the field $\mathscr{E}$. The first term in the above summation has a nonzero value when the electric field is applied, for two reasons. First, the numbers of electrons having the same energy but equal k_{xi} of opposite sign are different in the presence of the field. Second, when the scattering probability varies with electron energy, t_i for electrons with the same initial energy and equal k_{xi} of opposite sign would also be different, since electrons with negative k_{xi} would gain energy with time, and those with positive k_{xi} would loose energy with time. The average value of the term is, however, expected to be proportional to $\mathscr{E}$, and for low fields this average value would be much smaller than the individual terms. Since the individual terms vary statistically, the statistical fluctuations in the sum become comparable with or larger than the convergent value of the sum. It therefore becomes difficult, if not impossible, to obtain the low-field mobility from the Monte Carlo method, using (8.17) directly. The only method which has been successfully used to obtain the low-field mobility from the Monte Carlo trajectories consists in evaluation of the diffusion constant and then calculation of the mobility from the Einstein relation [8.5]

$$\mu_d = |e|D(k_BT)^{-1} \quad , \tag{8.18}$$

where μ_d is the drift mobility, D is the diffusion constant, and T is the lattice temperature.

The diffusion constant may be confidently evaluated from the Monte Carlo trajectories as it is related to the statistical fluctuations in the values of the first term of (8.17). We shall discuss the evaluation of the diffusion constant in Chap.11.

8.2 Drift Mobility for Anisotropic Bands

8.2.1 Ellipsoidal Band

The solution of the Boltzmann equation for ellipsoidal bands was discussed in Sect.7.7. The distribution function for electric field in the x direction was written as

$$f = f_0 - (e\hbar\mathscr{E}^*k^*/m_0)\partial_E f_0 \phi(\underline{k}^*) \quad , \tag{8.19}$$

and $\phi(\underline{k}^*)$ was expanded as

$$\phi(\underline{k}^*) = \sum a_{\ell x} P_{\ell x}(\cos\theta_{\underline{k}^*}) \quad , \tag{8.20}$$

where $\mathcal{E}^*$ and $\underline{k}^*$ are related to $\mathcal{E}$ and $\underline{k}$ by the Herring-Vogt transformation relation. Methods to be used for obtaining $a_{\ell x}$ have also been discussed and detailed expressions have been given for a_{1x}, a_{1y} for the conditions when relaxation-time approximation holds. The method of solution for polar optic phonon scattering has also been discussed.

Using the expression for the current density

$$J_x = e(4\pi^3\hbar)^{-1}\int(\partial E/\partial k_x)fd\underline{k} \quad , \tag{8.21}$$

it may be easily shown that

$$\begin{aligned} J_x &= -2n_v e^2\mathcal{E}_x(3m_x)^{-1}\int_0^\infty \partial_E f_0 a_{1x}(E)E^{3/2}dE\Big/\int_0^\infty f_0 E^{\frac{1}{2}}dE \\ &= n_v e^2\mathcal{E}_x\langle a_{1x}\rangle/m_x \quad , \end{aligned} \tag{8.22}$$

where n_v is the concentration of electrons in the particular valley under consideration, a_{1x} is as given by (8.20). It may thus be identified as the relaxation time when the electric field is applied in the x direction.

Similarly we get for the current density in the y and z directions

$$J_y = ne^2\mathcal{E}_y\langle a_{1y}\rangle/m_y \quad \text{and} \quad J_z = ne^2\mathcal{E}_z\langle a_{1z}\rangle/m_z \quad , \tag{8.23}$$

where a_{1y} and a_{1z} may be evaluated using (7.111), which should be rewritten after choosing y and z directions as the polar axes respectively. We recall that the constant-energy surfaces are spheroidal, and the major axis is along the <100> direction when the minimum lies on it, and along the <111> direction when the minimum lies on it. The scattering parameters would also be symmetric about these crystallographic directions. Hence if we choose the x direction in the direction of the major axis of the constant energy spheroids (i.e. in the <100> and <111> directions in the two cases), we are really required to evaluate two elements a_{1x} and a_{1y}, as a_{1z} will be equal to a_{1y}. We may identify these as the relaxation times for the electric field applied parallel, $\tau_\parallel$, and perpendicular, $\tau_\perp$, to the principal axes. The current density for ellipsoidal constant-energy surfaces may thus be written as

$$\underline{J} = n_v e^2\left(\hat{\underline{i}}\langle\tau_\parallel\rangle\mathcal{E}_x m_\parallel^{-1}+\hat{\underline{j}}\langle\tau_\perp\rangle\mathcal{E}_y m_\perp^{-1}+\hat{\underline{k}}\langle\tau_\perp\rangle\mathcal{E}_z m_\perp^{-1}\right) \quad , \tag{8.24}$$

where we have put $m_x = m_\parallel$, $m_y = m_\perp = m_z$, $a_{1x} = \tau_\parallel$, and $a_{1y} = \tau_\perp = a_{1z}$. The average $\langle\ \rangle$ is defined by (8.22). Expression (8.24) gives the current density due to electrons in a single valley. We find that in general, this current density will not be colinear with the electric field. However, most of the semiconductors in which the minima are away from the zone centre have cubic crystal structure. The band structure is of the multivalley kind and the minima are located symmetrically. The total current density is the sum of the contributions from electrons in different valleys. On evaluating these, taking into account the change in orientation of the constant-energy surfaces, we get for the current density

$$\underline{J} = (ne^2/3)\left(\langle\tau_\parallel\rangle m_\parallel^{-1} + 2\langle\tau_\perp\rangle m_\perp^{-1}\right)\underline{\mathscr{E}} \quad , \tag{8.25}$$

where n is the total carrier concentration. We note that the total current density is colinear with the electric field, and may be related to the electric field with an effective mobility given by

$$\mu_{\text{eff}} = (|e|/3)\left(\langle\tau_\parallel\rangle m_\parallel^{-1} + 2\langle\tau_\perp\rangle m_\perp^{-1}\right) \quad . \tag{8.26}$$

The expressions given above for $\tau_\parallel$ and $\tau_\perp$ are modified when we retain higher terms in the solution for f. For example, if we retain the second term in addition to the first, we find that a_{1x} is now given by (7.126). The current density being given by the same expression as (8.22), inclusion of the second term introduces a correction factor $1-A_{12}A_{31}/A_{33}A_{11}$. However, this correction, as mentioned, is of small order and involves a large amount of computation. It is hardly ever included as uncertainties in the values of the scattering parameters do not justify such accurate calculations. Most of the results reported in the literature are based on the expressions given earlier.

We have outlined in this section the principle of the method of solution for anisotropic band structures. We note the general result that the relaxation time becomes anisotropic and may be represented in general by two components $\tau_\parallel$ and $\tau_\perp$. The current density is in general given by

$$\underline{J} = ne^2 \begin{pmatrix} \tau_\parallel/m_\parallel & 0 & 0 \\ 0 & \tau_\perp/m_\perp & 0 \\ 0 & 0 & \tau_\perp/m_\perp \end{pmatrix} \cdot \underline{\mathscr{E}} \quad , \tag{8.27}$$

where the axes are required to be chosen parallel to the principal axes of the constant-energy ellipsoids.

We note that the anisotropy in the relaxation time is mixed up with the anisotropy in the effective mass and the general effect of relaxation-time anisotropy will appear as an alteration in the value of the ratio of $m_\parallel$ and

$m_\perp$ appearing in the transport coefficients. The usual band anisotropy ratio $m_\parallel/m_\perp$ will be modified to $(m_\parallel/m_\perp)(\tau_\perp/\tau_\parallel)$ in the expression for the transport coefficients under discussion.

8.2.2 Warped Band

Exact evaluation of the drift mobility for warped band structure involves complex algebra and is rarely attempted. In most of the studies the band structure is simplified by defining an average isotropic effective mass, so that the formulae of Table 8.2 may be used to analyse the transport of holes.

8.3 Galvanomagnetic Transport Coefficients

There are two galvanomagnetic transport coefficients commonly studied for semiconductors. These are the Hall coefficient and the magnetoresistance coefficient. The sample is usually taken in a rectangular form for the measurement of these coefficients, and the magnetic field is applied perpendicular to one plane of the sample (as shown in Fig.8.1). Due to the Lorentz force, electrons travel in curved paths in the presence of the magnetic field and accumulate on the sides of the sample to produce the so-called Hall voltage. Also, the resistance of the sample changes, and this change is the magnetoresistance phenomenon.

The Hall coefficient is defined as

$$R_H = \mathscr{E}_y (J_x B)^{-1} \quad , \tag{8.28}$$

where $\mathscr{E}_y$ is the field corresponding to the Hall voltage, B is the magnetic induction, and J_x is the current density in the sample. It may be shown that for an energy-independent scattering probability, the Hall coefficient is equal to $(ne)^{-1}$, where n is the carrier density. The electron mobility may

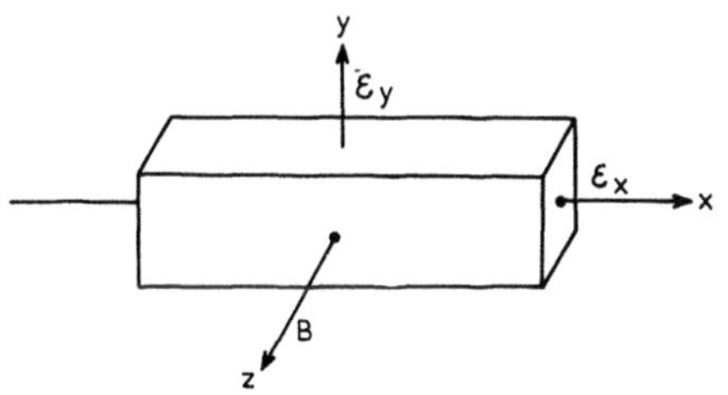

Fig. 8.1. Hall-effect experiment

hence be obtained from the results on Hall coefficient and conductivity using the relation

$$\mu_H = \sigma|R_H| \quad . \tag{8.29}$$

Note that, in general, the scattering probabilities are not energy independnet, and hence R_H is not exactly equal to $(ne)^{-1}$. The mobility obtained by using the above relation is therefore distinguished from the drift mobility and is called the Hall mobility.

The magnetoresistance coefficient R_m is defined by

$$\rho_B = \rho_0(1+R_m) \quad , \tag{8.30}$$

where ρ_B and ρ_0 represent respectively the resistivity in the presence and in the absence of the magnetic field. The coefficient is often measured by measuring the current density for a constant electric field. R_m is then given by

$$R_m = (J_{x0}-J_{xB})/J_{xB} \quad , \tag{8.31}$$

where J_{x0}, J_{xB} represent the current densities in the absence and in the presence of the magnetic field for an electric field applied in the x direction.

In materials with isotropic effective mass the magnetic field has no effect, on the average, when applied in the direction of the electric field in the absence of quantum effects. The magnetoresistance phenomenon is therefore observed only when the magnetic field is transverse to the electric field. On the other hand, in materials with anisotropic band structure, a magnetoresistance may be obtained even when the magnetic field is applied in the direction of the electric field. The latter kind of magnetoresistance is called longitudinal magnetoresistance, while the former kind is called transverse magnetoresistance. In such materials the magnitudes of the magnetoresistance coefficients depend also on the orientation of the fields in relation to the crystallographic axes.

We shall first confine our attention to materials with isotropic band structure, which is the case for a large number of compound semiconductors.

8.3.1 Formulae for Hall Coefficient, Hall Mobility, and Magnetoresistance

In the presence of a magnetic field, the carriers experience a force which is transverse to the field and to their velocity, given by $(e/\hbar)\underline{\nabla}_{k}E \times \underline{B}$. As a result, the electrons trace out curved trajectories and the drift velocity has components in directions other than that of the electric field. It may be recalled that the effective mass for nonspherical constant-energy surfaces is a tensor quantity, and the current density due to carriers in a single valley is not collinear with the electric field even when it acts alone. But at low fields in semiconductors with cubic crystal symmetry, the total current due to carriers in all the valleys is in the direction of the electric field as the cross terms cancel out due to symmetry. On the other hand, even in single-valley semiconductors with isotropic carrier effective mass, the current density in the presence of a magnetic field is related to the electric field by a tensor conductivity. It may then be written as

$$\underline{\underline{\sigma}} = \begin{pmatrix} \sigma_{xx} & \sigma_{xy} & 0 \\ \sigma_{yx} & \sigma_{yy} & 0 \\ 0 & 0 & \sigma_{zz} \end{pmatrix} , \tag{8.32}$$

assuming that the magnetic field has been applied in the z direction. We note that the conductivity component in the z direction is unaffected by the magnetic field, as it does not generate any force in its own direction. The other two diagonal components σ_{xx} and σ_{yy} are equal to each other due to symmetry. Also, the nondiagonal components σ_{xy} and σ_{yx} would be odd functions of the magnetic field as the forces in the cross directions due to the magnetic field are proportional to it. It may be shown from general thermodynamic arguments that [8.6]

$$\sigma_{xy}(B) = \sigma_{yx}(-B) \quad . \tag{8.33}$$

The above relation is one of the Onsager relations. Since σ_{xy} and σ_{yx} are odd functions of B,

$$\sigma_{xy} = -\sigma_{xy} \quad . \tag{8.34}$$

The current density may be written in terms of the conductivity tensor and the electric field components as follows:

$$J_x = \sigma_{xx}\mathscr{E}_x + \sigma_{xy}\mathscr{E}_y \quad , \tag{8.35}$$

$$J_y = -\sigma_{xy}\mathscr{E}_x + \sigma_{yy}\mathscr{E}_y \quad , \tag{8.36}$$

$$J_z = \sigma_{zz}\mathscr{E}_z \quad , \tag{8.37}$$

where $\mathscr{E}_x$, $\mathscr{E}_y$, $\mathscr{E}_z$ are the components of the electric field in the three directions. Since, as mentioned earlier, in an actual experiment the samples are rectangular in shape, J_y is zero. Hence,

$$\mathscr{E}_y = \sigma_{xy}\mathscr{E}_x/\sigma_{yy} \quad . \tag{8.38}$$

The Hall coefficient R_H and Hall mobility μ_H are thus given respectively by

$$R_H = \mathscr{E}_y(BJ_x)^{-1} = \sigma_{xy}\left[B\left(\sigma_{xx}\sigma_{yy}+\sigma_{xy}^2\right)\right]^{-1} \quad , \tag{8.39}$$

$$\mu_H = \sigma_{xx}(0)|R_H| = \sigma_{xx}(0)|\sigma_{xy}|\left[B\left(\sigma_{xx}\sigma_{yy}+\sigma_{xy}^2\right)\right]^{-1} \quad , \tag{8.40}$$

where $\sigma_{xx}(0)$ is the value of σ_{xx} in the absence of the magnetic field.

The magnetoresistance coefficient is given by

$$R_m = J_{x0}/J_{xB} - 1 = \sigma_{xx}\sigma_{xx}(0)/\left(\sigma_{xx}\sigma_{yy}+\sigma_{xy}^2\right) - 1 = (\sigma_{xx}/|\sigma_{xy}|)B\mu_H - 1 \quad . \tag{8.41}$$

The evaluation of the galvanomagnetic coefficients thus boils down to the calculation of the components of the conductivity tensor. The conductivity tensor may be evaluated assuming that the sample has no boundary and the electric field exists in one direction only. We shall follow this procedure in the following section.

8.3.2 Reduced Boltzmann Equation for the Galvanomagnetic Coefficients

The Boltzmann equation has the following form for conditions relevant to the evaluation of the galvanomagnetic coefficients:

$$(e/\hbar)\left(\underline{\mathscr{E}} + \hbar^{-1}\underline{\nabla}_k E\times\underline{B}\right)\cdot\underline{\nabla}_k f = \partial_c f \quad . \tag{8.42}$$

The sample is assumed to be uniform and homogeneous and the fields are steady. Let us now assume that the electric field is in the x direction and the magnetic field is in the z direction. Equation (8.42) may then be written as

$$\left\{e\mathscr{E}_x\hbar^{-1} + eBm^{*-1}k_y[\gamma'(E)]^{-1}\right\}(\partial f/\partial k_x) - eBm^{*-1}k_x[\gamma'(E)]^{-1}(\partial f/\partial k_y) = \partial_c f \quad . \tag{8.43}$$

The dispersion relation is assumed to be

$$\frac{\hbar^2 k^2}{2m^*} = E(1+\alpha E) = \gamma(E) \quad ,$$

$$\gamma'(E) = d\gamma/dE \quad ,$$

m^* being the band-edge effective mass, and $\alpha = E_g^{-1}$. It is evident that the distribution function is perturbed by the magnetic field in such a way that the perturbation part has components in the direction of the electric as well as in the direction transverse to both the electric and the magnetic fields. The distribution function may therefore be expanded as

$$f = f_0 - (e\hbar\mathscr{E}_x/m^*)\partial_E f_0\left[k_x\phi_x(E)-(eB/m^*)k_y\phi_y(E)\right] \quad , \tag{8.44}$$

where f_0 is the even part of the function which, as in the case of drift mobility will be the equilibrium function, i.e. either the Maxwellian function or the Fermi-Dirac function depending on the electron concentration in the material.

Substituting (8.44) in (8.43), neglecting small-order terms, and equating the coefficients of k_x and k_y on the two sides, we get

$$L_c\phi_x + \omega_B^2\gamma'^{-1}\phi_y = \gamma'^{-1} \quad , \tag{8.45}$$

$$L_c\phi_y - \phi_x\gamma'^{-1} = 0 \quad , \tag{8.46}$$

where

$$\omega_B = |e|B/m^* \quad , \tag{8.47}$$

and

$$\partial_c\left[k_x(\partial_E f_0)\phi_x\right] = -k_x(\partial_E f_0)L_c\phi_x \quad . \tag{8.48}$$

Note that ω_B is the same as the cyclotron resonance angular frequency for the band-edge electrons. Equations (8.45) and (8.46) are thus the equations resulting from the Boltzmann equation, which are required to be solved for obtaining the galvanomagnetic coefficients. The conductivity components may be easily evaluated when ϕ_x and ϕ_y are known. Using the expression for the current density (8.21), we find that the relevant formulae are

$$\sigma_{xx} = ne^2 \langle \phi_x \rangle / m^* \quad , \tag{8.49}$$

$$\sigma_{xy} = ne^3 \omega_B \langle \phi_y \rangle / m^* |e| \quad , \tag{8.50}$$

where

$$\langle \phi \rangle = -(2/3) \int_0^\infty \phi \gamma^{3/2} (\partial_E f_0) dE \Big/ \int_0^\infty f_0 \gamma^{\frac{1}{2}} \gamma' dE \quad . \tag{8.51}$$

8.3.3 Solution Using the Relaxation-Time Approximation Method

The collision term reduces for collision mechanisms obeying the relaxation-time approximation to

$$L_c \phi = \phi / \tau(E) \quad , \tag{8.52}$$

where $\tau(E)$ is the relaxation time.

Replacing $L_c \phi_x$ and $L_c \phi_y$ by ϕ_x/τ and ϕ_y/τ in (8.45) and (8.46) respectively, and solving the equations, we get

$$\phi_x = \tau_1 (1+\omega_B^2 \tau_1^2)^{-1} \quad , \tag{8.53}$$

$$\phi_y = \tau_1^2 (1+\omega_B^2 \tau_1^2)^{-1} \quad ; \tag{8.54}$$

where $\tau_1 = \tau/\gamma'(E)$. The conductivity components are

$$\sigma_{xx} = (ne^2/m^*) \left\langle \tau_1 (1+\omega_B^2 \tau_1^2)^{-1} \right\rangle \quad , \tag{8.55}$$

$$\sigma_{xy} = (ne^2/m^*) \omega_B \left\langle \tau_1^2 (1+\omega_B^2 \tau_1^2)^{-1} \right\rangle (e|e|^{-1}) \quad , \tag{8.56}$$

where $\langle \; \rangle$ denotes the average defined by (8.51).

We may now evaluate σ_{xx}, σ_{xy} when the expression for τ is available. Analytic expressions may be obtained only when $\tau(E)$ depends on E as E^r for parabolic nondegenerate materials. In general, however, the integrals are to be evaluated numerically and the Gaussian quadrature method discussed earlier is convenient for this purpose. We may calculate R_H, μ_H, and R_m after evaluating σ_{xx} and σ_{xy} using the formulae given earlier. It is of interest to note that in some cases we get simple results which are often used for the analysis of transport experiments. These are discussed below.

a) Energy-independent τ *and parabolic bands:* In this case,

$$\left\langle \tau_1(1+\omega_B^2\tau_1^2)^{-1} \right\rangle = \tau(1+\omega_B^2\tau^2)^{-1} \quad , \tag{8.57}$$

$$\left\langle \tau_1^2(1+\omega_B^2\tau_1^2) \right\rangle = \tau^2(1+\omega_B^2\tau^2)^{-1} \quad . \tag{8.58}$$

Hence, $R_H = (ne)^{-1}$, $\mu_H = |e|\tau/m^*$, $R_m = 0$. Thus, for energy-independent relaxation time and isotropic effective mass, the Hall mobility is the same as the drift mobility; the Hall coefficient is exactly the inverse of the product of electron charge and electron concentration; and the magnetoresistance is zero.

b) Low magnetic field: For very low magnetic fields, i.e. when $\omega_B\tau_1 << 1$, the expressions for σ_{xx} and σ_{xy} simplify to

$$\sigma_{xx} = (ne^2/m^*)\left(<\tau_1> - \left\langle \omega_B^2\tau_1^3 \right\rangle \right) \quad , \tag{8.59}$$

$$\sigma_{xy} = (ne^3/m^*)\left\langle \omega_B\tau_1^2 \right\rangle |e|^{-1} \quad , \tag{8.60}$$

where terms involving up to B^2 are retained. We therefore get

$$R_H = (ne)^{-1} <\tau_1^2><\tau_1>^{-2} \quad ; \quad \mu_H = (|e|/m^*) <\tau_1^2><\tau_1>^{-1} \quad ;$$

and

$$R_m = (eB/m^*)^2\left[\left(<\tau_1^3>/<\tau_1> \right) - \left(<\tau_1^2>^2/<\tau_1>^2 \right)\right] \quad . \tag{8.61}$$

It is seen that the Hall coefficient differs from $(ne)^{-1}$ by the factor $<\tau_1^2>/<\tau_1>^2$, which is also equal to the ratio of the Hall mobility and the drift mobility. This ratio r is often called the Hall ratio or the Hall factor. Its value is larger than unity. For parabolic bands it has a value of 1.18 for acoustic-phonon-scattering-limited τ and a value of 1.93 for ionised-impurity-scattering-limited τ if τ is assumed to vary as E^s. The value of r may thus be any where between 1 and 1.93 depending on the nature of nonparabolicity and the predominance of the various scattering mechanisms.

It is also found that the magnetoresistance coefficient is proportional to the square of the magnetic field at low fields. The value of the coefficient depends on the exact nature of the energy dependence of τ. But it varies directly as the square of the mobility and may be expected to have large values in high-mobility materials. A quantity is often defined to give the component of magnetoresistance, which is independent of the value of mobility. It is also called the magnetoresistance coefficient and is given by the ratio $R_m/(\mu_H B)^2$. Evidently, this coefficient would be independent of the magnetic field and the mobility of the material, but depend only on the functional form of $\tau_1(E)$. This ratio ξ is given by

$$\xi = R_m/(\mu_H B)^2 = <\tau_1^3><\tau_1>/<\tau_1^2>^2 - 1 \quad . \tag{8.62}$$

The value of ξ for acoustic phonon scattering and ionised impurity scattering in a material with parabolic band is given by 0.28 and 0.586 respectively.

c) High magnetic fields: In magnetic fields such that $\omega_B\tau_1 >> 1$ we may simplify σ_{xx} and σ_{xy} to

$$\sigma_{xx} = (ne^2/m^*)\left\langle \omega_B^{-2}\tau_1^{-1}\right\rangle \quad ; \quad \sigma_{xy} = (ne^3/m^*)<\omega_B^{-1}>|e|^{-1} \quad . \tag{8.63}$$

The quantities R_H, μ_H, and R_m are therefore

$$R_H \simeq (B\sigma_{xy})^{-1} = (ne)^{-1} \quad , \tag{8.64}$$

since

$$<\omega_B^{-1}> = -\omega_B^{-1}(2/3)\int_0^\infty \gamma^{3/2}\partial_E f_0 dE/\int_0^\infty \gamma^{\frac{1}{2}}\gamma' f_0 dE$$

$$= \omega_B^{-1}\left[-(2/3)\left(\gamma^{3/2}f_0\right)_0^\infty + \int_0^\infty \gamma^{\frac{1}{2}}\gamma' f_0 dE\right]\left[\int_0^\infty \gamma^{\frac{1}{2}}\gamma' f_0 dE\right]^{-1} = 1/\omega_B \quad .$$

Hence,

$$\mu_H = |e|<\tau_1>/m^* \quad , \tag{8.65}$$

and

$$R_m = <\tau_1><1/\tau_1> - 1 \quad . \tag{8.66}$$

We thus find that for high magnetic fields, the Hall factor has a value of unity and the Hall mobility is equal to the drift mobility. The ratio of Hall mobility for very high magnetic fields to that for very low magnetic fields thus gives the Hall ratio. This result is often used to obtain the drift mobility and Hall ratio in compound semiconductors. The Hall mobility being easily measurable, this quantity is directly measured experimentally. Electron mobility in compound semiconductors being very high, the high magnetic-field condition may be easily realised with available fields and the high-field mobility may be taken to represent the drift mobility.

We also find that the magnetoresistance coefficient at high fields reaches a saturation value, which is independent of the electron effective mass, but depends only on the functional form of τ_1. The limiting values for acoustic phonon scattering and ionised impurity scattering in parabolic bands are respectively $32/9\pi$ and $32/3\pi$, as may be easily evaluated from the expressions given earlier.

The expressions for the galvanomagnetic coefficients have been derived in this section assuming the relaxation-time expression to be valid. In compound semiconductors the assumption may be valid for low temperatures where polar optic phonon scattering is negligible. It is also valid at high temperatures at which the polar optic phonon scattering may be considered to be elastic. The formulae are not obeyed for the usual experimental temperatures. However, the general conclusions discussed above remain valid even when polar optic phonon scattering predominates and are therefore useful for an understanding of the behaviour of the coefficients. The methods which may be used to evaluate the coefficients accurately even when polar optic phonon scattering predominates are discussed in the following sections.

8.3.4 A Simple Formula for the Low-Field Hall Mobility

The evaluation of Hall mobility for arbitrary magnetic fields requires solution of (8.45) and (8.46) simultaneously and determination of both ϕ_x and ϕ_y. However, for low magnetic fields, the formula for Hall mobility may be simplified to such a form as to involve ϕ_x alone. For low values of the magnetic field $|\sigma_{xy}| \ll \sigma_{xx}$ and $\sigma_{xx} \simeq \sigma_{xx}(0)$. The formula for the Hall mobility hence simplifies to

$$\mu_H = |\sigma_{xy}|/\sigma_{xx}(0)B \quad . \tag{8.67}$$

Equation (8.45) may also be simplified to

$$L_c\phi_x = [\gamma'(E)]^{-1} \quad . \tag{8.68}$$

Now,

$$\sigma_{xy} = -(2/3)\left(ne^2/I_n\right)\int_0^\infty \phi_y\gamma^{3/2}f_0' dE \; eB/m^{*2} \quad , \tag{8.69}$$

$$f_0' = \partial f_0/\partial E \quad .$$

But,

$$\int_0^\infty \phi_y\gamma^{3/2}f_0' dE = \int_0^\infty \phi_y\gamma'^{-1}\gamma^{3/2}\gamma' f_0' dE \quad . \tag{8.70}$$

Replacing γ'^{-1} by $L_c\phi_x$ and using the self-adjoint property of the collision operator (i.e. $\int_0^\infty \psi L_c\phi\gamma^{3/2}\gamma' f_0' dE = \int_0^\infty \phi L_c\psi\gamma^{3/2}\gamma' f_0' dE$, see Sect.7.6.2), we get

$$\int_0^\infty \phi_y\gamma^{3/2}f_0' dE = \int_0^\infty \phi_x(L_c\phi_y)\gamma^{3/2}\gamma' f_0' dE = \int_0^\infty \phi_x^2\gamma^{3/2}f_0' dE \quad , \tag{8.71}$$

since $L_c\phi_y = \phi_x/\gamma'$ (8.46). The Hall mobility is thus given by [8.7,8]

$$\mu_H = (|e|/m^*)\int_0^\infty \phi_x^2\gamma^{3/2}f_0' dE \Big/ \int_0^\infty \phi_x\gamma^{3/2}f_0' dE \quad . \tag{8.72}$$

We thus find that the low-field Hall mobility may be obtained if the perturbation component ϕ_x in the absence of the magnetic field is known. The above formula may be used to obtain the Hall mobility and the Hall ratio by solving (8.68) alone with the help of the methods discussed in Sect.7.6. Since the solution of (8.45) and (8.46) simultaneously requires longer time, this formula is very convenient for the study of materials with complex scattering mechanisms, as we are interested mostly in the low-field Hall mobility, which is directly obtainable from experiments.

8.3.5 Numerical Methods for the Galvanomagnetic Coefficients for Arbitrary Magnetic Fields

As discussed above, for arbitrary magnetic fields we are required to solve (8.45) and (8.46) simultaneously. We note that the collision term has the following general form:

$$L_c\phi(E) = S_0(E)\phi(E) - S_+(E)\phi(E+\hbar\omega_\ell) - S_-(E)\phi(E-\hbar\omega_\ell) \quad , \tag{8.73}$$

where $S_0(E)$, $S_+(E)$, and $S_-(E)$ are as in (7.35), and ω_ℓ is the frequency of polar optic phonons. Equations (8.45) and (8.46) may thus be written as

$$S_0(E+\ell\hbar\omega_\ell)\phi_x(E+\ell\hbar\omega_\ell) - S_+(E+\ell\hbar\omega_\ell)\phi_x(E+(\ell+1)\hbar\omega_\ell)$$
$$- S_-(E+\ell\hbar\omega_\ell)\phi_x(E+(\ell-1)\hbar\omega_\ell)H(\ell-1) + \omega_B^2\left[\gamma'(E+\ell\hbar\omega_\ell)\right]^{-1}\phi_y(E+\ell\hbar\omega_\ell)$$
$$= \left[\gamma'(E+\ell\hbar\omega_\ell)\right]^{-1} \quad . \tag{8.74}$$

$$S_0(E+\ell\hbar\omega_\ell)\phi_y(E+\ell\hbar\omega_\ell) - S_+(E+\ell\hbar\omega_\ell)\phi_y(E+(\ell+1)\hbar\omega_\ell)$$
$$- S_-(E+\ell\hbar\omega_\ell)\phi_y(E+(\ell-1)\hbar\omega_\ell)H(\ell-1) - \left[\gamma'(E+\ell\hbar\omega_\ell)\right]^{-1}\phi_x(E+\ell\hbar\omega_\ell) = 0 \quad . \tag{8.75}$$

We have to solve an infinite number of equations obtained by giving ℓ integral values starting from zero. We note that the values of $\phi_x(E)$ and $\phi_y(E)$ are related to their values for energies corresponding to $E+\ell\hbar\omega_\ell$ where ℓ may have all the integral values with the limitation $E+\ell\hbar\omega_\ell > 0$. The functions ϕ_x and ϕ_y may therefore be obtained only by solving a set of infinite number of equations.

The variational method has been used in the past to evaluate the magnetoconductivity tensor [8.9]. The variational principle was similar to that used for the conductivity problem. The other numerical methods discussed with reference to the conductivity are however more suitable. The set of equations may be solved exactly by extending the matrix methods or the iterative methods. The matrix method of GRIGORIEV et al. [8.10] has been extended to the present problem by FORTINI et al. [8.11] and the DELVES [8.12] method has been extended by FLETCHER and BUTCHER [8.13]. However, as mentioned earlier, the iteration method developed by RODE [8.14] has some advantages and this method may be applied to this case with suitable modification. We shall describe this method in detail and refer the reader to the citations given above for the other methods. In the iteration method as developed by the present author [8.15], the equations are first solved in pairs to obtain $\phi_x(E+\ell\hbar\omega_\ell)$ and $\phi_y(E+\ell\hbar\omega_\ell)$ putting $\phi_x(E+(\ell\pm1)\hbar\omega_\ell)$ and $\phi_y(E+(\ell\pm1)\hbar\omega_\ell)$ equal to zero. The solutions are

$$\phi_x(E+\ell\hbar\omega_\ell) = S_0(E+\ell\hbar\omega_\ell)[\gamma'(E+\ell\hbar\omega_\ell)]^{-1}\left[[S_0(E+\ell\hbar\omega_\ell)]^2+\left\{\omega_B[\gamma'(E+\ell\hbar\omega_\ell)]^{-1}\right\}^2\right]^{-1}$$
$$= \tau_r(E+\ell\hbar\omega_\ell)[\gamma'(E+\ell\hbar\omega_\ell)]^{-1} \quad , \tag{8.76}$$

$$\phi_y(E+\ell\hbar\omega_\ell) = [\gamma'(E+\ell\hbar\omega_\ell)]^{-2}\left[[S_0(E+\ell\hbar\omega_\ell)]^2+\left\{\omega_B[\gamma'(E+\ell\hbar\omega_\ell)]^{-1}\right\}^2\right]^{-1}$$
$$= \tau_i(E+\ell\hbar\omega_\ell)[\gamma'(E+\ell\hbar\omega_\ell)]^{-1} \quad . \tag{8.77}$$

Values of ϕ_x and ϕ_y are worked out using the above formulae for values of ℓ up to some chosen value m.

In the next iteration step the values obtained in the first step are used for $\phi_x(E+(\ell\pm1)\hbar\omega_\ell)$ and $\phi_y(E+(\ell\pm1)\hbar\omega_\ell)$, and the equations are solved again to obtain new values of $\phi_x(E+\ell\hbar\omega_\ell)$ and $\phi_y(E+\ell\hbar\omega_\ell)$. Explicitly, the equations giving the values in the ith iteration step are

$$\begin{aligned}\phi_x^{(i)}(E+\ell\hbar\omega_\ell) = {} & \tau_r(E+\ell\hbar\omega_\ell)\Big\{[\gamma'(E+\ell\hbar\omega_\ell)]^{-1} \\ & + \sum_{+,-} S_\pm(E+\ell\hbar\omega_\ell)\phi_x^{(i-1)}(E+(\ell\pm1)\hbar\omega_\ell)\Big\} \\ & - \omega_B^2\tau_i(E+\ell\hbar\omega_\ell)\Big\{\sum_{+,-} S_\pm(E+\ell\hbar\omega_\ell)\phi_y^{(i-1)}(E+(\ell\pm1)\hbar\omega_\ell)\Big\} \quad . \quad (8.78)\end{aligned}$$

$$\begin{aligned}\phi_y^{(i)}(E+\ell\hbar\omega_\ell) = {} & \tau_r(E+\ell\hbar\omega_\ell)\Big\{\sum_{+,-} S_\pm(E+\ell\hbar\omega_\ell)\phi_y^{(i-1)}(E+(\ell\pm1)\hbar\omega_\ell)\Big\} \\ & +\tau_i(E+\ell\hbar\omega_\ell)\Big\{[\gamma'(E+\ell\hbar\omega_\ell)]^{-1}+\sum_{+,-} S_\pm(E+\ell\hbar\omega_\ell)\phi_x^{(i-1)}(E+(\ell\pm1)\hbar\omega_\ell)\Big\}. \\ & \qquad (8.79)\end{aligned}$$

These calculations are repeated till the values of ϕ_x and ϕ_y in successive steps agree to within the desired accuracy. It should be noted that in each step of the iteration we may obtain new iterated values of $\phi(E+\ell\hbar\omega_\ell)$ up to $\ell = n-1$, if n was the highest value of m in the last iteration step. Hence if the iteration converges in q steps and values of $\phi(E+\ell\hbar\omega_\ell)$ up to $\ell = p$ contribute significantly to the conductivity integrals, we should choose m in the first step as $m = p + q$. In a practical problem the iteration converges within 7-10 steps and the value of q is required to be about 6-10. Hence, the value of m may be chosen to be 20.

It is evident that the procedure described above gives values of ϕ_x and ϕ_y for some particular values of energy E, i.e. for E, $E + \hbar\omega_\ell$, $E + 2\hbar\omega_\ell$, ..., $E + \ell\hbar\omega_\ell$. In order to obtain ϕ_x and ϕ_y for other values of E, the procedure has to be repeated for that value of E. However, for the evaluation of conductivity tensor integrals, as discussed earlier (see Sect.8.1.3), we need to know the values of the functions at Gaussian quadrature points. The number of points required for sufficient accuracy is 10 to 16. On evaluating ϕ_x and ϕ_y, following the above procedure, the conductivity components are evaluated using the following formulae:

$$\begin{aligned}\sigma = {} & (2/3)ne^2(m^*k_BT)^{-1}\Big[\sum_{\ell=0}^{p}\sum_{i=1}^{16} \exp\Big\{-(E_i+\ell\hbar\omega_\ell)/k_BT\Big\}(E_i+\ell\hbar\omega_\ell)^{3/2}\phi(E+\ell\hbar\omega_\ell)B_i\Big] \\ & \times \Big[\sum_{\ell=0}^{p}\sum_{i=1}^{16} \exp\Big\{-(E_i+\ell\hbar\omega_\ell)/k_BT\Big\}(E_i+\ell\hbar\omega_\ell)^{\frac{1}{2}}B_i\Big]^{-1} \quad , \qquad (8.80)\end{aligned}$$

where the B_i are the Gaussian quadrature weighting numbers given earlier (Appendix C). Evidently, we get σ_{xx} by using ϕ_x for ϕ, and σ_{xy} by using ϕ_y for ϕ. Once σ_{xx} and σ_{xy} are obtained, the Hall mobility, Hall coefficient, and magnetoresistance may be evaluated using the formulae given earlier (8.39-41).

A computer program for the method described above is given in Appendix B. In the program, the scattering terms are written including the effects of nonparabolicity, overlap integrals, and screening. Calculated results for the parameter values of InSb are presented in Figs.8.2-4.

The perturbation functions ϕ_x and ϕ_y are found to show jumps in values at the energies equal to those of the polar optic phonons. These jumps are similar to those illustrated earlier for the electric field acting alone. It is also found that the components are markedly affected by the magnetic field and the magnetic field dependence is similar to that expected from the relaxation-time approximation as discussed earlier. At low fields, the magnetoresistance increases as the square of the magnetic field. At high fields the Hall ratio has a value of unity, while the magnetoresistance has a saturation value. It should, however, be noted that the value of the low-field Hall ratio and the saturation value of the magnetoresistance depends on the optic phonon temperature in relation to the lattice temperature. These values are often useful in the study of semiconductors with predominant polar optic phonon scattering. In Table 8.3 values of the Hall ratio and saturation magnetoresistance are given for various values of the ratio of the phonon temperature and the lattice temperature. Although results are presented here only for polar optic phonon scattering, the method may be used to include the effects of all kinds of scattering. The computer program given in Appendix B is quite general and may be used to calculate the coefficients for all kinds of complexity.

8.3.6 Evaluation of the Galvanomagnetic Transport Coefficients for Anisotropic Effective Mass

Galvanomagnetic transport coefficients for a material with anisotropic effective mass are evaluated following the same procedure as discussed earlier for the isotropic effective mass. We first develop the formulae for the relaxation-time formalism assuming that the relaxation time is isotropic. The modification in the formulae for anisotropic relaxation time will be next considered. The method to be used when the relaxation-time formalism is not applicable will be discussed at the end.

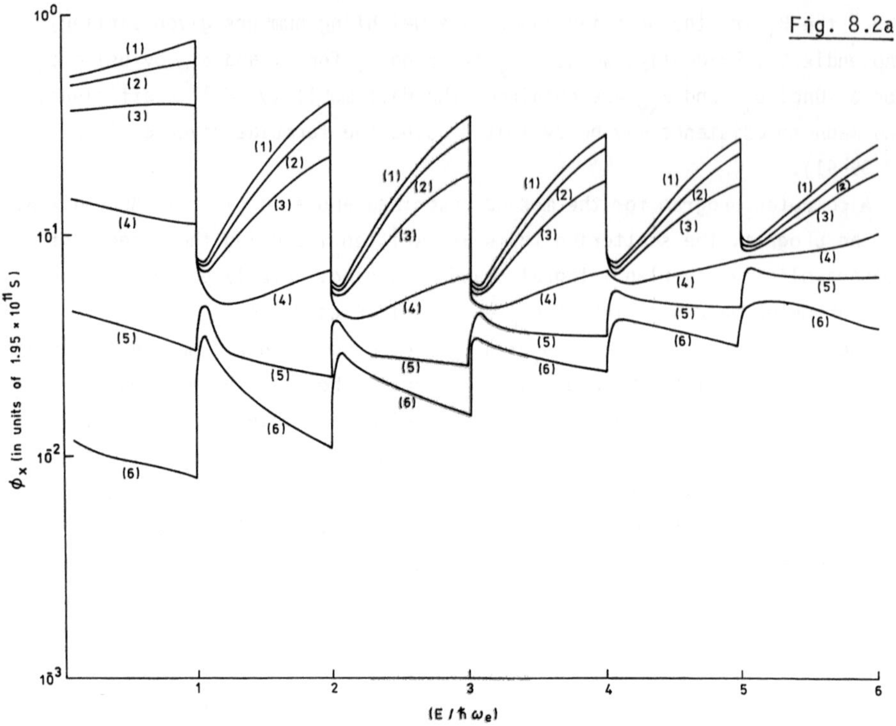

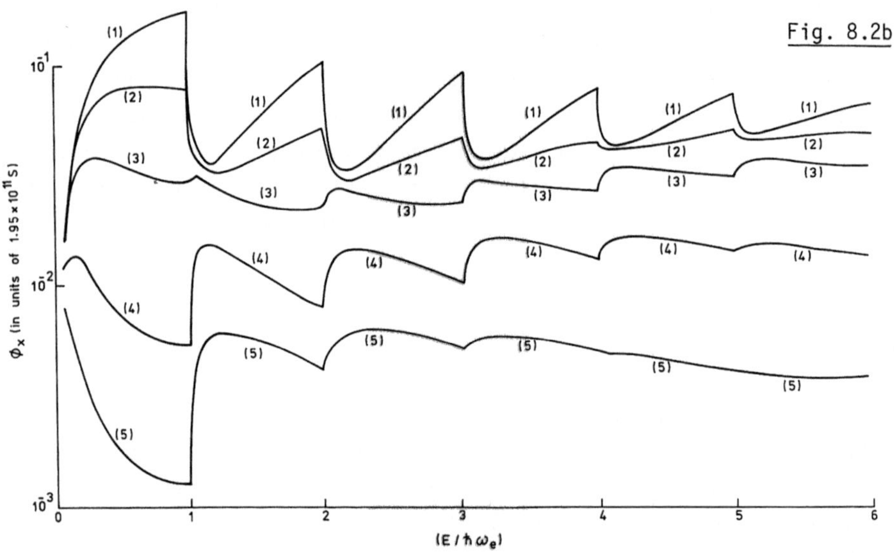

Fig. 8.2. Perturbation function ϕ_x. (a) Polar optic phonon scattering only (1) B = 0; (2) B = 0.01; (3) B = 0.02; (4) B = 0.005; (5) B = 0.1; (6) B = 0.2; (b) Polar optic phonon, acoustic phonon, and impurity atom scattering (1) B = 0; (2) B = 0.01; (3) B = 0.02; (4) B = 0.05; (5) B = 0.1

Fig. 8.3a

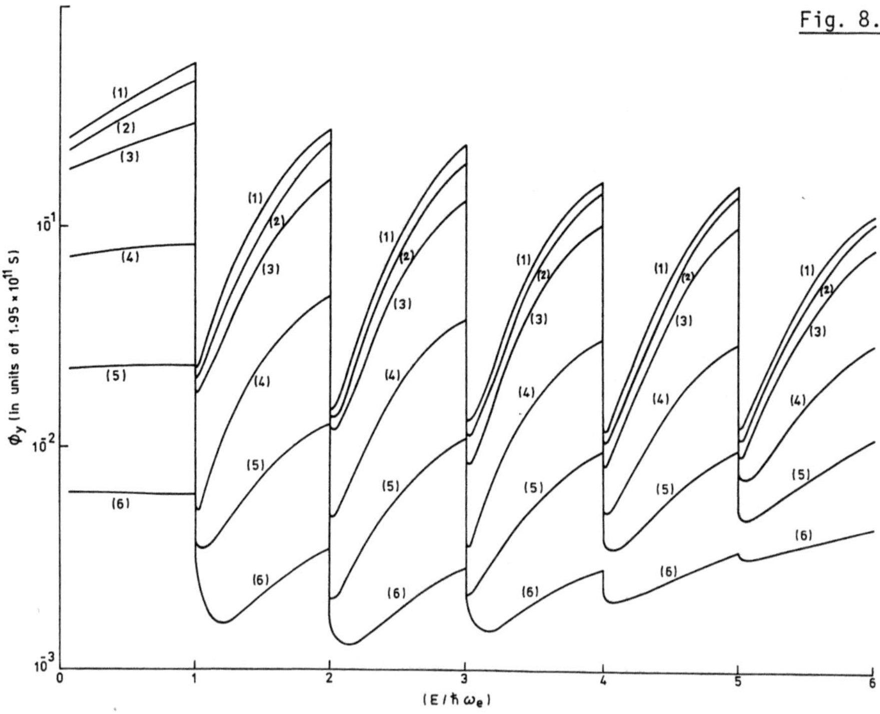

Fig. 8.3b

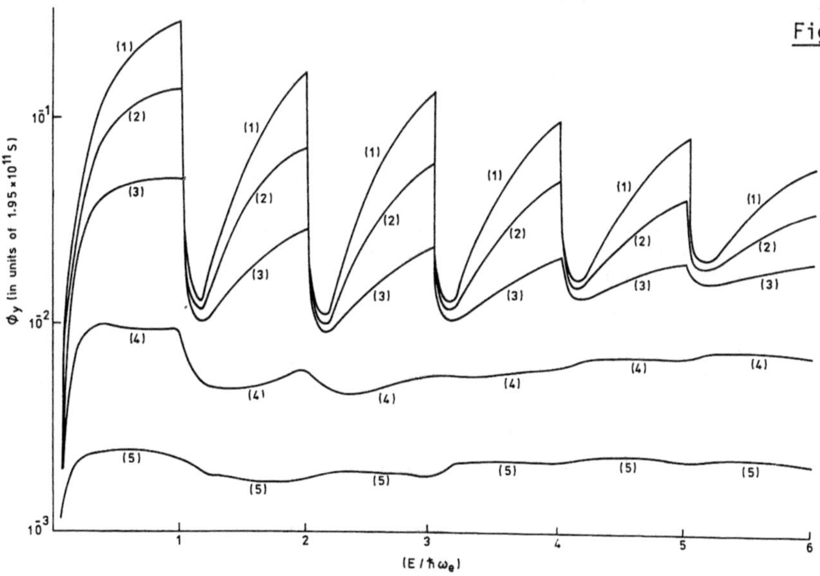

Fig. 8.3. Perturbation function ϕ_y. (a) Polar optic phonon scattering alone (1) B = 0; (2) B = 0.01; (3) B = 0.02; (4) B = 0.05; (5) B = 0.1; (6) B = 0.2; (b) Polar optic phonon, acoustic phonon, and impurity atom scattering (1) B = 0; (2) B = 0.01; (3) B = 0.02; (4) B = 0.05; (5) B = 0.1

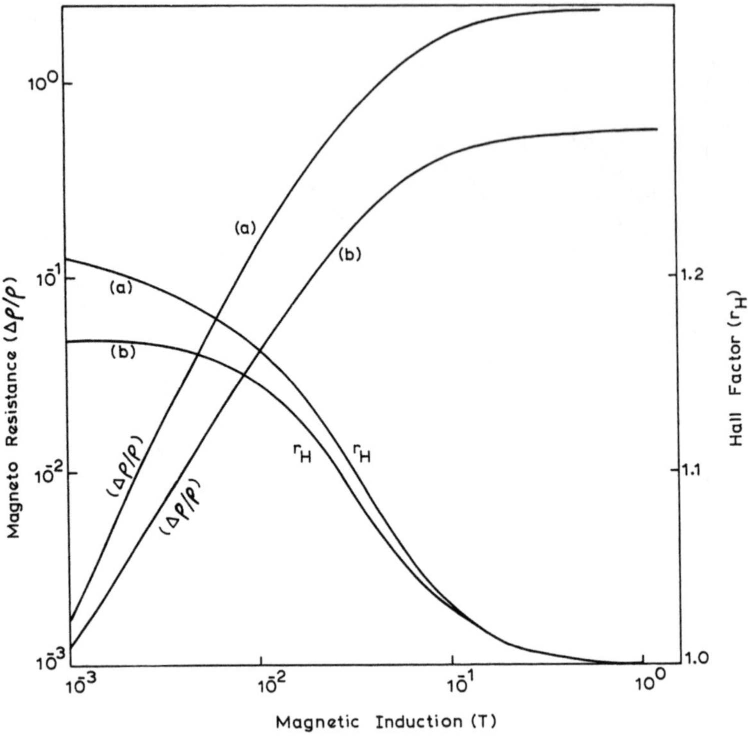

Fig. 8.4. Hall factor and magnetoresistance. (a) Polar optic phonon, acoustic phonon, and impurity atom scattering; (b) Polar optic phonon scattering only

Table 8.3. Hall factor and saturation magnetoresistance for polar optic phonon scattering

Lattice Temperature [K]	Optic phonon Temperature [K]	Hall factor	Saturation Magnetoresistance
77	100	1.11	0.23
	200	1.25	1.2
	300	1.14	2.5
	400	1.07	3.7
	500	1.03	4.8
300	100	1.1	0.08
	200	1.08	0.17
	300	1.09	0.17
	400	1.13	0.25
	500	1.21	0.42

We shall confine our attention to ellipsoidal parabolic band structure. As argued in connection with the conductivity (see Sect.8.2), the conductivity tensor for arbitrary directions of the fields may be obtained from the conductivity for electric and magnetic fields in the principal directions of the constant-energy ellipsoids. We assume that the electric field is in the x direction, which is one of the principal axes of the energy ellipsoids. The magnetic induction is, however, assumed to have all three components B_x, B_y, and B_z. Since the constant-energy ellipsoids are not, in general, orthogonally oriented, the magnetic field may not be completely transverse or longitudinal for all the ellipsoids and for the sake of generality it is convenient to assume that the magnetic induction is arbitrarity oriented. The Boltzmann equation for the assumed field directions has the form

$$(e\,\mathscr{E}_x/\hbar)(\partial f/\partial k_x) + (e/\hbar)^2(\underline{\nabla}_{\underline{k}}E\times\underline{B})\cdot\underline{\nabla}_{\underline{k}}f = \partial_c f \quad , \tag{8.81}$$

where

$$\underline{B} = \hat{\underline{i}}B_x + \hat{\underline{j}}B_y + \hat{\underline{k}}B_z \quad .$$

We transform the variables for the sake of simplification of the analysis by using the transformation relations

$$\mathscr{E}_x^* = (m_0/m_x)^{\frac{1}{2}}\mathscr{E}_x \quad , \tag{8.82}$$

$$\underline{k}^* = \underline{\underline{\alpha}}\cdot\underline{k} \quad , \quad \underline{B}^* = \underline{\underline{\alpha}}_j\cdot\underline{B} \quad , \tag{8.83}$$

where

$$\underline{\underline{\alpha}} = \begin{pmatrix} (m_0/m_x)^{\frac{1}{2}} & 0 & 0 \\ 0 & (m_0/m_y)^{\frac{1}{2}} & 0 \\ 0 & 0 & (m_0/m_z)^{\frac{1}{2}} \end{pmatrix} , \tag{8.84}$$

$$\underline{\underline{\alpha}}_j = \begin{pmatrix} m_0/(m_y m_z)^{\frac{1}{2}} & 0 & 0 \\ 0 & m_0/(m_z m_x)^{\frac{1}{2}} & 0 \\ 0 & 0 & m_0/(m_x m_y)^{\frac{1}{2}} \end{pmatrix} . \tag{8.85}$$

m_x, m_y, and m_z are the effective masses for the directions of the principal axes of the ellipsoids.

We also assume that the distribution function may be written as

$$f = f_0 - (e\mathscr{E}_x^*\hbar/m_0)\partial_E f_0\Big(k_x^*\phi_x(E)+k_y^*\phi_y(E)+k_z^*\phi_z(E)\Big) \quad . \tag{8.86}$$

It may be recalled that when the effective mass is anisotropic the perturbation in the distribution is such that a number of spherical harmonic terms are required to represent $\phi(k)$ in terms of functions involving energy alone. However, the contributions [8.16] of the higher-order terms is less than 10%, and in order to keep the analysis simple we shall not include these terms. They may be included if necessary by extending the method of analysis discussed in relation to conductivity. We should also point out that in the presence of a magnetic field, the first-order term in the perturbation has three components, since the magnetic field introduces additional displacements in transverse direction to the electric field. In contrast, in the presence of the electric field alone, we had to consider only one first-order perturbation term. These three components of the first-order spherical harmonic terms are proportional to $\cos\theta_{\underline{k}*}$, $\sin\theta_{\underline{k}*}\cos\phi_{\underline{k}*}$, and $\sin\theta_{\underline{k}*}\sin\phi_{\underline{k}*}$, and these terms are written in (8.86) multiplied by k_x^*, k_y^*, and k_z^*.

Substituting (8.86) in (8.81) we get

$$\begin{aligned}&+k_x^*(e\hbar\mathscr{E}_x^*/m_0)\partial_E f_0 - (k_y^*\omega_z\phi_x-k_z^*\omega_y\phi_x+k_z^*\omega_x\phi_y-k_x^*\omega_z\phi_y\\&\quad+k_x^*\omega_y\phi_z-k_y^*\omega_x\phi_z)(e\hbar\mathscr{E}_x^*/m_0)\partial_E f_0 = \partial_c f \quad ,\end{aligned} \tag{8.87}$$

where $\mu_i = eB_i^*/m_0$, and $i \equiv x,y,z$.

The collision term $\partial_c f$ may also be expressed as

$$\begin{aligned}\partial_c f &= -(8\pi^3)^{-1}(m_d/m_0)^{3/2}(e\hbar\mathscr{E}_x^*/m_0)\partial_E f_0 V_c\\&\times\int[1-f_0(\underline{k}'*)][1-f_0(\underline{k}*)]^{-1}S(\underline{k}*,\underline{k}'*)\Big\{k_x^*[\phi_x(\underline{k}*)-(k_x'^*/k_x^*)\phi_x(\underline{k}'*)]\\&+k_y^*[\phi_y(\underline{k}*)-(k_y'^*/k_y^*)\phi_y(\underline{k}'*)]+k_z^*[\phi_z(\underline{k}*)-(k_z'^*/k_z^*)\phi_z(\underline{k}'*)]\Big\}d\underline{k}'* \ .\end{aligned} \tag{8.88}$$

Equating the coefficients of k_x^*, k_y^*, and k_z^* on the two sides of (8.87) we get

$$\begin{aligned}&L_c\phi_x - \omega_z\phi_y + \omega_y\phi_z = 1 \quad ,\\&\omega_z\phi_x + L_c\phi_y - \omega_x\phi_z = 0 \quad ,\\&-\omega_y\phi_x + \omega_x\phi_y + L_c\phi_z = 0 \quad ,\end{aligned} \tag{8.89}$$

where

$$L_c\phi_i = (8\pi^3)^{-1}(m_d/m_0)^{3/2}V_c\int[1-f_0(k'^*)][1-f_0(k^*)]^{-1}S(\underline{k}^*,\underline{k}'^*)$$
$$\times\,[\phi_i(\underline{k}^*)-(k_i'^*/k_i^*)\phi_i(\underline{k}'^*)]d\underline{k}'^* \quad , \quad (i\equiv x,y,z) \quad .$$

Equation (8.89) may be easily solved when the collision mechanism is such that $L_x\phi_i$ may be written as ϕ_i/τ, i.e. when an isotropic relaxation time may be defined. We discuss the properties of the coefficients for this case first.

a) Transport Coefficients for Isotropic Relaxation Time: Replacing the collision operator by $1/\tau$ and solving the equations, we get

$$\phi_x = \left[\left(1+\omega_x^2\tau^2\right)/\left(1+\omega_B^2\tau^2\right)\right]\tau \quad ,$$
$$\phi_y = \left[(-\omega_z+\omega_x\omega_y\tau)/(1+\omega_B^2\tau^2)\right]\tau^2 \quad ,$$
$$\phi_z = \left[(\omega_y+\omega_x\omega_z\tau)/(1+\omega_B^2\tau^2)\right]\tau^2 \quad , \tag{8.90}$$

where $\omega_B^2=\omega_x^2+\omega_y^2+\omega_z^2$. The current densities in the three directions are given by

$$J_i = -(2/3)ne^2\mathscr{E}_x\left(I_n m_x^{\frac{1}{2}}m_i^{\frac{1}{2}}\right)^{-1}\int_0^\infty\phi_i E^{3/2}\partial_E f_0 dE \quad , \tag{8.91}$$

where $I_n=\int_0^\infty E^{\frac{1}{2}}f_0 dE$. The conductivity components are thus

$$\sigma_{xx} = \left(ne^2/m_x\right)\left\langle\left[(1+\omega_x^2\tau^2)/(1+\omega_B^2\tau^2)\right]\tau\right\rangle \quad ,$$
$$\sigma_{xy} = \left(ne^2/m_x^{\frac{1}{2}}m_y^{\frac{1}{2}}\right)\left\langle\left[(-\omega_z+\omega_x\omega_y\tau)/(1+\omega_B^2\tau^2)\right]\tau^2\right\rangle \quad ,$$
$$\sigma_{xz} = \left(ne^2/m_x^{\frac{1}{2}}m_z^{\frac{1}{2}}\right)\left\langle\left[(\omega_y+\omega_x\omega_z\tau)/(1+\omega_B^2\tau^2)\right]\tau^2\right\rangle \quad , \tag{8.92}$$

where $\langle\;\rangle$ denotes the average as defined below,

$$\langle x\rangle = -(2/3)\int_0^\infty xE^{3/2}\partial_E f_0 dE/\int_0^\infty E^{\frac{1}{2}}f_0 dE \quad . \tag{8.93}$$

Replacing ω_x, ω_y, ω_z by the corresponding expressions in terms of the magnetic induction and effective mass components, and generalising the expressions we get for the components of the conductivity tensor

$$\sigma_{ii} = \left(ne^2/m_i\right)\left\langle\left[\left(1+e^2B_i^2\tau^2/m_jm_k\right)/\left(1+\omega_B^2\tau^2\right)\right]\tau\right\rangle \quad ,$$

$$\sigma_{ij} = (ne^2)\left\langle\left[\left(-\varepsilon_{ij}eB_k/m_im_j+e^2B_iB_j/m_im_jm_k\right)/\left(1+\omega_B^2\tau^2\right)\right]\tau^2\right\rangle \quad , \qquad (8.94)$$

where $\varepsilon_{ij}=\varepsilon_{jk}=\varepsilon_{ki}=1$; $\varepsilon_{ji}=\varepsilon_{kj}=\varepsilon_{ik}=-1$; and i,j,k represent respectively x, y, and z. Also,

$$\omega_B^2 = e^2\left[B_x^2/(m_ym_z)+B_y^2/(m_xm_z)+B_z^2/(m_xm_y)\right] \quad . \qquad (8.95)$$

However, when we consider ellipsoidal constant-energy surfaces, we have to take into account multiple valleys, which are symmetrically oriented in a cubic crystal. The total conductivity should be obtained by summing the contribution of the electrons from all the valleys. Since the valleys are differently oriented, the components of the fields in the principal directions have different values in different valleys. For the purpose of this summation it is convenient to choose some general directions as the reference coordinate system. Let this be the x,y,z coordinate system (see Fig.8.5) and let us assume that the electric field is along the x direction. Let the direction cosines of the magnetic induction be ℓ_B, m_B, n_B with respect to the x,y,z coordinates. Let the direction cosines of the principal axes of one of the constant-energy ellipsoids be (ℓ_1,m_1',n_1), (ℓ_2,m_2',n_2), and (ℓ_3,m_3',n_3).

The components of the electric field and magnetic induction along these directions are thus given by

$$B_1 = (\ell_B\ell_1+m_Bm_1'+n_Bn_1)B = \ell_{B1}B \quad ,$$

$$B_2 = (\ell_B\ell_2+m_Bm_2'+n_Bn_2)B = \ell_{B2}B \quad ,$$

$$B_3 = (\ell_B\ell_3+m_Bm_3'+n_Bn_3)B = \ell_{B3}B \quad , \qquad (8.96)$$

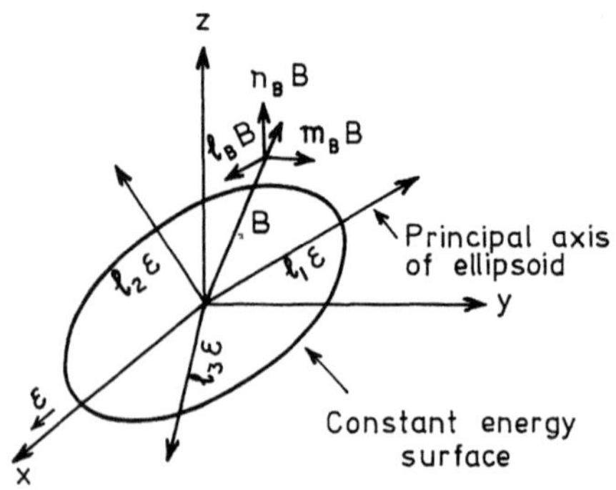

Fig. 8.5. Coordinate axes for the analysis of galvanomagnetic transport coefficients in a multivalley semiconductor

and

$$\mathscr{E}_1 = \ell_1\mathscr{E} \quad , \quad \mathscr{E}_2 = \ell_2\mathscr{E} \quad , \quad \mathscr{E}_3 = \ell_3\mathscr{E} \quad . \tag{8.97}$$

The current densities in the same directions are

$$J_i = \sum_j \sigma_{ij} \mathscr{E}_j \quad , \tag{8.98}$$

where the σ_{ij} are to be obtained using (8.94).

The current components may now be transformed to x,y,z coordinates to obtain

$$J_x = \sum_{i,j} \sigma_{ij}\ell_i\ell_j \mathscr{E}_x \quad ,$$
$$J_y = \sum_{i,j} \sigma_{ji}\ell_i m'_j \mathscr{E}_x \quad ,$$
$$J_z = \sum_{i,j} \sigma_{ji}\ell_i n_j \mathscr{E}_x \quad . \tag{8.99}$$

Thus the total conductivity tensor components are given by

$$\sigma_{xx} = \sum_v \left[\sum \sigma_{ij}\ell_i\ell_j\right]_v \quad ,$$
$$\sigma_{xy} = \sum_v \left[\sum \sigma_{ji}\ell_i m'_j\right]_v \quad ,$$
$$\sigma_{xz} = \sum_v \left[\sum \sigma_{ji}\ell_i n_j\right]_v \quad . \tag{8.100}$$

The subscript v indicates the value for a particular valley and the summation is to be carried out over all the valleys. It should be remembered that n, the carrier concentration in (8.94) should now represent the carrier concentration for a particular valley. If the total carrier concentration is n, the carrier concentration in each valley would be n/v (v is the number of valleys), as at low fields the carriers are equally distributed among the occupied valleys.

The method outlined above may be used to evaluate the conductivity tensor components, and then the transport coefficients, such as conductivity mobility, Hall mobility, and magnetoresistance using the formulae given earlier (see Sect.8.2.1) in connection with isotropic band structure. We may, however, note here some important properties of these parameters which are often useful in the analysis of experimental results of materials with ellipsoidal energy band structure. Confining our attention to low magnetic fields, so that a binomial expansion for $(1+\omega_B^2\tau^2)^{-1}$ is permissible, and retaining terms to second order in magnetic induction, we find that the current density may be expressed as

$$J_i = \sigma_{ij}\,\mathscr{E}_j + \sigma_{ij\ell}\,\mathscr{E}_j B_\ell + \sigma_{ij\ell m}\,\mathscr{E}_j B_\ell B_m \quad , \tag{8.101}$$

where

$$\sigma_{ij} = (\partial J_i/\partial\,\mathscr{E}_j)_{B\to 0} \quad ,$$
$$\sigma_{ij\ell} = \left(\partial^2 J_i/\partial\,\mathscr{E}_j \partial B_\ell\right)_{B\to 0} \quad ,$$
$$\sigma_{ij\ell m} = \left(\partial^3 J_i/\partial\,\mathscr{E}_j \partial B_\ell \partial B_m\right) \quad .$$

σ_{ij} is the zero-magnetic-field conductivity, $\sigma_{ij\ell}$ is the third-rank tensor associated with Hall effect and $\sigma_{ij\ell m}$ is the fourth-rank tensor giving the magnetoconductivity.

Following the method discussed earlier we may evaluate the components of each of these tensors. We find that for cubic materials with ellipsoidal band structure the conductivity mobility μ_d as well as the Hall mobility μ_H are scalar quantities given by

$$\mu_d = |e|\langle\tau\rangle(1/3)\left(m_1^{-1}+m_2^{-1}+m_3^{-1}\right) \quad ,$$
$$\mu_H = |e|(\langle\tau^2\rangle/\langle\tau\rangle)\left(m_1^{-1}m_2^{-1}+m_2^{-1}m_3^{-1}+m_1^{-1}m_3^{-1}\right)\left(m_1^{-1}+m_2^{-1}+m_3^{-1}\right)^{-1} \quad . \tag{8.102}$$

where m_1, m_2, and m_3 are the effective mass components along the directions of the principal axes of the constant-energy ellipsoids. In general, the band structure being spheroidal, we may reduce (8.102) to

$$\mu_d = |e|\langle\tau\rangle(3m_\perp)^{-1}(2+K^{-1}) \quad ,$$
$$\mu_H = |e|(\langle\tau^2\rangle/\langle\tau\rangle)m_\perp^{-1}(1+2K^{-1})/(2+K^{-1}) \quad , \tag{8.103}$$

where $m_\perp$ is the effective mass in the transverse direction and K is the ratio of the longitudinal and the transverse effective mass, often called the anisotropy ratio.

The magnetoresistance is, however, a tensor quantity, which may be expressed as [8.17]

$$M_{ijk}^{\ell mn} = \left[(\rho-\rho_0)/\rho_0 B^2\right]_{ijk}^{\ell mn} = b + c\Big(\sum_{\beta,\eta} \beta\eta\Big)^2 + d\sum_{\beta,\eta} \beta^2\eta^2 \quad , \tag{8.104}$$

where i,j,k and ℓ,m,n are respectively the directions of the current (in a constant-current magnetoresistance experiment) and the magnetic induction; β and η are the pairs of direction cosines of the current and the magnetic field for the same cubic axis.

The magnetoresistance coefficients for some particular directions of symmetry are then given by

$$M_{110}^{\bar{1}10} = b + d/2 \quad ; \quad M_{100}^{001} = b \quad ;$$
$$M_{\bar{1}10}^{111} = M_{\bar{1}11}^{110} = b + d/3 \quad ; \quad M_{\bar{1}10}^{11\bar{2}} = b + d/6 \quad ;$$
$$M_{112}^{112} = M_{110}^{110} = b + c + d/2 \quad ; \quad M_{100}^{100} = b + c + d \quad ;$$
$$M_{111}^{111} = b + c + d/3 \quad . \tag{8.105}$$

We may also show that for an isotropic material

$$d = 0 \quad ; \quad b + c = 0 \quad ;$$
$$b = \mu_H^2(A-1) \quad , \quad A = \langle\tau^3\rangle\langle\tau\rangle\langle\tau^2\rangle^{-2} \quad . \tag{8.106}$$

On the other hand, for [100]-type constant-energy spheroids (as in silicon)

$$b = \mu_H^2\left\{A[(K^2+K+1)(2K+1)K^{-1}(K+2)^{-2}]-1\right\} \quad ,$$
$$c = -\mu_H^2\left\{A[3(2K+1)/(K+2)^2]-1\right\} \quad ,$$
$$d = -\mu_H^2\left\{A[(K-1)^2(2K+1)K^{-1}(K+2)^{-2}]\right\} \quad . \tag{8.107}$$

Hence, $b+c+d=0$, $d<0$.
For [111]-type constant-energy spheroids (as in germanium),

$$b = \mu_H^2\left\{A[(2K+1)^2(1/3)K^{-1}(K+2)^{-1}]-1\right\} \quad ,$$
$$c = -\mu_H^2\left\{A[(2K+1)^2(1/3)K^{-1}(K+2)^{-1}]-1\right\} \quad ,$$
$$d = \mu_H^2\left\{A[2(K-1)^2(2K+1)(1/3)K^{-1}(K+2)^{-2}]\right\} \quad . \tag{8.108}$$

Hence, $b+c=0$, $d>0$.

We find from the above expressions that for a material with isotropic effective mass the longitudinal magnetoresistance is always zero and the transverse magnetoresistance is isotropic. For materials with [100] spheroids, both the transverse and the longitudinal magnetoresistance varies with the orientation of the current and the magnetic induction in relation to the crystallographic axes. The longitudinal magnetoresistance is zero for [100] directions and maximum for [111] directions. On the other hand, for [111] spheroid materials, there is no direction of zero magnetoresistance, and the longitudinal magnetoresistance is highest for the [111] directions.

b) Galvanomagnetic Transport Coefficients for Anisotropic Relaxation Time: We have seen earlier that in materials with anisotropic effective mass, the scattering probability is, in general, anisotropic due to the anisotropy of the coupling constants and also due to the dependence of $S(\underline{k},\underline{k}')$ on $|\underline{k}-\underline{k}'|$. The collision operator L_c would, therefore, be different even when relaxation-time formalism is applicable. If, however, the perturbation term in the distribution function is limited to $k_x\phi_x$, $k_y\phi_y$, and $k_z\phi_z$, we may express the collision operator as

$$L_c\phi_x = \phi_x/\tau_x \quad , \quad L_c\phi_y = \phi_y/\tau_y \quad \text{and} \quad L_c\phi_z = \phi_z/\tau_z \quad , \tag{8.109}$$

where now, for elastic collisions,

$$\frac{1}{\tau_x} = (3/4\pi)(m_d/m_0)^{3/2}(8\pi^3)^{-1}V_c\int\left\{\int[1-f_0(\underline{k}'^*)][1-f_0(\underline{k}^*)]^{-1}S(\underline{k}^*,\underline{k}'^*)\right.$$
$$\left.\times(1-k_x'^*/k_x^*)d\underline{k}'^*\right\}\cos^2\theta_{\underline{k}^*}d(\cos\theta_{\underline{k}^*})d\phi_{\underline{k}^*} \quad ,$$

and for randomising collision

$$\frac{1}{\tau_x} = (3/4\pi)(m_d/m_0)^{3/2}(8\pi^3)^{-1}V_c\int\left\{\int[1-f_0(\underline{k}'^*)][1-f_0(\underline{k}^*)]^{-1}S(\underline{k}^*,\underline{k}'^*)d\underline{k}'^*\right\}$$
$$\times\cos^2\theta_{\underline{k}^*}d(\cos\theta_{\underline{k}^*})d\phi_{\underline{k}^*} \quad .$$

Expressions for τ_y and τ_z are obtained by replacing $k_x'^*,k_x^*$ by $k_y'^*,k_y^*$ and $k_z'^*,k_z^*$ respectively, and replacing $\cos^2\theta_{\underline{k}^*}d(\cos\theta_{\underline{k}^*})d\phi_{\underline{k}^*}$ by $\sin^2\theta_{\underline{k}^*}d(\cos\theta_{\underline{k}^*})\cos^2\phi_{\underline{k}^*}d\phi_{\underline{k}^*}$ and $\sin^2\theta_{\underline{k}^*}d(\cos\theta_{\underline{k}^*})\sin^2\phi_{\underline{k}^*}d\phi_{\underline{k}^*}$ respectively.

On replacing the collision terms in (8.87) by these expressions, we get

$$\phi_x\tau_x^{-1} - \omega_z\phi_y + \omega_y\phi_z = 1 \quad ,$$
$$\omega_z\phi_x + \phi_y\tau_y^{-1} - \omega_x\phi_z = 0 \quad ,$$
$$-\omega_y\phi_x + \omega_x\phi_y + \phi_z\tau_z^{-1} = 0 \quad . \tag{8.110}$$

Solving the above equations and returning to B from B we find that the introduction of anisotropic relaxation time only replaces τ/m_i by τ_i/m_i. Thus the formulae derived in the preceding section may also be used in this case provided each τ/m_i term is replaced by τ_i/m_i. This modification complicates the evaluation of the final transport coefficients. However, when the energy dependence of the τ_i are identical but the magnitudes are different we find that the expressions given earlier remain generally valid but now K represents the ratio $(m_\parallel/m_\perp)(\tau_\perp/\tau_\parallel)$. The mass anisotropy factor is now multiplied by the scattering anisotropy factor to give the resultant anisotropy factor. It may be pointed out that analysis of the magnetoresistance results using (8.107) and (8.108) may be used to obtain K, but because of the above modification the value of K so obtained may not be the same as the mass anisotropy factor.

c) Galvanomagnetic Transport Coefficients for Inelastic and Nonrandomising Scattering: In polar materials, the predominant scattering mechanism being polar optic phonon scattering, the methods of analysis discussed above should not be relied upon to give accurate values of the galvanomagnetic transport coefficients. These coefficients are sensitive to the energy dependence of the scattering mechanism and, therefore, the effect of the collision term requires to be properly evaluated. The collision term for this case cannot be reduced to the simple forms. We get instead, following the procedure discussed in connection with conductivity,

$$S_{0x}\phi_x(E) - S_{+x}\phi_x(E+\hbar\omega_\ell) - S_{-x}\phi_x(E-\hbar\omega_\ell) - \omega_z\phi_y(E) - \omega_y\phi_z(E) = 1 \quad ,$$

$$-\omega_z\phi_x(E) + S_{0y}\phi_y(E) - S_{+y}\phi_y(E+\hbar\omega_\ell) - S_{-y}\phi_y(E-\hbar\omega_\ell) - \omega_x\phi_z(E) = 0 \quad ,$$

$$-\omega_y\phi_x(E) + \omega_x\phi_y(E) + S_{0z}\phi_z(E) - S_{+z}\phi_z(E+\hbar\omega_\ell) - S_{-z}\phi_z(E-\hbar\omega_\ell) = 0 \quad , \qquad (8.111)$$

where

$$S_{0x} = 3(4\pi)^{-1}(m_d/m_0)^{3/2}(8\pi^3)^{-1}V_c\int\Big\{\int[1-f_0(\underline{k}'^*)][1-f_0(\underline{k}^*)]^{-1}\Big(\sum_i S_i^{e\ell}(\underline{k}^*,\underline{k}'^*)$$
$$\times(1-k_x'^*/k_x^*)+\sum_{i=r,p}S_i(\underline{k}^*,\underline{k}'^*)\Big)d\underline{k}'^*\Big\}\cos^2\theta_{\underline{k}^*}\,d(\cos\theta_{\underline{k}^*})\,d\phi_{\underline{k}^*} \quad .$$

($S_i^{e\ell}$ indicates scattering probability for the ith kind of elastic scattering and the first summation is over all elastic scattering mechanisms. The second summation is over randomising and polar optic scattering.)

$$S_+(x) = 3(4\pi)^{-1}(m_d/m_0)^{3/2}(8\pi^3)^{-1}V_c\int\left\{\int[1-f_0(\underline{k}'^*)][1-f_0(\underline{k}^*)]^{-1}\right.$$
$$\left.\times S_{pe}(\underline{k}^*,\underline{k}'^*)(k_x'^*/k_x^*)d\underline{k}'^*\right\}\cos^2\theta_{\underline{k}^*}d(\cos\theta_{\underline{k}^*})d\phi_{\underline{k}^*} \quad ,$$

$$S_-(x) = 3(4\pi)^{-1}(m_d/m_0)^{3/2}(8\pi^3)^{-1}V_c\int\left\{\int[1-f_0(\underline{k}'^*)][1-f_0(\underline{k}^*)]^{-1}\right.$$
$$\left.\times S_{pa}(\underline{k}^*,\underline{k}'^*)(k_x'^*/k_x^*)d\underline{k}'^*\right\}\cos\theta_{\underline{k}^*}d(\cos\theta_{\underline{k}^*})d\phi_{\underline{k}^*} \quad .$$

S_{pe} and S_{pa} are the scattering probabilities for the emission and absorption respectively of a polar optic phonon.

The corresponding expressions with subscripts y and z for the two other directions are obtained by replacing $k_x^*,k_x'^*$ by $k_y^*,k_y'^*$ and $k_z^*,k_z'^*$ and replacing $\cos^2\theta_{\underline{k}^*}d(\cos\theta_{\underline{k}^*})d\phi_{\underline{k}^*}$ by $\sin^2\theta_{\underline{k}^*}d(\cos\theta_{\underline{k}^*})\cos^2\phi_{\underline{k}^*}d\phi_{\underline{k}^*}$ and $\sin^2\theta_{\underline{k}^*}d(\cos\theta_{\underline{k}^*})\sin^2\phi_{\underline{k}^*}d\phi_{\underline{k}^*}$ respectively. As in the case of conductivity or galvanomagnetic coefficients of isotropic polar materials we get a set of infinite equations relating the values of $\phi_i(E+\ell\hbar\omega_\ell)$ with $\phi_i(E+(\ell\pm1)\hbar\omega_\ell)$ and in this case we have three such unknowns. We may, however, solve these equations by extending the iteration method discussed earlier. In this method we first assume that $\phi_x(E\pm\hbar\omega_\ell)$, $\phi_y(E\pm\hbar\omega_\ell)$, and $\phi_z(E\pm\hbar\omega_\ell)$ are zero and solve the three equations simultaneously to obtain $\phi_x(E)$, $\phi_y(E)$, and $\phi_z(E)$. This method is repeated for the next set of equations to obtain $\phi_x(E+\hbar\omega_\ell)$, $\phi_y(E+\hbar\omega_\ell)$ and $\phi_z(E+\hbar\omega_\ell)$ assuming $\phi_x(E+2\hbar\omega_\ell)$, $\phi_y(E+2\hbar\omega_\ell)$, $\phi_z(E+2\hbar\omega_\ell)$ zero and substituting for $\phi_x(E)$, $\phi_y(E)$ and $\phi_z(E)$ the values obtained in the earlier step. This process is repeated for n steps to obtain values of ϕ_i up to $\phi_i(E+n\hbar\omega_\ell)$. The whole procedure is followed again, but this time while evaluating the $\phi_i(E+\ell\hbar\omega_\ell)$, the $\phi_i(E+(\ell+1)\hbar\omega_\ell)$ are not put equal to zero, but given the values obtained in the previous step of the iteration. It is usually found that after 8-10 iteration steps convergent values are obtained, which may be taken as solutions for ϕ_i and used in (8.91) to calculate the current density and then the various transport coefficients may also be calculated. This method may be extended to include higher-order terms in the perturbation term in the distribution function. For example, if we retain the third Legendre polynomial terms we get six equations instead of three, and we have to solve these equations following the procedure outlined above. One may readily understand that the computational work becomes prohibitive for the application of these methods. It is no wonder, therefore, that although the method of analysis is known, very few studies with this kind of analysis are reported. The interested reader may consult [8.13], where this kind of analysis, but using the matrix method, is reported for GaSb.

8.4 Transport Coefficients for Nonuniform Conditions

Transport coefficients have been discussed in previous sections assuming the semiconductor sample to be homogeneous and at uniform temperature. The position-dependent term $(1/\hbar)\nabla_k E \cdot \nabla_r f$ is absent from the Boltzmann equation under these conditions. On the other hand, for inhomogeneous samples or in samples in which the temperature is not uniform, the distribution function will be different in different parts of the sample and electron transport of a kind not yet considered may occur. The nonuniform concentration of carriers and the nonuniform temperature give rise to forces which make the carriers drift even in the absence of an external electric field. These forces originate from the kinetic pressure of the charge carriers as explained below.

It is well known that the pressure of an assembly of particles is related to the particle concentration and temperature. The pressure for particles obeying nondegenerate statistics is nk_BT, where n is the particle concentration, k_B is Boltzmann's constant, and T is the temperature. The expression for the pressure of particles obeying Fermi-Dirac statistics is

$$P = (2/3)n\langle E\rangle = \left[F_{3/2}(\eta)/F_{\frac{1}{2}}(\eta)\right]nk_BT \quad , \tag{8.112}$$

where

$$F_j(\eta) = [\Gamma(j+1)]^{-1}\int_0^\infty x^j[\exp(x-\eta)+1]^{-1}dx \quad ; \quad \eta = E_F(k_BT)^{-1} \quad .$$

In an inhomogeneous material E_F and hence η will vary with position and in a material with nonuniform temperature, T, and as a result E_F also, varies with position. In such material, therefore, P will be a function of position. Let us consider a variation in the direction x only. Then considering the particles in a volume Adx of the material (see Fig.8.6), we find that these particles experience a force

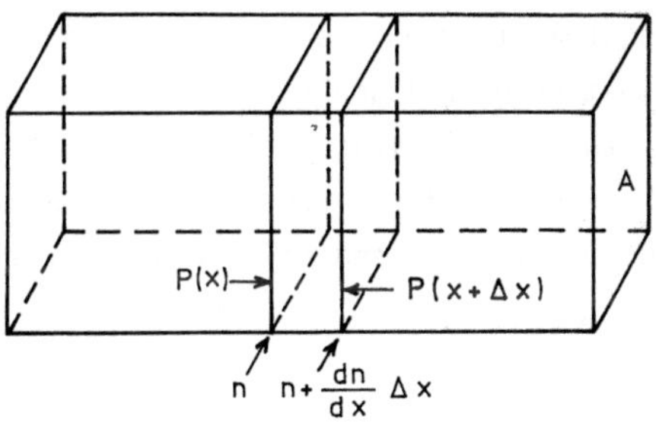

Fig. 8.6. Semiconductor sample with nonuniform carrier concentration

$$F = A[P(x)-P(x+dx)] = -A\frac{dP(x)}{dx}dx \quad . \tag{8.113}$$

Each individual particle, therefore, experiences a force

$$F_p = F/(nAdx) = -(1/n)\frac{dP}{dx} \quad . \tag{8.114}$$

We may also express dP/dx as

$$\frac{dP}{dx} = \left.\frac{\partial P}{\partial n}\right|_T \frac{dn}{dx} + \left.\frac{\partial P}{\partial T}\right|_n \frac{dT}{dx} \quad . \tag{8.115}$$

The force has, therefore, one component proportional to the concentration gradient and a second component proportional to the temperature gradient. The drift of carriers caused by the first component of the force is called diffusion. The second component of the force gives rise to effects, generally grouped together as thermoelectric effects. We shall derive in this section the formulae for transport coefficients related to these two effects.

8.4.1 Diffusion

The variation of concentration gradient may be produced due to inhomogeneity of the material, such as difference in impurity concentration. It may also be produced by external injection of charge carriers. As discussed above, if there is a variation of concentration, irrespective of how it is produced, the carriers at any position experience a force proportional to the concentration gradient at that point. The drift velocity resulting from this force may be written as

$$v_d = -D(\partial n/\partial x)n^{-1} \quad . \tag{8.116}$$

The constant of proportionality is called the diffusion coefficient or the diffusion constant. It is our object to derive a formula for it starting from the Boltzmann equation. We should note that when the charge carriers diffuse, an electric field appears as the charge neutrality condition is disturbed. This field is related to the excess charge through Poisson's equation. We shall assume, for the sake of generality, that an electric field also exists along with the concentration gradient. The electric field representing the sum total of the internal field generated by diffusion and any field applied externally is assumed to be $\mathscr{E}_x$.

The Boltzmann equation for diffusion should thus be taken as

$$-(1/\hbar)\underline{\nabla}_{\underline{k}}E \cdot \underline{\nabla}_{\underline{r}}f - (e\underline{\mathscr{E}}/\hbar) \cdot \underline{\nabla}_{\underline{k}}f + \partial_c f = \partial_t f \quad . \tag{8.117}$$

For the sake of simplicity, we shall confine our attention to isotropic bands which may be nonparabolic. We also assume, as stated earlier, that the carrier concentration varies in the x direction, and also that the external electric field is applied in the same direction. Equation (8.117) thus simplifies to

$$-[\hbar k_x/m^*(\underline{k})]\partial_x f - (e\mathscr{E}_x/\hbar)(\partial f/\partial k_x) + \partial_c f = \partial_t f \quad , \tag{8.118}$$

where $m^*(\underline{k})$ is the effective mass corresponding to the wave vector $\underline{k}$ being given by $m^*(k) = [(1/\hbar^2 k)(\partial E/\partial k)]^{-1}$. We note that the distribution function will be a function of E as well as of x and t. We also note that the electron concentration $n(x,t)$ at any point x at a time t is given by

$$n(x,t) = (4\pi^3)^{-1}\int f d\underline{k} \quad . \tag{8.119}$$

The electric field as well as the concentration gradient, as argued earlier, causes a perturbation in the distribution function, which has the maximum magnitude in the direction of the field and the concentration gradient, i.e. the x direction. As in the earlier sections, we may therefore seek a solution by expanding the distribution function in spherical harmonics, retaining terms up to the second, i.e. we assume that

$$f(x,\underline{k},t) = f_0(E,x,t) + k_x\phi(E,x,t) \quad . \tag{8.120}$$

Substituting in (8.118), and equating the even and odd terms, we get

$$-(2/3)\left(\frac{k}{2}\frac{\partial E}{\partial k}/\hbar\right)(\partial\phi/\partial x) - (2/3)(e\mathscr{E}_x/\hbar)E^{-\frac{1}{2}}\frac{\partial}{\partial E}(E^{3/2}\phi) + \partial_c f_0 = \partial_t f_0 \quad , \tag{8.121}$$

$$-[e\mathscr{E}_x\hbar/m^*(k)](\partial f_0/\partial E) - [\hbar/m^*(k)](\partial f_0/\partial x) + \partial_c\phi = \partial_t\phi \quad . \tag{8.122}$$

Further progress in the analysis may be made if we introduce simplifying assumptions. First, for small concentration gradients and small fields, we may assume that the form of f_0 is determined by the collision term alone and hence it has the same form as in the absence of concentration gradient. The functional form of f_0 is therefore either Maxwellian or Fermi-Dirac. Equation

(8.121) has to be satisfied on the average and f_0 is also required to satisfy the equation

$$\partial_t[(4\pi^3)^{-1}\int f_0 d\underline{k}] = -\partial_x[(1/4\pi^3\hbar)\int (2/3)E\phi d\underline{k}] \quad , \tag{8.123}$$

since the integrals of the second and third term on the left-hand side vanish identically. We also note that the average velocity of the carriers in the x direction is

$$\langle v_x \rangle = (1/n)(1/4\pi^3)\int (1/\hbar)(\partial E/\partial k_x) f d\underline{k} = (4\pi^3\hbar n)^{-1}(2/3)\int E\phi d\underline{k} \quad . \tag{8.124}$$

Equation (8.123) thus reduces to

$$\frac{\partial n}{\partial t} + \frac{\partial}{\partial x}(n\langle v_x \rangle) = 0 \quad , \tag{8.125}$$

which is the continuity equation. The diffusion problem has to be finally solved by solving this equation. It would require knowledge of $\langle v_x \rangle$ which again depends on $\mathscr{E}_x$ and $(\partial n/\partial x)$. Since $\mathscr{E}_x$ has an internal field component, we are also required to consider the Poisson equation along with this equation. In many problems the excess charge of both the kinds exist at the same time and the diffusion of the charge carriers of each kind affects that of the other. For a complete solution, we are therefore required to consider the continuity equation for the different kinds of charge carriers and Poisson equation simultaneously. We shall not consider this aspect of the problem, but refer the reader to [Ref.8.2, p.326] and [8.18], where a detailed discussion of the problem is given. We shall confine our attention here only to the relation between $\langle v_x \rangle$ and $\partial n/\partial x$.

An expression for the diffusion coefficient may be obtained by assuming open-circuit conditions in which the current is zero when steady-state conditions (i.e. $\partial n/\partial t$ or $\partial f/\partial t = 0$) are produced. However, in experiments the diffusion current flows along with the current produced by the field. We shall consider this dynamic condition to keep our derivation quite general. We shall, however, introduce another assumption, that $\partial_t\phi \ll \partial_c\phi$. The time scale of diffusion and hence of $\partial_t\phi$ is of the same order as the life time of minority carriers [8.2,18], which is not usually smaller than 10^{-9} s. The time scale of the collision term is, on the other hand, of the same order as that of the momentum relaxation time, which is not larger than 10^{-10} s. The assumption that $\partial_t\phi \ll \partial_c\phi$ is therefore satisfied for most experimental conditions. Let us also write

$$\phi(x,E) = -\phi_{\mathscr{E}}\,[e\hbar\,\mathscr{E}_x/m^*(0)](\partial f_0/\partial E) - \phi_D[\hbar/m^*(0)](\partial f_0/\partial x) \quad . \tag{8.126}$$

Equation (8.122) may therefore be separated into

$$L_c\phi = m^*(0)m^{*-1}(\underline{k}) \quad , \tag{8.127}$$

$$\partial_c\phi_D(\partial f_0/\partial x) = -m^*(0)m^{*-1}(\underline{k})(\partial f_0/\partial x) \quad , \tag{8.128}$$

where the operator L_c is the same as defined earlier. We note, however, that

$$\frac{\partial f_0}{\partial x} = \frac{\partial f_0}{\partial E_F}\frac{\partial E_F}{\partial x} = -\frac{\partial f_0}{\partial E}\frac{\partial E_F}{\partial x} \quad , \tag{8.129}$$

since $\partial f_0/\partial E_F = -\partial f_0/\partial E$ and the sample being at uniform temperature, only E_F in f_0 may vary with x and t.
Hence (see Sect.7.4),

$$\partial_c[\phi_D(\partial f_0/\partial x)] = -\frac{\partial E_F}{\partial x}\frac{\partial f_0}{\partial E}L_c\phi_D \quad . \tag{8.130}$$

Equation (8.128) thus also reduces to

$$L_c\phi_D = m^*(0)m^{*-1}(\underline{k}) \quad . \tag{8.131}$$

Thus ϕ_D is identical to $\phi_{\mathscr{E}}$ and need not be evaluated separately. It is the same function as ϕ, discussed earlier in connection with the conductivity mobility.

The component of current arising from $\partial f_0/\partial x$ is given by

$$\begin{aligned} J_d &= -(e/4\pi^3)\int m^{*-1}(0)(\partial E/\partial k_x)k_x\phi_D(\partial f_0/\partial x)d\underline{k} \\ &= ne(\partial E_F/\partial x)\int m^{*-1}(0)(\partial E/\partial k_x)k_x\phi_D(\partial f_0/\partial E)d\underline{k}(\int f_0 d\underline{k})^{-1} \\ &= -n\mu_d(\partial E_F/\partial x)(e/|e|) \quad , \end{aligned} \tag{8.132}$$

since

$$\mu_d = -|e|\int m^{*-1}(0)(\partial E/\partial k_x)k_x\phi_{\mathscr{E}}\,(\partial f_0/\partial E)d\underline{k}(\int f_0 d\underline{k})^{-1} \tag{8.133}$$

and

$$\phi_D = \phi_{\mathscr{E}} \quad . \tag{8.134}$$

The diffusion velocity may thus be written as

$$\langle v_d \rangle = J_d(ne)^{-1} = -(\mu_d k_B T)|e|^{-1}\,(\partial n/\partial \eta)^{-1}(\partial \eta/\partial x) \quad . \tag{8.135}$$

The diffusion constant is thus [see (8.116)]

$$D = \mu_d k_B T|e|^{-1} n(\partial n/\partial \eta)^{-1} \quad . \tag{8.136}$$

It should be evident from the deduction given above that the above relation should be valid for all scattering mechanisms and for all conditions of inhomogeneity provided the temperature is uniform. We also note that it is not necessary to assume that the scattering mechanism is such that a relaxation time is applicable, as has been done in the literature [8.19] to arrive at this result in the presence of an electric field.

Relation (8.136) may be reduced to the following form for nondegenerate semiconductors:

$$D/\mu_d = k_B T/|e| \quad , \tag{8.137}$$

since $n(\partial n/\partial \eta)^{-1}$ is equal to unity. The relation in this form was first deduced in connection with the diffusion of gas particles and is usually termed the NERNST-TOWNSEND-EINSTEIN [8.20] or simply the Einstein relation.

We may also obtain a simple form of the relation for degenerate semiconductors with isotropic parabolic bands. Since in this case [see (5.20)]

$$n = N_c F_{\frac{1}{2}}(\eta)$$

$$D/\mu_d = (k_B T/|e|)\left[F_{\frac{1}{2}}(\eta)/F_{-\frac{1}{2}}(\eta)\right] \quad . \tag{8.138}$$

This form of the relation was first given by LANDSBERG [8.21], and is often referred to as the generalised Einstein relation.

We note that the value of the ratio depends on the lattice temperature and the relation between the electron concentration and the Fermi level which again depends only on the band structure. The exact form of the relation for different band structures have been studied extensively [8.22]. We give below the forms for two nonparabolic band structures which will be of interest in our future discussion.

We get for nonparabolic bands obeying the KANE relation [8.23]

$$D/\mu_d = (k_BT/|e|)\left[F_{\frac{1}{2}}(\eta)+(15/4)k_BT\alpha F_{3/2}(\eta)\right]\left[F_{-\frac{1}{2}}(\eta)+(15/4)k_BT\alpha F_{\frac{1}{2}}(\eta)\right]^{-1} . \qquad (8.139)$$

On the other hand, the relation for superlattice structures is [8.24]

$$\frac{D}{\mu_d} = \frac{k_BT}{|e|}\int_0^1 \ln\left[1+\exp(\eta-\varepsilon_0+\varepsilon_1 \cos\pi y)\right]dy \Big/ \int_0^1\left[1+\exp(\varepsilon_0-\eta-\varepsilon_1 \cos\pi y)\right]^{-1}dy \quad , \qquad (8.140)$$

where $\varepsilon_0 = E_{01}/k_BT$ and $\varepsilon_1 = E_{11}/k_BT$, ε_0 and ε_1 being the band-centre energy and the bandwidth of the first miniband. It is interesting to note that if D/μ_d could be determined experimentally, we could get information about the band gap in nonparabolic materials for superlattice structures. However, no such experimental results are as yet available.

We find from the relations given above that the conductivity mobility and the diffusivity are given by the same transport integral. If the band structure is known one constant may be obtained from the knowledge of the other without knowing the details of the scattering mechanism. Diffusion plays a major role in many semiconductor devices. Since the diffusion coefficient cannot be easily determined experimentally, its value is often obtained from the experimental values of mobility using the relation given here. On the other hand, theoretical values of the diffusion constant may be obtained conveniently using the Monte Carlo method (see Sect.11.7.3). The values of mobility may then be obtained from the diffusion constant using the same relation.

8.4.2 Thermal Transport Coefficients

We consider a homogeneous sample with a temperature gradient. It has been explained earlier that the carriers experience a force due to the pressure gradient produced by the temperature gradient of magnitude $(1/n)(\partial P/\partial T)(\partial T/\partial x)$. This force is balanced in open circuit by the internal field produced by the redistribution of charges. On the other hand, when an external voltage is applied, the resulting current flow is associated with a heat flow. These effects are usually discussed in terms of three thermoelectric transport coefficients: the Seebeck coefficient, the Peltier coefficient, and the Thompson coefficient, which are related to effects bearing the same names. We first introduce these coefficients.

a) Seebeck Coefficient - Thermoelectric Power: Let us consider a semiconductor sample to the ends of which two identical metallic wires are connected through ohmic contacts. Let the ends of the semiconductor sample (Fig.8.7) be main-

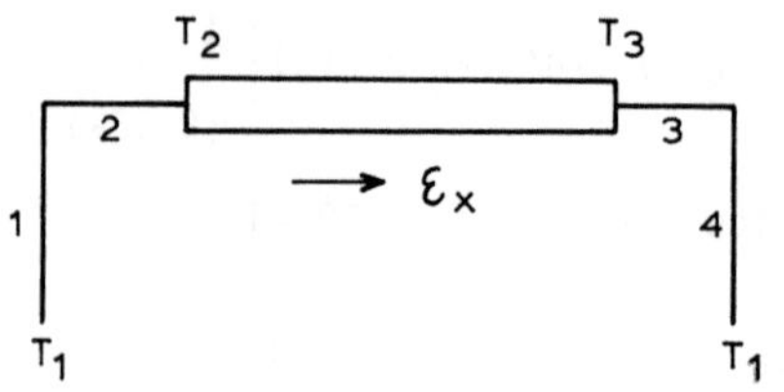

Fig. 8.7. Temperature distribution in the sample for thermoelectric power experiment

tained at temperatures T_2 and T_3 and let the temperature of the open ends of the metallic wire be T_1. It is evident from the preceding discussion that due to the temperature gradient a position dependent electric field is produced in addition to a gradient in the Fermi energy. All these three driving forces, the electric field, gradient in Fermi energy and the temperature gradient cause the carriers to drift and hence a current. The total current may thus be phenomenologically described by

$$J = \sigma\left[\mathscr{E}_x-(1/e)(\partial E_F/\partial x)-Q(dT/dx)\right] \quad , \tag{8.141}$$

where σ is the conductivity, $\mathscr{E}_x$ is the electric field along the semiconductor sample and the metal wires, E_F is the Fermi energy at any point x, and dT/dx is the temperature gradient. Under open-circuit conditions, $J=0$ and we get for the electric field

$$\mathscr{E}_x = (1/e)(\partial E_F/\partial x) + Q(dT/dx) \quad . \tag{8.142}$$

A potential difference results between the ends of the metal wires due to this field and this effect, i.e. the production of electromotive force by the temperature difference between the junctions of two materials is the Seebeck effect.

The emf between the ends of the wires is

$$\begin{aligned} V_{14} &= -\int_1^4 \mathscr{E}_\ell d\ell = \int_1^4 -\left[(1/e)(\partial E_F/\partial \ell)+Q(dT/d\ell)\right]d\ell \\ &= -e^{-1}\left[E_F(\ell)\right]_1^4 + \int_1^4 Q(T)(dT/d\ell)d\ell = \int_{T_2}^{T_3} \left[Q_1(T)-Q_2(T)\right]dT \quad , \end{aligned} \tag{8.143}$$

where 1,2,3,4 denote the values of ℓ for the points marked in Fig.8.7 with these numbers; Q_1 and Q_2 represent the values of Q (a function of temperature) respectively for the semiconductor and the metal. We note that the Fermi-level at the points 1 and 4 being identical, the first term is zero.

The Seebeck coefficient, more commonly called the thermoelectric power, is defined as

$$\alpha_{ms} = V_{14}/(T_3-T_2) \quad . \tag{8.144}$$

It is taken as positive if the high-temperature end is positive.

The thermoelectric power is defined above in relation to the two materials forming the junctions. We may, however, define an absolute thermoelectric force characteristic of a particular material. It is defined as

$$\Theta = \int_0^T Q(T)dT \quad . \tag{8.145}$$

The thermoelectric power between two materials is the difference between the absolute thermoelectric forces of the two materials. We also find that the absolute thermoelectric power is given by

$$\frac{d\Theta}{dT} = Q \quad . \tag{8.146}$$

Thus, the thermoelectric power of a material is the current density for a unit temperature gradient normalised by the conductivity.

It should be mentioned that the thermoelectric power of semiconductors obtained by using junctions between a semiconductor and a metal may be taken to represent the absolute thermoelectric power of a semiconductor, since the thermoelectric power of metals is negligible in comparison to that of semiconductors.

b) Peltier Coefficient: Let us assume that a source of voltage is connected between the terminals 1 and 4 of the experimental circuit shown in Fig.8.7. It will be found that due to the current produced by this voltage, heat will be absorbed from the ambient at one junction and in the other junction heat will be given out to the ambient. There is cooling at one junction and heating at the other. This effect is known as the Peltier effect. The Peltier coefficient is defined as

$$\Pi_{12} = W/J \quad , \tag{8.147}$$

where W is the heat flux due to a current density J for a temperature difference $T_1 - T_2$ between the junctions.

Although the Peltier and Seebeck effects were discovered at different times and are known individually, the effects are results of the same electromotive force generated by the temperature gradient. A general relation between the two coefficients was derived thermodynamically by Thompson. The relation is

$$\Pi_{12} = T\alpha_{12} \quad , \tag{8.148}$$

where T is the temperature at which the two coefficients are measured. In deriving this result, Thompson predicted also a new effect, now known by his name.

c) Thompson Effect: It was predicted that when a current flows through a semiconductor in which a temperature gradient is maintained, a heat flux is also produced. The Thompson coefficient is defined as the heat flux for unit current density and unit temperature gradient. It is given by

$$\mu = W/J(dT/dx) \quad , \tag{8.149}$$

where μ is the heat flux, J is the current density, and dT/dx is the temperature gradient.

The Thompson coefficient is also related to the thermoelectric power by the relation

$$\mu = -T\frac{d}{dT}\left(\frac{d\Theta}{dT}\right) \quad . \tag{8.150}$$

We find that although three coefficients are related to thermal transport, the coefficients are interrelated. Knowledge of one coefficient is sufficient to derive the others. The thermoelectric power is most easily measured, and most of the available results are about this coefficient. We shall, therefore, focus our attention to this coefficient alone.

8.4.3 Formula for Thermoelectric Power

The thermoelectric power, as discussed in the preceding section, is given by the current density for unit temperature gradient normalised by the conductivity. In order to evaluate the current density for a unit temperature gradient we take the Boltzmann equation as

$$-(1/\hbar)\underline{\nabla}_{\underline{k}}E \cdot \underline{\nabla}_{\underline{r}}f - (e\underline{\mathscr{E}}/\hbar) \cdot \underline{\nabla}_{\underline{k}}f + \partial_c f = \partial_t f \quad . \tag{8.151}$$

The thermoelectric power is measured by producing null conditions and in equilibrium we may assume that the current is zero. The time derivative $\partial_t f$ may, therefore, be put equal to zero. The temperature gradient is assumed to be along the x direction. The electric field which develops due to the transport of the electrons is also in the x direction. The Boltzmann equation may, therefore, be simplified to

$$-(1/\hbar)(\partial E/\partial k_x)(\partial f/\partial x) - (e\,\mathcal{E}_x/\hbar)(\partial f/\partial k_x) + \partial_c f = 0 \quad . \tag{8.152}$$

We now assume that the solution for f is given by

$$f = f_0(E,E_F,T) + k_x\phi(E,E_F,T) \quad , \tag{8.153}$$

where f_0 is the even part and $k_x\phi$ is the odd part of f.

Let us confine our attention to isotropic bands, which may, however, be nonparabolic. The derivatives with respect to k_x may thus be written as

$$\partial E/\partial k_x = \hbar^2 k_x/m^*(E) \quad \text{and} \quad \partial f_0/\partial k_x = \left[\hbar^2 k_x/m^*(E)\right](\partial f_0/\partial E) \quad , \tag{8.154}$$

where $m^*(E)$ is the effective mass corresponding to the energy E and the wave vector $\underline{k}$, the x component of which is k_x. It is given for the Kane-type nonparabolic bands by

$$m^*(E) = m^*(0)(1+2\alpha E) = m^*(0)\gamma'(E) \quad ; \quad \alpha = E_g^{-1} \quad ; \quad \gamma = E(1+\alpha E) \quad , \tag{8.155}$$

where $m^*(0)$ is band edge effective mass and E_g is the band gap energy.

Substituting (8.154) in (8.152) and equating the even and odd parts on the two sides we get

$$\left[\hbar k_x^2/m^*(E)\right](\partial\phi/\partial x) + (e\,\mathcal{E}_x/\hbar)\left\{\phi_0+\hbar^2 k_x^2[m^*(E)]^{-1}(\partial\phi/\partial E)\right\} = \partial_c f_0 \quad , \tag{8.156}$$

$$[\hbar/m^*(E)](\partial f_0/\partial x) + e\,\mathcal{E}_x\hbar[m^*(E)]^{-1}(\partial f_0/\partial E) = \partial_c\phi \quad . \tag{8.157}$$

The temperature gradient is assumed to be small so that the first two terms of (8.156) may be considered negligible and the equation may be simplified to $\partial_c f_0 = 0$. The even part of the distribution function is then the equilibrium Fermi function, but the temperature and consequently the Fermi energy will vary with position. The x derivative of f_0 may be expanded as

$$\partial_x f_0 = (\partial_T f_0)_{E_F}\partial_x T + (\partial f_0/\partial E_F)_T \partial_x E_F \quad . \tag{8.158}$$

But,

$$\partial_T f_0 = -(E-E_F)/T \cdot \partial_E f_0 \quad , \tag{8.159}$$

$$\partial f_0/\partial E_F = -\partial_E f_0 \quad . \tag{8.160}$$

Equation (8.157) may thus be written as

$$\partial_c \phi - e\hbar[m^*(E)]^{-1}[\mathscr{E}_x-(1/e)(\partial E_F/\partial x)]\partial_E f_0$$
$$+ \hbar[m^*(E)]^{-1}[(E-E_F)/T](\partial_E f_0)(\partial_x T) = 0 \quad . \tag{8.161}$$

We may separate this equation into two equations putting

$$\phi = -e\hbar[m^*(0)]^{-1}[\mathscr{E}_x-(1/e)(\partial E_F/\partial x)](\partial_E f_0)\phi_1$$
$$- \hbar[m^*(0)]^{-1}(1/T)(\partial T/\partial x)(\partial_E f_0)\phi_2 \quad . \tag{8.162}$$

We get

$$L_c\phi_1 = 1/\gamma'(E) \quad , \tag{8.163}$$

$$L_c\phi_2 = (E-E_F)/\gamma'(E) \quad , \tag{8.164}$$

where L_c is the collision operator defined earlier (see Sect.7.6).

The current density may also be expressed in terms of ϕ_1 and ϕ_2 as follows:

$$J = e(4\pi^3)^{-1}(1/\hbar)\int(\partial E/\partial k_x) f d\underline{k}$$
$$= -(2/3)(ne^2/I_n)[m^*(0)]^{-1}\left\{\left[\mathscr{E}_x - e^{-1}(\partial E_F/\partial_x)\right]\int_0^\infty \phi_1\gamma^{3/2}\partial_E f_0 dE\right.$$
$$\left. - (eT)^{-1}(\partial T/\partial x)\int_0^\infty \phi_2\gamma^{3/2}\partial_E f_0 dE \right. \quad , \tag{8.165}$$

where

$$I_n = \int_0^\infty f_0 \gamma^{\frac{1}{2}}\gamma' dE \quad . \tag{8.166}$$

Comparing (8.165) with (8.141) we find that the thermoelectric power

$$Q = (eT)^{-1}\left[\int_0^\infty \phi_2\gamma^{3/2}\partial_E f_0 dE \Big/ \int_0^\infty \phi_1\gamma^{3/2}\partial_E f_0 dE\right] \quad . \tag{8.167}$$

We note that (8.163) is the same as that obtained for the evaluation of conductivity mobility. It may, therefore, be solved by the techniques already discussed. We are, however, required to solve (8.164) in addition, for evaluating Q. The same methods as discussed for (7.35) may be used for this purpose. However, the integral $\int_0^\infty \phi_2\gamma^{3/2}\partial_E f_0 dE$ may be expressed in terms of ϕ_1 as follows:

$$\int_0^\infty \phi_2\gamma^{3/2}\partial_E f_0 dE = \int_0^\infty (L_c\phi_1)\phi_2\gamma'\gamma^{3/2}\partial_E f_0 dE$$

$$= \int_0^\infty \phi_1(L_c\phi_2)\gamma'\gamma^{3/2}\partial_E f_0 dE = \int_0^\infty \phi_1(E-E_F)\gamma^{3/2}\partial_E f_0 dE \quad .$$

We obtain the second step by using the self-adjoint property of the collision operator L_c. It is seen that the thermoelectric power may be evaluated from the values of ϕ_1 which is the same as the perturbation function for the drift mobility and we are not required to solve any additional equation.

We discuss below the formulae for thermoelectric power for different kinds of scattering.

a) Relaxation-Time Scattering: Solutions for this case may be obtained straightaway since

$$L_c\phi_2 = \phi_2/\tau \quad . \tag{8.168}$$

We obtain from (8.164)

$$\phi_2 = \tau(E-E_F)/\gamma'(E) \quad . \tag{8.169}$$

Note also that

$$\phi_1 = \tau/\gamma'(E) \quad . \tag{8.170}$$

The thermoelectric power Q is thus given by

$$Q = (eT)^{-1}\left[\langle E\tau\rangle\langle\tau\rangle^{-1} - E_F\right] \quad , \tag{8.171}$$

where $\langle\tau\rangle$ is defined as

$$< \tau > = I_n^{-1} \int_0^{\infty} \tau \gamma'^{-1} \gamma^{3/2} \partial_E f_0 dE \quad . \tag{8.172}$$

In degenerate semiconductors we may approximate $\partial_E f_0$ by $(k_B T)^{-1} \delta(E-E_F)$ for extreme degeneracy. The thermoelectric power turns out to be zero in the first approximation, since $< E\tau >/< \tau > = E_F$. If however, we retain the second-order terms, we get a nonzero value given by [Ref.8.25, p.315]

$$Q \simeq (\pi^2/3) k_B^2 T/(eE_F) = -2.45 \times 10^{-8} T/E_F \quad , \tag{8.173}$$

where E_F is in units of eV.

It is of interest to note that the value of E_F is of the order of a few electron volts in metals. Assuming a value of $E_F = 10$ eV, we find that the thermoelectric power is about 10^{-6} eV/K at 300 K. The thermoelectric power of metals, as mentioned earlier may thus be considered negligible compared to that of semiconductors, which would be of the order of 10^{-2} eV/K. We may also conclude from the above discussion that, in general, the thermoelectric power of a semiconducting material will decrease with increasing degeneracy.

For nondegenerate semiconductors the averages may be easily worked out for relaxation time τ varying as E^{-s}. We then get

$$Q = (k_B/e)[(5/2)-s-E_F/(k_B T)] \quad . \tag{8.174}$$

It is evident from the formula that Q has the same sign as that of e, since in nondegenerate materials E_F is negative and $|s|$ is less than 5/2. This result is often used to determine the sign of the majority carriers, i.e. the type of a semiconductor.

Relation (8.174) is also used to determine the Fermi level. The effective mass of the carriers is then obtained by combining the knowledge of the Fermi level with that of the carrier concentration. The value of s is usually taken to be 1/2 for this purpose. This assumption would be valid if the predominant scattering mechanism is deformation potential acoustic phonon scattering, which is not however generally true.

We should note that $(E_F/k_B T)$ at temperatures near room temperature is of the same order as (5/2-s). Hence, the nature of scattering which determines the value of s, significantly affects the thermoelectric power of nondegenerate semiconductors. Theoretical interpretation of the results on thermoelectric power, therefore, requires a more elaborate analysis. In the case of mixed scattering, described by relaxation time, such an analysis may be based

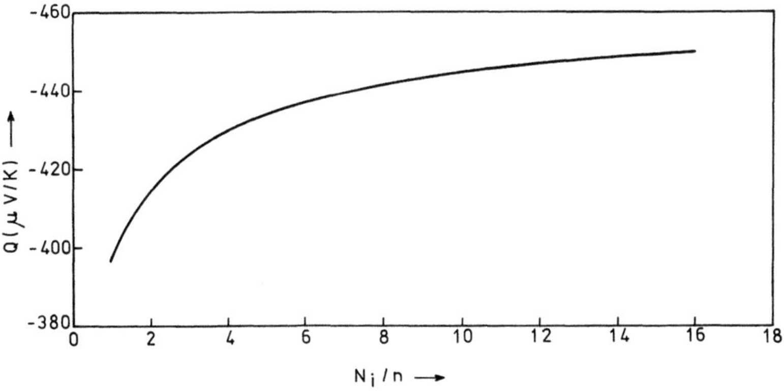

Fig. 8.8. Thermoelectric power for a sample with a fixed electron concentration but different impurity content

on (8.171). We are only required to evaluate the averages by considering the actual scattering mechanisms. We illustrate in Fig.8.8 the variation of thermoelectric power with the variation of the impurity concentration for a material with fixed electron concentration, i.e. the value of E_F, assuming acoustic phonon and impurity scattering. It makes evident the dependence of Q on the relative importance of the two kinds of scattering. In compound semiconductors, the polar optic phonon scattering being often the dominant scattering mechanism formula (8.174) cannot be directly applied. For the evaluation of Q for this case we are required to extend the iterative and other methods of analysis discussed for electric transport. We discuss in the following section how the iterative method may be extended for this case.

b) Polar Optic Phonon Scattering: Equation (8.164) has to be solved to obtain ϕ_2 when polar optic phonon scattering predominates. We may split the equation into two equations putting $\phi_2 = \phi_3 + \phi_4$ where

$$L_c\phi_3 = -[\gamma'(E)]^{-1}E_F \quad , \tag{8.175}$$

$$L_c\phi_4 = [\gamma'(E)]^{-1}E \quad . \tag{8.176}$$

Evidently (8.175) is the same as $-E_F\phi_1$ where ϕ_1 is the solution of (8.163). Equation (8.176) may also be solved by using the same iteration technique. Expressing the collision operator in detail we get (see Sect.7.5.2)

$$S_0\phi_4(E+\ell\hbar\omega_\ell) - S_+\phi_4(E+(\ell+1)\hbar\omega_\ell) - S_-\phi_4(E+(\ell-1)\hbar\omega_\ell) = (E+\ell\hbar\omega_\ell)[\gamma'(E+\ell\hbar\omega_\ell)]^{-1} \quad . \tag{8.177}$$

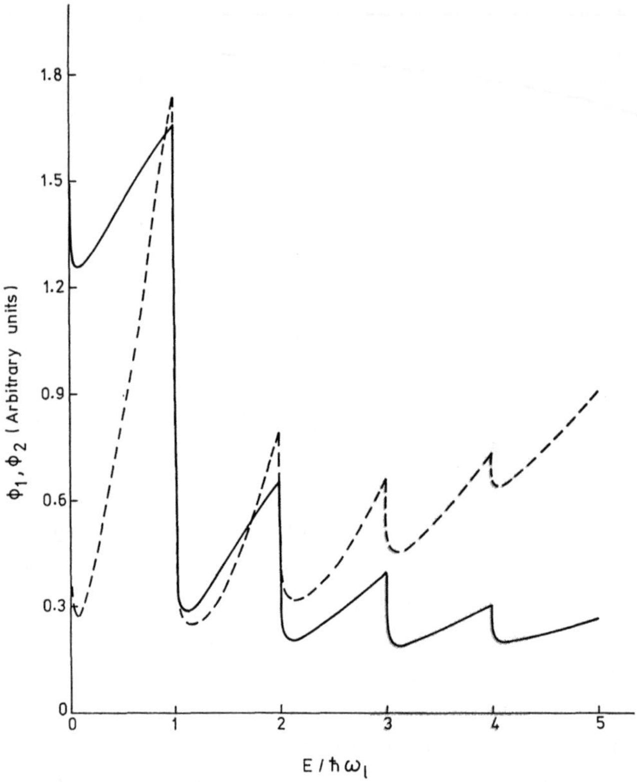

Fig. 8.9. Perturbation functions for thermoelectric power experiment. Solid line — ϕ_1; dotted line — ϕ_2

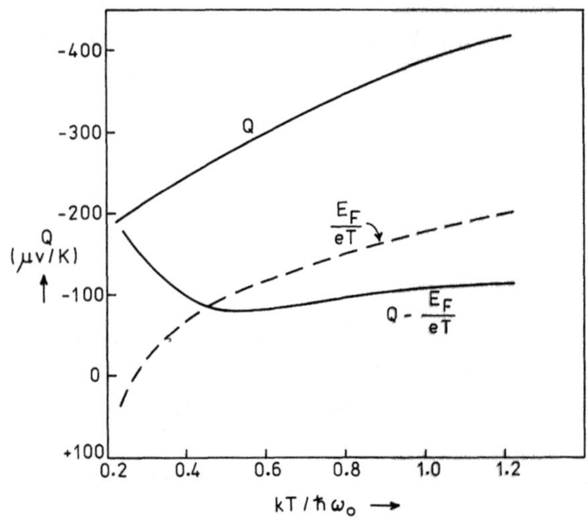

Fig. 8.10. Thermoelectric power for polar optic phonon scattering

The above equation is a typical member of the set of infinite equations obtained by giving ℓ integral values starting from zero. The set has to be solved simultaneously to obtain the function ϕ_4. The iterative procedure as discussed earlier for solving (7.35) may be used for this purpose. Values for the first iterative step are obtained by putting $\phi_4(E+(\ell\pm1)\hbar\omega_\ell)$ equal to zero in each equation. The first set of values are thus given by

$$\phi_4(E+\ell\hbar\omega_\ell) = (E+\ell\hbar\omega_\ell)[S_0(E+\ell\hbar\omega_\ell)]^{-1}[\gamma'(E+\ell\hbar\omega_\ell)]^{-1} \quad . \tag{8.178}$$

Values of $\phi_4(E+\ell\hbar\omega_\ell)$ in successive steps are then obtained by putting in the equations values of $\phi_4(E+(\ell\pm1)\hbar\omega_\ell)$ obtained in the previous step. We illustrate in Fig.8.9 the function obtained by applying this procedure for polar optic phonon scattering acting alone. The function ϕ_3 is also shown for comparison. It may be pointed out that the method is applicable equally well when other elastic scattering mechanisms are effective along with polar optic phonon scattering. The expression for S_0 is then only modified as it should include contributions of all other elastic scattering mechanisms.

The thermoelectric power in the present case is given by

$$Q = -E_F(eT)^{-1}\left[\int_0^\infty \phi_4\gamma^{3/2}\partial_E f_0 \Big/ \int_0^\infty \phi_3\gamma^{3/2}\partial_E f_0 dE + 1\right] \quad . \tag{8.179}$$

It has, however, been shown earlier [see (8.167)] that we may express the above expression also as

$$Q = (k_B/e)\left[(k_BT)^{-1}\int_0^\infty E\phi\gamma^{3/2}\partial_E f_0 dE \Big/ \int_0^\infty \phi\gamma^{3/2}\partial_E f_0 dE - E_F(k_BT)^{-1}\right] \quad ,$$

and Q may be obtained from the knowledge of ϕ although the perturbation function ϕ_4 is very different from ϕ.

We present in Fig.8.10 the values of $Q + E_F/eT$ for polar optic phonon scattering at different temperatures. We find that the value lies between 1.2 and 1.9, whereas the corresponding value for deformation potential scattering is 2.

8.4.4 Electronic Thermal Conductivity

The transport of electrons due to a temperature gradient also causes a transport of energy. The heat flux thus produced contributes to the heat conduction. Electrons, therefore, contribute also to the thermal conductivity. We illustrate in Fig.8.11 the variation of the thermal conductivity with the conductivity in a semiconductor. We find that although the electronic contribu-

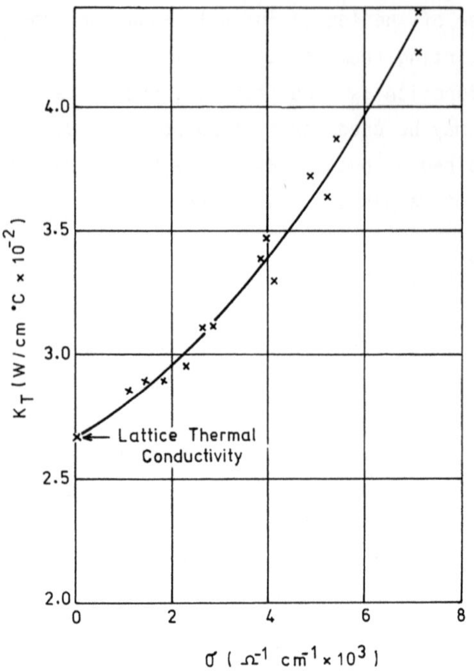

Fig. 8.11. Thermal conductivity for samples with different electrical conductivity [8.26]

tion in nondegenerate semiconductors is not usually as large as the lattice thermal conductivity, the contribution may be significant in heavily doped semiconductors. The contribution may be evaluated by solving the Boltzmann equation following the procedure discussed in the earlier section.

The energy flux is given by

$$W = (4\pi^3)^{-1} \int E(1/\hbar)(\partial E/\partial k_x) f d\underline{k} \quad . \tag{8.180}$$

Replacing f by (8.153) and (8.162), we get

$$W = -2ne(3I_n m^*)^{-1}\Big[\Big(\mathscr{E}_x - e^{-1}\partial E_F/\partial x\Big)\int_0^\infty E\gamma^{3/2}\phi_1(\partial f_0/\partial E)dE$$

$$-(1/eT)(dT/dx)\int_0^\infty E\gamma^{3/2}\phi_2(\partial f_0/\partial E)dE\Big] \quad , \tag{8.181}$$

where ϕ_1 and ϕ_2 are the solutions of (8.163) and (8.164). Putting

$$-2ne(3m^*I_n)^{-1}\int_0^\infty E\gamma^{3/2}\phi_1\partial_E f_0 dE = \sigma_E \quad , \tag{8.182}$$

and

$$-2n(3m^*I_n)^{-1}\int_0^\infty E\gamma^{3/2}\phi_2\partial_E f_0 dE = Q_E \quad , \tag{8.183}$$

we get

$$W = \sigma_E\left[\mathscr{E}_x-(1/e)(\partial E_F/\partial x)\right] - Q_E(1/T)(dT/dx) \quad . \tag{8.184}$$

We remember that the expression for the current density in the presence of temperature gradient (see Sect.8.4.2) is

$$J = \sigma\left[\mathscr{E}_x-(1/e)(\partial E_F/\partial x)\right] - Q\sigma(dT/dx) \quad . \tag{8.185}$$

For an open-circuited sample, $J=0$. We may use this condition to eliminate $\left[\mathscr{E}_x-(1/e)(\partial E_F/\partial x)\right]$ and obtain

$$W = (TQ\sigma_E-Q_E)(1/T)(dT/dx) \quad . \tag{8.186}$$

The thermal conductivity K_T is defined to be the negative of the energy flux per unit temperature gradient. It is thus given by

$$K_T = (1/T)(Q_E-TQ\sigma_E) \quad . \tag{8.187}$$

For the relaxation-time approximation

$$\phi_1 = \tau(E) \quad \text{and} \quad \phi_2 = (E-E_F)\tau(E) \quad , \tag{8.188}$$

where $\tau(E)$ is the relaxation time for electrons of energy E.

On evaluating σ_E and Q_E, we get

$$\begin{aligned} K_T/\sigma &= (1/T)\left[\left\langle(E-E_F)E\tau\right\rangle - \langle E\tau\rangle(\langle E\tau\rangle/\langle\tau\rangle - E_F)\right](e^2\langle\tau\rangle)^{-1} \\ &= (1/T)\left[(\langle\tau\rangle\langle E^2\tau\rangle - \langle E\tau\rangle^2)/\langle\tau\rangle\right](e^2\langle\tau\rangle)^{-1} \quad , \end{aligned} \tag{8.189}$$

where

$$\langle x\rangle = -(2/3)\int_0^\infty x\gamma^{3/2}\partial_E f_0 dE\Big/\int_0^\infty f_0\gamma^{1/2}\gamma' dE \quad .$$

We find that for degenerate semiconductors under extreme conditions of degeneracy, i.e. when $\partial_E f_0$ is approximated by the delta function $\delta(E-E_F)$, the above expression yields a value of zero for K_T. However, when the integral is properly evaluated for finite temperatures we get [Ref.8.25, p.306]

$$K_T/(\sigma T) = (\pi^2/3)(k_B^2/e^2) \quad . \tag{8.190}$$

The ratio is independent of scattering mechanisms and temperatures. This relation is the Wiedemann-Franz law and the constant is called the Lorentz number.

For nondegenerate materials, on the other hand, the ratio depends on the scattering mechanisms. If the relaxation time τ varies with E as E^s, we get for parabolic bands,

$$K_T/(\sigma T) = (k_B^2/e^2)(5/2+s) \quad . \tag{8.191}$$

The ratio has thus a value lying between $2(k_B^2/e^2)$ and $4(k_B^2/e^2)$ as s varies from -1/2 to 3/2 for deformation acoustic potential phonon and ionised impurity scattering.

The method of solution of the equations giving ϕ_1 and ϕ_2 for polar optic phonon or mixed scattering has been discussed earlier. These solutions may be used in (8.162) and (8.183) to obtain K_T and then $K_T/\sigma T$ for the general

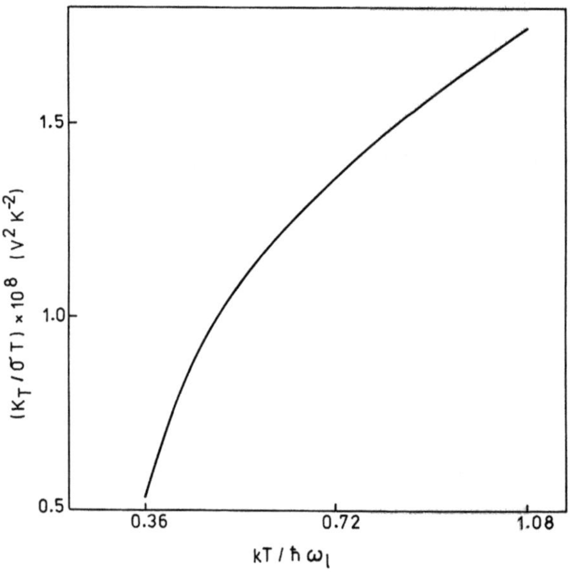

Fig. 8.12. Lorentz number for polar optic phonon scattering

kind of scattering. As an illustration, we have shown in Fig.8.12 the value of this ratio for pure polar optic phonon scattering for different ratios of ambient temperature. It is found that the ratio changes from a value of 0.53×10^{-8} at low temperatures to a value of 1.76×10^{-8} V^2/K^2 at high temperatures.

We have discussed the formalism for obtaining the electronic thermal conductivity in this section. We find that it may be evaluated along with the thermoelectric power using the same solutions. The thermal conductivity is similar to that in metals. However, in nondegenerate semiconductors, the ratio of thermal and electrical conductivity is intimately related to the nature of the scattering mechanism. If the electronic thermal conductivity in such materials is separable from the lattice conductivity, we may investigate the nature of scattering mechanism. Such studies, however, would require very accurate methods of measurement of thermal conductivity.

8.5 Conclusions

We have presented in this chapter various methods of solution of the Boltzmann equation and evaluation of the various DC electron transport coefficients. The methods used for the evaluation of the galvanomagnetic and thermal transport coefficients have been discussed for all kinds of complexity of band structure, and scattering mechanisms. In the past the transport coefficients were mostly evaluated by using the relaxation-time formalism and by assuming a parabolic E-$\underline{k}$ relation. The numerical methods for the evaluation of the coefficients taking into account the effects of the polar optic phonon scattering, nonparabolic dispersion relation and the admixture of the wave functions have been developed in the past few years. These methods have been used extensively to obtain the conductivity mobility. The values of other transport coefficients have been evaluated for a few materials. We may expect that our understanding of electron transport in compound semiconductors will be much improved when the available experimental results for the various transport coefficients are examined in the light of these new developments in theory.

9. Low-Field AC Transport Coefficients

The drift velocity attained by the carriers in the presence of an external field is determined by the free time available to the electrons between collisions during which they are accelerated by the field. The effect of a time-dependent field may therefore be expected to be different from that of a DC field if the magnitude of the field changes appreciably in the time between collisions. The mean collision time for electrons in semiconductors ranges between 10^{-9} and 10^{-12} seconds and hence a difference between AC and DC transport coefficients may be expected to start from frequencies in the microwave range. Semiconductor devices applicable for generating and amplifying microwave signals have been developed in the last few years [9.1]. An interest in the electron transport properties in semiconductors at microwave frequencies has grown with the emergence of these devices. However, even before the development of microwave semiconductor devices, transport coefficients at microwave frequencies had been extensively studied. These studies were undertaken mainly for two reasons. First, the microwave transport coefficients may be measured without attaching contact leads to the samples [9.2]. For samples in which the production of an ohmic contact is technologically difficult, microwave methods therefore provide a good alternative to DC methods for the measurement of the parameters. The sensitivity of the measuring instruments may also be much improved with microwave signals by using the high-Q value of microwave cavities. Hence, the conductivity of small-sized samples or the Hall mobility of low-mobility samples may also be conveniently measured by the microwave methods [9.3]. These advantages have prompted extensive application of microwaves for the study of the semiconductor transport coefficients. Second, the changes in the transport coefficients at microwave frequencies cause interesting phenomena, such as cyclotron resonance [9.4], plasma resonance [9.5], and electron susceptibility change [9.6]. The electron effective mass and the relaxation time may be accurately determined from the study of the phenomena. Interest in the AC transport properties of semiconductors has also been prompted by the need for understanding the results of such experiments.

Transport coefficients of semiconductors are also studied at infrared and optical signal frequencies. Many semiconducting materials are used for making light-emitting diodes and laser diodes. The efficiency of such diodes is dependent on the absorption coefficient of light signals. For signals having wavelengths beyond the absorption edge, the absorption is due to free carriers and is related to the conductivity of the material at the frequency of the signal. Estimation of the absorption coefficient therefore requires knowledge of the processes limiting the conductivity at high frequencies. The study of absorption of high frequency (infrared or optical) signals, or the so-called free-carrier absorption (often abbreviated as FCA) is also very fruitful for the understanding of the scattering mechanisms. It turns out that the contributions of various scattering mechanisms are directly additive in FCA and hence the individual contribution to the total FCA may be separately studied. The unknown coupling coefficients of some scattering mechanisms have been determined through such analysis. The wavelength dependence of FCA for the various scattering mechanisms have their characteristic index like the temperature-dependence index of mobility (see Sect.8.1.1). Knowledge of this index may also be readily used to identify the predominant scattering mechanisms.

We present in this chapter the theory of AC transport coefficients of semiconductors, both for the microwave and the infrared and optical signals. The microwave transport coefficients may be evaluated by extending the classical theory of transport presented in the previous chapter. The classical theory, however, is not adequate for the understanding of the FCA, and we have to turn to the quantum theory at the end of this chapter.

9.1 Classical Theory of AC Transport Coefficients

The classical theory of AC transport coefficients is developed starting with the Boltzmann equation as in DC transport theory. The Boltzmann equation including all the terms is (see Sect.7.2)

$$\partial_t f + (1/\hbar)\nabla_{\underline{k}} E \cdot \nabla_{\underline{r}} f + (\underline{F}/\hbar) \cdot \nabla_{\underline{k}} f - \partial_c f = 0 \quad , \tag{9.1}$$

where $\underline{F}$ is the force on the electrons due to the applied fields. In order to simplify the development, we shall first confine our attention to a uniform homogeneous material with an isotropic band structure. We shall also assume that only a sinusoidal electric field has been applied and that there is no magnetic field.

The electric field may be expressed as

$$\underline{\mathcal{E}} = \underline{\mathcal{E}}_0 \, \mathrm{Re}\{\exp(i\omega t)\} \quad , \tag{9.2}$$

where $\underline{\mathcal{E}}_0$ is the magnitude of the field, ω is its angular frequency, and Re denotes the real part of the complex quantity.

For the assumptions stated above, the Boltzmann equation may be simplified to

$$\partial_t f + (e/\hbar)\underline{\mathcal{E}}_0 \, \mathrm{Re}\{\exp(i\omega t)\} \cdot \underline{\nabla}_{\underline{k}} f - \partial_c f = 0 \quad . \tag{9.3}$$

On examining the above equation we find that the field being time dependent in this case, even in the steady state, the distribution function will be time dependent. We may expand f as

$$f(\underline{k},t) = f_0(E) - k \cos\theta \; \phi_1(E,t)(\partial f_0/\partial E) \quad , \tag{9.4}$$

where $\underline{k}$ is the wave vector of the electron and θ is the angle between the wave vector and the electric field. For small fields, as shown for DC fields, the even part of the distribution function remains unaffected, and is the same as the Fermi-Dirac function. Replacing f in (9.3) by the expression (9.4), we obtain

$$\partial_t \phi_1 + L_c \phi_1 = (e\hbar\mathcal{E}_0/m^*)\mathrm{Re}\{\exp(i\omega t)\} \quad , \tag{9.5}$$

where L_c is the collision operator defined by (7.35), and m* is the effective mass of the electron. Evidently in the steady state, the time dependence of ϕ_1 will be identical to that of $\mathcal{E}$, and we may express ϕ_1 as

$$\phi_1 = (e\hbar\mathcal{E}_0/m^*)\mathrm{Re}\{\phi \, \exp(i\omega t)\} \quad . \tag{9.6}$$

The equation for ϕ is

$$(L_c + i\omega)\phi = 1 \quad . \tag{9.7}$$

The AC transport coefficients may now be obtained by solving the above equation for ϕ. The mobility is given by

$$\mu_d = -(2|e|)/(3m^*)\left[\int_0^\infty \phi E^{3/2}\partial_E f_0 dE / \int_0^\infty f_0 E^{\frac{1}{2}} dE\right] \quad . \tag{9.8}$$

9.1.1 Solution for the Relaxation-Time Approximation

For the relaxation-time formalism, $L_c\phi$ is replaced by ϕ/τ, and the solution for ϕ is

$$\phi = (i\omega+1/\tau)^{-1} \quad . \tag{9.9}$$

The expression for the conductivity mobility is then

$$\mu_d = |e|\left\langle (i\omega+1/\tau)^{-1}\right\rangle/m^* \quad , \tag{9.10}$$

where the average is as defined by (8.189). We find that the mobility has an out-of-phase component, signifying that the velocity of the electron will have a phase difference with the applied field. This behaviour introduces an interesting effect as discussed below.

Separating the mobility into its real and imaginary components we get

$$\mu_r = (|e|/m^*)\left\langle \tau/(1+\omega^2\tau^2)\right\rangle \quad ,$$

and

$$\mu_i = -(|e|/m^*)\omega\left\langle \tau^2/(1+\omega^2\tau^2)\right\rangle \quad . \tag{9.11}$$

The real part of the mobility gives the conductivity. We find that the conductivity decreases with increasing frequency. The exact nature of this variation depends on the energy dependence of τ, since the averages have to be evaluated considering the distribution of the electrons in the energy space. Often, however, the energy dependence is ignored and the frequency dependence of μ_r is utilised to estimate the order of magnitude of τ [9.7]. Such an estimate may be obtained on the basis of the formula

$$\tau = |e|(m^*\omega^2)^{-1}\left[\mu_r^{-1}(\omega)-\mu_r^{-1}(0)\right] \quad , \tag{9.12}$$

where ω is the angular frequency of the signal for which the real part of the mobility is $\mu_r(\omega)$. In the particular case, when $\mu_r(\omega)$ is $(1/2)\mu_r(0)$ we get

$$\tau = 1/\omega \quad . \tag{9.13}$$

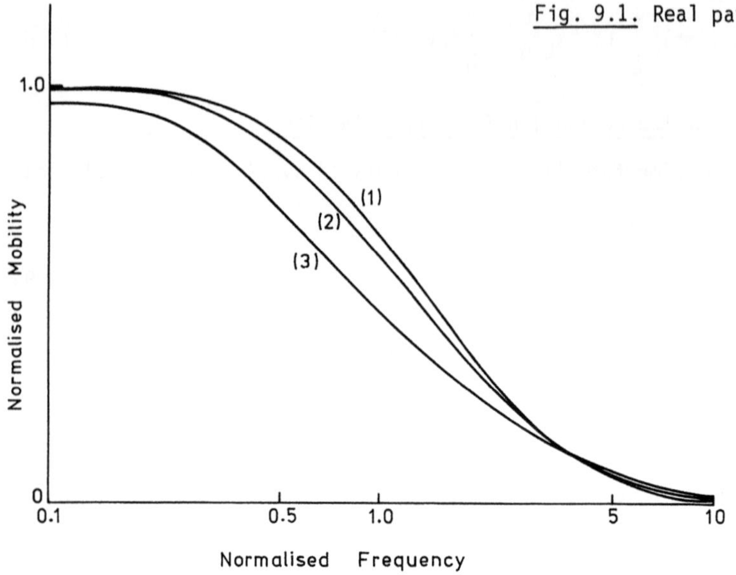

Fig. 9.1. Real part of mobility

The formula is not however exactly valid when τ is energy dependent. In order to illustrate the effect of the energy dependence we have plotted in Fig.9.1 the variation of μ_r with frequency assuming that τ varies with energy as E^{-s}, and that the material is nondegenerate.

For small values of frequency, we may expand $1/(1+\omega^2\tau^2)$ binomially, and express $\mu_r(\omega)$ as

$$\mu_r(\omega) = (|e|/m^*)(\langle\tau\rangle-\omega^2\langle\tau^3\rangle) \quad . \tag{9.14}$$

Thus, the plot of $\mu_r(\omega)/\mu_r(0)$ against ω^2 would be a straight line, the slope m of which gives $\langle\tau^3\rangle/\langle\tau\rangle$. In nondegenerate materials, if τ varies as E^{-s}, we may use the above result to estimate $\langle\tau\rangle$, since (see Sect.8.1.1)

$$\langle\tau\rangle = a\Gamma(5/2-s)/\Gamma(5/2) \quad , \tag{9.15}$$

$$\langle\tau^3\rangle = a^3\Gamma(5/2-3s)/\Gamma(5/2) \quad , \tag{9.16}$$

where a is a constant determined by the coupling parameters for the scattering mechanisms.

Thus,

$$\langle\tau^3\rangle/\langle\tau\rangle = a^2\Gamma(5/2-3s)/\Gamma(5/2-s) \quad , \tag{9.17}$$

and

$$\langle\tau\rangle = [\Gamma(5/2-s)/\Gamma(5/2)][m\Gamma(5/2-s)/\Gamma(5/2-3s)]^{\frac{1}{2}} = cm^{\frac{1}{2}} \quad , \tag{9.18}$$

where

$$c = [\Gamma(5/2-s)]^{3/2}\left\{\Gamma(5/2)[\Gamma(5/2-3s)]^{\frac{1}{2}}\right\}^{-1} \quad .$$

The value of c lies between $4/(3\sqrt{\pi})$ and $4/(30\pi)^{\frac{1}{2}}$ for s varying from +1/2 to -3/2, corresponding to acoustic phonon and ionised impurity scattering.

The current resulting from the imaginary part of μ_d is identical in nature to the dielectric current. The apparent dielectric constant of the material will therefore be different. The apparent change in the dielectric constant ΔK is given by

$$\Delta K = +n|e|\mu_i/\omega\varepsilon_0 \quad , \tag{9.19}$$

where n is the carrier concentration and ε_0 ($=8.854\times10^{-12}$ F/m) is the free-space permittivity. Replacing μ_i by the expression (9.11) we get

$$\Delta K = -(ne^2/m^*\varepsilon_0)\left\langle\tau^2/(1+\omega^2\tau^2)\right\rangle \quad . \tag{9.20}$$

Note that $(ne^2/m^*\varepsilon_0)$ is K times the square of the plasma frequency ω_p of the carriers, K being the dielectric constant of the material. For frequencies such that $\omega^2\tau^2 \ll 1$, we may simplify the expression to

$$\Delta K = -K\omega_p^2 < \tau^2 > \quad . \tag{9.21}$$

Using the expression for the low-field Hall mobility and DC conductivity mobility we may also express ΔK as

$$\Delta K = -K\omega_p^2(m^*/e)^2\mu_d\mu_H = -(nm^*/\varepsilon_0)\mu_d\mu_H = -(m^*/\varepsilon_0)(\sigma/|e|)\mu_H \quad . \tag{9.22}$$

Thus, by measuring ΔK, σ and μ_H we may obtain m* from the above expression. This method is particularly useful for materials with low-values of τ (due to large concentration of impurities or at high temperature) for which other methods (e.g. cyclotron resonance) cannot be applied. In fact, this possibility was first pointed out by BENEDICT and SHOCKLEY [9.6] and the suggestion prompted a number of studies of microwave transport coefficients [9.8]. Initially, τ was considered independent of energy and m* was obtained by using the relation [9.6]

$$\Delta K = (m^*\sigma^2/ne^2\varepsilon_0) = (m^*n/\varepsilon_0)\mu_d^2 \quad . \tag{9.23}$$

The results obtained in the early experiments had been confusing. But this was more due to the inaccuracy in the microwave method of measurement than due to the error introduced by the above assumption. We find that the above assumption only introduces a factor of μ_H/μ_d, which in most of the cases is lower than 1.5. The discrepancy in the values of m* was much larger in the initial measurements. However, the agreement between theory and experiment has much improved with time and the method has been used to study the effect of temperature on the effective mass.

The change in the dielectric constant introduced by the free carriers also leads to another interesting physical phenomenon. We note that $|\Delta K|$ has a value of $K\omega_p^2<\tau^2>$ at low frequencies. It decreases with frequency and at very high frequencies such that $\omega^2\tau^2 \gg 1$, it falls with frequency as $1/\omega^2$. Now, the propagation constant for an AC signal of frequency ω is

$$\beta = (2\pi/\lambda)(K-|\Delta K|)^{\frac{1}{2}} \quad , \tag{9.24}$$

where λ is the free-space wavelength of the signal of frequency ω.

If the carrier concentration is such that the low-frequency value of $|\Delta K|$ is larger than K, then the signal cannot propagate in the material. This condition is satisfied if $\omega_p^2<\tau^2>$ is larger than unity. The AC signal will be totally reflected from the material. With increase in the frequency of the signal, the material remains totally reflecting till $|\Delta K| < K$. For higher frequencies the material becomes partially reflecting and the reflection coefficient becomes zero when $K - |\Delta K| = 1$, i.e. when

$$\omega_p^2\left\langle\tau^2/(1+\omega^2\tau^2)\right\rangle = 1 - K^{-1} \quad . \tag{9.25}$$

For further increase in frequency, the reflection coefficient increases again and at very high frequency, it attains the asymptotic value $(\sqrt{K}-1)/(\sqrt{K}+1)$. We illustrate in Fig.9.2 the variation of ΔK with frequency and the consequent variation in the reflection coefficient. The condition of no reflection may be simplified further noting that the particular kind of variation occurs if $\omega_p^2\tau^2 \gg 1$. Then the condition also requires that ω^2 should be equal to $\omega_p^2/(1-K^{-1})$ for no reflection. No reflection thus occurs when the signal fre-

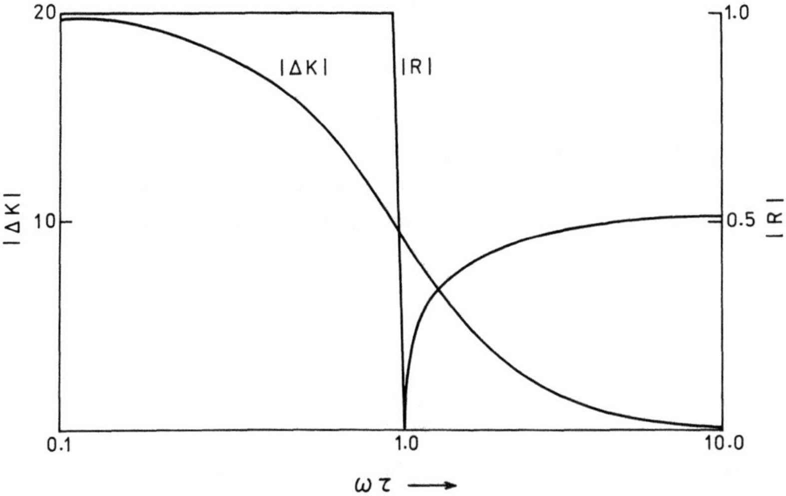

Fig. 9.2. (a) Change in dielectric constant; (b) Reflection coefficient

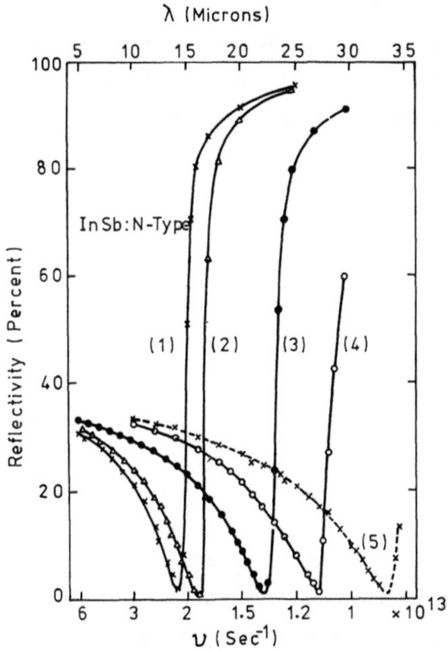

Fig. 9.3. Reflection coefficient for different plasma frequencies [9.5a]

quency approximately resonates with the plasma frequency, and the phenomenon is often characterised as plasma resonance [9.5]. It provides a direct means for the determination of the plasma frequency. We also find that we may determine the value of m* from ω_p, provided n is determined by other means.

Study of the plasma resonance phenomenon may thus be used to study the effective mass. In principle, the phenomenon will be observable if the low frequency value of $|\Delta K|$ is larger than K-1. The required value of ω_p is to be such that $\omega_p^2\tau^2 > (1-K^{-1})$. We show in Fig.9.3 the nature of the reflection coefficient versus frequency curves as ω_p^2 decreases. In the limiting case when $\omega_p^2\tau^2$ is close to 1, the value of the plasma resonance frequency ω_{pr} is not exactly identical to ω_p, but is given for an energy-independent relaxation time by

$$\omega_{pr}^2 = \omega_p^2/(1-K^{-1}) - 1/\tau^2 \quad , \tag{9.26}$$

and ω_p^2 may be determined only when τ is also known. Since τ is usually energy dependent, the above simplified formula does not apply and a more detailed analysis is necessary for the evaluation of ω_p from ω_{pr} under the above conditions.

9.1.2 Solution for Polar Optic Phonon and Mixed Scattering

The reduced transport equation (9.7) for polar optic phonon scattering mixed with other kinds of scattering may be transformed to a form identical to that for the galvanomagnetic transport equation (8.45,46) by putting

$$\phi = \phi_r + i\omega\phi_i \quad , \tag{9.27}$$

where ϕ_r and ϕ_i are respectively the real and imaginary parts of ϕ. Replacing ϕ by the expression (9.27) and separating the equation into real and imaginary parts, we get

$$L_c\phi_r = 1 + \omega^2\phi_i \quad , \tag{9.28}$$

$$L_c\phi_i = -\phi_r \quad . \tag{9.29}$$

The above set of equations is identical to (8.45,46) if ϕ_r and ϕ_i are identified with ϕ_x and ϕ_y, respectively. The real and imaginary parts of the mobility, μ_r and μ_i, are therefore given by the same expressions as for $\mu_{xx} = \sigma_{xx}n^{-1}|e|^{-1}$ and $\mu_{xy} = \sigma_{xy}n^{-1}|e|^{-1}$. The methods discussed ealier for obtaining the galvanomagnetic coefficients may hence be used to evaluate the AC conductivity components.

In order to illustrate the shape of the conductivity versus frequency curves for polar optic phonon scattering we have plotted in Fig.9.4 the real

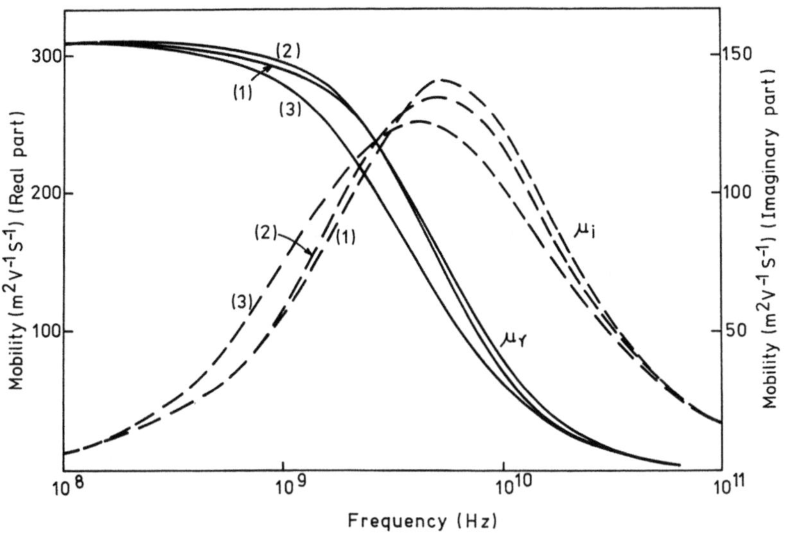

Fig. 9.4. Mobility-frequency curves. Solid line — real part; dotted line imaginary part. (1) Polar optic; (2) Acoustic; (3) Ionised impurity

and imaginary parts of the conductivity for polar optic phonon scattering acting alone. We have also plotted for comparison the curves for deformation potential acoustic phonon and ionised impurity scattering. The magnitude of the DC mobility has been adjusted in each case to obtain the same value of DC mobility. It is evident that the shapes of the curves are significantly dependent on the nature of the scattering. In fact, the frequency at which the conductivities is halved differs by about 20% for the different kinds of scattering. These results only emphasise the fact that for the interpretation of the AC conductivity along the lines discussed in the preceding section we are required to evaluate the conductivity by using the numerical techniques described earlier. Use of the relaxation time formalism and assumption of an energy-independent relaxation time may introduce large amounts of error. The error would be particularly large in the analysis of compound semiconductors in which polar optic phonon scattering predominates.

9.1.3 Solution for Nonparabolic and Anisotropic Bands

We have seen that the reduced transport equation has the same form for AC signals as for the DC galvanomagnetic fields. The methods of solution of the equation relevant to galvanomagnetic coefficients for anisotropic nonparabolic bands have been discussed. These methods are applicable for the evaluation of the AC coefficients. However, the equation for the AC signals for anisotropic bands are somewhat simpler and the detailed form of the method is discussed below. As in the earlier analysis we choose the coordinate axes in the direction of the principal axes of the constant-energy ellipsoids and resolve the applied field $\mathcal{E}_0 \exp(i\omega t)$ into its components along the directions of the coordinate axis. Let the magnitude of the components be $\mathcal{E}_{0x}$, $\mathcal{E}_{0y}$, and $\mathcal{E}_{0z}$, the corresponding components of the effective mass, m_x, m_y, and m_z. We shall illustrate the method for the evaluation of the mobility for one of the components, $\mathcal{E}_{0x}$. We shall also use the transformed quantities; k_x^*, k_y^*, k_z^*, $\mathcal{E}_{0x}^*$ given by

$$k_i^* = (m_0/m_i)^{\frac{1}{2}} k_i \quad (i \equiv x,y,z) \quad ; \text{ and } \quad \mathcal{E}_{0x}^* = (m_0/m_x)^{\frac{1}{2}} \mathcal{E}_{0x} \quad .$$

As explained in Sect.7.7, the θ_{k^*} dependence of the perturbation in the distribution function is in general more complex than that of $\cos\theta_{k^*}$ (θ_{k^*} is the angle between $\underline{k}^*$ and $\mathcal{E}_{0x}^*$). The steady-state distribution function f is therefore written as

$$f = f_0 - (e\hbar\mathcal{E}_{0x}^* \exp i\omega t) m_0^{-1} k^* (\partial f_0/\partial E) \sum \phi_{\ell x}(\underline{k}^*) P_\ell(\cos\theta_{\underline{k}^*}) \quad , \tag{9.30}$$

where $P_\ell(\cos\theta_{\underline{k}^*})$ is the Legendre polynomial of order ℓ.

Substituting the expression for f in (9.1) and noting that f_0 is the equilibrium distribution function, we get for the steady state

$$i\omega \sum \phi_{\ell x}(\underline{k}^*) P_\ell(\cos\theta_{\underline{k}^*}) + L_c\left(\sum_\ell \phi_\ell(\underline{k}^*) P_\ell(\cos\theta_{\underline{k}^*}) \right) = [\gamma'(E)]^{-1} \cos\theta_{\underline{k}^*} \quad , \tag{9.31}$$

where L_c is the collision operator.

It has been assumed (see Sect.4.5.4) that the nonparabolicity is identical for the different directions of $\underline{k}$, and that

$$(\hbar^2/2) \sum_i \left(k_i^2/m_i \right) = E(1+\alpha E) = \gamma(E) \quad (i \equiv x,y,z) \quad . \tag{9.32}$$

We may now obtain equations relating the different components of the perturbation function using the property of the orthogonality of the Legendre polynomials. We have seen, however, that the contribution of the higher-order perturbation terms is small in comparison to that of the first-order terms. Keeping only the first order terms we get

$$i\omega\phi_{1x}(\underline{k}^*) + L_{c1}\phi_{1x}(\underline{k}^*) = [\gamma'(E)]^{-1} \quad , \tag{9.33}$$

where

$$L_{c1}\phi_{1x}(k^*) = 3(4\pi)^{-1}(m_d/m_0)^{3/2}(8\pi^3)^{-1}V_c\int\left\{\int[1-f_0(\underline{k}'^*)][1-f_0(\underline{k}^*)]\right.$$
$$\left.\times\ S(\underline{k}^*,\underline{k}'^*)[\phi_{1x}(\underline{k}^*)\cos\theta_{\underline{k}^*}-\phi_{1x}(\underline{k}'^*)k'^*\cos\theta_{\underline{k}'^*}/k^*]d\underline{k}'^*\right\}$$
$$\times\ \cos\theta_{\underline{k}^*}d\cos\theta_{\underline{k}^*}\,d\phi_{\underline{k}^*} \quad . \tag{9.34}$$

The equation giving the perturbation term for AC coefficients in materials with ellipsoidal energy bands may thus be reduced to the same form as that of the equation for the isotropic bands. The method of solution needs, therefore, no further discussion.

The mobility for the ith direction is given by

$$\mu_i = -(2/3)(|e|/m_i)\int_0^\infty \phi_{1i}\partial_E f_0\gamma^{3/2}dE/\int_0^\infty f_0\gamma^{\frac{1}{2}}\gamma' dE \quad . \tag{9.35}$$

Considering only the collision processes for which the relaxation-time approximation holds, we get

$$\phi_{1i} = \gamma'^{-1}\tau_i(1+i\omega\tau_i)^{-1} \quad . \tag{9.36}$$

The real and the imaginary parts of the mobility are hence given by

$$\mu_{ri} = -(2/3)(|e|/m_i)I_n^{-1}\int_0^\infty \gamma'^{-1}\tau_i(1+\omega^2\tau_i^2)^{-1}\partial_E f_0\gamma^{3/2}dE \quad , \tag{9.37}$$

$$\mu_{ii} = (2/3)(|e|/m_i)I_n^{-1}\int_0^\infty \gamma'^{-1}\omega\tau_i^2(1+\omega^2\tau_i^2)^{-1}\partial_E f_0\gamma^{3/2}dE \quad , \tag{9.38}$$

where $I_n = \int_0^\infty f_0\gamma^{\frac{1}{2}}\gamma' dE$.

The conductivity may be obtained by adding the contributions to the current density from the different energy valleys, following the procedure outlined in connection with DC conductivity. We get on summing the contributions

$$\sigma = (ne^2/3)\sum_i m_i^{-1}\left\langle\tau_i(1+\omega^2\tau_i^2)^{-1}\right\rangle \quad . \tag{9.39}$$

It has been shown earlier that the effective mass of electrons may be obtained from the analysis of low-frequency AC conductivity by using the formula

$$\Delta K = -(m^*/\varepsilon_0)(\sigma/|e|)\mu_H \quad , \tag{9.40}$$

where ΔK is the effective change in the dielectric constant, σ is the DC conductivity, and μ_H is the Hall mobility. In the present case we find that the change in dielectric constant for $\omega\tau >> 1$ is

$$|\Delta K| = (1/3)(ne^2/\varepsilon_0) \sum_i < \tau_i^2/m_i > \quad , \tag{9.41}$$

and the conductivity is

$$\sigma = (ne^2/3) \sum_i < \tau_i/m_i > \quad . \tag{9.42}$$

It is thus evident that the formula for isotropic bands is not strictly applicable for anisotropic effective mass. The anisotropy in the relaxation time is usually about 20% in most of the materials. Hence, although the error arising from the use of the formula may not be important in the estimation of the effective mass, it should be taken into account when temperature dependence of the effective mass is studied by this method.

9.2 AC Galvanomagnetic Coefficients

The AC galvanomagnetic coefficients like the corresponding DC coefficients are studied to obtain information about the carrier concentration and mobility. The AC methods of measurement may be made more sensitive than DC methods and are found suitable for measuring the mobility of low-mobility samples. In recent years the AC methods are being increasingly used for the study of organic and bioelectronic samples [9.3a,10]. Further, the characteristics of the phenomena of cyclotron resonance and Faraday rotation are related to the AC coefficients. The interpretation of these experiments also require knowledge of these coefficients. We discuss in this section the methods used for the evaluation of AC galvanomagnetic coefficients. We limit the discussion to isotropic bands and uniform homogeneous materials.

We assume that the electric field is in the x direction given by $\mathscr{E}_{0x} \exp(i\omega t)$ and that the magnetic induction is B, in the z direction. The corresponding Boltzmann equation is

$$\begin{aligned} &\partial_t f + (e/\hbar)\left[\mathscr{E}_{0x} \exp i\omega t + (B/\hbar)(\partial E/\partial k_y)\right](\partial f/\partial k_x) \\ &\quad - (eB/\hbar^2)(\partial E/\partial k_x)(\partial f/\partial k_y) - \partial_c f = 0 \quad . \end{aligned} \tag{9.43}$$

We note that as the bands are isotropic for the assumed field configuration the electrons experience forces in the x and y directions. The distribution function is therefore perturbed in the same directions. The solution for the equation is assumed to be

$$f = f_0 - (e\hbar/m^*)\mathscr{E}_{0x} \exp i\omega t(\partial f_0/\partial E)(k_x\phi_x - k_y\phi_y) \quad . \tag{9.44}$$

Substituting in (9.43) and equating the coefficients of k_x and k_y, we obtain

$$i\omega\phi_x + \omega_B(E)\phi_y + L_c\phi_x = [\gamma'(E)]^{-1} \quad , \tag{9.45}$$

$$i\omega\phi_y - \omega_B(E)\phi_x + L_x\phi_y = 0 \quad , \tag{9.46}$$

where $\omega_B(E) = eB/[m^*\gamma'(E)]$, and L_c is the collision operator defined by (7.35).

For elastic and for randomising collisions, the solution may be obtained straightaway.

Replacing $L_c\phi$ by $\phi/\tau(E)$, $\tau(E)$ being the relaxation time, and solving we get

$$\phi_x = \tau(1+i\omega\tau)[\gamma'(E)]^{-1}\left\{1+i2\omega\tau+\left[\omega_B(E)^2-\omega^2\right]\tau^2\right\}^{-1} \quad , \tag{9.47}$$

$$\phi_y = \tau^2\omega_B(E)[\gamma'(E)]^{-1}\left\{1+i2\omega\tau+\left[\omega_B(E)^2-\omega^2\right]\tau^2\right\}^{-1} \quad . \tag{9.48}$$

In the case of polar optic phonon scattering, the collision term has the form

$$L_c\phi(E) = S_0(E)\phi(E) - S_+(E)\phi(E+\hbar\omega_\ell) - S_-(E)\phi(E-\hbar\omega_\ell) \quad . \tag{9.49}$$

The equation cannot then be solved analytically. A numerical method of solution was developed by the author [9.10] by extending RODE's method. This method is outlined below. It may be noted that the perturbation functions ϕ_x and ϕ_y are, in general, complex. Separating the functions into real and imaginary parts, we get the following set of difference equations for the above collision operator:

$$\begin{aligned} &S_0(E_i+\ell\hbar\omega_\ell)\phi_{xr}(E_i+\ell\hbar\omega_\ell) - \omega\phi_{xi}(E_i+\ell\hbar\omega_\ell) + \omega_B(E_i+\ell\hbar\omega_\ell)\phi_{yr}(E_i+\ell\hbar\omega_\ell) \\ &= \left[\gamma'(E_i+\ell\hbar\omega_\ell)\right]^{-1} + S_+(E_i+\ell\hbar\omega_\ell)\phi_{xr}(E_i+(\ell+1)\hbar\omega_\ell) \\ &\quad + S_-(E_i+\ell\hbar\omega_\ell)\phi_{xr}(E_i+(\ell-1)\hbar\omega_\ell) \quad , \end{aligned}$$

$$S_0(E_i+\ell\hbar\omega_\ell)\phi_{xi}(E+\ell\hbar\omega_\ell) + \omega\phi_{xr}(E_i+\ell\hbar\omega_\ell) + \omega_B(E_i+\ell\hbar\omega_\ell)\phi_{yi}(E_i+\ell\hbar\omega_\ell)$$
$$= S_+(E_i+\ell\hbar\omega_\ell)\phi_{xi}(E_i+(\ell+1)\hbar\omega_\ell) + S_-(E_i+\ell\hbar\omega_\ell)\phi_{xi}(E_i+(\ell-1)\hbar\omega_\ell) \quad ,$$

$$S_0(E_i+\ell\hbar\omega_\ell)\phi_{yr}(E_i+\ell\hbar\omega_\ell) - \omega\phi_{yi}(E_i+\ell\hbar\omega_\ell) - \omega_B(E_i+\ell\hbar\omega_\ell)\phi_{xr}(E_i+\ell\hbar\omega_\ell)$$
$$= S_+(E_i+\ell\hbar\omega_\ell)\phi_{yr}(E_i+(\ell+1)\hbar\omega_\ell) + S_-(E_i+\ell\hbar\omega_\ell)\phi_{yr}(E_i+(\ell-1)\hbar\omega_\ell) \quad ,$$

$$S_0(E_i+\ell\hbar\omega_\ell)\phi_{yi}(E_i+\ell\hbar\omega_\ell) + \omega\phi_{yr}(E_i+\ell\hbar\omega_\ell) - \omega_B(E_i+\ell\hbar\omega_\ell)\phi_{xi}(E_i+\ell\hbar\omega_\ell)$$
$$= S_+(E_i+\ell\hbar\omega_\ell)\phi_{yi}(E_i+(\ell+1)\hbar\omega_\ell) + S_-(E_i+\ell\hbar\omega_\ell)\phi_{yi}(E_i+(\ell-1)\hbar\omega_\ell) \quad . \qquad (9.50)$$

Here E_i is a chosen value of the energy. The values of $\phi_x(E_i)$ and $\phi_y(E_i)$ are determined by a set of infinite number of equations, which are obtained by giving ℓ integral values starting from zero. In the method of solution suggested by the author $\phi_x(E_i+\ell\hbar\omega_\ell)$ and $\phi_y(E_i+\ell\hbar\omega_\ell)$ are first evaluated by putting $\phi_{x,y}(E_i+(\ell\pm1)\hbar\omega_\ell)$ on the right-hand sides of the corresponding equations equal to zero and solving simultaneously the four equations in (9.50) linking the components of ϕ_x and ϕ_y. Solutions are obtained for $\phi_{x,y}(E_i+\ell\hbar\omega_\ell)$ up to a chosen value of ℓ. In the next iteration step $\phi_{x,y}(E_i+(\ell\pm1)\hbar\omega_\ell)$ on the right-hand side are given the values obtained in the previous step, and $\phi_{x,y}(E_i+\ell\hbar\omega_\ell)$ are obtained again. This process is repeated till the values of $\phi_{x,y}(E_i+\ell\hbar\omega_\ell)$ change less than a predetermined value. At the end of the iteration, we thus get the values of $\phi_{x,y}(E_i+\ell\hbar\omega_\ell)$ for different values of ℓ. In order to obtain the solution for other values of E_i, a new value of E_i is chosen and the values of $\phi_{x,y}(E_i+\ell\hbar\omega_\ell)$ are obtained by repeating the above procedure. For the sake of convenience E_i is chosen to lie in the interval $0 < E_i < \hbar\omega_\ell$ and the particular values for which ϕ is evaluated are chosen so as to fit the integration scheme to be used in the evaluation of the conductivities. Usually, since the Gaussian quadrature method is preferred for this purpose, the values are chosen according to the Gaussian quadrature scheme.

The conductivity components are evaluated, when ϕ_x and ϕ_y are known, using the following formulae:

$$\sigma_x = -(2/3)(ne^2/I_n m^*)\int_0^\infty \phi_x \partial_E f_0 \gamma^{3/2} dE \quad , \qquad (9.51)$$

$$\sigma_y = -(2/3)(ne^2/I_n m^*)\int_0^\infty \phi_y \partial_E f_0 \gamma^{3/2} dE \quad , \qquad (9.52)$$

where $I_n = \int_0^\infty f_0 \gamma^{1/2} \gamma' \, dE$. Since ϕ_x and ϕ_y are complex, σ_x and σ_y are also complex.

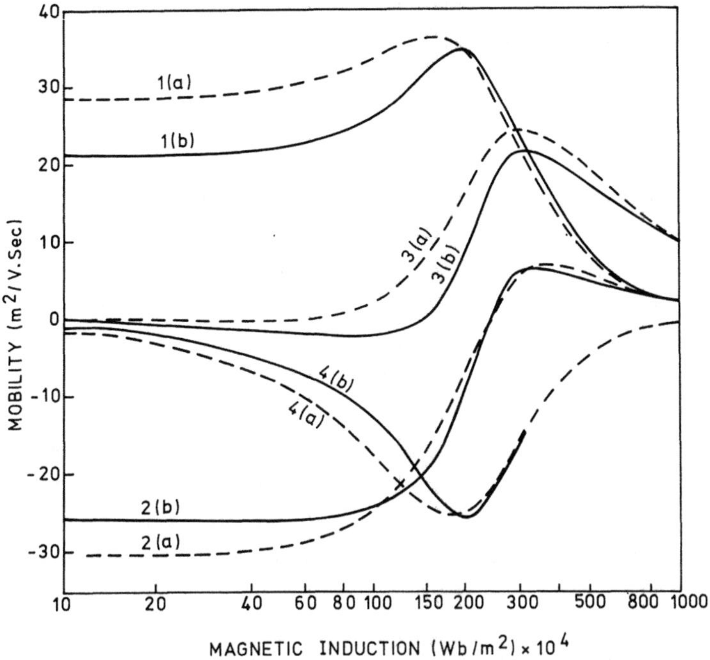

Fig. 9.5. Microwave magnetoconductivity of InSb at 35 GHz. Solid curves, present theory; dotted, Drude theory. 1,2,3,4 refer respectively to μ_{xxr}, μ_{xxi}, μ_{xyr}, and μ_{xyi}

In order to illustrate the kind of deviations to be expected from an exact analysis, we present in Fig.9.5 the values of conductivity components obtained by the author [9.11] for InSb. The full curves represent the values obtained by using the numerical method of evaluation described above. The dotted curves, on the other hand, were obtained by using the relaxation-time approximation method, the relaxation being obtained from the DC mobility using the formula $\tau = (m^*/|e|)\mu$. It is seen that the frequency dependences obtained by the two methods are identical in nature, but the exact values are slightly different. The critical frequencies at which the components change sign or attain maximum values are somewhat different. The magnetoconductivity tensor is often measured to obtain information about the relaxation time and the mobility from the measured values of critical frequencies [9.12]. In such interpretation of the results on compound semiconductors, it appears that an accurate analysis may be needed to obtain proper values.

The method of analysis presented here may in principle be extended to the case of anisotropic effective masses, but we refrain from presenting the details as the corresponding algebra is a little involved. There is also no report so far of an experiment on or analysis of magnetoconductivity of a compound semiconductor with anisotropic effective mass.

9.3 Cyclotron Resonance and Faraday Rotation

We find from the analysis of the preceding section that the conductivity in the presence of a magnetic field is frequency dependent and shows peaks at some frequencies. These peaks lead to the phenomenon of cyclotron resonance [9.4], which has been used extensively for the evaluation of electron effective mass. The tensor character of the conductivity leads also to another interesting phenomenon, namely, Faraday rotation [9.13], which has also been used extensively in semiconductor diagnostics. We shall briefly describe these two phenomena in this section. We first consider the characteristics of propagation of a plane electromagnetic wave in a semiconductor in the presence of a magnetic field.

9.3.1 Propagation of Electromagnetic Waves in a Semiconductor in the Presence of a Magnetic Field

We consider a plane electromagnetic harmonic wave of frequency ω propagating in a semiconductor. A magnetic field is also assumed to be applied in the direction of propagation of the wave which is taken as the z direction. Maxwell's equations for the problem are

$$\underline{\nabla} \times \underline{\mathscr{E}} = -(\partial \underline{B}/\partial t) = -i\omega\mu_0 \underline{H} \quad , \tag{9.53}$$

$$\underline{\nabla} \times \underline{H} = \underline{J} + (\partial \underline{\mathscr{D}}/\partial t) = \underline{\underline{\sigma}} \cdot \underline{\mathscr{E}} + i\omega K\varepsilon_0 \underline{\mathscr{E}} \quad , \tag{9.54}$$

where ε_0 and μ_0 are the free-space permittivity and permeability, $\underline{\mathscr{E}}$ is the electric field vector, $\underline{H}$ is the magnetic field intensity vector, $\underline{B}$ is the magnetic induction vector, $\underline{\mathscr{D}}$ is the electric displacement vector, and $\underline{J}$ is the conduction current density vector. The semiconductor is assumed to be nonmagnetic and to have a dielectric constant K and a tensor conductivity $\underline{\underline{\sigma}}$. The tensor conductivity for the present problem is

$$\underline{\underline{\sigma}} = \begin{pmatrix} \sigma_{xx} & \sigma_{xy} & 0 \\ \sigma_{yx} & \sigma_{yy} & 0 \\ 0 & 0 & \sigma_{zz} \end{pmatrix} \quad , \tag{9.55}$$

where $\sigma_{xx} = \sigma_{yy} = \sigma_x$ [given by (9.51)], $\sigma_{xy} = -\sigma_{yx} = \sigma_y$ [given by (9.52)], and σ_z is the AC conductivity of the sample in the absence of the magnetic field [given by (9.8)].

It may be easily shown from (9.55) and (9.54) that the electric field components $\mathscr{E}_x$, $\mathscr{E}_y$, and $\mathscr{E}_z$ satisfy the equations

$$(\partial^2\mathscr{E}_x/\partial z^2) + \omega^2\mu_0\varepsilon_0 K(1-i\sigma_{xx}/\omega\varepsilon_0 K)\mathscr{E}_x - i\omega\mu_0\sigma_{xy}\mathscr{E}_y = 0 \quad ,$$

$$(\partial^2\mathscr{E}_y/\partial z^2) + \omega^2\mu_0\varepsilon_0 K(1-i\sigma_{yy}/\omega\varepsilon_0 K)\mathscr{E}_y - i\omega\mu_0\sigma_{yx}\mathscr{E}_x = 0 \quad ,$$

$$(\partial^2\mathscr{E}_z/\partial z^2) + \omega^2\mu_0\varepsilon_0 K(1-i\sigma_{zz}/\omega\varepsilon_0 K)\mathscr{E}_z = 0 \quad . \tag{9.56}$$

Since we are considering uniform plane waves and $\underline{\nabla}\cdot\underline{\mathscr{E}}=0$, the component of the electric field in the direction of propagation, $\mathscr{E}_z$, is zero. However, we find that in the presence of the magnetic field, $\mathscr{E}_x$, $\mathscr{E}_y$ are coupled to each other, and the coordinate axes cannot be chosen to make one of them zero. The electrons moving under the action of one of the electric field components generate current and hence field in the transverse direction by interacting with the magnetic field and the presence of one field component requires presence of a transverse component.

Let us now assume that the propagation constant of the wave is γ so that the field components $\mathscr{E}_x$ and $\mathscr{E}_y$ may be written as

$$\mathscr{E}_x = \mathscr{E}_{x0}\exp i(\omega t-\gamma z) \quad , \quad \mathscr{E}_y = \mathscr{E}_{y0}\exp i(\omega t-\gamma z) \quad . \tag{9.57}$$

Substituting in (9.56), we get

$$\left\{-\gamma^2+\omega^2\mu_0\varepsilon_0 K[1-i\sigma_{xx}/(\omega K\varepsilon_0)]\right\}\mathscr{E}_{x0} - i\omega\mu_0\sigma_{xy}\mathscr{E}_{y0} = 0 \quad ,$$

$$-i\omega\mu_0\sigma_{xy}\mathscr{E}_{x0} + \left\{-\gamma^2+\omega^2\mu_0\varepsilon_0 K[1-i\sigma_{yy}/(\omega\varepsilon_0 K)]\right\}\mathscr{E}_{y0} = 0 \quad . \tag{9.58}$$

The characteristic equation giving the propagation constant γ is

$$\left\{-\gamma^2+\omega^2\mu_0\varepsilon_0 K[1-\sigma_{xx}/(\mu\varepsilon_0 K)]\right\}\left\{-\gamma^2+\omega^2\mu_0\varepsilon_0 K[1-\sigma_{yy}/(\omega\varepsilon_0 K)]\right\} + \omega^2\mu_0^2\sigma_{xy}\sigma_{yx} = 0 \quad . \tag{9.59}$$

Putting

$$\sigma_{xx} = \sigma_{yy} = \sigma_x \quad ,$$

$$\sigma_{xy} = -\sigma_{yx} = \sigma_y \quad ,$$

we get

$$\left\{-\gamma^2+\omega^2\mu_0\varepsilon_0K\left[1-i\sigma_x(\omega\varepsilon_0K)^{-1}\right]\right\}^2 = \omega^2\mu_0^2\sigma_y^2 \quad . \tag{9.60}$$

The propagation constant is

$$\gamma = \pm\omega(\mu_0\varepsilon_0K)^{\frac{1}{2}}\left[1-i\sigma_x(\omega\varepsilon_0K)^{-1}\mp\sigma_y(\omega\varepsilon_0K)^{-1}\right]^{\frac{1}{2}} \quad . \tag{9.61}$$

We also find that

$$\mathscr{E}_y/\mathscr{E}_x = i\omega\mu_0\sigma_y\left\{-\gamma^2+\omega^2\mu_0\varepsilon_0K\left[1-i\sigma_x(\omega\varepsilon_0K)^{-1}\right]\right\}^{-1} = \pm i \quad . \tag{9.62}$$

The characteristic waves of the medium are thus circularly polarised and the two characteristic waves have different propagation constants. We also find that the propagation constant has a real and also an imaginary part, the values of which are different for the two waves. The two characteristic waves thus suffer different amounts of attenuation and propagate with different phase velocities. The difference in the phase velocities leads to the effect known as Faraday rotation. The conductivity components are strongly dependent on frequency and it is found that one of the components suffers high attenuation for a particular frequency. This is the so-called cyclotron resonance phenomenon. We explain these phenomena using the expressions for σ_x and σ_y based on the relaxation-time approximation. For the sake of simplicity we shall neglect the energy dependence of τ and ω_B and consider a parabolic band.

9.3.2 Cyclotron Resonance Effect

Using the expressions for σ_x and σ_y for the relaxation-time approximation and neglecting the energy dependence of $\omega_B(E)$, m* and τ we may write for γ

$$\gamma = \omega(\mu_0\varepsilon_0K)^{\frac{1}{2}}\left\{1-ine^2\tau(m\omega\varepsilon_0K)^{-1}[1+i(\omega\pm\omega_B)\tau]^{-1}\right\}^{\frac{1}{2}} \quad , \quad \omega_B = |\omega_B(E)| \quad . \tag{9.63}$$

Let us assume that the conductivity is small and let us expand binomially the term in the square brackets. Retaining only the first and the second terms in the expression, we get

$$\beta = \omega(\mu_0\varepsilon_0K)^{\frac{1}{2}}\left\{1-ne^2(2m^*\omega\varepsilon_0K)^{-1}\tau^2(\omega\pm\omega_B)\left[1+(\omega\pm\omega_B)^2\tau^2\right]^{-1}\right\} \quad , \tag{9.64}$$

$$\alpha = (1/2)(\mu_0/\varepsilon_0K)^{\frac{1}{2}}ne^2\tau/\left\{m^*\left[1+(\omega\pm\omega_B)^2\tau^2\right]\right\} \quad , \tag{9.65}$$

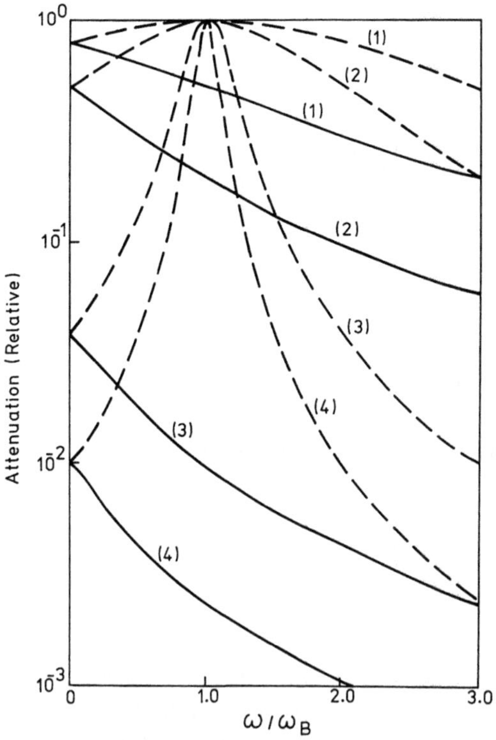

Fig. 9.6. Microwave attenuation in a magnetic field. Solid line, clockwise; dotted line, anticlockwise. (1) $\omega\tau = 0.5$; (2) $\omega\tau = 1$; (3) $\omega\tau = 5$; (4) $\omega\tau = 10$

where β and α are respectively the real and imaginary parts of γ. We consider first the attenuation constant. The constant is plotted against ω for different values of $\omega\tau$ in Fig.9.6. It is seen that the attenuation constant for the wave corresponding to the plus sign (the electric vector rotates clockwise when looking along the direction of propagation) decreases monotonically with frequency. On the other hand, the attenuation of the wave corresponding to the negative sign (the electric vector rotates anticlockwise when looking along the direction of propagation) passes through a maximum at a frequency of $\omega = \omega_B$. The sharpness of the maximum increases with increase in the value of τ.

The maximum in absorption thus occurs when the frequency of the signal ω is in resonance with the frequency of cyclotron motion of the charge carriers. The phenomenon is usually referred to as cyclotron resonance absorption. It is widely used to determine the effective mass of the carriers and also to determine the momentum relaxation time τ [9.14]. These applications are discussed below.

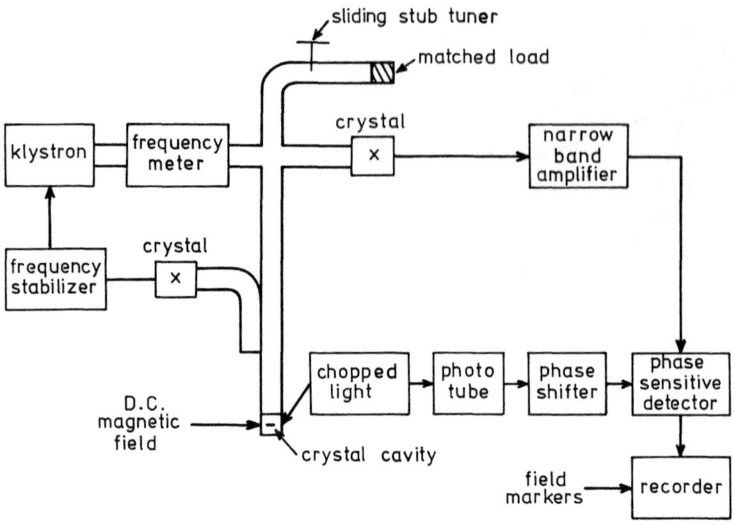

Fig. 9.7. Experimental arrangement for cyclotron resonance [9.15]

a) Determination of Effective Mass

The experimental arrangement for the observation of cyclotron resonance is shown in Fig.9.7. The semiconductor sample is mounted in a cavity placed in a magnetic field. The signal reflected from the cavity is detected and displayed on an oscilloscope or a recorder, the horizontal deflection in which is made proportional to the magnetic field. As the magnetic field is varied, due to variation in ω_B, the power absorbed by the sample changes and thereby alters the magnitude of the reflected signal. The sample usually being small, the power absorbed by the sample is given by P, where

$$P = P_0[1-\exp(-2\alpha d)] \simeq 2\alpha dP_0 \quad . \tag{9.66}$$

P_0 is the incident power, d is the thickness of the sample, and α is the attenuation constant for the signal in the semiconductor. The absorbed power and the reflected signal thus varies as the attenuation constant. For circularly polarised waves, a resonance peak is observed when the sense of rotation of the electric field vector is the same as the direction of cyclotron rotation of the charge carriers and the frequency of rotation is the same as the signal frequency, since then the conductivity and hence α becomes maximum. Hence, for circularly polarised waves, the direction of the magnetic field must be chosen in accordance with the nature of the charge carriers to obtain resonance.

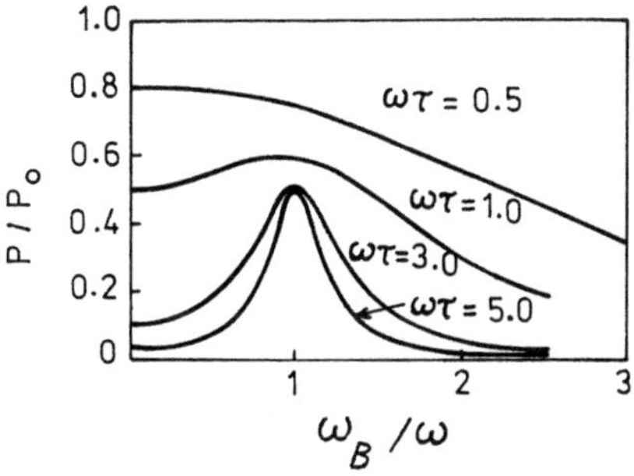

Fig. 9.8. Absorbed power in a cyclotron resonance experiment [9.16]

The experiment is often performed with linearly polarised waves. A linearly polarised wave is, however, made up of two circularly polarised waves with two opposite senses of rotation. The absorbed power for the linearly polarised wave is hence given by

$$P = (\alpha_+ + \alpha_-) dP_0 \quad , \tag{9.67}$$

where α_+ and α_- represent respectively the attenuation constants for the two components of the circularly polarised waves. Substituting the expressions for α_+ and α_- derived earlier, we get

$$P = A\left[1+(\omega^2+\omega_B^2)\tau^2\right]\left[1+2(\omega^2+\omega_B^2)\tau^2+(\omega^2-\omega_B^2)^2\tau^4\right]^{-1} \quad , \tag{9.68}$$

where $A = (\mu_0/K\varepsilon_0)^{\frac{1}{2}} ne^2\tau m^{*-1}$. We have plotted in Fig.9.8 P/A against ω_B/ω for different values of $\omega\tau$. It may be easily shown [9.12] that the curve has a maximum if $\omega\tau$ is larger than $1/\sqrt{3}$. The maximum occurs at

$$\omega_B = \omega\left[2(1+\omega^{-2}\tau^{-2})^{\frac{1}{2}}-(1+\omega^{-2}\tau^{-2})\right]^{\frac{1}{2}} \quad . \tag{9.69}$$

The resonance is obtained for either of the signs of the charge of the carriers. The sharpness of resonance depends on the value of $\omega\tau$, being sharper with increase in its value. We also find that the maximum occurs at a magnetic field slightly smaller than the resonance field. The difference varies as $\omega^{-2}\tau^{-2}$ and would be negligible for large values of $\omega\tau$.

We note that when the maximum is clearly identifiable, the effective mass of the carriers is obtained from the value of the magnetic field B_r at resonance from the formula

$$m^* = \omega^{-1}|e|B_r \quad , \tag{9.70}$$

ω being the signal frequency. The value of $\omega\tau$ has to be large enough for the above method to apply to the determination of m*. The experiment in silicon and germanium was initially carried out using signals at a frequency of about 10 GHz. The values of τ are usually such that the experiment has to be done near the temperature of liquid helium, 4.2 K, and pure samples free of crystalline imperfections must be used. Since free carriers are not available in such samples, carriers are excited either by shining light [9.15] or through breakdown and excitation from the impurity levels. In order to apply the method at higher temperatures, the frequency has to be higher to obtain the required value of $\omega\tau$, since τ usually decreases with increase in temperature. The highest usable value of ω is, however, limited by the highest practical value of the magnetic induction, which is about 50 T. The method cannot be used therefore at temperatures above about 20 K, for elemental semiconductors, in which the effective mass is of the order of 0.1 m_0. In compound semiconductors, the value of the effective mass of the electrons is one order lower. Signals in the infrared region may therefore be used to obtain resonance with the highest available magnetic fields. The method may be used in such materials for lower values of τ and to explore the temperature dependence of the effective mass.

The discussion above assumes an isotropic band structure. Resonance is observed for such a band structure at the same magnetic field, irrespective of its orientation with respect to the crystallographic axes. In the case of materials with anisotropic band structure the magnitude of the resonance field, on the other hand, depends on the direction of the magnetic field. As mentioned earlier, essentially two kinds of anisotropic band structures must be considered.

i) Band Structure with Ellipsoidal Constant Energy Surfaces. The cyclotron resonance frequency for this band structure is given by [see (8.90)]

$$\omega_B = |e|B\left(m_x\alpha^2+m_y\beta^2+m_z\gamma^2\right)^{\frac{1}{2}}(m_x m_y m_z)^{-\frac{1}{2}} \quad , \tag{9.71}$$

where α, β, and γ are respectively the direction cosines of the principal axes of the constant-energy ellipsoids, the effective mass components along which are m_x, m_y, and m_z respectively. We should remember that such constant-energy surfaces are spheroidal and two of the effective mass components are equal in value. Further, minima with such surfaces occur either along the <111> directions or along the <100> directions and are oriented so that the principal axes are also along the same directions. It is then evident that in materials with minima along the <100> directions, a single resonance field is obtained for the magnetic field along the <111> direction. It is given by

$$\omega = |e|B_r\left(3m_\ell m_t^2\right)^{-\frac{1}{2}}(m_\ell+2m_t)^{\frac{1}{2}} \quad , \tag{9.72}$$

where m_ℓ and m_t are respectively the longitudinal and transverse effective masses. When the magnetic field is along the <100> directions, we get instead two resonance fields B_{r1} and B_{r2} given by

$$\omega = |e|B_{r1}(m_\ell m_t)^{-\frac{1}{2}} \quad \text{and} \quad \omega = |e|B_{r2}m_t^{-1} \quad . \tag{9.73}$$

On the other hand, in materials with minima along the <111> directions, one resonance field given by (9.72) is obtained when the magnetic field is in the <100> directions. But, when the magnetic field is in the <111> directions, two resonance fields B_{r1} and B_{r2} given by

$$\omega = |e|B_{r1}m_t^{-1} \quad \text{and} \quad \omega = |e|B_{r2}m_\ell^{-\frac{1}{2}}m_t^{-1}(m_\ell+8m_t)^{\frac{1}{2}}/3 \quad , \tag{9.74}$$

are obtained.

Considering the above results, we find that the components of the effective mass tensor and the directions of the location of the minima may also be determined from the cyclotron resonance experiment in materials with ellipsoidal constant energy surface band structures, but two experiments with the magnetic field along the <100> and <111> directions are required to be performed.

ii) Warped Constant-Energy Surfaces. The analysis of the motion of the carriers with warped constant-energy surfaces is more complicated. It may be shown by using the E-$\underline{k}$ relation for such surfaces that the resonance magnetic field for a direction in the (110) plane and making an angle θ with the [001] direction is given by [9.15]

$$\omega = |e|B_r\left\{A\pm[B^2+(C/2)^2]^{\frac{1}{2}}\right\}$$
$$\times\left\{1\pm(C^2/64)(1-3\cos^2\theta)^2[B^2+(C/2)^2]^{-\frac{1}{2}}[A\pm(B^2+C^2/4)]^{-\frac{1}{2}}\right\}^{-1} \quad . \tag{9.75}$$

The values of the resonance magnetic fields when applied along the <111> direction are, thus,

$$\omega = |e|B_{r1}\left\{A+[B^2+(C/2)^2]^{\frac{1}{2}}\right\} \quad , \tag{9.76}$$

and

$$\omega = |e|B_{r2}\left\{A-[B^2+(C/2)^2]^{\frac{1}{2}}\right\} \quad . \tag{9.77}$$

Similarly, for the magnetic field along the <100> direction, the resonance fields are given by

$$\omega = |e|B_{r3}\left\{A+[B^2+(C/2)^2]^{\frac{1}{2}}+(1/16)[B^2+(C/2)^2]^{-\frac{1}{2}}C^2\right\} \quad , \tag{9.78}$$

$$\omega = |e|B_{r4}\left\{A-[B^2+(C/2)^2]^{\frac{1}{2}}-(1/16)[B^2+(C/2)^2]^{-\frac{1}{2}}C^2\right\} \quad . \tag{9.79}$$

The values of the constants A,B,C involved in the dispersion relation for warped constant-energy surfaces, may thus be obtained by determining the values of B_{r1}, B_{r2}, and B_{r3}.

We have considered so far a parabolic band structure, so that the carrier effective mass and hence the cyclotron resonance frequency could be taken to be independent of carrier energy. The nonparabolic nature of the band structure, however, alters these results. In order to estimate the magnitude of the effect of the nonparabolicity we consider an isotropic band with the dispersion relation

$$\hbar^2k^2/(2m^*) = \gamma(E) \quad . \tag{9.80}$$

It may be remembered that this relation is applicable to most of the compound semiconductors. Following the procedure outlined earlier for the evaluation of DC transport coefficients it may be shown that the components of the conductivity $\sigma_\pm$ is now given by

$$\sigma_\pm = -(2/3)(n/I_n)(e^2/m^*)\int_0^\infty \tau\gamma'^{-1}\gamma^{3/2}\partial_E f_0\left\{1+i[\omega_B\gamma'^{-1}\pm\omega]\tau\right\}^{-1}dE \quad , \tag{9.81}$$

where f_0 is the carrier distribution function, $\omega_B = |e|B/m^*$ is the cyclotron resonance frequency for the band-edge effective mass and I_n is the normalisation integral.

If the experiment is carried out with circularly polarised waves, we find by maximising the above expression that the real part of σ_- becomes a maximum for a value of magnetic field given by

$$|e|Bm^{*-1}\int_0^\infty \tau^3\gamma'^{-3}\gamma^{3/2}\partial_E f_0\left[1+(\omega_B\gamma'^{-1}-\omega)^2\tau^2\right]^{-2}dE$$

$$= \omega\int_0^\infty \tau^3\gamma^{3/2}\gamma'^{-1}\partial_E f_0\left[1+(\omega_B\gamma'^{-1}-\omega)^2\tau^2\right]^{-2}dE \quad . \tag{9.82}$$

The above equation may be solved only numerically. It is found on detailed computations that the deviation of the effective mass value may be appreciable and the effect of nonparabolicity may be important only at a high temperature. However, the resonance cannot be obtained at higher temperatures using signals in the microwave range, for which the classical theory of cyclotron resonance presented in this section applies. The value of τ being smaller at high temperatures, infrared signals must be used to make $\omega\tau$ large. For such signals, $\hbar\omega$ is larger than k_BT or the energy of the carriers, and the experiment may be interpreted only by applying the quantum theory of transport. This theory is discussed in the following chapter.

b) Determination of the Relaxation Time

The plots of the absorption against ω_B given in Fig.9.8 make it evident that the shapes of the curves are very much dependent on the value of the relaxation time τ. The value of τ may hence be determined from the analysis of the shape of the curves. When a distinct peak is obtained in the absorption curves, the value of τ is obtained from the following expressions for experiments with circularly polarised and linearly polarised waves.

i) Circular Polarised Waves

$$\tau = (2m^*/|e|)(B_{r1}-B_{r2})^{-1} \quad . \tag{9.83}$$

ii) Linearly Polarised Waves

$$\begin{aligned}\tau &= \sqrt{2}\left[\left(e^2B_{r1}^2m^{*-2}-\omega^2\right)-\left(\omega^2-e^2B_{r2}^2m^{*-2}\right)\right]^{-\frac{1}{2}} \\ &= \frac{\sqrt{2}\,m^*}{|e|}\left[\left(B_{r1}^2-B_0^2\right)-\left(B_0^2-B_{r2}^2\right)\right]^{-\frac{1}{2}} \quad ,\end{aligned} \tag{9.84}$$

where B_{r1} and B_{r2} represent the magnetic fields for which the absorption is half the peak value, and B_0 is the resonance value of the magnetic field. Thus, knowing B_{r1}, B_{r2}, and B_0, we may evaluate τ.

Even when a distinct resonance peak is not obtained, the value of τ may be evaluated from the analysis of the absorption curves [Ref.9.16, p.252]. One may theoretically evaluate the absorption coefficient for different magnetic fields assuming different values of τ and m^* and determine the values which give the best fit to the experimental curves. Such an analysis is often done when the materials are impure, and $\omega\tau$ cannot be made large enough to obtain a resonance. It should, however, be remembered that results obtained from such an analysis are not as accurate as those obtained when resonance is observed. The shape of curves is not sensitive enough to the values of m* and τ to give accurate values.

We should also remember that, in general, τ may be expected to be a function of energy. The value obtained from the analysis presented above should therefore be considered as an effective value. However, as the experiment is mostly carried out at low temperatures, the predominant scattering mechanism is neutral impurity atom scattering. The relaxation time for such scattering, as discussed earlier, is independent or weakly dependent on electron energy, so that the values obtained at very low temperatures may be taken to represent the correct values of τ in the material. In infrared cyclotron resonance experiments, the temperature being often higher, the value of τ should be obtained by taking into account the energy dependence. This may be done by evaluating the conductivities from the expressions given earlier, taking for τ the energy-dependent expressions appropriate to the predominant scattering mechanisms. But in infrared experiments $\hbar\omega$ is often greater than k_BT, so that the classical theory of cyclotron resonance presented here does not apply. The quantum theory of transport has to be applied for such analyses (see [9.17]).

9.3.3 Faraday Rotation

We have shown earlier that the characteristic waves in a semiconductor in the presence of a magnetic field are circularly polarised. A linearly polarised wave when incident in the medium excites those two characteristic waves. Since the propagation constants for the two waves are different, the two components suffer different amounts of attenuation and phase shift. The emergent wave from the semiconductor is, thus, in general elliptically polarised, with the major axis of the ellipse, however, rotated with respect to the original plane of polarisation. In the case when the attenuation suffered is small, the emergent wave may be considered as linearly polarised, but the plane of polarisation is rotated. This rotation is similar to the rotation of plane of polarisation of light signals observed by Faraday in glass in the presence of a magnetic field and is usually referred to as Faraday rotation. In dielectrics, the rotation is due to the anisotropy in the dielectric property caused by the magnetic field by altering the motion of the bound electrons. In semiconductors a similar rotation may be observed. For signals having wavelengths smaller than the absorption edge, a Faraday rotation results from the anisotropy introduced by magnetic field in the susceptibility due to interband absorption effects [Ref.9.16, p.259; 9.18]. The effect being considered here is, however, due to free electrons and may be observed at all frequencies. It will be shown that the analysis of this component of rotation allows us to evaluate the effective mass and also the relaxation time. The phenomenon has, therefore, been studied in great detail in compound semiconductors at microwave and at far-infrared frequencies [9.19], at which the rotation is solely due to the free carriers. We first derive the relation between the angle of rotation and the material parameters.

Let us consider a plane polarised wave of amplitude $\mathscr{E}_0$ and angular frequency ω, polarised in the x direction. The electric field of the wave is $\underline{\mathscr{E}} = \underline{i}\mathscr{E}_0 \exp(i\omega t) = (1/2)[(\hat{\underline{i}}+i\hat{\underline{j}})+(\hat{\underline{i}}-i\hat{\underline{j}})]\mathscr{E}_0\exp(i\omega t)$. The wave after propagation through a length ℓ in a semiconductor sample in the presence of a magnetic field in the x direction is given by

$$\underline{\mathscr{E}} = (\mathscr{E}_0/2)\left[(\hat{\underline{i}}+i\hat{\underline{j}})\exp(-\alpha_+\ell)\exp(i\beta_+\ell)+(\hat{\underline{i}}-i\hat{\underline{j}})\exp(-\alpha_-\ell)\exp(-i\beta_-\ell)\right]\exp(i\omega t) \quad , \tag{9.85}$$

where $\alpha_\pm$ and $\beta_\pm$ represent the attenuation and phase constant for the two circularly polarised characteristic waves. These are given by (9.63) and by (9.64) and (9.65) when the conductivity is small.

After some algebraic manipulation we find that the x and y components of the emergent waves are

$$\begin{aligned}\mathscr{E}_x &= \mathscr{E}_0 \exp[-(\alpha_++\alpha_-)\ell/2]\Big\{\cos\Delta\beta\ell\,\cos(\omega t-[\beta_++\beta_-]\ell/2)\cosh\Delta\alpha\ell\\ &\quad -\sinh\Delta\alpha\ell\,\sin\Delta\beta\ell\,\sin(\omega t-[\beta_++\beta_-]\ell/2)\Big\} \quad ,\\ \mathscr{E}_y &= \mathscr{E}_0 \exp[-(\alpha_++\alpha_-)\ell/2]\Big\{\cosh\Delta\alpha\ell\,\sin\Delta\beta\ell\cos(\omega t-[\beta_++\beta_-]\ell/2)\\ &\quad +\sinh\Delta\alpha\ell\,\cos\Delta\beta\ell\,\sin(\omega t-[\beta_++\beta_-]\ell/2)\Big\} \quad ,\end{aligned} \tag{9.86}$$

where $\Delta\alpha = (\alpha_+-\alpha_-)/2$ and $\Delta\beta = (\beta_++\beta_-)/2$. It may be easily shown that the resultant wave is elliptically polarised with the major axis and minor axis given by

$$A = (\cosh\Delta\alpha\ell) \quad , \quad B = (\sinh\Delta\alpha\ell) \quad . \tag{9.87}$$

The major axis makes an angle θ ($\equiv\Delta\beta\ell$) with the x direction, i.e. the direction of polarisation of the incident wave. It is evident that when the attenuation is negligible, $B = 0$ and the emergent wave is plane polarised, with the plane of polarisation rotated by the angle $\Delta\beta\ell$ from the incident wave.

Determination of Effective Mass and Relaxation Time

Substituting the expression for β_+ and β_- from (9.64) assuming that the conductivity is small, we find that

$$\theta = \Delta\beta\ell = (\ell/2)(\mu_0\varepsilon_0)^{\frac{1}{2}}\sigma_{yr}/(\varepsilon_0\sqrt{K}) \quad , \tag{9.88}$$

where σ_{yr} is the real part of σ_y [defined by (8.50)]. We find from the above equation that the Faraday rotation angle really gives the values of σ_{yr}. As σ_{yr} is related to m* and τ, the experimental results of Faraday rotation are often used to evaluate these parameters. For this purpose the experiments are performed under conditions for which certain approximations may be made and the expressions for σ_{yr} may be simplified. These are discussed below

i) $\omega\tau >> 1$, *Parabolic Bands*

$$\theta = \ell ne^3\omega_B\left[2c\sqrt{K}\, m^*\varepsilon_0\left(\omega_B^2-\omega^2\right)|e|\right]^{-1} \quad , \quad c = (\mu_0\varepsilon_0)^{-\frac{1}{2}} \quad . \tag{9.89}$$

If, further, $\omega_B << \omega$,

$$\theta = -\ell Bne^3\lambda^2\left(8\pi^2c^3\sqrt{K}\, m^{*2}\varepsilon_0\right)^{-1} \quad . \tag{9.90}$$

Under these conditions we find that the Faraday rotation angle is proportional to the magnetic field, square of the wavelength and inverse of the square of the effective mass. It is also evident from (9.89) that θ, the angle of Faraday rotation, will increase as ω_B approaches ω, show a resonance peak when $\omega_B = \omega$, and will change sign for larger values of ω.

ii) $\omega\tau >> 1$, *Nonparabolic Bands*. Using the expression for σ_{yr} for nonparabolic bands, we find that when $\omega\tau >> 1$,

$$\theta = \ell ne^3B\left(2c\sqrt{K}\,\omega^2\varepsilon_0 m^{*2}\right)^{-1}(2/3)\int_0^\infty \gamma^{3/2}\gamma'^{-2}\partial_E f_0 dE\left[\int_0^\infty f_0\gamma^{\frac{1}{2}}\gamma'\, dE\right]^{-1} \quad , \tag{9.91}$$

where f_0 is the Fermi-Dirac function, m* is the band-edge mass, and the dispersion relation is

$$\hbar^2k^2/(2m^*) = \gamma(E) \quad . \tag{9.92}$$

The effective mass, determined from the Faraday rotation results, using the formula

$$(1/m_F^*)^2 = -2\theta c\sqrt{K}\,\varepsilon_0\omega^2(\ell ne^3B)^{-1} \quad , \tag{9.93}$$

is thus related to the band-edge mass by the relation

$$(1/m_F^*)^2 = (1/m^*)^2(2/3)\int_0^\infty \gamma^{3/2}\gamma'^{-2}\partial_E f_0 dE\left[\int_0^\infty f_0\gamma^{\frac{1}{2}}\gamma' \, dE\right]^{-1} \quad . \tag{9.94}$$

In the case of extreme degeneracy, $\partial_E f_0$ is unity on the Fermi surface and zero elsewhere. Also, f_0 is unity for k less than k_F and, zero for k greater than k_F, k_F being the value of the wave vector on the Fermi surface. Expression (9.94) may then be simplified to

$$m_F^{*-1} = \left(\hbar^2 k_F\right)^{-1}(\partial E/\partial k_F) \quad . \tag{9.95}$$

The rotation effective mass thus gives essentially the effective mass corresponding to the Fermi energy under extremely degenerate conditions. The method may hence be used to explore the shape of nonparabolic bands. This has been done for materials like InSb and the results found to agree with the Kane relation.

If, however, the material is degenerate, but the conditions of extreme degeneracy does not apply, then expression (9.94) may be simplified to the following relation, assuming that $\alpha E_F \ll 1$.

$$m_F^{*-1} = m^{*-1}\left\{1-5\alpha k_B T \, F_{3/2}(E_F/k_B T)\left[F_{\frac{1}{2}}(E_F/k_B T)\right]^{-1}\right\} \quad , \tag{9.96}$$

where $F_j(E_F/k_B T)$ are the Fermi integrals given by

$$F_j(E_F/k_B T) = [\Gamma(j+1)]^{-1}\int_0^\infty x^j\left\{\exp\left[x-(E_F/k_B T)\right]+1\right\}^{-1}dx \quad . \tag{9.97}$$

It may be noted that under conditions of extreme degeneracy, since

$$F_{3/2}(E_F/k_B T)/F_{\frac{1}{2}}(E_F/k_B T) = 2/5 \quad , \tag{9.98}$$

the above expression reduces to (9.95).

iii) $\omega\tau \ll 1$, $\omega_B \ll \omega$, *Parabolic Bands*. Neglecting terms containing $\omega^2\tau^2$ in comparison to unity, we get

$$\begin{aligned}\theta &= \ell n e^3 B\left(2c\sqrt{K}\, m^{*2}\varepsilon_0\right)^{-1}(2/3)\int_0^\infty \tau^2 E^{3/2}\partial_E f_0 dE\Big/\int_0^\infty f_0 E^{\frac{1}{2}}dE \\ &= -\ell n e^3 B\langle\tau^2\rangle\left(2c\sqrt{K}\,\varepsilon_0 m^{*2}\right)^{-1} \quad . \end{aligned} \tag{9.99}$$

In the present case Faraday rotation angle is related both to the effective mass and the relaxation time. We may, therefore, determine one knowing the other. Further, we find that

$$|\theta|/\sigma = \ell B|e| <\tau^2>\left(2c\sqrt{K}\ \varepsilon_0 m^*<\tau>\right)^{-1} = \ell B\mu_H(2c\sqrt{K}\ \varepsilon_0)^{-1} \quad , \qquad (9.100)$$

where σ is the conductivity and μ_H is the Hall mobility. The Faraday rotation experiment carried out at frequencies such that $\omega\tau << 1$ thus gives a means of determining μ_H. Since the condition is valid in most of the semiconductors at room temperature for microwave frequencies, the method may be gainfully employed to obtain μ_H by microwaves.

The formulae given above may be generalised to include the anisotropic bands. However, this generalisation does not involve any new principles (see [9.20]).

Before concluding this section, it should be mentioned that we may determine $\Delta\alpha\ell$ in reasonably conductive samples also from the Faraday rotation experiment by measuring the ratio of the minimum and maximum amplitude of the emrgent elliptically polarised wave using the relation

$$B/A = \tanh \Delta\alpha\ell \quad . \qquad (9.101)$$

Knowing $\Delta\alpha\ell$ we may again determine m* and τ, since $\Delta\alpha\ell$ is given by

$$\Delta\alpha\ell = (\alpha_+ - \alpha_-)\ell/2 = \ell\sigma_{yi}(2c\sqrt{K}\ \varepsilon_0)^{-1} \quad . \qquad (9.102)$$

The value of $\Delta\alpha\ell$ may, therefore, be used to obtain σ_{yi} and hence the value of τ. Experiments of this kind are not, however, much reported in the literature [9.21].

9.4 Free-Carrier Absorption (FCA)

Electromagnetic radiation is absorbed in semiconductors through two distinct processes. In one process, electrons are excited from the valence band to the conduction band by absorbing the radiation. Such absorption may take place without the participation of phonons if the band gap is direct. If, however, the band gap is indirect, participation by phonons is required for momentum balance. This process of absorption is called the interband absorption. We shall not discuss this kind of absorption in this book.

In the second process, the electromagnetic radiation is absorbed by free electrons in the conduction band or quasi-free holes in the valence band. Electrons or holes are transferred to higher energy levels in the same band by the absorption of the radiation. This process of absorption is called free-carrier absorption. It is essentially an intraband transition phenomenon. We discuss this phenomenon in this section. We first present the classical theory of the phenomenon.

9.4.1 Classical Theory of FCA

It has been shown in the earlier section that an electromagnetic wave when propagating through a semiconductor sample suffers attenuation, the attenuation constant being given by

$$\alpha = -\mathscr{I}m\left\{\omega(\mu_0\varepsilon_0 K)^{\frac{1}{2}}[1-i\sigma(\omega\varepsilon_0 K)^{-1}]^{\frac{1}{2}}\right\} , \quad (9.103)$$

where σ is the conductivity for the signal frequency ω. If $\sigma(\omega\varepsilon_0 K)^{-1}$ is small compared to unity, we may simplify the above expression. The attenuation constant α is then given by

$$\alpha = \sigma_r/(2c\varepsilon_0\sqrt{K}) \quad , \quad (9.104)$$

where c is the velocity of light in free space and σ_r is the real part of the conductivity σ.

The electromagnetic radiation thus suffers attenuation if σ_r has a nonzero value. The classical theory of AC conduction developed in the earlier section thus provides an explanation for the phenomenon. We shall first discuss the qualitative features of free-carrier absorption on the basis of the above formula.

We assume for the purpose of the present discussion a parabolic isotropic energy band. We also assume that the relaxation-time approximation is applicable. The real part of the conductivity is hence given by

$$\sigma_r = (ne^2/m^*)\left\langle \tau(E)\left\{1+\omega^2[\tau(E)]^2\right\}^{-1}\right\rangle \quad , \quad (9.105)$$

where m* is the effective mass of the carriers and < > indicates average value as defined by

$$<x> = -(2/3)\int_0^\infty xE^{3/2}\partial_E f_0 dE\Big/\int_0^\infty f_0 E^{\frac{1}{2}} dE \quad , \quad (9.106)$$

f_0 being, in general, the Fermi function. $\tau(E)$ is the resultant relaxation time for the scattering mechanisms causing changes in the electron state. It is related to the individual relaxation times τ_i by the relation

$$1/\tau(E) = \sum_i 1/\tau_i(E) \quad . \tag{9.107}$$

Free-carrier absorption refers to the absorption of signals at wavelengths smaller than the far infrared. The largest value considered is about 30 μm. The corresponding frequency being 10^{13} Hz, we are concerned with signals having angular frequencies larger than $2\pi \times 10^{13}$ rad/s. In most semiconducting materials, τ is larger than 10^{-13} s, up to room temperature. We may, hence, assume that $\omega^2\tau^2$ is much larger than unity. Expression (8.295) for σ_r thus simplifies to

$$\sigma_r = ne^2(m^*\omega^2)^{-1} < 1/\tau(E) > \quad . \tag{9.108}$$

Replacing $[\tau(E)]^{-1}$ by $\sum_i [\tau_i(E)]^{-1}$, we get

$$\sigma_r = ne^2(m^*\omega^2)^{-1}\left\langle \sum_i [\tau_i(E)]^{-1} \right\rangle = ne^2(m^*\omega^2)^{-1} \sum_i < \tau_i^{-1} > \quad . \tag{9.109}$$

The expression for the absorption coefficient of a high-frequency signal is thus

$$A = ne^2\lambda^2\left(4\pi^2 c^3 \sqrt{K}\, m^*\varepsilon_0\right)^{-1} \sum_i \left\langle [\tau_i(E)]^{-1} \right\rangle \quad , \tag{9.110}$$

A being equal to 2α.

We find from the above formula that free-carrier absorption (abbreviated hereafter as FCA) has the following characteristic features:

a) FCA increases as the square of the wavelength of the absorbed signal.

b) The contribution of the various scattering mechanisms to the free-carrier absorption is directly additive. We may evaluate the FCA for the individual scattering mechanisms, and obtain the total absorption by adding the individual contributions. This feature simplifies the analysis of the contributions of various scattering mechanisms to FCA.

c) FCA in nondegenerate materials is directly proportional to the carrier concentration. In degenerate material, however, the proportionality will not be linear.

d) FCA varies with the effective mass of the carriers. The index of variation of FCA with m* are given below.

Scattering mechanism		Index of mass dependence
Acoustic phonon:	Deformation potential	1/2
	Piezoelectric	- 1/2
Optic phonon :	Polar	- 1/2
	Nonpolar	1/2
Ionised impurity		- 3/2

The characteristic features of FCA obtained from the classical theory are in approximate agreement with the experiments. A major discrepancy occurs between the magnitude of the FCA obtained from the classical theory and that given by the experiments. The index of the wavelength dependence is also not exactly 2 but varies between the limits [9.5a,22] of 1.5 and 3 depending on the scattering mechanism. Although in some analyses the classical expression has been used, quantitative analysis of FCA, or derivation of physical parameters from the FCA by using the classical expression, can hardly be justified in view of the above discrepancies.

FCA experiments, as mentioned earlier, are carried out at frequencies higher than 10^{13} Hz. The corresponding photon energy is about 0.04 eV. On the other hand, average value of the electron energies at room temperature is 0.037 eV. Photon energies thus being larger or comparable with the electron energy, the assumption (which is basic in the classical theory) that the electron gains energy from the radiation field quasi-continuously is not applicable. The classical theory is based on the Boltzmann equation, and the change in the distribution function caused by the electric field associated with the radiation is assumed to be essentially of the same form as for the steady fields. The only modification that is taken into account in calculating the AC conductivity is that the perturbation in the distribution function introduced by the AC field (proportional to the electric field) is time varying. As a result, the $\partial_t f$ term in the Boltzmann equation has a nonzero value. However, as discussed above for the frequencies relevant to FCA, photon energies being comparable with the electron energy, this classical description of the interaction between the electron and the radiation field is not applicable. We are required to treat this interaction on the basis of the quantum theory, which is developed in the following section.

9.4.2 Quantum-Mechanical Theory of FCA

a) General Theory

In the quantum-mechanical theory, each mode of electromagnetic radiation of frequency $\omega/2\pi$ and wave number κ, is characterised by particles called photons, having energy $\hbar\omega$ and momentum $\hbar\kappa$. A semiconductor sample, kept at a finite temperature T, is permeated by electromagnetic radiation of all frequencies. Electrons interact with the photons associated with this radiation all the time. In some interactions photons are absorbed by the atoms and in others photons are emitted. Through such interactions a statistical equilibrium is established, in which the photon number [9.23] associated with a mode per unit volume of phase space becomes

$$n_{po} = [\exp(\hbar\omega/k_BT)-1]^{-1} \quad . \tag{9.111}$$

When external radiation is incident on the sample, the photon number increases and the rate of absorption and emission of photons changes. A new steady-state condition is established in which the number of photons absorbed is larger than the number emitted. Let $-dn_p/dt$ represent the rate of the absorption of the photons. The absorption coefficient, defined as the ratio of the power absorbed per unit length of the sample and incident power, is then given by

$$\alpha = -(dn_p/dt)[(n_p-n_{po})(c/n_r)]^{-1} \quad , \tag{9.112}$$

since the ratio of $-(dn_p/dt)\hbar\omega$ and $(n_p-n_{po})\hbar\omega$ represents the ratio of the incident and absorbed power, and c/n_r represents the distance travelled by the wave in unit time, c being the velocity of light in vacuum and n_r the refractive index of the medium. For small values of the absorption coefficient, the refractive index of the sample may be taken as $K^{\frac{1}{2}}$.

Evaluation of the absorption coefficient in the quantum-mechanical theory thus requires an expression for dn_p/dt. The required expression is obtained by considering the interaction between the electrons and the photons quantum mechanically. Photons interact with the electrons through the field associated with the radiation. The radiation field adds a term H_1 to the Hamiltonian in the Schrödinger equation [9.24]:

$$H_1 = i(e\hbar/m_0)\underline{A}\cdot\underline{\nabla}_{\underline{r}} \quad , \tag{9.113}$$

where $\underline{A}$ is the vector potential operator associated with the photon field. We have not included the term arising from the spin of the electron, since in the final expression for the interaction matrix element it cancels out. We also note that the vector potential operator may be expressed as

$$\underline{A} = A_0\hat{\underline{a}}\left[a_\kappa \exp(i\underline{k}\cdot\underline{r})+a_\kappa^\dagger \exp(-i\underline{k}\cdot\underline{r})\right] \quad , \tag{9.114}$$

where $\hat{\underline{a}}$ is the unit vector in the direction of the polarisation of the wave and $\underline{\kappa}$ is the wave vector of the wave, a_κ and $a_\kappa^\dagger$ are respectively the annihilation and creation operators,

$$A_0^2 = (\hbar^2/4\pi\varepsilon_0)(2\pi/\hbar\omega)\left(1/V_c n_r^2\right) \tag{9.115}$$

(this factor differs from that given elsewhere, as rationalised MKS units are used and the radiation inside a solid is being considered) and V_c is the volume of the sample.

The effect of the additional term in the Hamiltonian may be evaluated by perturbation theory. The electron wave vector changes from $\underline{k}$ to $\underline{k}'$ due to the interaction. The transition probability is given by

$$T^{\pm}_{\underline{k}\to\underline{k}'} = (2\pi/\hbar)|M^{\pm}(\underline{k},\underline{k}')|^2\delta(E_{\underline{k}'}-E_{\underline{k}}\pm\hbar\omega) \ , \tag{9.116}$$

where $M(\underline{k},\underline{k}')$ represents the matrix element for such transitions. The plus and minus signs correspond respectively to the interactions in which a photon is emitted and is absorbed. The matrix element $|M^{\pm}(\underline{k},\underline{k}')|$ is given by

$$M(\underline{k},\underline{k}') = \int\psi^*_{\underline{k}'}U_f H_1 \psi_{\underline{k}} U_i d\underline{r} \ , \tag{9.117}$$

where $\psi_{\underline{k}'}$ and $\psi_{\underline{k}}$ are the unperturbed wave functions of the electron corresponding respectively to the final and the initial state, and U_f, U_i the wave functions for the radiation oscillator.

Replacing H_1 by the expression (9.113) and $\underline{A}$ by the expression (9.114), we get

$$|M^{\pm}(\underline{k},\underline{k}')| = (e\hbar A_0/m_0)|\hat{\underline{a}}\cdot\int\left[U^*_{\underline{k}'}\exp(-i\underline{k}'\cdot\underline{r})\exp(i\underline{k}\cdot\underline{r})\nabla_{\underline{r}}\left(U_{\underline{k}}\exp(i\underline{k}\cdot\underline{r})\right)\right]d\underline{r}| \times\left(n_p+\frac{1}{2}\pm\frac{1}{2}\right)^{1/2} \ . \tag{9.118}$$

Expression (9.118) for $|M^{\pm}(\underline{k},\underline{k}')|$ may be simplified further noting that

$$(H_0-E_{\underline{k}})\psi_{\underline{k}} = 0 \ , \quad \partial(H_0-E_{\underline{k}})\psi_{\underline{k}}/\partial k_x = 0 \ ,$$

$$(H_0-E)\left\{\partial\left[U_{\underline{k}}\exp(i\underline{k}\cdot\underline{r})\right]/\partial k_x\right\} - \left(i\hbar^2/m_0\right)\partial_x\psi_{\underline{k}} - (\partial E_{\underline{k}}/\partial k_x)\psi_{\underline{k}} = 0 \ .$$

Hence,

$$|M(\underline{k},\underline{k}')| = |(e\hbar A_0/m_0)\hat{\underline{a}}\cdot\int[\psi^*_{\underline{k}'}\exp(\mp i\underline{\kappa}\cdot\underline{r})\underline{\nabla}_r\psi_{\underline{k}}]d\underline{r}|\left(n_p+\frac{1}{2}\pm\frac{1}{2}\right)^{\frac{1}{2}}$$
$$= |(eA_0/\hbar)\underline{a}\cdot\int\left\{\psi^*_{\underline{k}'}\exp(\mp i\underline{\kappa}\cdot\underline{r})\left[(H_0-E)\underline{\nabla}_{\underline{k}}U_{\underline{k}}\exp(i\underline{k}\cdot\underline{r})\right.\right.$$
$$\left.\left.-(\underline{\nabla}_{\underline{k}}E)\psi_{\underline{k}}\right]\right\}d\underline{r}| \quad . \qquad (9.119)$$

The first term may be shown to be equal to zero. We then have

$$|M(\underline{k},\underline{k}')| = (eA_0/\hbar)\underline{a}\cdot\int\psi^*_{\underline{k}}\exp(\mp i\underline{\kappa}\cdot\underline{r})(\underline{\nabla}_{\underline{k}}E)\psi_{\underline{k}'}\left(n_p+\frac{1}{2}\pm\frac{1}{2}\right)^{\frac{1}{2}}d\underline{r} \quad . \qquad (9.120)$$

It is evident from (9.120) that the transition probability is nonzero only when the following equations are satisfied:

$$\underline{k}' = \underline{k}\pm\underline{\kappa} \quad , \qquad (9.121)$$

and

$$E(\underline{k}') = E(\underline{k})\pm\hbar\omega \quad . \qquad (9.122)$$

We note, however, that the magnitude of the electron wave vector corresponding to the energy $\hbar\omega$ is about 10^8 m^{-1}, whereas the wave number κ of the radiation is about 10^6 m^{-1} for the wavelengths of 30 μm (since the refractive index of semiconductors is about 4). The wave vector of the electron in the final state $\underline{k}'$ is thus almost the same as in the initial state $\underline{k}$. The matrix element $M(\underline{k},\underline{k}')$ may thus be simplified to

$$M^{\pm}(\underline{k},\underline{k}') = (eA_0/\hbar)\hat{\underline{a}}\cdot(\underline{\nabla}_{\underline{k}}E)\left(n_p+\frac{1}{2}\pm\frac{1}{2}\right)^{\frac{1}{2}} \quad , \qquad (9.123)$$

as $\int U^*_{\underline{k}'}U_{\underline{k}}d\underline{r}=1$, k' being approximately equal to k. We, however, find that as $k'\approx\underline{k}$, the energy conservation condition cannot be satisfied, $\hbar\omega$ being comparable or larger than E(k). Hence, the probability of absorption of photons by the quasi-free electrons is zero. FCA does not, therefore, take place if we consider only the photons and electrons. This result is the same in classical theory. We have seen that according to the classical theory, FCA is proportional to $<\tau^{-1}>$. When electrons and photons only are considered, scattering of the electrons by phonons or impurity atoms is neglected. Hence, the value of τ is infinitely large and the FCA is zero.

A finite value of FCA is obtained only when we consider the effect of phonons and impurity centres on the interaction of the electrons and the photons. In order not to complicate our discussion we shall first consider only the phonons. The effect of impurity centres may be included in the final formula by inspection. The simultaneous interaction of the electrons, the photons and the phonons may be dealt with by the perturbation theory since $\omega\tau$, as discussed earlier, is much larger than unity. We may hence neglect linewidth broadening effects [9.26] and consider the collision events as rather infrequent compared to the period of oscillation of the radiation.

The transition probability of the electron from the $\underline{k}$ state to $\underline{k}'$ state is given by

$$T^{\pm}_{\pm(\underline{k}\rightarrow\underline{k}')} = (2\pi/\hbar)|M^{\pm}_{\pm}(\underline{k},\underline{k}')|^2\delta[E_{\underline{k}'}-E_{\underline{k}}\mp\hbar\omega\mp\hbar\omega_q] \quad . \tag{9.124}$$

The superscript corresponds to the photon interaction and subscript to phonon interaction; $\hbar\omega_q$ is the energy of a phonon, and n_q is the average phonon occupation number given by

$$n_q = [\exp(\hbar\omega_q/k_BT)-1]^{-1} \quad . \tag{9.125}$$

The matrix element $M^{\pm}_{\pm}(\underline{k},\underline{k}')$ is expressed as

$$\begin{aligned}|M^{\pm}_{\pm}(\underline{k},\underline{k}')| = {} & |M^{\pm}_{pt}(\underline{k},\underline{k}_{int})M^{\pm}_{pn}(\underline{k}_{int},\underline{k}')/(E_0-E_{int}) \\ & +M^{\pm}_{pt}(\underline{k}_{int},\underline{k}')M^{\pm}_{pn}(\underline{k},\underline{k}_{int})/(E_0-E'_{int})| \quad ,\end{aligned} \tag{9.126}$$

where $M^{\pm}_{pt}$ and $M^{\pm}_{pn}$ represent respectively the matrix elements for the transitions through the interaction with the photons and the phonons. The plus and minus signs correspond respectively to the absorption and the emission of a photon or a phonon. The transition from the $\underline{k}$ state to the $\underline{k}'$ state may be envisaged to occur in two stages.

In the initial stage the energy of the system consisting of the photon, the electron and the phonon is E_0. In the first stage of transition the electron absorbs or emits a photon to reach an intermediate stage in which the electron wave vector is given by

$$\underline{k}_{int} = \underline{k} \pm \underline{\kappa} \approx \underline{k} \quad . \tag{9.127}$$

The energy in the intermediate stage is E_{int} and

$$E_0 - E_{int} = E(\underline{k}) - E(|\underline{k}\pm\underline{\kappa}|) \pm \hbar\omega \approx \pm\hbar\omega \quad . \tag{9.128}$$

This intermediate stage is evidently a virtual state and the energy is not conserved in this transition. In the second stage, the electron passes from the $\underline{k}_{int}$ state to the $\underline{k}'$ state by absorbing or emitting a phonon, so that

$$\underline{k}' = \underline{k}_{int} \pm \underline{q} \approx \underline{k} \pm \underline{q} \quad , \tag{9.129}$$

and

$$E(\underline{k}') = E(\underline{k}) \pm \hbar\omega_q \pm \hbar\omega \quad , \tag{9.130}$$

$\underline{q}$ being the wave vector of the phonon.

The second term in the matrix element corresponds to the same transition, in which the order of absorption or emission of the photons and the phonons is different. In this case, in the first stage the electron absorbs or emits a phonon to reach a virtual stage such that

$$\underline{k}_{int} = \underline{k} \pm \underline{q} \quad , \tag{9.131}$$

and

$$E_0 - E'_{int} = E(\underline{k}) - E(\underline{k}_{int}) \pm \hbar\omega_q \quad . \tag{9.132}$$

In the second stage, the electron absorbs or emits a photon, so that

$$\underline{k}' = \underline{k}_{int} \pm \underline{\kappa} \approx \underline{k}_{int} \quad , \tag{9.133}$$

and

$$E(\underline{k}') = E(\underline{k}) \pm \hbar\omega_q \pm \hbar\omega \quad . \tag{9.134}$$

Using relations (9.130) and (9.134) we may write the transition probability as

$$T^{\pm}_{\pm}(\underline{k}\to\underline{k}') = (2\pi/\hbar^3\omega^2)|M^{\pm}_{pt}(\underline{k},\underline{k})M^{\pm}_{pn}(\underline{k},\underline{k}') - M^{\pm}_{pt}(\underline{k}',\underline{k}')M^{\pm}_{pn}(\underline{k},\underline{k}')|^2\delta(E_{\underline{k}'}-E_{\underline{k}}\pm\hbar\omega\pm\hbar\omega_q) \quad ,$$

since $E_0 - E'_{int} = \mp\hbar\omega$. Replacing the matrix element for the photon interactions by the expression (9.123), we get

$$T^{\pm}_{\pm(\underline{k}\to\underline{k}')} = (2\pi/\hbar^3\omega^2)(eA_0/\hbar)^2\left\{\hat{\underline{a}}\cdot\left[(\nabla_{\underline{k}}E)_{\underline{k}}-(\nabla_{\underline{k}}E)_{\underline{k}'}\right]\right\}^2$$
$$\times\ |M^{\pm}_{pn}(\underline{k},\underline{k}')|^2\left(n_p+\frac{1}{2}\pm\frac{1}{2}\right)\delta(E_{\underline{k}'}-E_{\underline{k}}\mp\hbar\omega\mp\hbar\omega_q)$$
$$= S^{\pm}_{\pm}(\underline{k},\underline{k}')\left(n_p+\frac{1}{2}\pm\frac{1}{2}\right)\left(n_q+\frac{1}{2}\pm\frac{1}{2}\right)\delta(E_{\underline{k}'}-E_{\underline{k}}\mp\hbar\omega\mp\hbar\omega_q)\quad, \tag{9.135}$$

where the subscripts ± correspond to the photons and the superscript ± to the phonons.

We may now write the expression for dn_p/dt, using the above expression for the transition probability. We find that dn_p/dt is given by

$$-(dn_p/dt) = 2V_c^2(2\pi)^{-6}\iint\left[S^{+}_{+}(\underline{k},\underline{k}')\delta(E_{\underline{k}'}-E_{\underline{k}}-\hbar\omega-\hbar\omega_q)n_pn_q\right.$$
$$+S^{-}_{+}(\underline{k},\underline{k}')\delta(E_{\underline{k}'}-E_{\underline{k}}-\hbar\omega+\hbar\omega_q)n_p(n_q+1)$$
$$-S^{+}_{-}(\underline{k},\underline{k}')\delta(E_{\underline{k}'}-E_{\underline{k}}+\hbar\omega-\hbar\omega_q)(n_p+1)n_q$$
$$\left.-S^{-}_{-}(\underline{k},\underline{k}')\delta(E_{\underline{k}'}-E_{\underline{k}}+\hbar\omega+\hbar\omega_q)(n_p+1)(n_q+1)\right]f(\underline{k})\,[1-f(\underline{k}')]d\underline{k}d\underline{k}'\quad. \tag{9.136}$$

The four terms on the right-hand side of (9.136) arise from the interactions in which an electron is scattered from the $\underline{k}$ state to the $\underline{k}'$ state by absorbing a photon and a phonon, by absorbing a photon and emitting a phonon, by emitting a photon and absorbing a phonon and by emitting a photon and also a phonon. The rate of transition being determined by the occupancy of the $\underline{k}$ states and vacancies in the $\underline{k}'$ states we have the factor $f(\underline{k})[1-f(\underline{k}')]$. Integration over all the $\underline{k}'$ states is carried out to determine the rate of change of photon population due to the transition from a $\underline{k}$ state to all the possible $\underline{k}'$ states. The final integration over the $\underline{k}$ states gives the total rate of change of population caused by all the electrons. The factor $2V_c(2\pi)^{-6}$ accounts for the density of states in the $\underline{k}$ and $\underline{k}'$ space and for the spin-degeneracy of each state. It is assumed that the spin is not changed in any transition.

Since $\delta(-x)=\delta(x)$, we may rearrange the terms in dn_p/dt as follows by interchanging k and k' in the third and fourth terms:

$$-(dn_p/dt) = 2V_c^2(2\pi)^{-6}\iint\left\{n_pn_qS^{+}_{+}(\underline{k},\underline{k}')f(\underline{k})[1-f(\underline{k}')]\right.$$
$$-(n_p+1)(n_q+1)S^{-}_{-}(\underline{k},\underline{k}')f(\underline{k}')[1-f(\underline{k})]\delta(E_{\underline{k}'}-E_{\underline{k}}-\hbar\omega-\hbar\omega_q)$$
$$+\left[n_p(n_q+1)S^{-}_{+}(\underline{k},\underline{k}')f(\underline{k})(1-f(\underline{k}'))-(n_p+1)n_qS^{+}_{-}(\underline{k},\underline{k}')f(\underline{k}')\right.$$
$$\left.\left.\times(1-f(\underline{k}))\right]\delta(E_{\underline{k}'}-E_{\underline{k}}-\hbar\omega+\hbar\omega_q)\right\}d\underline{k}d\underline{k}'\quad. \tag{9.137}$$

In the absence of the external radiation dn_p/dt is equal to zero, and therefore the two terms should be individually equal to zero. In addition,

$$n_{po}n_q f(\underline{k})[1-f(\underline{k}')] = \left[\exp(\hbar\omega/k_BT)-1\right]^{-1}\left[\exp(\hbar\omega_q/k_BT)-1\right]^{-1}$$
$$\times\left[\exp\left((E_{\underline{k}}-E_F)/k_BT\right)+1\right]^{-1}\left[1+\exp\left((E_F-E_{\underline{k}'})/k_BT\right)\right]^{-1} \tag{9.138}$$

$$(n_{po}+1)(n_q+1)f(\underline{k}')[1-f(\underline{k})] = \exp(\hbar\omega/k_BT)n_{po}\,\exp(\hbar\omega_q/k_BT)n_q$$
$$\times\left[\exp\left((E_{\underline{k}'}-E_F)/k_BT\right)+1\right]^{-1}\left[1+\exp\left((E_F-E_{\underline{k}})/k_BT\right)\right]^{-1}$$
$$= n_{po}n_q\left[\exp\left((E_{\underline{k}}-E_F)/k_BT\right)+1\right]^{-1}\left[\exp\left((E_{\underline{k}'}-E_F)/k_BT\right)+1\right]^{-1}\left[\exp\left((E_{\underline{k}}+\hbar\omega+\hbar\omega_q-E_F)/k_BT\right)\right]$$
$$= n_{po}n_q\left[\exp\left((E_{\underline{k}}-E_F)/k_BT\right)+1\right]^{-1}\left[\exp\left((E_F-E_{\underline{k}'})/k_BT\right)+1\right]^{-1} = n_{po}n_q f(\underline{k})[1-f(\underline{k}')] \quad . \tag{9.139}$$

We have used the energy conservation condition, i.e. $E_{\underline{k}'} = E_{\underline{k}} + \hbar\omega + \hbar\omega_q$, in deriving the above identity. It may be shown similarly that

$$n_{po}(n_q+1)f(\underline{k})[1-f(\underline{k}')] = (n_{po}+1)n_q f(\underline{k}')[1-f(\underline{k})] \quad . \tag{9.140}$$

Utilising the above identities and equating the two terms to zero, we get

$$S_+^+(\underline{k},\underline{k}') = S_-^-(\underline{k}',\underline{k}) \quad \text{and} \quad S_+^-(\underline{k},\underline{k}') = S_-^+(\underline{k}',\underline{k}) \quad . \tag{9.141}$$

The wave vectors $\underline{k}'$ and $\underline{k}$ in the above two equations are related by the energy conservation conditions of the relevant delta functions.

Using the above equilibrium relations, we get

$$-(dn_p/dt) = 2V_c^2(2\pi)^{-6}(n_p-n_{po})\sum_{+,-}\iint\left\{\left[S_+^\pm(\underline{k},\underline{k}')\left(n_q+\frac{1}{2}\mp\frac{1}{2}\right)f(\underline{k})\right.\right.$$
$$\left.\left.\times(1-f(\underline{k}'))-\left(n_q+\frac{1}{2}\pm\frac{1}{2}\right)f(\underline{k}')(1-f(\underline{k}))\right]\right\}\delta(E_{\underline{k}'}-E_{\underline{k}}-\hbar\omega\mp\hbar\omega_q)d\underline{k}d\underline{k}' \quad . \tag{9.142}$$

Further,

$$\left(n_q+\frac{1}{2}\mp\frac{1}{2}\right)f(\underline{k})(1-f(\underline{k}')) - \left(n_q+\frac{1}{2}\pm\frac{1}{2}\right)f(\underline{k}')(1-f(\underline{k}))$$
$$= \left(n_q+\frac{1}{2}\mp\frac{1}{2}\right)[f(\underline{k})-f(\underline{k}')] \mp f(\underline{k}')[1-f(\underline{k})] \quad . \tag{9.143}$$

Combining the above identity with the equilibrium condition

$$n_{po}\left(n_q+\frac{1}{2}\mp\frac{1}{2}\right)f(\underline{k})[1-f(\underline{k}')] - (n_{po}+1)\left(n_q+\frac{1}{2}\pm\frac{1}{2}\right)f(\underline{k}')[1-f(\underline{k})] = 0 \quad , \tag{9.144}$$

and noting that

$$f(\underline{k})[1-f(\underline{k}')] - f(\underline{k}')[1-f(\underline{k})] = f(\underline{k}) - f(\underline{k}') \quad , \tag{9.145}$$

we get

$$\begin{aligned}&\left(n_q+\frac{1}{2}\mp\frac{1}{2}\right)f(\underline{k})[1-f(\underline{k}')] - \left(n_q+\frac{1}{2}\pm\frac{1}{2}\right)f(\underline{k}')[1-f(\underline{k})]\\&= [\sinh(\hbar\omega/2k_BT)]\left\{2\sinh(\omega_q/2k_BT)\sinh[(\omega\pm\omega_q)/2k_BT]\right\}^{-1}[f(\underline{k})-f(\underline{k}')]\\&= F_\pm(\omega,\omega_q)[f(\underline{k})-f(\underline{k}')] \quad .\end{aligned} \tag{9.146}$$

It should be noted that in (9.143-146), $\underline{k}'$ and $\underline{k}$ are such that

$$E(\underline{k}') = E(\underline{k}) + \hbar\omega \pm \hbar\omega_q \quad . \tag{9.147}$$

Using the above results and the expression for the absorption coefficient (9.112) given earlier, we get finally

$$\alpha = 2V_c^2(2\pi)^{-6}(n_r/c)\sum_{+,-}\iint S_+^\pm(\underline{k},\underline{k}')[f(\underline{k})-f(\underline{k}')]F_\pm(\omega,\omega_q)\delta(E_{\underline{k}'}-E_{\underline{k}}-\hbar\omega\mp\hbar\omega_q)d\underline{k}d\underline{k}' \ . \tag{9.148}$$

Expressions for the absorption coefficient due to the different scattering mechanisms may now be obtained by using for the matrix element $M_{pn}^\pm(\underline{k},\underline{k}')$ in $S_+^\pm(\underline{k},\underline{k}')$ the relevant expressions given in Table 6.1.

b) Formulae for the Absorption Coefficient

We present in this section formulae for the absorption coefficient due to the various scattering mechanisms. In order to indicate the steps involved in the derivation of these formulae, we shall derive that for acoustic phonon scattering. We shall assume the energy-wave-vector relation to be isotropic and parabolic. The electron concentration is also assumed to be small so that the effects of electron screening may be neglected.

Putting the expression for $M_{pn}^\pm(\underline{k},\underline{k}')$ from Table 6.1 into the expression (9.148) for the absorption coefficient, we get

$$\begin{aligned}\alpha = \sum_{+,-} C\int&\left\{\hat{\underline{a}}\cdot\left[(\underline{\nabla}_{\underline{k}}E)_{\underline{k}}-(\underline{\nabla}_{\underline{k}}E)_{\underline{k}'}\right]\right\}^2F_\pm(\omega,\omega_q)(\hbar\omega_q/2k_BT)\\&\times[f(\underline{k})-f(\underline{k}')]\delta(E_{\underline{k}}-E_{\underline{k}'}+\hbar\omega\mp\hbar\omega_q)d\underline{k}d\underline{k}' \quad ,\end{aligned} \tag{9.149}$$

where

$$C = e^2E_1^2k_BT(2\pi)^{-5}\left[\varepsilon_0\hbar^4\omega^3\rho s^2cn_r\right]^{-1} \quad ,$$

$\underline{k}$ and $\underline{k}'$ are respectively the wave vectors of the electron before and after scattering and $\underline{q}$ that of the phonon. These are related by

$$\underline{k}' = \underline{k} \pm \underline{q} \quad . \tag{9.150}$$

The value of the quantity within the braces depends on the orientation of the polarisation vector in relation to the phonon wave vector. For unpolarised signals, it has the average value

$$\left\{\hat{\underline{a}}\cdot\left[(\underline{\nabla}_kE)_{\underline{k}}-(\underline{\nabla}_kE)_{\underline{k}'}\right]\right\}^2_{av} = (1/3)|\underline{k}-\underline{k}'|^2(\hbar^2/m^*)^2 \quad . \tag{9.151}$$

We also note that for acoustic phonons, $\omega_q \ll \omega$ and $\hbar\omega_q \ll k_BT$. Hence,

$$F_\pm(\omega,\omega_q) \approx k_BT/\omega_q \tag{9.152}$$

and

$$\delta(E_{\underline{k}}-E_{\underline{k}'}+\hbar\omega\pm\hbar\omega_q) \approx \delta(E_{\underline{k}}-E_{\underline{k}'}+\hbar\omega) \quad . \tag{9.153}$$

For the integration over $d\underline{k}'$ we choose $\underline{k}$ as the polar axis. We then have

$$\alpha = C_1\iint(k^2-2kk'\cos\theta+k'^2)[f(\underline{k})-f(\underline{k}')] \times \delta(E_{\underline{k}}-E_{\underline{k}'}+\hbar\omega)E_{\underline{k}'}^{\frac{1}{2}}dE_{\underline{k}'}\sin\theta\, d\theta d\underline{k} \quad , \tag{9.154}$$

where

$$C_1 = (\pi/3)(\hbar^2/m^*)^2(2m^*/\hbar^2)^{3/2}C = (\sqrt{2}/3)(2\pi)^{-4}e^2E_1^2k_BT\left(m^{*\frac{1}{2}}\varepsilon_0\hbar^3\omega^3\rho s^2cn_r\right)^{-1} \quad .$$

Integrating over $dE_{\underline{k}'}$ (using the property of the delta function) and over $d\underline{k}$, we get

$$\alpha = C_2\int(2E_{\underline{k}}+\hbar\omega)[f(E_{\underline{k}})-f(E_{\underline{k}}+\hbar\omega)](E_{\underline{k}}+\hbar\omega)^{\frac{1}{2}}d\underline{k} \quad , \tag{9.155}$$

where

$$C_2 = 2C_1(2m^*)/\hbar^2 = (4\sqrt{2}/3)(2\pi)^{-4}e^2E_1^2m^{*\frac{1}{2}}k_BT\left(\varepsilon_0\hbar^5\omega^3\rho s^2cn_r\right)^{-1} \quad .$$

We may proceed further analytically if we assume the sample to be nondegenerate so that the distribution function

$$f(E_{\underline{k}}) = n\hbar^3\sqrt{2}(\pi/m^*k_BT)^{3/2}\exp(-E_{\underline{k}}/k_BT) \quad , \tag{9.156}$$

and

$$f(E_{\underline{k}}+\hbar\omega) = \exp(-\hbar\omega/k_BT)f(E_{\underline{k}}) \quad , \tag{9.157}$$

where n is the electron concentration. Putting in the expression (9.155) and carrying out the integration over the azimuthal and polar angles, we get

$$\alpha = C_3\int_0^\infty(2E_{\underline{k}}+\hbar\omega)\exp(-E_{\underline{k}}/k_BT)(E_{\underline{k}}+\hbar\omega)^{\frac{1}{2}}E_{\underline{k}}^{\frac{1}{2}}dE_{\underline{k}} \quad , \tag{9.159}$$

where

$$C_3 = n8\pi^{5/2}[1-\exp(-\hbar\omega/k_BT)]C_2(k_BT)^{-3/2}$$
$$= (2\sqrt{2}/3)\pi^{-3/2}ne^2E_1^2m^{*\frac{1}{2}}[1-\exp(-\hbar\omega/k_BT)]\left(\varepsilon_0\hbar^5\omega^3\rho s^2cn_r\right)^{-1}(k_BT)^{-\frac{1}{2}} \quad .$$

We may express the above integral in terms of modified Bessel functions [9.25] to obtain

$$\alpha = C_4K_2(\hbar\omega/2k_BT)\exp(\hbar\omega/2k_BT) \quad , \tag{9.159}$$

where

$$C_4 = 2(\hbar\omega/2)^2k_BTC_3 = ne^2(4\pi\varepsilon_0)^{-1}(4\sqrt{2}/3)\pi^{-\frac{1}{2}}(k_BT)^{\frac{1}{2}}E_1^2m^{*\frac{1}{2}}$$
$$\times[1-\exp(-\hbar\omega/k_BT)]\left(\hbar^3\omega\rho s^2cn_r\right)^{-1} \quad .$$

The above expression for α may be simplified for the two limiting cases. (i) $\hbar\omega \ll k_BT$ and (ii) $\hbar\omega \gg k_BT$.

We get in the first case [9.26]

$$K_2(\hbar\omega/2k_BT) \approx 8k_B^2T^2/\hbar^2\omega^2 \quad , \quad \text{since} \quad K_2(z) \cong 2/z^2 \quad . \tag{9.160}$$

Also, $\exp(\hbar\omega/2k_BT)\approx 1$, $1-\exp(-\hbar\omega/k_BT)\approx\hbar\omega/k_BT$. Hence,

$$\alpha = (8\sqrt{2}/3)\pi^{-3/2}ne^2E_1^2m^{*\frac{1}{2}}(k_BT)^{3/2}\left(\varepsilon_0\hbar^4\omega^2\rho s^2cn_r\right)^{-1} \quad . \tag{9.161}$$

It is of interest to compare the expression given by the classical theory. Putting the expression for the relaxation time due to acoustic phonon scattering (see Table 8.1)

$$\tau^{-1} = \sqrt{2}\ E_1^2m^{*3/2}(k_BT)^{3/2}(\pi\rho\hbar^4s^2)^{-1}(E/k_BT)^{\frac{1}{2}} \quad . \tag{9.162}$$

in the classical expression for α (9.110), and evaluating the average for the Maxwellian energy distribution we get the same expression as (9.161). Thus, in the limit when $\hbar\omega \ll k_BT$, the quantum theory and classical theory give identical results. This condition may therefore be used as a criterion to judge whether to use the classical theory or the quantum theory.

It may be mentioned in this connection that the classical expression for α has often been used in the literature [9.22e] to analyse FCA even when this criterion is not satisfied, for the sake of simplicity. The criterion used is the index of wavelength dependence, which is 2 for the classical theory. But this criterion is not good enough to justify the use of the classical theory. Further, in the formula for α, $\langle\tau^{-1}\rangle$ is often replaced by $e/m^*\mu_d$, where μ_d is the DC mobility. If we use the expression for μ_d for acoustic phonon scattering (see Table 8.2), we get for α

$$\begin{aligned}\alpha &= ne^2\langle\tau^{-1}\rangle(m^*\omega^2cn_r\varepsilon_0)^{-1} = ne^2(m^*\omega^2cn_r\varepsilon_0)^{-1}|e|(m^*\mu_{ac})^{-1}\\ &= 3\sqrt{2}\ ne^2E_1^2m^{*\frac{1}{2}}(k_BT)^{3/2}(4\sqrt{\pi}cn_r\varepsilon_0\rho\hbar^4s^2\omega^2)^{-1} \quad .\end{aligned} \tag{9.163}$$

The above expression is smaller than the exact expression by a factor of $9\pi/32$. This difference arises from the inequality $\langle\tau^{-1}\rangle \neq \langle\tau\rangle^{-1}$ and may account for discrepancies in the values of the parameters obtained from the analysis of the FCA even when $\hbar\omega \ll k_BT$ and the classical theory is expected to apply.

In the second limiting case, i.e. when $\hbar\omega \gg k_BT$,

$$K_2(\hbar\omega/2k_BT) \simeq (\pi/2)^{\frac{1}{2}}(\hbar\omega/2k_BT)^{-\frac{1}{2}}\exp(-\hbar\omega/2k_BT) \quad , \tag{9.164}$$

and we get

$$\alpha = \sqrt{2}\; ne^2 m^{*\frac{1}{2}} k_B T E_1^2 \left(3\pi\varepsilon_0 c n_r \rho s^2 \omega^{3/2} \hbar^{7/2}\right)^{-1} \quad . \tag{9.165}$$

We find that the expression is qualitatively different from that for the classical case. Both the temperature and the wavelength dependences are different. In this case the absorption varies as T and as $\lambda^{1.5}$ whereas in the classical case it varies as $T^{1.5}$ and as λ^2. Thus when $\hbar\omega$ is comparable to k_BT, we find that the FCA should always be analysed using the quantum theory. In order to illustrate the magnitude of difference we have plotted in Fig.9.9 the absorption coefficient obtained from the quantum theory as a function of $\hbar\omega/k_BT$, normalised by the value for the classical theory. We find that the discrepancy increases with $\hbar\omega/k_BT$ and approaches a value $(4/\sqrt{\pi})(\hbar\omega/k_BT)^{\frac{1}{2}}$, which may be substantial for the experimental conditions.

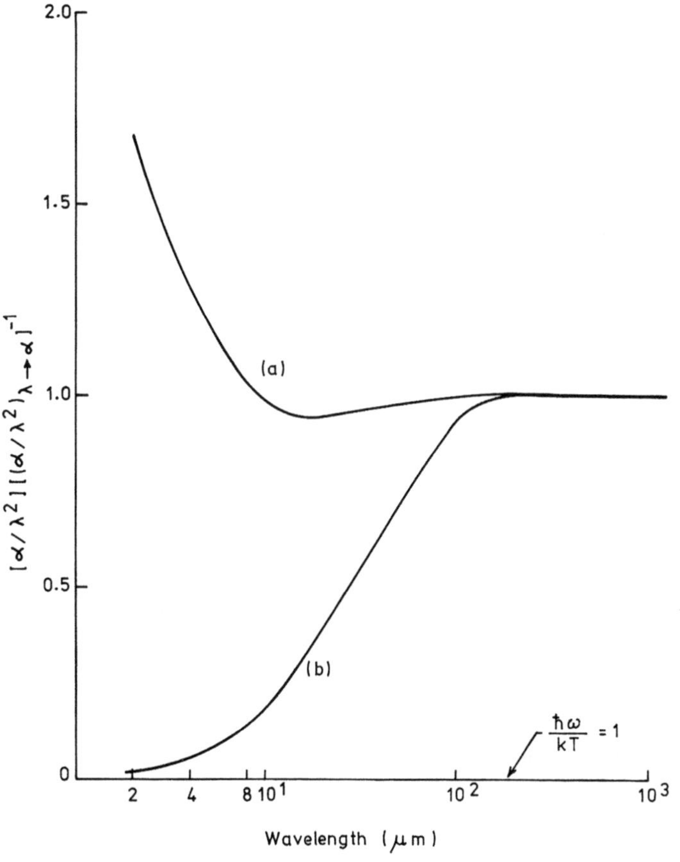

Fig. 9.9. Deformation potential acoustic phonon and ionised impurity scattering-limited FCA. (a) Acoustic phonon; (b) Ionised impurity

Table 9.1. Expressions for the absorption coefficients

Scattering mechanisms	Expression for α/n for parabolic nondegenerate material	Index of wavelength dependence in the quantum limit
Deformation potential acoustic phonon	$\dfrac{2\sqrt{2}e^2E_1^2(m^*k_BT)^{\frac{1}{2}}\sinh(\hbar\omega/2k_BT)K_2(\hbar\omega/2k_BT)}{3\pi^{3/2}\varepsilon_0\hbar^3\rho s^2cn_r\omega}$	1.5
Piezoelectric acoustic phonon	$\dfrac{\sqrt{2}e^4\left(h_{pz}^2/\varepsilon\rho s^2\right)(k_BT)^{\frac{1}{2}}\sin(\hbar\omega/2k_BT)K_1(\hbar\omega/2k_BT)}{3\pi^{3/2}\varepsilon_0\varepsilon cn_rm^{*\frac{1}{2}}\omega^2\hbar^2}$	2.5
Ionised impurity (neglecting screening)	$\dfrac{N_iZ^2e^6\ \sinh(\hbar\omega/2k_BT)K_0(\hbar\omega/2k_BT)}{3\sqrt{2}\ \pi^{3/2}\hbar\varepsilon^2\varepsilon_0cn_rm^{*3/2}\omega^3(k_BT)^{\frac{1}{2}}}$	3.5
Nonpolar optic phonon	$\dfrac{2\sqrt{2}e^2D_0^2m^{*\frac{1}{2}}(k_BT)^{3/2}F_{op}(z,z_+,z_-)}{3\pi^{3/2}\varepsilon_0\hbar^4\rho cn_r\omega_0\omega^3}$	1.5
	$F_{op}(z,z_+,z_-)=\left[\sinh(z_+-z)z_+^2K_2(z_+)+\sinh(z_-+z)z_-^2K_2(\lvert z_-\rvert)\right]/\sinh z$;	
	$z=(\hbar\omega_0/2k_BT)$; $z_+=\dfrac{\hbar(\omega_0+\omega)}{2k_BT}$; $z_-=\dfrac{\hbar(\omega-\omega_0)}{2k_BT}$.	
Polar optic	$\dfrac{\left(K^{-1}-K_S^{-1}\right)\sqrt{2}e^4\hbar\omega_\ell(k_BT)^{\frac{1}{2}}F_{pop}(z,z_+,z_-)}{4\pi\varepsilon_0 3\pi^{\frac{1}{2}}\varepsilon_0\hbar^3\omega^3cn_rm^{*\frac{1}{2}}}$	2.5
	$F_{pop}(z,z_+,z_-)=[\sinh(z_+-z)z_+K_1(z_+)+\sinh(z_-+z)K_1(\lvert z_-\rvert)z_-]/\sinh z$;	
	$z=\dfrac{\hbar\omega_\ell}{2k_BT}$, $z_+=\dfrac{\hbar(\omega+\omega_\ell)}{2k_BT}$, $z_-=\dfrac{\hbar(\omega-\omega_\ell)}{2k_BT}$.	

Expressions for α for the various scattering mechanisms may be obtained following the procedure outlined above. Extension of the formulae to cover the case of nonparabolic bands also does not present any special problem. We have collected in Table 9.1 the available expressions for the nondegenerate materials with isotropic parabolic energy bands.

c) Formulae for Isotropic Nonparabolic Energy Bands

Formulae for absorption coefficients in materials with nonparabolic energy bands may be derived following the procedure outlined in the preceding section. The final expressions cannot, however, be reduced to analytic forms and the integration over energy has to be carried out numerically. We shall not derive the formulae but only quote those available in the literature.

Expression for the FCA for isotropic nonparabolic bands may be written [9.29] as

$$\alpha = \sum_{s=+,-} C_{\nu s} \int_0^\infty (F_s X_\nu - J_s Y_\nu)[f(y)-f(y_s')]dy \quad , \tag{9.166}$$

where the subscript ν indicates the kind of scattering,

$$y = E/k_BT \quad , \quad y_\pm' = y + \hbar\omega/k_BT \qquad \text{for } \nu = \text{ac, imp.}$$

$$y_\pm' = y + (\hbar\omega \pm \hbar\omega_\ell)/k_BT \quad \text{for } \nu = \text{op,pop.}$$

$$f(y) = [\exp(y-z)+1]^{-1} \quad , \quad z = E_F/k_BT \quad ;$$

$$F_\pm = \frac{xy}{(x+y)^2} + \frac{x_\pm' y_\pm'}{(x_\pm'+y_\pm')^2} \frac{(x+y)(x_\pm'+y_\pm')}{(x+y+\delta)(x_\pm'+y_\pm'+\delta)} \quad ;$$

$$J_\pm = P_\pm/[(x+y+\delta)(x_\pm'+y_\pm'+\delta)] \quad ;$$

$$x = y + \beta \; ; \quad x_\pm' = y_\pm' + \beta \quad ; \quad \beta = E_g/k_BT \quad ,$$

$$P_\pm = [xx_\pm' yy_\pm']^{\frac{1}{2}} \quad ; \quad A = yy_\pm' \quad , \quad B = xx_\pm' \quad ;$$

$$\delta = [E_g\Delta/D]\beta \quad ; \quad D = 6E_g^2 + 9E_g\Delta + 4\Delta^2 \quad .$$

The energy-wave-vector dispersion relation as given by (4.77) has been assumed to apply. The constants C_ν and the functions X_ν and Y_ν are given by the following expressions for the different scattering mechanisms:

i) Deformation Potential Acoustic Phonon Scattering

$$C_{ac\pm} = \frac{2e^2E_1^2 m^{*2}(k_BT)^4}{3\pi^2\hbar^8 4\pi\varepsilon_0 c n_r \rho s^2 \beta^2 \omega^3} \quad ,$$

$$X_{ac\pm} = \left[\frac{1}{2}(b_2+b_4)A_\pm + b_1 2B_\pm\right]P_\pm \quad ,$$

$$Y_{ac\pm} = b_3 8P^2/3 \quad ,$$

$$b_1 = 4(E_g+\Delta)^2(3E_g+2\Delta)^2/D^2 \quad ,$$

$$b_2 = 16\Delta^2(2E_g+\Delta)^2/D^2 \quad ,$$

$$b_3 = 4\left(3E_g^2+4E_g\Delta+2\Delta^2\right)\left(3E_g^2+5E_g\Delta+2\Delta^2\right)/D^2 \quad . \tag{9.167}$$

ii) Nonpolar Optic Phonon Scattering

$$C_{nop\pm} = \frac{e^2}{3\pi^2}\frac{m^{*2}D_0^2(k_BT)^3}{\hbar^7\rho c n_r\omega^3\omega_0\beta^2 4\pi\varepsilon_0}\cdot\frac{\sinh(\hbar\omega/2k_BT)}{\sinh(\hbar\omega_0/2k_BT)\sinh(\hbar|\omega\pm\omega_0|/k_BT)} \quad . \tag{9.168}$$

Expressions for X_{nop} and Y_{nop} are identical to those for acoustic phonon scattering but y' is given by

$$y'_\pm = y + \hbar(\omega\pm\omega_0)/k_BT \quad . \tag{9.169}$$

iii) Ionised Impurity Atom Scattering

$$C_{imp\pm} = N_i e^6 k_BT\left(24\hbar^4 c n_r\pi^3\varepsilon^2\varepsilon_0\omega^3\right)^{-1} \quad ,$$

$$X_{imp\pm} = b_4(3A_\pm)/(4P_\pm) + 2P_\pm R'_\pm/D_\pm - (1/2)[b_3+b_4(3S_\pm)/(8B_\pm)]L_\pm \quad ,$$

$$Y_{imp\pm} = 2b_3 + b_4 9S_\pm/8B_\pm + 2S_\pm R'_\pm/D_\pm$$
$$- (1/2P_\pm)(b_1B_\pm+b_2A_\pm/4+2b_3S_\pm+b_4 9S_\pm^2/16B_\pm)L_\pm \quad ,$$

$$b_4 = 16(E_g+\Delta)^2\left(3E_g^2+2E_g\Delta+\Delta^2\right)/D^2 \quad ,$$

$$S_\pm = xy + x'_\pm y'_\pm + a \quad ,$$

$$R'_\pm = (1/4)(b_2+3b_4)A_\pm + b_1B_\pm + b_3S_\pm \quad ,$$

$$D_\pm = S_\pm^2 - 4P_\pm^2 \quad ,$$

$$L_\pm = \ell n[(S_\pm+2P_\pm)/(S_\pm-2P_\pm)] \quad ,$$

$$a = \frac{e^2}{\sqrt{2}\pi^2\varepsilon}\left(\frac{m^*}{\hbar^2\beta k_BT}\right)^{\frac{1}{2}}\int_0^\infty (xy)^{\frac{1}{2}}(x+y)f(y)[1-f(y)]dy \quad . \tag{9.170}$$

iv) Polar Optic Phonon Scattering

$$C_{pop\pm} = e^4m^*(k_BT)^2\omega_\ell\left(24\hbar^5 c n_r\beta\omega^3\pi^3\varepsilon_0^2\right)^{-1}\sinh(\hbar\omega/2k_BT)\left(K_\alpha^{-1}-K_S^{-1}\right)$$
$$\times\left[\sinh(\hbar\omega_\ell/2k_BT)\sinh(\hbar|\omega\pm\omega_\ell|/2k_BT)\right]^{-1} \quad ,$$

$$X_{pop\pm} = \left\{(1/2)\left[b_1B_\pm+b_3(S_\pm+a)\right]+(1/8)\left[b_2A_\pm+b_4 3S_\pm(S_\pm+2a)/4B_\pm\right]\right\}L_\pm$$
$$- \left[2b_3P_\pm+(3/8)b_4(S_\pm+2a)A_\pm/P_\pm+2P_\pm R'_\pm a/D_\pm\right] \quad ,$$

$$Y_{pop\pm} = \Big\{(1/2)\big[(S_\pm+a)(b_3S_\pm+b_1B_\pm+b_2A_\pm/4+b_3S_\pm a\big]+(3/32)b_4S_\pm^2(S_\pm+3a)/B_\pm\Big\}L_\pm/P_\pm$$
$$-\big[2b_3(S_\pm+a)+(1/2)(b_2+b_4)A_\pm+2b_1B_\pm+3b_4S_\pm(S_\pm+3a)/8B_\pm+2S_\pm R'_\pm/D_\pm\big] \quad . \tag{9.171}$$

The formulae given above have been used to calculate the FCA in InSb [9.27] and InAs [9.28], in which the effects of nonparabolicity are large, as the band-gaps are small. However, in the calculation of the FCA, the effect of nonparabolicity is important even in a material with a comparatively larger band gap, for which the nonparabolic effect may be neglected in the calculation of DC mobility. This is because in mobility calculations nonparabolic effects become significant when k_BT is comparable with E_g, whereas in FCA, the effect may be important when $\hbar\omega$ is comparable with E_g even though k_BT may be much smaller than E_g. In order to illustrate the effect of nonparabolicity in a large-band-gap material, we have plotted in Fig.9.10 the calculated values of the FCA in InP at room temperature, both including and excluding the effects of nonparabolicity. We find that nonparabolicity contributes about 40% in the experimental wavelength range even though k_BT is only 10^{-2} times E_g.

d) Formulae for Anisotropic Energy Bands

Formulae for FCA in materials with ellipsoidal constant-energy surfaces have also been derived [9.29]. We illustrate below the derivation for the nonpolar

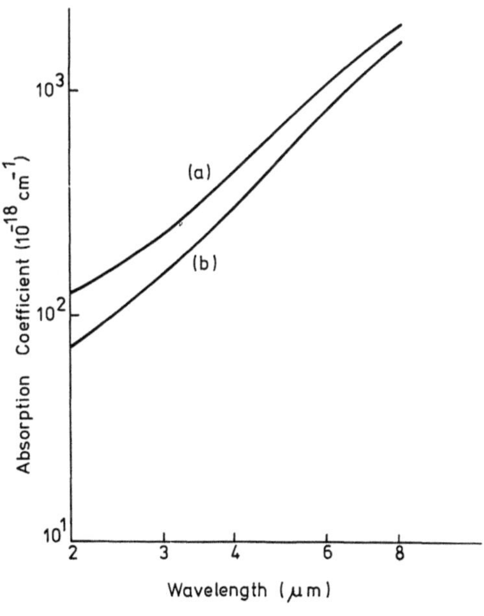

Fig. 9.10. Free-carrier absorption in indium phosphide for electron concentration 3.3×10^{17} [cm^{-3}]. (a) Parabolic; (b) Nonparabolic

optic phonon scattering. Putting the expression for the electron-phonon interaction matrix element in the expression for the absorption coefficient, we get

$$\alpha = \sum_{+,-} C\int\left\{\hat{\underline{a}}\cdot\left[(\underline{\nabla}_k E)_{\underline{k}}-(\underline{\nabla}_k E)_{\underline{k}'}\right]\right\}^2 F_\pm(\omega,\omega_0)$$
$$\times\ [f(\underline{k})-f(\underline{k}')]\delta(E_{\underline{k}}-E_{\underline{k}'}+\hbar\omega\pm\hbar\omega_0)d\underline{k}d\underline{k}' \quad , \qquad (9.172)$$

where

$$C = e^2 D_0^2 (2\pi)^{-4}(4\pi\varepsilon_0 c n_r \hbar^3\omega^3\rho\omega_0)^{-1} \quad ,$$

and the energy-wave-vector dispersion relation is

$$E = (\hbar^2/2)\left[(k_x^2/m_x)+(k_y^2/m_y)+(k_z^2/m_z)\right] \quad . \qquad (9.173)$$

Introducing the transformation

$$\underline{k}^* = \underline{\underline{\alpha}}^{\frac{1}{2}}\cdot\underline{k} \quad , \quad \underline{k}'^* = \underline{\underline{\alpha}}^{\frac{1}{2}}\cdot\underline{k}' \quad ,$$

where

$$\underline{\underline{\alpha}}^{\frac{1}{2}} = \begin{pmatrix} (m_c/m_x)^{\frac{1}{2}} & 0 & 0 \\ 0 & (m_c/m_y)^{\frac{1}{2}} & 0 \\ 0 & 0 & (m_c/m_z)^{\frac{1}{2}} \end{pmatrix} \quad , \qquad (9.174)$$

m_c is the conductivity effective mass [see (8.102)], and taking the average for the polarisation direction of the signal, we get

$$\alpha = \sum_{s=+,-} C_1\int [\underline{k}^*\cdot\underline{\underline{\alpha}}\cdot\underline{k}^*-\underline{k}'^*\cdot\underline{\underline{\alpha}}\cdot\underline{k}'^*]^2\delta(E_{\underline{k}}-E_{\underline{k}'}-\hbar\omega\pm\hbar\omega_0)$$
$$\times\ F_s(\omega,\omega_0)[f(\underline{k}^*)-f(\underline{k}'^*)]d\underline{k}^*d\underline{k}'^* \quad , \qquad (9.175)$$

where

$$C_1 = \hbar^4 m_d^3 C(3m_c^5)^{-1} = D_0^2 e^2 m_d^3\left(4\pi\varepsilon_0 c n_r\rho\omega_0 m_c^5\omega^3\right)^{-1}(1/3)(2\pi)^{-4}\hbar \quad ,$$
$$m_d = (m_x m_y m_z)^{1/3} \quad .$$

Expressing $\underline{k}^*$ and $\underline{k}'^*$ in polar coordinates with the x direction as the polar axis and carrying out the integrations over θ', ϕ', E_f, θ, and ϕ we obtain

$$\alpha = \sum_{+,-} C_2 \int_0^{\infty} (E_{\underline{k}}+E_{\underline{k}}+\hbar\omega\pm\hbar\omega_0)E_{\underline{k}}^{\frac{1}{2}}(E_{\underline{k}}+\hbar\omega\pm\hbar\omega_0)^{\frac{1}{2}}$$
$$\times F_{\pm}(\omega,\omega_0)\left[f(E_{\underline{k}})-f(E_{\underline{k}}+\hbar\omega\pm\hbar\omega_0)\right]dE_{\underline{k}} \quad , \tag{9.176}$$

where

$$C_2 = (4\pi)^2(1/4)(1/3)(2m_c/\hbar^2)^4\left(m_x^{-1}+m_y^{-1}+m_z^{-1}\right)m_c C_1$$
$$= 16D_0^2e^2m_d^3(4\pi\varepsilon_0 cn_r\rho\omega_0 3\hbar^7\omega^3 m_c)^{-1}(2\pi)^{-2} \quad ,$$

since $m_c^{-1} = (1/3)\left(m_x^{-1}+m_y^{-1}+m_z^{-1}\right)$. We shall now assume that the sample is non-degenerate and that the distribution function is Maxwellian. Expressing the integrals in terms of modified Bessel functions, we get finally

$$\alpha = \sum_{+,-} C_3 z_{\pm}^2 \sinh(z_{\pm}\mp z)K_2(z_{\pm})/\sinh z \quad , \tag{9.177}$$

where

$$C_3 = \sqrt{2}\pi^{3/2}n\hbar^3(4\pi)^2 2(k_BT)^3 C_2 m_d^{-3/2}(k_BT)^{-3/2}$$
$$= D_0^2e^2n8\sqrt{2}m_d^{3/2}(k_BT)^{3/2}/(4\pi\varepsilon_0 cn_r\rho\omega_0 3\sqrt{\pi}\hbar^4\omega^3 m_c) \quad ,$$
$$z = \hbar\omega_0/2k_BT \quad , \quad z_{\pm} = \hbar(\omega\pm\omega_0)/2k_BT \quad .$$

Comparing this expression with that given earlier for isotropic effective mass we find that $m^{*\frac{1}{2}}$ is replaced by $m_d^{3/2}/m_c$. This result could also be envisaged from the formulae for the conductivity mobility given earlier. A similar expression has also been derived for the deformation potential acoustic phonon scattering. In this case, however, the deformation potential constant as well as the acoustic velocity is anisotropic. Hence, when the integrations over θ', ϕ', and θ and ϕ are carried out we get an average value for the coupling constant. The final expression [9.29a] may be written as

$$\frac{\alpha}{n} = 2\sqrt{2}e^2m_d^{3/2}(k_BT)^{\frac{1}{2}} \sinh(\hbar\omega/2k_BT)K_2(\hbar\omega/2k_BT)$$
$$\times \left(3\pi^{3/2}\varepsilon_0\hbar^3\rho cn_r\omega m_c\right)^{-1}\left(E_{1av}^2/S_{av}^2\right) \quad , \tag{9.178}$$

where

$$E_{1av}^2 = (1/2)\left[1.07E_u^2 + 1.66E_uE_d + 1.34E_d^2\right](m_c/m_\perp) \quad ,$$

$$\rho S_{av}^2 = (2/5)(3C_{11}+2C_{12}+4C_{44}) \quad ,$$

for a material having mass and elastic constant anisotropy as in germanium. E_u and E_d are the deformation potential constants for the spheroidal band structure, and C_{11}, C_{12}, C_{44} are the usual elastic stiffness coefficients. The expression for S_{av}^2 remains the same for materials with the cubic crystal structure. The expression for E_{1av}^2, on the other hand, depends on the mass anisotropy and the anisotropy of the elastic stiffness coefficients. These anisotropies being different in different materials, the expression for E_{1av}^2 has to be worked out for each individual material. An approximate expression is often used for E_{1av}:

$$E_{1av} = E_d + E_u/3 \quad . \tag{9.179}$$

However, as E_u is not easily obtained from experiments, E_1 is often used as an empirical parameter. For such an analysis we find that we may use for the acoustic deformation potential scattering the expression for isotropic effective mass by replacing $m^{*\frac{1}{2}}$ by $m^{3/2}/m_c$, as indicated by the classical theory.

Analytic expressions of the above forms cannot be obtained for the other scattering mechanisms as the matrix element for the electron-phonon scattering depends on $|\underline{k}-\underline{k}'|$ in all other cases. Integrations over the polar and azimuthal angles must then be worked out numerically. A discussion of the effect of the mass anisotropy for polar optic phonon scattering has been given by DAS and NAG [9.30].

Expressions for the FCA in nonparabolic anisotropic bands may be worked out from expression (9.148). However, as very little simplification may be made analytically and as the algebra is rather involved, we shall not present these formulae here. The relevant expressions have been worked out for all the scattering mechanisms and the results have been discussed by DAS and NAG [9.31].

We should mention that we have given the formulae for the common scattering mechanisms, assuming that the intermediate state lies in the same energy band. It has been suggested by DUMKE [9.26] that the intermediate stage may lie in a secondary conduction band, and by JENSEN [9.32] that it may lie in the valence band. Contributions of such processes to the total FCA are not very significant for the experimental wavelength ranges and hence we shall exclude these scattering mechanisms. It has also been suggested that two-phonon scattering may account for discrepancies between experiment and theory for some materials [9.33]. However, as the values of the deformation potential constant for this scattering mechanism are not yet determined from independent experiments for most of the materials, it is not possible at this stage to assess its actual importance. A detailed expression for this scattering mechanism is not also available. However, some estimates have been made using the expression for conductivity which appear to suggest that the discrepancies in some materials may be ascribed to this scattering mechanism using reasonable values for the deformation potential constant.

9.5 Concluding Remarks

We have presented in this chapter the methods to be used for the evaluation of the AC transport coefficients. It has been shown that by analysing these coefficients we may obtain information on the scattering mechanisms and the energy band structure. Of the various low-frequency (i.e. microwave) transport phenomena, cyclotron resonance absorption has been extensively studied in various materials as it gives the effective mass directly. Very few experiments, however, have been reported on the frequency dependence of conductivity or magnetoconductivity. Such experiments may be expected to throw light on the processes controlling the electron transport. Free-carrier absorption, on the other hand, has been very extensively studied experimentally in many semiconducting materials. The data obtained from some of the experiments have also been analysed to determine the strength of the different scattering mechanisms. HAGA and KIMURU analysed the results for InSb and InAs taking fully into account the effects of band nonparabolicity. Analyses have also been reported for GaAs [9.22e], GaP [9.34], PbSe [9.33c], and PbTe [9.35], but in none of these have all the complexities of band structure, such as mass anisotropy and nonparabolicity been fully incorporated. A more careful study of the experimental results of these and other materials is needed.

10. Electron Transport in a Strong Magnetic Field

The effect of a magnetic field on the transport coefficients has been analysed in the earlier chapters by including a Lorentz force term in the Boltzmann equation and by assuming that the energy-wave-vector dispersion relations and the scattering probabilities are not altered by the magnetic field. These assumptions are not valid when the magnetic field is strong. It was shown in Sect.4.6.2 that the energy-wave-vector dispersion relation is altered in magnetic fields. The component of the electron energy corresponding to the electron wave vector transverse to the magnetic field is quantised. The conduction band thereby breaks into Landau levels. This quantisation affects the electron statistics and affects also the scattering probabilities. As a result, the theories of electron transport developed in the earlier chapters cannot be applied to explain electron motion in strong magnetic fields.

Experimental results on strong-field transport coefficients are also very much different from those obtained from the theory presented earlier. Magnetoresistance [10.1,2], thermoelectric power [10.3], optical absorption [10.4], and almost all the transport coefficients are found to pass through maxima and minima as the magnetic field is varied. The saturation in the magnetoresistance predicted from the classical theory is also not observed experimentally. It continues to increase with the magnetic field and the nature of the variation with the magnetic field depends on the scattering mechanism which predominates.

The peculiarities of the behaviour of the transport coefficients in strong magnetic fields have been of great interest to semiconductor physicists in the last decade. Understanding these phenomena required generalisation of the transport theory to the quantised dispersion relations. Also, information could be obtained from experiments in quantising fields about important physical parameters such as the Fermi level, the electron effective mass, and the optic phonon frequency.

We discuss these phenomena in this chapter and present the methods available for their analysis. It is evident from the dispersion relation of Sect. 4.6.2 that the components of a momentum in the transverse direction vanish in strong magnetic fields. The method of analysis developed in the earlier chapters has been based essentially on a particle picture for the electron, whose motion is described by Newton's law. Since a momentum cannot be associated with the transverse motion, we cannot evaluate the effect of an electric field on this motion by using the particle description. Electron transport transverse to a magnetic field is studied by using a different formalism. The density matrix approach has been found to be convenient for this purpose [10.5]. We shall not, however, discuss this method. It requires for its development background material which is not covered in this book. On the other hand, even in strong magnetic fields the component of the wave function corresponding to the longitudinal direction is of the same form as that of a free electron. We may define an electron wave vector for this direction of motion and treat the motion by using the particle picture as in the earlier chapters. We should, however, remember that the dispersion relation being very different, the scattering probabilities assume different forms. These therefore must be worked out before the theory for the evaluation of the transport coefficients is presented. We work out the scattering probabilities in the following section and present the results on longitudinal transport coefficients in the following sections. We shall confine our attention to isotropic parabolic bands, for the sake of simplicity. The methods may be extended straightaway to deal with nonparabolic bands.

10.1 Scattering Probabilities

The wave function and eigenvalues of energy of the electron in the presence of a magnetic field as derived in Sect.4.6.2 are respectively

$$\psi = (L_y L_z)^{-\frac{1}{2}} \phi_\ell (x - \hbar k_y/|e|B) \exp(ik_y y + ik_z z) \quad , \tag{10.1}$$

and

$$E_\ell = \hbar^2 k_z^2/(2m^*) + (\ell + 1/2)\hbar|e|B/m^* \pm (1/2) g\mu_B B \quad , \tag{10.2}$$

where $(L_y L_z)^{-\frac{1}{2}}$ is the normalisation factor corresponding to a sample of dimensions L_y and L_z in the y and z directions; ϕ_ℓ is the normalised harmonic oscillator wave function corresponding to the ℓth eigenvalue; k_y and k_z are analogues of the wave vectors for the y and z directions. The magnetic field is assumed to be in the z direction; x and y are two arbitrarily chosen orthogonal directions.

The scattering probabilities are worked out by applying perturbation theory, which involves the assumption that the duration of a collision is negligible when compared with the time between collisions.

Expressions for the scattering probabilities are to be worked out from the expression for the matrix element given in Sect.6.6. In the present case, however, the state of the electron is characterised by ℓ, k_y, and k_z. The matrix element may hence be written as

$$M(\ell k_y k_z, \ell' k_y' k_z') = \int_\Omega \psi^*_{\ell' k_y' k_z'} V \psi_{\ell k_y k_z} d\underline{r} \quad , \tag{10.3}$$

where V is the potential energy of the electron due to the scattering potential. The scattering probability for the electron being scattered from the $\ell k_y k_z$ state to the $\ell' k_y' k_z'$ state is

$$S(\ell k_y k_z, \ell' k_y' k_z') = (2\pi/\hbar)|M(\ell k_y k_z, \ell' k_y' k_z')|^2 \delta\left[E(\ell k_z) - E(\ell' k_z') \pm \Delta E\right] \quad , \tag{10.4}$$

where $\pm\Delta E$ represents the change in energy of the electron. We have used only ℓ and k_z in the subscript to the electron energy E to denote the energy corresponding to the $\ell k_y k_z$ state, as E depends only on these two variables. The scattering potential V may be caused by lattice vibrations or by ionised impurities. In the former case, we may expand V in terms of normal modes as

$$V = \sum c_q \left[a_q \exp(i\underline{q}\cdot\underline{r}) + a_q^\dagger \exp(-i\underline{q}\cdot\underline{r})\right] \quad . \tag{10.5}$$

In the latter case, it may be expanded in a Fourier series as

$$V = \sum c_q \exp(i\underline{q}\cdot\underline{r}) \quad . \tag{10.6}$$

Using the above expressions in (10.3), we get for the matrix element for the lattice scattering

$$M(\ell k_y k_z, \ell' k_y' k_z') = c_q \left[\int \phi_\ell \exp(\pm i q_x x) \phi_{\ell'} dx\right] \delta(k_y - k_y' + q_y) \delta(k_z - k_z' \pm q_z) \times (n_q + 1/2 \pm 1/2)^{\frac{1}{2}} \quad , \tag{10.7}$$

where n_q is the phonon occupation number for phonons of wave vector $\underline{q}$, and the plus and minus signs correspond respectively to the emission and the absorption terms. For ionised impurity scattering, we get

$$M(\ell k_y k_z, \ell' k_y' k_z') = c_q \left[\int \phi_\ell \exp(i q_x x) \phi_{\ell'} dx\right] \delta(k_y - k_y' + q_y) \delta(k_z - k_z' + q_z) \quad . \tag{10.8}$$

The integrals over y and z vanish for all values of q_y and q_z (the y and z component of $\underline{q}$) except for the values $k'_y - k_y$ and $k'_z - k_z$ because of the periodicity of the integrands in these two directions. The integral within the square bracket may be evaluated in terms of Laguerre polynomials [10.6]. We refer the reader to the article by TETEICA [10.7], where the complete derivation is given. The integral is expressed as

$$J_{\ell\ell'} = \int \phi_\ell \exp(iq_x x)\phi_{\ell'} dx = \int \phi_\ell(x-\hbar k_\perp/|e|B)\phi_{\ell'}(x-\hbar k'_\perp/|e|B)dx \quad . \qquad (10.9)$$

The x direction being arbitrary, it has been chosen so that $q_x = 0$ and $q_y = q$. It may be shown [10.2] that

$$J_{\ell\ell'}(u) = (\ell!\ell'!)^{\frac{1}{2}} \exp(-u/2)u^{(\ell'-\ell)/2}L_\ell^{\ell-\ell'}(u) \quad , \qquad (10.10)$$

where $u = q_\perp^2(\hbar/2|e|B)$, and $L_\ell^{\ell'-\ell}(u)$ is the associated Laguerre polynomial [10.6] given by

$$L_\ell^{\ell'-\ell}(u) = \exp(u)u^{-(\ell'+\ell)}d^\ell(\exp(-u)u^{\ell'})/du^\ell \quad . \qquad (10.11)$$

Note that

$$\int |J_{\ell\ell'}(u)|^2 du = 1 \quad . \qquad (10.12)$$

The scattering probability for lattice scattering may now be expressed as

$$S(\ell k_y k_z, \ell' k'_y k'_z) = \left(V_c/8\pi^3\right)(2\pi/\hbar)\sum_{+,-}\left[|J_{\ell\ell'}|^2 c_q \delta(k_y - k'_y + q_y)\delta(k_z - k'_z + q_z)\right]$$
$$\times (n_q + 1/2 \pm 1/2)\delta\left[E(\ell k_z) - E(\ell' k'_z) \pm \hbar\omega_q\right]d\underline{k}' \quad . \qquad (10.13)$$

For ionised scattering we get the same expression with the $(n_q + 1/2 \pm 1/2)$ and $\hbar\omega_q$ omitted.

Expressions for the out-scattering rate for the various scattering mechanisms may be derived now, using the relevant expressions for c_q from Table 6.1. The procedure of derivation is illustrated below taking polar optic phonon scattering as an example. For this scattering mechanism,

$$|c_q|^2 = 2\pi\hbar^2|e|E_0/(m^* V_c q^2) \quad , \qquad (10.14)$$

where

$$E_0 = m^*|e|\omega_\ell(4\pi\hbar\varepsilon_0)^{-1}\left(K_\infty^{-1}-K_S^{-1}\right) \quad . \tag{10.15}$$

Substituting (10.14) in (10.13), we get

$$\begin{aligned} S_{pop}(\ell k_y k_z) = V_c(8\pi^3)^{-1}(2\pi/\hbar)\sum_{+,-}\int |J_{\ell\ell'}|^2(2\pi\hbar^2|e|E_0)(m^*V_c q^2)^{-1} \\ \times\delta(k_y-k_y'+q_y)\delta(k_z-k_z'\pm q_z)(n+1/2\pm 1/2) \\ \times\delta\left[E(\ell k_z)-E(\ell' k_z')\pm\hbar\omega_\ell\right]d\underline{k}' \quad . \end{aligned} \tag{10.16}$$

With

$$E(\ell' k_z') - E(\ell k_z) \mp \hbar\omega_\ell = x \quad , \tag{10.17}$$

we get

$$x = \left[\hbar^2/(2m^*)\right]\left(k_z'^2-k_z^2-k_{\ell\ell'}^2\mp k_0^2\right) \quad , \tag{10.18}$$

where

$$k_{\ell\ell'}^2 = (2m^*/\hbar)(\ell-\ell')\omega_B \quad , \tag{10.19}$$

$$k_0^2 = (2m^*/\hbar)\omega_\ell \quad . \tag{10.20}$$

The minus and plus signs correspond respectively to the absorption and emission processes.

We note that

$$dk_z' = (m^*/\hbar^2)(1/k_z')dx \quad , \tag{10.21}$$

and when $x = 0$

$$k_z' = \pm\left(k_z^2+k_{\ell\ell'}^2\pm k_0^2\right)^{1/2} \quad . \tag{10.22}$$

Replacing dk_z' by dx and integrating over x, using the property of the delta function, (10.16) may be reduced to

$$S_{pop}(\ell k_y k_z) = (|e|E_0/\hbar)\sum_{\ell'}\sum_{+,-}\sum_i\int_0^\infty |J_{\ell\ell'}|^2 q_\perp\left(q_\perp^2+q_{zi}^2\right)^{-1}|k_{zi}|^{-1}\left(n_\ell+\frac{1}{2}\pm\frac{1}{2}\right)dq_\perp \quad , \tag{10.23}$$

where $|k_{zi}|$ are the possible values of k'_z for $x=0$ given by (10.22). The values of q_z corresponding to the two possible values of $|k_{zi}|$ are

$$q_{zi} = k_z \pm \left(k_z^2+k_{\ell\ell'}^2 \pm k_0^2\right)^{\frac{1}{2}} \quad . \tag{10.24}$$

The integral over $q_\perp$ may also be expressed as

$$\int_0^\infty |J_{\ell\ell'}|^2 q_\perp \left(q_\perp^2+q_z^2\right)^{-1} dq_\perp = (1/2)\int_0^\infty |J_{\ell\ell'}(u)|^2 \left[u+\hbar/(2|e|Bq_z^{-2})\right]^{-1} du = (1/2) I_{\ell\ell'}(q_z) . \tag{10.25}$$

When the different terms in (10.23) are written in terms of $I_{\ell\ell'}$, the expression for the out-scattering rate for polar optic phonons assumes the form

$$\begin{aligned} S_{pop}(\ell k_y k_z) = (|e|E_0/2\hbar)\sum_{\ell'} \Big\{ & n_\ell \left(k_z^2+k_{\ell\ell'}^2+k_0^2\right)^{-\frac{1}{2}} \sum_{+,-} I_{\ell\ell'}\left[k_z \pm \left(k_z^2+k_{\ell\ell'}^2+k_0^2\right)^{\frac{1}{2}}\right] \\ & + H\left(k_z^2+k_{\ell\ell'}^2-k_0^2\right)(n_\ell+1)\left(k_z^2+k_{\ell\ell'}^2-k_0^2\right)^{-\frac{1}{2}} \\ & \times \sum_{+,-} I_{\ell\ell'}\left[k_z \pm \left(k_z^2+k_{\ell\ell'}^2-k_0^2\right)^{\frac{1}{2}}\right]\Big\} \quad . \end{aligned} \tag{10.26}$$

Expressions for the scattering probabilities for all the other scattering mechanisms may be derived following the same procedure. We give below the expressions for the deformation potential acoustic phonon, piezoelectric phonon, ionised impurity, nonpolar optic phonon, and intervalley phonon scattering.

Deformation Potential Acoustic Phonon:

$$S_{ac}(\ell k_y k_z) = C_{acB} \sum_{\ell'} \left(k_z^2+k_{\ell\ell'}^2\right)^{-\frac{1}{2}} \quad , \tag{10.27}$$

$$C_{acB} = E_1^2 m^* k_B T |e| B (\pi\rho s^2 \hbar^4)^{-1} \quad . \tag{10.28}$$

Piezoelectric Phonon:

$$S_{pz}(\ell k_y k_z) = C_{pzB} \sum_{\ell'} \left(k_z^2+k_{\ell\ell'}^2\right)^{-\frac{1}{2}} \sum_{+,-} I_{\ell\ell'}\left[k_z \pm \left(k_z^2+k_{\ell\ell'}^2\right)^{\frac{1}{2}}\right] \quad , \tag{10.29}$$

$$C_{pzB} = m^* k_B T e^2 h_{pz}^2 (4\pi\hbar^3 \rho\varepsilon^2 s^2)^{-1} \quad . \tag{10.30}$$

Ionized Impurity:

$$S_{impB}(\ell k_y k_z) = C_{impB} \sum_{\ell'} \left(k_z^2+k_{\ell\ell'}^2\right)^{-\frac{1}{2}} \sum_{+,-} I_{1\ell\ell'}\left[k_z \pm \left(k_z^2+k_{\ell\ell'}^2\right)^{\frac{1}{2}}\right] \quad , \tag{10.31}$$

where

$$I_{1\ell\ell'}(x) = \int_0^\infty |J_{\ell\ell'}(y)|^2 \left[y+\hbar(2|e|B)^{-1}(x^2+\lambda_s^2)\right]^{-2} dy \quad , \tag{10.32}$$

$$C_{impB} = N_i e^4 m^* (4\pi\varepsilon^2\hbar^2|e|B)^{-1} \quad , \tag{10.33}$$

λ_s is the screening wavelength given by $\lambda_s = \left[ne^2(\varepsilon k_B T)^{-1}\right]^{\frac{1}{2}}$. The screening has been assumed to be isotropic. It should be, however, mentioned that in the presence of a quantising magnetic field, anisotropies are introduced due to the restricted motion of the electrons in the transverse plane. The values of λ_s may be significantly altered by the anisotropy, but this complication may be omitted for the present discussion.

Nonpolar Optic Phonon:

$$S_{opB}(\ell k_y k_z) = C_{opB} \sum_{\ell'} \left[\left(k_z^2+k_{\ell\ell'}^2+k_0^2\right)^{-\frac{1}{2}} + H\left(k_z^2+k_{\ell\ell'}^2-k_0^2\right)(n_0+1)n_0^{-1}\left(k_z^2+k_{\ell\ell'}^2-k_0^2\right)^{-\frac{1}{2}}\right] , \tag{10.34}$$

where

$$C_{opB} = m^* D_0^2 |e| B (2\pi\hbar^3 \rho\omega_0)^{-1} n_0 \quad . \tag{10.35}$$

Expression for the intervalley scattering is obtained by replacing D_0 by D_i, n_0 by n_i, and ω_0 by ω_i.

10.2 Mobility in Strong Magnetic Fields

Electron mobility in strong magnetic fields may be worked out using the expressions for the scattering probabilities derived in the previous section. As mentioned earlier, the effect of a magnetic field applied in the direction of the electric field may be studied by using the particle formalism and the Boltzmann equation. The effect of a transverse magnetic field cannot, however, be studied in the same way. We derive in this section expressions for the longitudinal mobility using the Boltzmann equation approach.

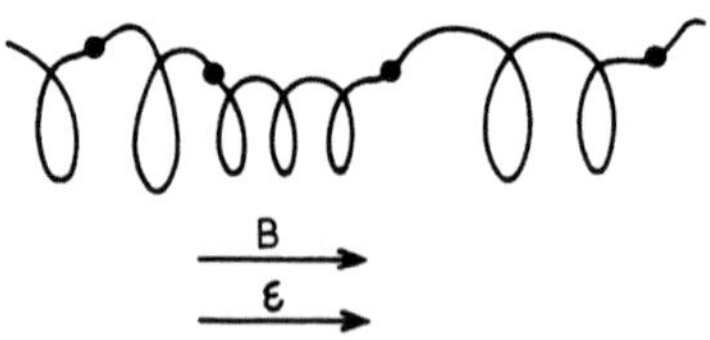

Fig. 10.1. Electron trajectories in a strong magnetic field and a longitudinal electric field

We may define a distribution function for electrons in each Landau level and set up a Boltzmann equation for this function. In strong magnetic fields, the motion of the electron due to an electric field in the direction of the magnetic field (z direction) may be visualised as shown in Fig.10.1. The electron executes circular motion in the x-y plane while drifting in the k_z direction. The radius of the orbit is unaltered but the rate of drift of the centre of the orbit is altered by the electric field as the electron drifts. Thus only the k_z component is changed in time. When the electron suffers a collision, the orbit of the electron may be changed and at the same time the sign and magnitude of k_z may be altered. We may hence consider the electrons to belong essentially to a one-dimensional system and the drift term of the Boltzmann equation will have only one component for the z direction. The term introduced by the magnetic field in the classical formulation, i.e. $\hbar^{-1}\underline{\nabla}_k E \times \underline{B}$, is absent. The effect of the magnetic field, however, would be introduced through the collision term, which should include the possibility of the change of the electron orbit, i.e. of the transfer of the electron from one Landau level to another, and of a change of k_z to k_z'. The Boltzmann equation thus has the form

$$\begin{aligned}(e\mathscr{E}/\hbar)(\partial f_{\ell\pm}/\partial k_z) + \sum_{\ell'}\int\Big[&S(\ell k_y k_z, \ell' k_y' k_z') f_{\ell\pm}(1-f_{\ell\pm}') \\ &- S(\ell' k_y' k_z', \ell k_y k_z) f_{\ell\pm}'(1-f_{\ell\pm})\Big] d\underline{k}' = 0 \quad ,\end{aligned} \tag{10.36}$$

where plus and minus signs correspond to the two possible spins, $f_{\ell\pm}$ represents the distribution function for the ℓth level with the two spins. It has been assumed that the electron may be scattered from one level to another but without spin reversal.

We shall try a solution of the form

$$f_{\ell\pm} = f_{\ell\pm}^0 - (e\hbar\mathscr{E}/m^*) k_z \phi_{\ell\pm}(E_z)\left(\partial f_{\ell\pm}^0/\partial E\right) \quad , \tag{10.37}$$

where $f_{\ell\pm}^0$ is the distribution function in the absence of the electric field, and

$$E_z = \hbar^2 k_z^2 (2m^*)^{-1} \quad . \tag{10.38}$$

The distribution function $f_{\ell\pm}^0$ has the form of the Fermi-Dirac function for a degenerate material and may be written as

$$f^0_{\ell\pm} = \left\{\exp\left[\hbar^2 k_z^2(2m^*)^{-1}+(\ell+1/2)\hbar\omega_B \pm (1/2)g\mu_B B - E_F\right](k_B T)^{-1}+1\right\}^{-1} \quad , \qquad (10.39)$$

since the energy is

$$E = \hbar^2 k_z^2 (2m^*)^{-1} + (\ell+2^{-1})\hbar\omega_B \pm (1/2)g\mu_B B \quad .$$

The distribution function for nondegenerate materials may also be expressed in terms of the carrier concentration as follows:

$$f^0_{\ell\pm} = (n_\pm/I_{n\pm})\exp\left\{-\left[(\ell+2^{-1})\hbar\omega_B+\hbar^2 k_z^2(2m^*)^{-1}\pm(1/2)g\mu_B B\right](k_B T)^{-1}\right\} \quad , \qquad (10.40)$$

where

$$\begin{aligned} I_{n\pm} &= \sum_{\ell=0}^{\infty} eB(4\pi^2\hbar)^{-1}\int_{-\infty}^{\infty} f^0_{\ell\pm}dk_z \\ &= \hbar\omega_B(2k_B T)^{-1}N_c\left[1-\exp(-\hbar\omega_B/k_B T)\right]^{-1}\exp\left[-(\hbar\omega_B \pm g\mu_B B)(2k_B T)^{-1}\right] \quad . \qquad (10.41) \end{aligned}$$

Substituting (10.40) in (10.36), using the condition of detailed balance,

$$f^0_{\ell\pm}S(\ell k_y k_z, \ell' k'_y k'_z) = f^0_{\ell'\pm}S(\ell' k'_y k'_z, \ell k_y k_z) \quad , \qquad (10.42)$$

and neglecting small order terms, we get

$$\begin{aligned} &\sum_{\ell'}\int S(\ell k_y k_z, \ell' k'_y k'_z)\left[1-f^0_{\ell\pm}(E')\right] \\ &\quad \times\left[1-f^0_{\ell\pm}(E)\right]\left(\phi_{\ell\pm}-k'_z\phi'_{\ell\pm}k_z^{-1}\right)d\underline{k}' = L\phi_{\ell\pm} = 1 \quad . \qquad (10.43) \end{aligned}$$

We have presented in the earlier section the integrated expression for $\phi_{\ell\pm}$. The integration for $\phi'_{\ell\pm}$ may be carried out following the same procedure. We note, however, that due to the presence of the delta function, implying the balance of energy, k_z is restricted to have either of the two values

$$k'_z = \pm\left(k_z^2+k_{\ell\ell'}^2 \pm k_0^2\right)^{\frac{1}{2}} \quad . \qquad (10.44)$$

The second term in the scattering integral (representing the in-scattering) may, thus, be obtained by inspection of the out-scattering term, replacing k_z by the relevant expression from (10.44). We give below the expressions for the different scattering mechanisms.

Deformation Potential Acoustic Phonon:

$$L_{acB}\phi_{\ell\pm} = C_{acB}\sum_{\ell'}\phi_{\ell\pm}(E_z)\left(k_z^2+k_{\ell\ell'}^2\right)^{-\frac{1}{2}} \quad . \tag{10.45}$$

Piezoelectric Acoustic Phonon:

$$\begin{aligned} L_{pzB}\phi_{\ell\pm} = C_{pzB}\sum_{\ell'} & \left\{\left(k_z^2+k_{\ell\ell'}^2\right)^{-\frac{1}{2}}\phi_{\ell\pm}(E_z)\sum_{+,-} I_{\ell\ell'}\left(k_z+\left(k_z^2\pm k_{\ell\ell'}^2\right)^{\frac{1}{2}}\right)\right. \\ & -\phi_{\ell'\pm}(E_z+(\ell'-\ell)\hbar\omega_B)k_z^{-1}\left[I_{\ell\ell'}\left(k_z-\left(k_z^2+k_{\ell\ell'}^2\right)^{\frac{1}{2}}\right)\right. \\ & \left.\left.-I_{\ell\ell'}\left(k_z+\left(k_z^2+k_{\ell\ell'}^2\right)^{\frac{1}{2}}\right)\right]\right\} \quad . \end{aligned} \tag{10.46}$$

Ionised Impurity Scattering:

$$\begin{aligned} L_{impB}\phi_{\ell\pm} = C_{impB}\sum_{\ell'} & \left\{\left(k_z^2+k_{\ell\ell'}^2\right)^{-\frac{1}{2}}\phi_{\ell\pm}(E_z)\sum_{+,-} I_{\ell\ell'}\left(k_z+\left(k_z^2+k_{\ell\ell'}^2\right)^{\frac{1}{2}}\right)\right. \\ & -k_z^{-1}\phi_{\ell\ell'}(E_z+(\ell'-\ell)\hbar\omega_B)\left[I_{1\ell\ell'}\left(k_z-\left(k_z^2+k_{\ell\ell'}^2\right)^{\frac{1}{2}}\right)\right. \\ & \left.\left.-I_{1\ell\ell'}\left(k_z+\left(k_z^2+k_{\ell\ell'}^2\right)^{\frac{1}{2}}\right)\right]\right\} \quad . \end{aligned} \tag{10.47}$$

We note that in the above expressions the scattering has been assumed to be elastic. The factor $\left[1-f_\ell^0(E')\right]\left[1-f_{\ell'}^0(E)\right]^{-1}$ is therefore unity and the expressions are applicable to both degenerate and nondegenerate materials.

The optic scattering mechanism being inelastic, the factor remains in the expressions for the collision terms for this kind of scattering. We write below the expressions for degenerate materials; the corresponding expressions for nondegenerate materials are obtained by replacing this factor by unity.

Nonpolar Optic Phonon:

$$\begin{aligned} L_{opB}\phi_{\ell\pm} = C_{opB}\sum_{\ell'} & \left[1-f_{\ell\pm}^0(E_z)\right]^{-1}\phi_\ell(E_z)\left\{\left[(1-f_{\ell'\pm}E_{z1}))k_{z1}^{-1}+(n_0+1)n_0^{-1}\right.\right. \\ & \left.\left.\times H(E_z+(\ell-\ell')\hbar\omega_B-\hbar\omega_0)(1-f_{\ell'\pm}(E_{z2}))k_{z2}^{-1}\right]\right\} \quad , \end{aligned} \tag{10.48}$$

where

$$k_{z1} = \left(k_z^2+k_{\ell\ell'}^2+k_0^2\right)^{\frac{1}{2}} \quad , \tag{10.49}$$

$$k_{z2} = \left(k_z^2+k_{\ell\ell'}^2-k_0^2\right)^{\frac{1}{2}} \quad , \tag{10.50}$$

$$E_{z1} = E_z + \hbar\omega_0 + \hbar\omega_B(\ell'-\ell) \quad , \tag{10.51}$$

$$E_{z2} = E_z - \hbar\omega_0 + \hbar\omega_B(\ell'-\ell) \quad . \tag{10.52}$$

Polar Optic Phonon:

$$L_{popB}\phi_{\ell\pm} = (|e|E_0/2\hbar)n_\ell \sum_{\ell'} \left[1-f^0_{\ell\pm}(E_z)\right]\phi_{\ell\pm}(E_z)$$
$$\times\left\{\left[\left(1-f^0_{\ell'\pm}(E_{z1})\right)k_{z1}^{-1}\sum_{+,-} I_{\ell\ell'}(k_z\pm k_{z1})+(n_\ell+1)n_\ell^{-1}\right.\right.$$
$$\left.\left.\times H(E_z+\hbar\omega_B(\ell-\ell')-\hbar\omega_\ell)\left(1-f^0_{\ell'\pm}(Ez_2)\right)k_{z2}^{-1}\sum_{+,-} I_{\ell\ell'}(k_z\pm k_{z2})\right]\right\}$$
$$-\phi_{\ell'\pm}(E_{z1})k_z^{-1}\left(1-f^0_{\ell'\pm}(E_{z1})\right)\left[I_{\ell\ell'}(k_z-k_{z1})-I_{\ell\ell'}(k_z+k_{z1})\right]$$
$$\times(n_\ell+1)n_\ell^{-1}-k_z^{-1}\phi_{\ell'\pm}(E_{z2})\left(1-f^0_{\ell'\pm}(E_{z2})\right)H(E_z+(\ell'-\ell)\hbar\omega_B-\hbar\omega_\ell)$$
$$\times\left[I_{\ell\ell'}(k_z-k_{z2})-I_{\ell\ell'}(k_z+k_{z2})\right]\Big\}\quad . \tag{10.53}$$

The perturbation function ϕ_ℓ may now be obtained by solving (10.41) using the above expressions for $L_c\phi_\ell$. The mobility is then obtained from the expressions for the current density. In the present case, the current density

$$J = -(e/\hbar)\sum_{+,-}\int_{-\infty}^{\infty}\sum_{\ell}(\hbar^2k_z/m^*)(e\hbar\mathscr{E}/m^*)(|e|B/4\pi^2\hbar)k_z\phi_{\ell\pm}(\partial f_{\ell\pm}/\partial E)dk_z$$
$$= -ne^2|e|B(k_BT)^{\frac{1}{2}}\mathscr{E}\left(2^{\frac{1}{2}}\pi^2\hbar^2 I_n m^{*\frac{1}{2}}\right)^{-1}\sum_{\ell}\sum_{+,-}\int_0^{\infty}\left[W-(2\ell+1)\theta/2\pm(1/2)g\mu_BB/(k_BT)\right]^{\frac{1}{2}}$$
$$\times\phi_{\ell\pm}\left(\partial f^0_{\ell\pm}/\partial W\right)dW\quad , \tag{10.54}$$

where

$$I_n = \sum_{+,-}\sum_{\ell}\int_{-\infty}^{\infty} f^0_{\ell\pm}|e|B(4\pi^2\hbar)^{-1}dk_z$$
$$= (\sqrt{\pi}2)^{-1}N_c\theta\sum_{\ell}\sum_{+,-}\int\left[W-(2\ell+1)\theta/2\pm g\mu_BB(2k_BT)^{-1}\right]^{-\frac{1}{2}}f^0_{\ell\pm}dW\quad ; \tag{10.55}$$

$$W = E(k_BT)^{-1}\quad ,\quad \theta = \hbar\omega_B(k_BT)^{-1}\quad ,\quad N_c = 2(2\pi m^*k_BT)^{3/2}h^{-3}\quad ,$$

$|e|B/4\pi^2\hbar$ is the density of states per unit length of k_z [see (5.18)] in each Landau level.

The mobility is given by

$$\mu_d = -e^2B(k_BT)^{\frac{1}{2}}\left(\sqrt{2}\pi^2\hbar^2 I_n m^{*\frac{1}{2}}\right)^{-1}\sum_{+,-}\sum_{\ell}\int_0^{\infty}\left[W-\theta(2\ell+1)/2\right.$$
$$\left.\pm(1/2)g\mu_BB(k_BT)^{-1}\right]^{\frac{1}{2}}\left(\partial f^0_{\ell\pm}/\partial W\right)dW\quad . \tag{10.56}$$

10.3 Electron Mobility in the Extreme Quantum Limit (EQL)

Examining the different expressions for $L_c\phi_\ell$ we find that a straightforward solution in the relaxation-time formalism may be obtained only for randomising collisions, i.e. for deformation potential acoustic phonon and for nonpolar optic phonon scattering. But even for these scattering mechanisms simple expressions for the mobility cannot be obtained in the general case. Only for the special case of very strong magnetic fields manageable expressions for the mobility may be obtained. We may assume for very strong magnetic fields, i.e. for $\hbar\omega_B \gg k_BT$ or E_F, electrons occupy only the lowest Landau level, the level corresponding to $\ell = 0$. Electrons are confined to this level even when scattered and hence ℓ' is also 0. This condiction is usually referred to as the extreme quantum limit or EQL [10.8]. In the EQL, ϕ_ℓ for the two other elastic scattering mechanisms may also be expressed in terms of a relaxation time. We may hence obtain expressions for the mobility in the EQL for all the scattering mechanisms except for the polar optic phonon scattering. We have collected in Table 10.1 expressions for the mobility in the EQL for the various scattering mechanisms for extremely degenerate as well as for nondegenerate materials. It has been assumed for the degenerate material that the carrier density is not altered by the magnetic field. The Fermi energy in the presence of a magnetic field may then be expressed in terms of the energy in the absence of the field as follows:

$$
\begin{aligned}
n &= (1/2)N_c\theta F_{-\frac{1}{2}}\left[E_{FB}/k_BT-\hbar\omega_B(2k_BT)^{-1}-g\mu_BB(2k_BT)^{-1}\right] \\
&= N_cF_{\frac{1}{2}}(E_F/k_BT) \quad . \qquad (10.57)
\end{aligned}
$$

For the case of extreme degeneracy, we may assume that the distribution function is unity up to the Fermi level and zero above. We then get

$$
(E_{FB}-\hbar\omega_B/2-g\mu_BB/2)^{\frac{1}{2}}(k_BT)^{-\frac{1}{2}} = (E_F/k_BT)^{3/2}4(3\theta)^{-1} \quad . \qquad (10.58)
$$

We note that the expressions for the scattering terms for acoustic phonon scattering were obtained assuming that $k_BT \gg \hbar\omega_q$, i.e. the collisions are elastic. However, quantum magnetic effects are usually studied at very low temperatures so that this assumption is not often satisfied. The expressions for the mobility when the collisions are regarded as inelastic has been worked out in [10.8,9]. As the formalism is somewhat different from that presented here we shall not discuss the method of derivation but only present the expressions.

Table 10.1. Electron mobility in longitudinal magnetic fields in the EQL

Nature of scattering	Expressions for the mobility: Nondegenerate	Degenerate
1. Deformation potential acoustic phonon:		
a) elastic	$(2\sqrt{2\pi}\ \rho s^2\hbar^3)\left(E_1^2 m^{*3/2}\sqrt{k_BT}\ B\right)^{-1}$	$4\sqrt{2}\ \rho s^2\pi\hbar^2E_F^{3/2}\left(m^{*\frac{1}{2}}3E_1^2k_BT\|e\|B^2\right)^{-1}$
b) inelastic	$\|e\|8\sqrt{2\pi}\ \rho s^4\hbar^4\left(3E_1^2m^{*3/2}k_B^{5/2}T^{5/2}C_2\right)^{-1}$	$\sqrt{2}\ \rho s^4E_F^{3/2}\pi\hbar^3\left(m^{*\frac{1}{2}}E_1^2k_B^3T^3C_1B\right)^{-1}$
2. Piezoelectric:		
a) elastic	$4\sqrt{2}\|e\|\pi^{3/2}\rho s^2\hbar^2(eh_{pz}/\varepsilon)^{-2}(k_BT)^{-\frac{1}{2}}m^{*-3/2}$ $\times\left\{\ln\left[\hbar\omega_B/(E_s+4k_BT)\right]\right\}^{-1}$	$8\sqrt{2}\ \rho s^2\pi\hbar E_F^{3/2}m^{*-\frac{1}{2}}3^{-1}(eh_{pz}/\varepsilon)^{-2}$ $\times(k_BT)^{-1}B^{-1}\left\{\ln\left[\hbar\omega_B/(4E_F+E_s)\right]\right\}^{-1}$.
b) inelastic	$\|e\|4\sqrt{2}\ \pi^{3/2}\rho s^2\hbar^2(eh_{pz}/\varepsilon)^{-2}(k_BT)^{-\frac{1}{2}}$ $\times\left\{m^{*3/2}\ln\left[(k_BT/4m^*s^2)(1+E_s/4k_BT)\right]\right\}^{-1}$	$8\sqrt{2}\ \rho s^2\pi\hbar E_F^{3/2}m^{*-\frac{1}{2}}3^{-1}(eh_{pz}/\varepsilon)^{-2}$ $\times(k_BT\ B)^{-1}\left\{\ln\left[(k_BT)^2/m^*s^2(E_s+4E_F)\right]\right\}^{-1}$.
3. Ionised impurity	$64\|e\|\pi^{\frac{1}{2}}(k_BT)^{3/2}m^{*-\frac{1}{2}}(1+E_s/4k_BT)N_i^{-1}$ $\times(Ze^2/\varepsilon)^{-2}\sqrt{2}$	$16\sqrt{2}\pi E_F^{3/2}(E_s+4E_F)m^{*\frac{1}{2}}$ $\times\left[3(Ze^2/\varepsilon)^2N_i\hbar B\right]^{-1}$.
4. Nonpolar optic	$2\sqrt{2}\ \pi\hbar\rho(\hbar\omega_0)^{3/2}\left[D_0^2m^{*3/2}n_0B\right]^{-1}$.	

where

$$C_2^{-1} = (3/2)\int_0^\infty x^{\frac{1}{2}}e^{-x}\left\{\int_{-x}^\infty y^2\left[(x+y)^{-\frac{1}{2}}/\|e^{-y}-1\|\right]^{-1}dy\right\}^{-1}dx \approx 1 \ , \qquad E_s = (\hbar^2/2m^*)(2\pi/\lambda)^2 \ ,$$

$$C_1^{-1} = (16/3)\int_0^\infty (e^x+1)^{-1}(1+\exp -x)^{-2}\left\{\int_{-\infty}^\infty y^2[\|\exp y-1\|(1-\exp(-y)x\ \exp(-x))]^{-1}dy\right\}^{-1}dx \approx 1 \ .$$

10.3.1 Electron Mobility for Polar Optic Phonon Scattering in the EQL

The perturbation in the distribution function for polar optic phonon scattering cannot be expressed as a relaxation time even in the EQL. The Boltzmann equation may, however, be solved by the iterative method discussed in Sect.7.5.3. The equation reduces to the following form for polar optic phonon scattering in the EQL:

$$a(E_z+\ell\hbar\omega_0)\phi_0(E_z+\ell\hbar\omega_0) - b(E_z+\ell\hbar\omega_0)\phi_0(E_z+(\ell+1)\hbar\omega_0)$$
$$- c(E_z+\ell\hbar\omega_0)\phi_0(E_z+(\ell-1)\hbar\omega_0) = 1 \quad , \tag{10.59}$$

where

$$a(E_z) = (|e|E_0/2\hbar)n_0\Big\{k_{z1}^{-1}\Big[I_{00}(k_z+k_{z1})+I_{00}(k_z-k_{z1})$$
$$+k_{z2}^{-1}(n_\ell+1)n_\ell^{-1}H(E_z-\hbar\omega_0)\Big[I_{00}(k_z+k_{z2})+I_{00}(k_z-k_{z2})\Big]\Big]\Big\} \quad , \tag{10.60}$$

$$b(E_z) = (|e|E_0/2\hbar)(n_\ell+1)\Big\{k_z^{-1}\Big[I_{00}(k_z-k_{z1})-I_{00}(k_z+k_{z1})\Big]\Big\} \quad , \tag{10.61}$$

$$c(E_z) = (n_\ell)H(E_z-\hbar\omega_0)\Big\{k_z^{-1}\Big[I_{00}(k_z-k_{z2})-I_{00}(k_z+k_{z2})\Big]\Big\} \quad , \tag{10.62}$$

$$k_{z1} = \left(k_z^2+k_0^2\right)^{\frac{1}{2}} \quad , \quad k_{z2} = \left(k_z^2-k_0^2\right)^{\frac{1}{2}} \quad .$$

We get a series consisting of an infinite number of equations by giving ℓ positive integral values starting from zero. All these equations have to be solved simultaneously to get the perturbation functions. We find that the equations have the same form as (7.87). Hence, the same iteration technique may be used. In the first iteration step we put in each equation $\phi(E_z+(\ell\pm1)\hbar\omega_0)$ equal to zero and evaluate $\phi(E_z+\ell\hbar\omega_0)$ from

$$\phi(E_z+\ell\hbar\omega_0) = \left[a(E_z+\ell\hbar\omega_0)\right]^{-1} \quad . \tag{10.63}$$

In the second and succeeding iteration steps we use for $\phi(E_z+(\ell\pm1)\hbar\omega_0)$ the values obtained in the earlier step and evaluate $\phi(E_z+\ell\hbar\omega_0)$ again. These iterations are to be continued till steady values are obtained. The method has been used by MAGNUSSON [10.10] and BASU and NAG [10.11]. We show in Fig.10.2 the form of the distribution function obtained by MAGNUSSON. The mobility is evaluated with this function using (10.54). The mobilities obtained for some characteristic values of ω_0 using the physical constants for InSb are illustrated in Fig.10.3.

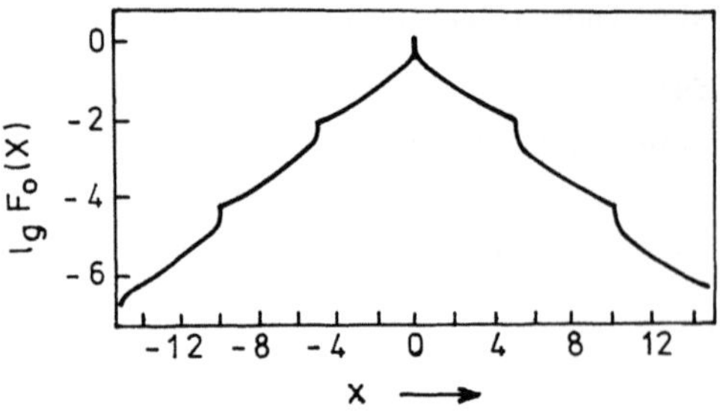

Fig. 10.2

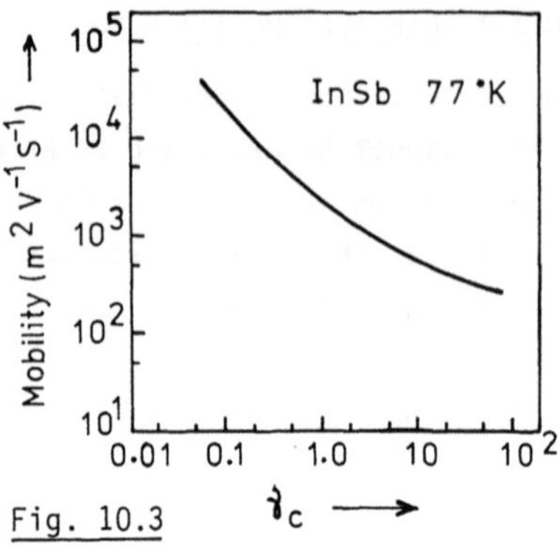

Fig. 10.3

Fig. 10.2. Electron distribution function in quantising magnetic fields with an electric perturbation [10.10]

Fig. 10.3. Electron mobility in InSb in the EQL [10.11]. $x = (E_Z/k_BT) \times \mathrm{Sign}(k_Z)$; $\log F_0(a) \equiv$ Distribution function for $\ell = 0$ in log scale; $\gamma_C = \hbar\omega_B/k_BT$

10.4 Oscillatory Effects in the Magnetoresistance

The study of longitudinal magnetoresistance in quantising magnetic fields is undertaken mostly because the magnetoresistance shows distinct oscillations in amplitude, which allows us to determine important physical constants of semiconductors. These oscillations occur through two distinct physical processes and are referred to by two different names. The first kind of oscillation is observed only in degenerate materials at very low temperatures and is called the SHUBNIKOV-DE HAAS effect after the discoverers [10.1]. The second kind occurs also in nondegenerate materials. However, the temperature must not be either too high or too low. These oscillations are often referred to as GUREVICH-FIRSOV oscillations after the discoverers [10.2], but are more commonly called magnetophonon oscillations. The theory of oscillations may be developed in the framework of Boltzmann equation using the expressions for the scattering probabilities given earlier. However, analytic expressions for the oscillatory conditions are difficult to obtain. At the same time, exact expressions are hardly necessary for the interpretation of experiments. The oscillatory effects arise from the singularities in the density of states (see Fig.5.4) which go to infinity for idealised conditions. Scattering of electrons broadens the Landau levels and makes the density of states finite. As yet, proper theories taking account of all the relevant effects accurately are not available. Even though a quantitative comparison between theory and experiment may not be fully justified, we may explain all the qualitative characteristics from the idealised theory. Our discussion in the following paragraphs will be confined to the general features of the oscillations which may be so explained. It should, however, be mentioned that even these qualitative characteristics of the phenomenon yield a wealth of information about the electron parameters.

10.4.1 Shubnikov-de Haas Effect

The Shubnikov-de Hass effect, as mentioned earlier, is observed in degenerate materials at very low temperatures (1.7-4 K). At these temperatures, impurity atom scattering is mostly dominant and acoustic phonon scattering also makes some contribution. Optic phonon scattering may be considered negligible as optic phonons are not excited significantly at the experimental temperatures. We also recall that we may treat only the deformation potential acoustic phonon scattering using a relaxation time constant. We shall discuss the characteristics of the oscillations for this kind of scattering. The treatment is, however, applicable to ionised impurity scattering for large electron concentrations when $\lambda^{-1} \gg k$ as we may then neglect q and consider the scattering to be randomising.

The perturbation in the distribution function

$$\begin{aligned}\phi_{\ell\pm} &= C_{acB}^{-1}\left[\sum_{\ell'}\left(k_z^2+k_{\ell\ell'}^2\right)^{-\frac{1}{2}}\right]^{-1}\\ &= (2m^*)^{\frac{1}{2}}\hbar^{-1}C_{acB}^{-1}\left\{\sum_{\ell'}\left[E-(\ell'+1/2)\hbar\omega_B\mp(1/2)g\mu_B B\right]^{-\frac{1}{2}}\right\}^{-1} \quad . \qquad (10.64)\end{aligned}$$

Replacing $\phi_{\ell\pm}$ in the expression for the current density and carrying out the integration by assuming that the material is extremely degenerate, i.e. $E_F \gg k_BT$, we get

$$\begin{aligned}J &= \mathscr{E}e^2|e|B(k_BT)^{\frac{1}{2}}\left(\pi^2\hbar^3C_{acB}\right)^{-1}\sum_{+,-}\left\{\sum_{\ell}\left[W_F-(\ell+1/2)\theta\mp(1/2)g\mu_B B(k_BT)^{-1}\right]^{\frac{1}{2}}\right.\\ &\quad\left.\times\left[\sum_{\ell'}\left(E_F-(\ell'+1/2)\hbar\omega_B\mp(1/2)g\mu_B B\right)^{-\frac{1}{2}}\right]^{-1}\right\} \quad . \qquad (10.65)\end{aligned}$$

But,

$$n_\pm = \pi^{-\frac{1}{2}}N_c\theta\sum_{\ell}\left[W_F-(\ell+1/2)\theta\mp g\mu_B B(2k_BT)^{-1}\right]^{\frac{1}{2}} \quad , \qquad (10.66)$$

and

$$J_{B\to 0} = ne^2\mathscr{E}\hbar E_{F0}^{-\frac{1}{2}}\left(C_{ac}m^{*3/2}2^{\frac{1}{2}}\right)^{-1} \quad , \qquad (10.67)$$

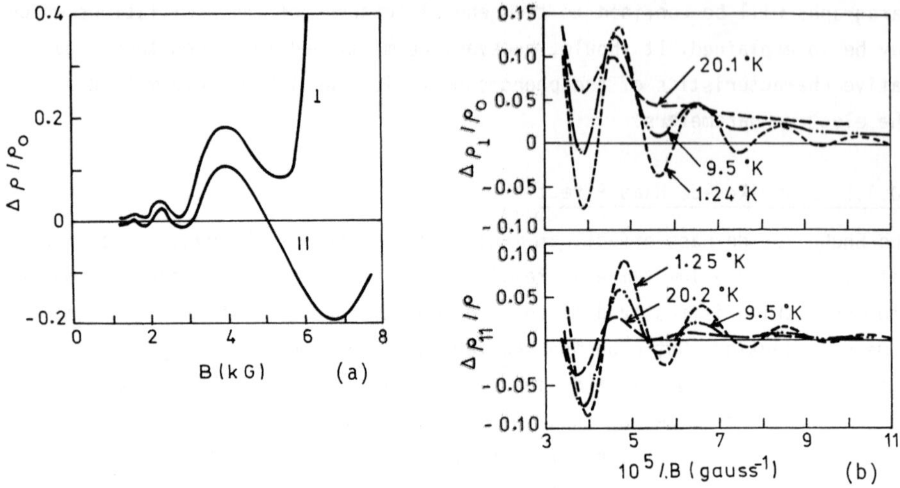

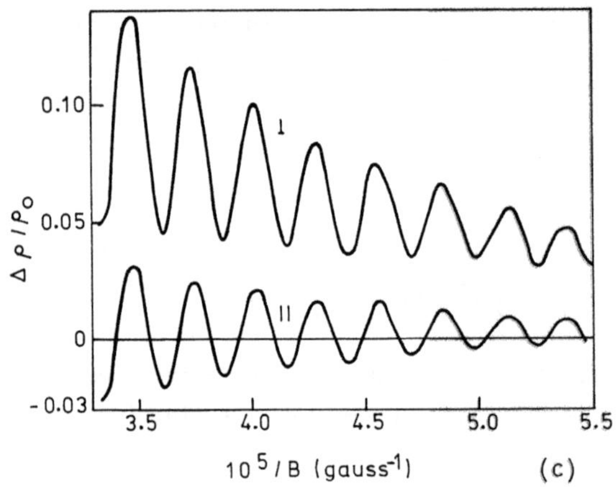

Fig. 10.4. Shubnikov-de Haas oscillations. (a) Indium antimonide [10.20]. (b) Indium arsenide [10.21]. (c) Gallium antimonide [10.22]

C_{ac} being defined in (6.62) and E_{F0} is the Fermi level for $B=0$. Eliminating $\sum_{\ell}[W_F-(\ell+1/2)\theta\mp g\mu_B B/2k_B T]^{\frac{1}{2}}$ by using $n_{\pm}$ and expressing the other constants in terms of the conductivity in the absence of the magnetic field, we get for the conductivity in the magnetic field

$$\sigma = \sigma_0 2E_{F0}^{\frac{1}{2}}(n\hbar\omega_B)^{-1}\sum_{+,-} n_{\pm}\left\{\sum_{\ell}\left[E_F-(\ell+1/2)\hbar\omega_B\mp g\mu_B B\right]^{-\frac{1}{2}}\right\}^{-1} \quad . \tag{10.68}$$

It is evident from (10.68) that for some values of B, the sum in braces has very large values and consequently the conductivity becomes very small for these values of B. Hence the resistance of a sample when measured by applying a magnetic field will pass through infinitely large values as the field is varied. These oscillations were discovered by SHUBNIKOV and DE HAAS in bismuth in 1930 [10.1] and are referred to as Shubnikov-de Haas effect. The effect has also been observed in many compound semiconductors. We present in Fig.10.4 some experimental characteristics showing the effect in InSb, InAs, and GaSb. A characteristic feature of these oscillations is that these are periodic in 1/B and the period is determined by the carrier concentration. It is independent of the effective mass and is not affected by the band nonparabolicity for spherical constant-energy surface band structure. This feature may be directly explained from (10.68).

We neglect, for the sake of simplicity, the effect of spin splitting. The term within braces is then large when

$$E_F = (\ell+1/2)\hbar\omega_B \quad . \tag{10.69}$$

We further assume that E_F is not significantly altered by the magnetic field and hence it may be expressed in terms of the carrier concentration in the case of extreme degeneracy as

$$E_F = (3\pi^2)^{2/3}\hbar^2(2m^*)^{-1}n^{2/3} \quad , \tag{10.69a}$$

Substituting in (10.69),

$$1/B = 2(3\pi^2)^{-2/3}|e|(\hbar n^{2/3})^{-1}(\ell+1/2) \quad . \tag{10.70}$$

Thus the oscillations will be periodic in 1/B with a period of

$$\Delta(1/B) = 2(3\pi^2)^{-2/3}|e|(\hbar n^{2/3})^{-1} \quad . \tag{10.71}$$

This simple relation remains valid even for the nonparabolic bands. Using the nonparabolic dispersion relation (4.68) and assuming that $\alpha^2E_F^2$ and higher-order terms are negligible, we get

$$n = 2^{3/2}m^{*3/2}(3\pi^2)^{-1}\hbar^{-3}E_F^{3/2}(1+3\alpha E_F/2) \quad . \tag{10.72}$$

The fields for large values of resistance are also given by [see (5.31) for the density of states in nonparabolic bands]

$$1/B = |e|\hbar(\ell+1/2)\left[E_F(1+\alpha E_F)m^*\right]^{-1} \quad . \tag{10.73}$$

On eliminating E_F using (10.72), we find that the period has the same value as given by (10.70). The periods are found to be given by the same expression even when higher-order terms of αE_F are included, so that we may generally conclude that the period of Shubnikov-de Haas oscillations is related only to the carrier concentration in the sample and does not depend on other physical parameters.

The expression given for $\sigma_{\parallel}$ indicates that the conductivity becomes zero when the condition (10.69) is satisfied. In actual experiments such zeros are not observed, for a number of reasons. The perturbation theory on which the expression is based is not valid for this extreme condition as $\omega_B\tau$ is then not much larger than unity. The broadening of the levels due to collisions mentioned earlier as well as the effects of carrier-carrier scattering are not also taken into account in the theory. Further, we have to combine the contribution of the electrons with two spins taking the proper values of n_+ and n_-, which vary with the magnetic field. An analytic expression taking these effects into account has been given by ROTH and ARGYRES [10.12] assuming that $E_F \gg \hbar\omega_B$, so that the relaxation times for the two spins could be averaged. This expression is quoted below:

$$\sigma = \sigma_0\left\{1+\sum_{r=1}^{\infty} b_r \cos\left[2\pi E_F r(\hbar\omega_B)^{-1}-\pi/4\right]\right\}^{-1} \quad , \tag{10.74}$$

where

$$\begin{aligned} b_r = {} & (-1)^r r^{-\frac{1}{2}}\left[\hbar\omega_B(2E_F)^{-1}\right]^{\frac{1}{2}} 2\pi^2 r k_B T(\hbar\omega_B)^{-1}\left\{\sinh\left(2\pi^2 r k_B T(\hbar\omega_B)^{-1}\right)\right\}^{-1} \\ & \times \cos(r\pi m^* g/2m_0)\exp\left[-2\pi\Gamma r(\hbar\omega_B)^{-1}\right] \quad . \end{aligned} \tag{10.75}$$

This expression gives the same period of oscillations as derived earlier. We find in addition that the amplitude of oscillation should vary with the temperature and the magnetic field. It should thus be possible to determine the cyclotron resonance frequency by analysing the temperature dependence of the oscillation. For not too low temperatures, all terms for $r > 1$ may be neglected. We then find that the log of amplitude normalised by

$$(\hbar\omega_B/E_F)^{\frac{1}{2}} k_B T(\hbar\omega_B)^{-1}/\sinh(2\pi k_B T/\hbar\omega_B)$$

should be a straight line, the slope of which gives Γ and hence the Dingle temperature $\Gamma' = \Gamma(\pi k_B)^{-1}$. All these characteristics of the Shubnikov-de Haas oscillations have been extensively studied in many semiconductors. These experimental results are also reviewed in detail by ROTH and ARGYRES [10.12].

10.4.2 Magnetophonon Oscillations

GUREVICH and FIRSOV [10.2] predicted that the transverse magnetoresistance of nondegenerate semiconductors with predominant optic phonon scattering should show oscillations with a period

$$\Delta(1/B) = |e|(m^*\omega_0)^{-1} \quad , \tag{10.76}$$

where ω_0 is the optic phonon frequency, and m^* is the effective mass. This prediction has been confirmed by experiments. Oscillations are observed both in the transverse and in the longitudinal magnetoresistance. Similar oscillations are also observed in the thermal transport coefficients, e.g. in thermoelectric power [10.3]. These oscillations arise from the variation in the scattering probabilities with the magnetic induction. When the magnetic induction is such that the separation between the Landau levels is larger than the optic phonon energy the scattering probability is small. As the induction is decreased and the separation approaches the phonon energy, the scattering probability increases, and the magnetoresistance suffers a large alteration. The same kind of change occurs as the separation between the Landau levels assumes values such that the optic phonon energy is equal to the separation between two levels (consecutive or not). The oscillatory feature is immediately evident from the expression for the transverse magnetoresistance [10.13]:

$$\sigma_{xx} = \int \sum_{\ell\ell'} c_{\ell\ell'} f(E)\left[1-f(E+\hbar\omega_0)\right]\left[E-\hbar\omega_B(\ell+1/2)\right]^{-\frac{1}{2}}\left[E+\hbar\omega_0-(\ell'+1/2)\hbar\omega_B\right]^{-\frac{1}{2}} \quad , \tag{10.77}$$

where $c_{\ell\ell'}$ is a smoothly varying function of E. It is evident that σ_{xx} will become large when $\omega_0 = p\omega_B$, (p is an integer). For such values of the magnetic induction the two singularities $E = (\ell+1/2)\hbar\omega_B$ and $E = (\ell'+1/2)\hbar\omega_B - \hbar\omega_0$ $=(\ell'-p+1/2)\hbar\omega_B$ may coincide. The resistivity ρ_{xx} is given by

$$\rho_{xx} = \sigma_{xx}\left(\sigma_{xx}^2+\sigma_{xy}^2\right)^{-1} \quad . \tag{10.78}$$

Since $\sigma_{xy} \gg \sigma_{xx}$ for the quantising fields, $\rho_{xx} \propto \sigma_{xx}$, and for the same values of the magnetic induction we may expect maxima to occur in the resistivity. The transverse magnetoresistance cannot, however, be analysed by using the effective mass formalism. The transverse component of the energy being quantised, we cannot relate it to a transverse wave vector by a simple dispersion relation. The problem has been solved by applying the density matrix formalism, which is beyond the scope of this treatise. Some experimental results are presented in the next section.

The longitudinal magnetoresistance, on the other hand, may be analysed by using the Boltzmann equation and the effective mass formalism. We present below the analysis for nonpolar and polar optic phonon scattering.

Nonpolar Optic Phonon Scattering

The matrix element for the nonpolar optic phonon scattering being independent of $(|\underline{k}'-\underline{k}|)$, the incoming terms cancel out. The current is hence obtained from expression (10.54) substituting for the relaxation time the expression for nonpolar optic phonon scattering given in (10.48). We get accordingly

$$J = -ne^2|e|B(k_BT)^{\frac{1}{2}}\left(2^{\frac{1}{2}}\pi^2\hbar^2 I_n m^{*\frac{1}{2}}\right)^{-1}\mathscr{E}\sum_{\ell}\int_{(1/2)(2\ell+1)\theta}^{\infty}[W-(2\ell+1)\theta/2]^{\frac{1}{2}}$$
$$\times (C_{opB})^{-1}\left[\sum_{\ell'}k_{z1}^{-1}+\sum_{\ell'}H\left(E_z+(\ell'-\ell)\hbar\omega_B-\hbar\omega_0\right)(n_0+1)n_0^{-1}k_{z2}^{-1}\right]^{-1}(\partial f_\ell^0/\partial W)\,dW \quad , \tag{10.79}$$

where

$$k_{z1} = \left(k_z^2+k_{\ell\ell'}^2+k_0^2\right)^{\frac{1}{2}} \quad ,$$
$$k_{z2} = \left(k_z^2+k_{\ell\ell'}^2-k_0^2\right)^{\frac{1}{2}} \quad .$$

Using the above expression for the current density, the conductivity σ_{zz} may be expressed as [10.14]

$$\sigma_{zz}(B) = \sigma(0)[1-\exp(-\theta)]\theta^{-1}I(\theta)/I(0) \quad , \tag{10.80}$$

where $\sigma(0)$ is the conductivity in the absence of the magnetic induction and

$$I(0) = (1/3)\int_0^{\infty}\left[(W+\theta')^{\frac{1}{2}}+n_0^{-1}(n_0+1)H(W-\theta')(W-\theta')^{\frac{1}{2}}\right]^{-1}W^{3/2}(\partial f_0/\partial W)\,dW \quad , \tag{10.81}$$

and

$$I(\theta) = \int_0^{\infty}\sum_{\ell}[W-(2\ell+1)\theta/2]^{\frac{1}{2}}\times\left\{\sum_{\ell'}[W+\ell'\theta+\theta']^{-\frac{1}{2}}+\sum_{\ell'}H(W+\ell'\theta-\theta')\right.$$
$$\left.\times(n_0+1)n_0^{-1}[W+\ell'\theta-\theta']^{-\frac{1}{2}}\right\}^{-1}(\partial f_\ell^0/\partial W)\,dW \quad , \tag{10.82}$$

$$\theta' = \hbar\omega_0(k_BT)^{-1} \quad .$$

Although the oscillatory nature of σ_{zz} is not immediately evident from the above expression we find on detailed computations that the first term I(0) varies monotonically with θ, but the second term $I(\theta)$ varies periodically with θ. Further the derivative of $I(\theta)$ has discontinuities for values of θ such that $\ell\theta = \theta'$ and therefore extrema will be present for these values of θ.

Polar Optic Phonon Scattering

The perturbation function for the polar optic phonon scattering must be obtained by solving the following equation:

$$\begin{aligned}(|e|E_0/2\hbar)n_\ell\left[1-f_\ell^0(E_z)\right]^{-1}\Big(\sum_{\ell'}\phi_{\ell\pm}(E_z)\Big\{&\left[1-f_{\ell'\pm}^0(E_{z1})\right]k_{z1}^{-1}\sum_{+,-}I_{\ell\ell'}(k_z\pm k_{z1})\\ &+\left[1-f_{\ell'\pm}^0(E_{z2})\right]H(E_z+\hbar\omega_B(\ell-\ell')-\hbar\omega_\ell)(n_\ell+1)n_\ell^{-1}k_{z2}^{-1}\sum_{+,-}I_{\ell\ell'}(k_z\pm k_{z2})\Big\}\\ &-\phi_{\ell'\pm}(E_{z1})\Big\{\left[1-f_{\ell'\pm}^0(E_{z2})\right]k_z^{-1}(n_\ell+1)n_\ell^{-1}\left[I_{\ell\ell'}(k_z-k_{z1})-I_{\ell\ell'}(k_z+k_{z1})\right]\Big\}\\ &-\phi_{\ell'\pm}(E_{z2})\Big\{k_z^{-1}H(E_z+\hbar\omega_B(\ell-\ell')-\hbar\omega_\ell)\left[1-f_{\ell'\pm}^0(E_{z2})\right]\Big[I_{\ell\ell'}(k_z-k_{z2})\\ &-I_{\ell\ell'}(k_z+k_{z2})\Big]\Big\}\Big) = 1 \quad . \end{aligned} \tag{10.83}$$

This equation is obtained by substituting the detailed expressions for the collision term in the Boltzmann equation. We find that each perturbation term is coupled to all other terms and an analytic solution is impossible to obtian. The iterative method discussed earlier for the EQL may be extended to obtain solutions when two levels are occupied and in principle also when larger number of levels are occupied. But the calculations become more and more complicated as larger number of terms are included, and no calculation by this method including more than two terms is reported in the literature. An iteration scheme has, however, been given by MAGNUSSON [10.15] which is very similar to this method but more convenient to apply. In this scheme, values of $\phi_\ell(E_z)$ for chosen values of k_z are obtained in the first step by putting $\phi_{\ell\pm}(E_{z1})$ and $\phi_{\ell\pm}(E_{z2})$ equal to zero in (10.83), i.e. $\phi_\ell(k_z)$ are evaluated as

$$\phi_{\ell\pm}^0(k_z) = \tau_\ell \quad , \tag{10.84}$$

where τ_ℓ^{-1} is the coefficient of $\phi_\ell(E_z)$ in (10.83). In the second and subsequent iteration steps, $\phi_\ell(E_z)$ is evaluated from the equation

$$\phi_{\ell\pm}^{\mu+1}(E_z) = \tau_\ell\left[1+L_1\phi_{\ell\pm}^{\mu}(E_z)\right] = \tau_\ell\sum_{\nu=0}^{\mu}[L_1(\tau_\ell)]^\nu \quad , \tag{10.85}$$

where L_1 is the operator which gives the collision term excluding those involving $\phi_\ell(k_z)$.

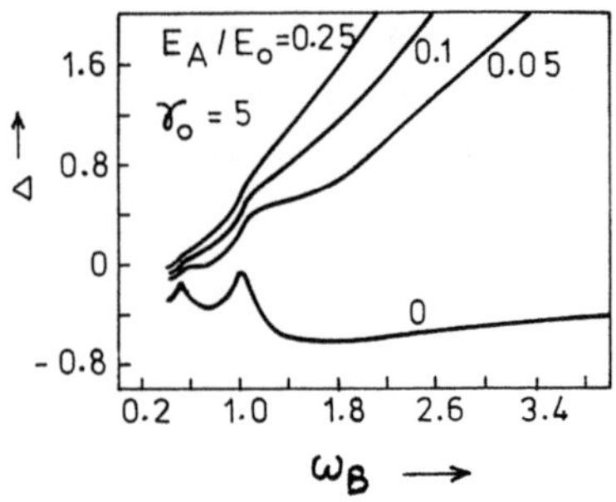

Fig. 10.5. Theoretical magnetophonon oscillation curves [10.15]. $\gamma_0 = \hbar\omega_\ell/k_BT$; E_A/E_0 = Ratio of the coupling constants for acoustic phonon and optic phonon scattering

That the magnetoresistance even in the longitudinal configuration would be oscillatory may be surmised from (10.83) since there are certain singularities for values of ω_B satisfying the resonance relation. However, any further detailed features of the resonance curves cannot be predicted from the equation. Certain features have been predicted from approximate analytic solutions of the equation neglecting the incoming terms [10.16]. MAGNUSSON [10.15], however, presented some magnetoresistance curves obtained by the iteration technique and including elastic scattering, which may be considered to represent exact solutions of (10.83). We reproduce in Fig.10.5 some of these curves. These curves bring out certain characteristics which have bearings on the experimental results. It is found that maxima are produced for magnetic fields satisfying the relation $\omega_B = \omega_0/p$. The resonance points shift to higher values of the magnetic field with increase in the strength of elastic scattering, but there remain discontinuities in the slope at the resonance fields. It is also found that if the lattice temperature is high and elastic scattering is very strong minima may be produced at the resonance points. A pseudoresonance series also appears for magnetic inductions satisfying the relation

$$\omega_B = 2\omega_0/(2n+1) \quad (n=0,1,2,\dots) \quad . \tag{10.86}$$

The theory as discussed above is really good enough to predict at this stage only the broad qualitative features. But, even these, as we shall see in the next section, do not fully agree with experiments.

Attempts to develop a quantitative theory are not much reported in the literature. The collisional broadening and the effects of electron-electron collisions are important and must be taken into account if the theory is to agree quantitatively with experiments. A theory including the broadening effects has been presented recently [10.17] and some encouraging results have been obtained. Further progress along this line may be expected in the future.

10.5 Experimental Results on Magnetophonon Resonance

The magnetophonon oscillations, although not quantitatively explained as yet, have been very useful in the study of energy band structures and phonon spectra. In the early years, the effective mass was determined from the resonance relation

$$|e|B/m^{*} = \omega_0/n \quad , \tag{10.87}$$

using the values of ω_0 determined by Raman scattering or other techniques. The resonance being observable over a wide range of temperature, the variation of effective mass with temperature could be studied with a precision of a few per cent. Lately the resonance phenomenon has been used to explore the phonon spectrum using the knowledge of the band structure [10.18]. Experiments have been carried out with indium antimonide, gallium arsenide, indium phosphide, indium arsenide, and cadmium telluride and some of the experimental records are presented in Fig.10.6. A vast amount of useful information has emerged from these experiments. These results have been reviewed in a number of articles by the experimentalists themselves. We shall briefly summarise here the basic aspects of these results [10.13].

The magnetophonon oscillations are observed above a temperature of about 40 K and reach the peak amplitude around 100 K. The amplitudes of transverse magnetoresistance oscillation are found to satisfy the empirical relation

$$\rho_{xx} \sim \exp(\gamma\omega_0/\omega_B)\,\cos(2\pi\omega_0/\omega_B) \quad , \tag{10.88}$$

where γ is an empirical factor. Unlike Shubnikov-de Haas oscillations, γ in this case is not very strongly related to the sample mobility, but it varies inversely as the ionised impurity concentration. This is mainly due to the fact that the magnetophonon oscillations are mostly controlled by electrons with $k_z = 0$. The concentration of ionised impurities control the rate of scattering but the energy dependence of scattering is not reflected in the results due to the predominance of the electrons with $k_z = 0$. The temperature dependence of the amplitude is also easily explained. We get low amplitudes below 40 K as optic vibrations are not excited in large numbers at low temperatures. The amplitude decreases at high temperatures as the Landau levels are broadened by collisions. A theoretical justification for the above empirical formula has been given by BARKER [10.17], where γ is found to be related to the electron self-energy arising from optic phonon and impurity

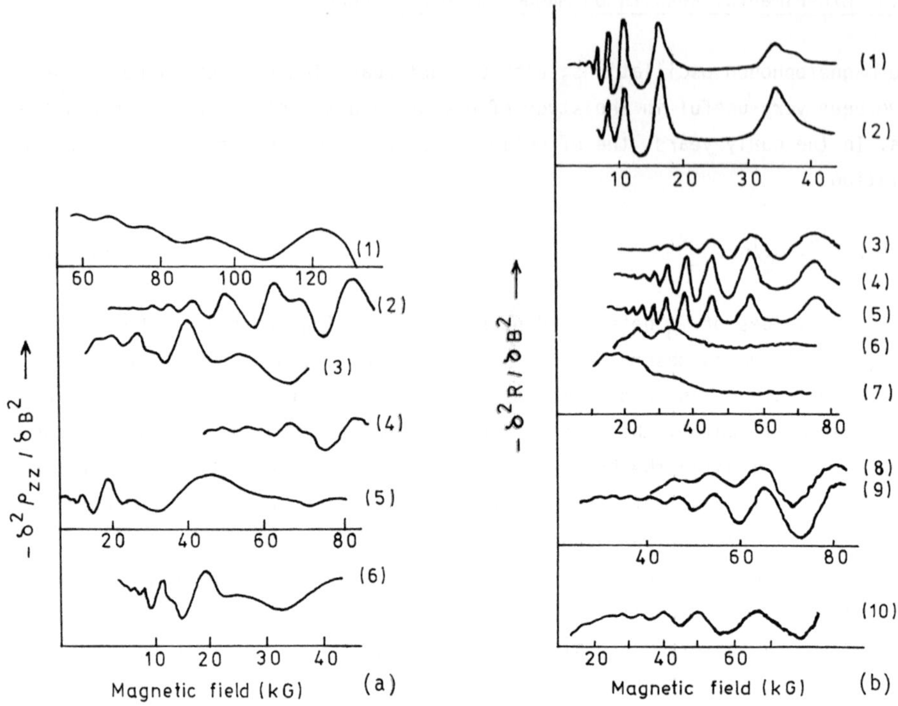

Fig. 10.6. Magnetophonon oscillations 10.13 . (a) Longitudinal: (1) GaAs (240 K); (2) GaAs (200 K); (3) InAs (290 K); (4) InP (270 K); (5) InSb (180 K). (b) Transverse: (1) InSb (120 K); (2) InSb (77 K); (3) GaAs (290 K); (4) GaAs (150 K); (5) GaAs (77 K); (6) InAs (290 K); (7) InAs 110 K); (8) InP (290 K); (9) InP (175 K); (10) CdTe (60 K)

scattering. No such formula for the amplitude of longitudinal magnetoresistance oscillations has yet been given. However, from the arguments given earlier we may expect (as found experimentally) that the amplitude will increase with temperature, pass through a maximum, and decrease at high temperatures.

The period of magnetophonon resonance is independent of the carrier concentration to first order. It is observed in both nondegenerate and degenerate materials. In degenerate materials, the peaks satisfy the relation [10.19]

$$(\ell+1/2)\hbar\omega_B \pm \hbar\omega_0 = E_F \quad . \tag{10.89}$$

In nondegenerate materials, the peaks in transverse magnetoresistance are observed at

$$\omega_B = \omega_0/p \quad . \tag{10.90}$$

Table 10.2. Effective mass obtained from magnetophonon experiments

Material	Temperature	Effective mass in units of electron mass
InSb	40	0.0133
	260	0.01210
InAs	110	0.0240
	290	0.0223
GaAs	60	0.0665
	290	0.0636
InP	77	0.0806
	290	0.0765

However, the mass involved in ω_B has a strong polaron correction, which may be as large as 13% in some materials. This correction is experimentally found to be about $0.8\alpha/3$, α is as defined by (3.10a). The position of the peaks is periodic in $1/B$ as indicated by the relation (10.90), but in nonparabolic materials, the period changes with the order of the peak due to the energy dependence of the effective mass. The energy dependence of the effective mass obtained from the analysis of the variation of this period is found to be in general agreement with the Kane theory. The effective mass values obtained experimentally are also found to vary with temperature. The change is found to be close to that predicted from the dilatational change in energy band gap and not to that calculated from the change in optical energy gap.

The positions of the extrema in longitudinal magnetoresistance, on the other hand, do not agree with any theoretical prediction. In some materials, minima occur near the fields for which maxima are observed in transverse magnetoresistance. The positions of these minima are shifted to lower fields by 0-30%. In some materials, maxima are obtained at the corresponding positions but shifted to higher fields by 3-15%. The theoretical prediction that the minima should change to maxima with the predominance of polar optic phonon scattering over elastic scattering has not been confirmed experimentally. Although the positions of the extrema do not agree with theoretical expectations, their positions in some materials form a good harmonic series, so that effective mass values may also be obtained by analysing the longitudinal magnetoresistance results.

Experimental curves of magnetophonon oscillations have been used to obtain the values of band-edge effective mass at different temperatures. We present the available results [10.13] in Table 10.2.

10.6 Conclusions

We have discussed in this chapter the transport properties of compound semiconductors in the presence of a quantising magnetic field. The analytical methods as available in the literature for the longitudinal orientation, i.e. for electric field parallel to the magnetic field, have been presented. For the transverse configuration, i.e. for electric field perpendicular to the magnetic field, only the final expressions have been quoted. Experimental results for both the cases have, however, been discussed. It is seen that the transport properties have many interesting features in quantising fields. But, due to the absence of a quantitative theory, experimental results cannot be exploited to their full potential in obtaining the constants related to physical processes. The possibilities of diagnosing the scattering processes from the saturation values of magnetoresistance in the extreme quantum limit have not been fully realised. On the other hand, the oscillatory phenomena, although quantitatively not yet understood, have been very useful in the estimation of the electron concentration, the Fermi level, and the effective mass. The magnetophonon effect, in particular, has opened the way for a new range of experiments, among the latest of which is the study of the resonance spectrum by applying a moderately high electric field [10.18]. It is explained in the following chapter that a high electric field alters the electron distribution. This excitation of the electrons to high-energy states when coupled by the magnetophonon effect provides a means for exploring the phonon spectrum and the parameters of collision processes responsible for energy loss of the electrons. Such studies have been very fruitfully made in elemental semiconductors, and are likely to be extended to compound semiconductors. It may be expected that in the near future, many new features of electron conduction will be revealed by experiments of this kind in compound semiconductors.

11. Hot-Electron Transport

11.1 Phenomenon of Hot Electrons

Electron transport is caused by an electric field, a thermal gradient, or a concentration gradient. In the latter two cases, a uniform condition is established as a result of the transport unless external sources are used to maintain the nonuniform conditions. The transport of electrons under these conditions may cause transfer of energy from one part of the sample to the other, but the electrons do not gain energy from the external sources in the process of transport. On the other hand, when the transport is caused by an electric field, electrons are continuously supplied with energy from the source of the electric field at a rate $\underline{J} \cdot \underline{\mathscr{E}}$ ($\underline{J}$ is the current density and $\underline{\mathscr{E}}$ is the electric field), and it would appear that the total energy of the electron system should go on increasing indefinitely. This, however, does not happen as this gain of energy is balanced by transfer of energy to the lattice atoms through the process of collisions. It has been seen that an electron is scattered by the lattice either by emitting or by absorbing a phonon. The lattice absorbs energy from the electron when a phonon is emitted and it delivers energy to the electron when a phonon is absorbed. In the absence of the electric field the absorption and emission processes are so balanced that there is no net transfer of energy from the electron system to the lattice system or the vice versa. It means in effect that the temperature, the thermodynamic coeffícent determining transfer of energy from one system to another, of the electron and that of the lattice system are identical. One may verify easily from the Boltzmann equation that in the field-free condition,

$$\begin{aligned}&\int E\int\left[S(\underline{k},\underline{k}')_{ab}f_0(\underline{k})(1-f_0(\underline{k}'))+S(\underline{k}',\underline{k})_{ab}f_0(\underline{k}')(1-f_0(\underline{k}))\right]d\underline{k}'d\underline{k}\\ &=\int E\int\left[S(\underline{k},\underline{k}')_{em}f_0(\underline{k})(1-f_0(\underline{k}'))+S(\underline{k}',\underline{k})_{em}f_0(\underline{k}')(1-f_0(\underline{k}))\right]d\underline{k}'d\underline{k} \quad . \qquad (11.1)\end{aligned}$$

for the same value of temperature in the electron distribution function and in the phonon occupation number. On the other hand, in the presence of an external electric field, the energy of the electron system starts to increase immediately after the application of the field. But as the energy increases, electrons emit more phonons than they absorb, and there is a net transfer of energy from the electron system to the lattice system. Referring back again to the Boltzmann equation, we find that if the temperature T_e of the electron system is larger than the lattice temperature T_L,

$$\int E(\partial f/\partial t)_c d\underline{k} > 0 \quad , \tag{11.2}$$

and this transfer continues to increase as the difference between T_e and T_L increases with time. A new equilibrium may therefore be expected instead of a continuous increase of the energy of the electrons. This equilibrium is established when the difference between the electron and the lattice temperature is such that the rate of gain of energy of the electrons from the electric source ($\underline{J}\cdot\underline{\mathcal{E}}$) is balanced by the rate of loss of energy to the lattice atoms, i.e. when

$$\underline{J}\cdot\underline{\mathcal{E}} = (4\pi^3)^{-1}\int E(\partial f/\partial t)_c d\underline{k} \quad . \tag{11.3}$$

We thus find that when electron transport occurs under the action of an electric field, the electron temperature is required to be higher than the lattice temperature to produce a steady state condition. The statement should remain valid even when the electric field is low. But, for low fields the required rise in electron temperature is so small that the difference between the electron and the lattice temperature does not produce any measurable change in the values of the transport coefficients. We may hence neglect this rise for low electric fields and take the electron temperature to be identical to the lattice temperature even when the electric field is applied. The theory of electron transport has been developed in the earlier chapters on this assumption. A consequence of this was that the symmetric part of the electron distribution function could be taken to be identical to the equilibrium function and the current density was found to vary linearly with the electric field. In other words, it is due to the validity of this assumption that we find the current-voltage relation to be given by Ohm's law at low fields.

As the electric field is increased, the difference between the electron and the lattice temperature becomes significant. The current density starts to vary nonlinearly with the electric field as it increases, even if the lattice temperature and the electron concentration are kept unaltered. The nonlinearity may be approximated by a square law,

$$J = \sigma_0 \mathscr{E}(1+\beta\mathscr{E}^2) \quad , \tag{11.4}$$

for moderate values of the field. The electron temperature is not too high for such fields and such conditions are referred to as warm-electron conditions [11.1]. The coefficient β is called the warm-electron transport coefficient.

For very large electric fields, however, the electron temperature may become an order higher than the lattice temperature. For such high electron temperatures, the whole character of the electron transport may change radically. In addition to the failure of Ohm's law, many of the low-field results become invalid. The isotropic and scalar character of conductivity and Hall mobility for cubic semiconductors is changed [11.2]. The current density, instead of increasing linearly with the electric field, attains a saturation value [11.3]. New phenomena occur which could not be expected from the low-field results. One such phenomenon is the transfer of electrons from one valley of the energy band structure to another valley, the so-called transferred electron effect [11.4]. A consequence of this effect is the production of a negative differential mobility which has been of great technical importance in realising solid-state microwave and millimeter wave sources. In addition to the phenomena already observed there are suggestions of the existence of even more unexpected phenomena than those observed. In the preceding discussion it was implied that more and more energy is transferred from the electron system to the lattice system as the electron temperature rises. This should be true if the electrons are predominantly scattered by deformation potential acoustic or nonpolar optic or intervalley phonons. However, it is not true for piezoelectric and polar optic phonon scattering. In fact, it is expected that if these scattering processes predominate, one may observe a cooling of the electrons [11.5] or a run-away condition at high electric fields [11.6].

The differential conductivity has been proved to be negative experimentally. There are predictions that the diffusion constant may also be negative [11.7]. Also the conductivity may have an absolute negative character [11.8]. Because of all these expectations and for the novelty of the observed phenomena electron transport in high fields has been a very active subject of study for the last three decades. As the electron temperature is very high for these fields transport for these fields is often called the hot-electron transport. We present in this chapter the theory of hot-electron transport and we review the results so far obtained on this kind of transport in compound semiconductors.

11.2 Experimental Characteristics

The change in electronic conductivity caused by the electric field had been studied theoretically as early as 1930 in relation to ionised gases [11.9]. It was in fact shown that at high electric fields the energy distribution function for nondegenerate gases would be quite different from the Maxwellian function, which applies at low fields. The consequent conductivity was shown to decrease with electric field as the inverse of the square root of the field for collisions characterised by a constant mean free path. The distribution function for this case was named Druyvestyn distribution to distinguish it from the low-field Maxwellian function [11.10]. The physics of electron conduction in conductors or semiconductors being similar to that in ionised gases, the variation of conductivity with electric field could be theoretically predicted in these materials also. However, experimental evidence showing this variation was not available till 1951 [11.3]. Apparently, this was due to experimental difficulties. First, the hot-electron condition would be experimentally evident only when the electron energy corresponding to the rise in temperature is comparable with the average energy of the electrons under field-free condition. In metals or degenerate semiconductors, the required value of the electric field for producing this condition is so high and the corresponding power delivered to the sample is so large that the sample is likely to disintegrate physically before the effect is observed. In fact no definite experimental evidence of the effect is available for metals even today, although one or two preliminary reports appeared some years ago.

The experiment became feasible when techniques of preparation of pure semiconductors with small electron concentration were established. Experiments could be planned with high resistivity samples in which the required field for producing the hot-electron effect was only a few hundred volts. The experiment is essentially very simple. One is only required to measure the current density for different fields. The experiment may be carried out by applying a steady voltage. But in that case the lattice temperature varies with the applied field. As explained earlier, when an electric field is applied, the energy supplied to the electrons from the source of the electric field is balanced by the energy transferred to the lattice atoms through collisions. The energy so gained by the lattice causes its temperature to rise above that of the ambient or the heat sink on which the sample may be mounted. However, with increase in the lattice temperature, the lattice starts to lose energy to the heat sink and a steady lattice temperature is established when the energy gained by the lattice atoms is balanced by the

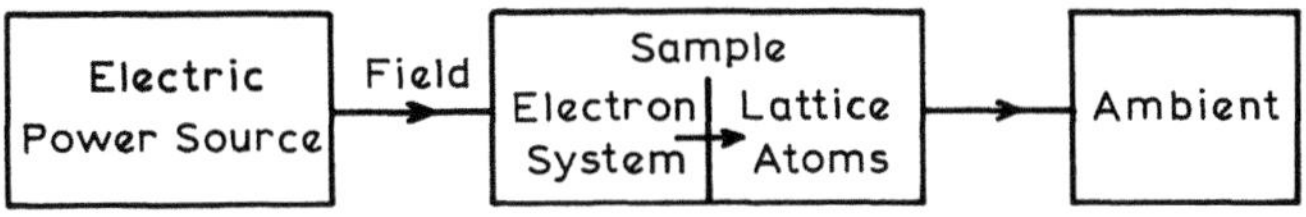

Fig. 11.1. Energy flow in hot-electron experiment

amount lost to the sink through the processes of radiation or conduction. In the steady state, therefore, the electron temperature is higher than that of the lattice and the lattice temperature is higher than that of the sink. The energy flow is illustrated in Fig.11.1. The variation of conductivity observed with a steady voltage is thus partly due to the electric field and partly due to the change in lattice temperature. The contribution of the electric field, in other words the change brought about solely by the hot-electron condition, may be determined by carrying out the experiment at different heat sink temperatures. But, the analysis is complicated by the non-uniformity of the lattice temperature as the part of the sample in direct contact with the heat sink will have a lower temperature than the remote parts. A more serious complication is introduced by the difficulty of making ohmic and non-injecting contacts to high resistivity semiconductor samples. The possibility of injection increases with the increase in the contact field. Hence, in the experiment under discussion the electron concentration may also be expected to change with the electric field. This introduces a complexity which is almost impossible to sort out and the determination of the true contribution of the hot-electron effect to the variation of the conductivity with the field becomes correspondingly awkward.

The difficulties discussed above for an experiment with a steady voltage were removed in the pioneering experiments by using a pulse voltage and preparing the sample in dumbbell form. The electron temperature starts to increase as soon as the electric field is applied. The energy transferred to the lattice through collisions also causes the lattice temperature to start increasing immediately. However, the heat capacity of the electrons being much smaller than that of the lattice atoms, electron temperature rises much faster. The typical time taken by the electrons to reach steady state is about 10^{-9}-10^{-12} s. On the other hand, the lattice temperature does not increase significantly in a time of the order of few hundred nanoseconds. The condition is illustrated schematically in Fig.11.2. The lattice temperature may therefore be taken to be unaltered and the hot-electron condition to have been established if the voltage is applied in the form of a pulse of duration less than about 1 μs. It should, however, be ensured that sufficient time

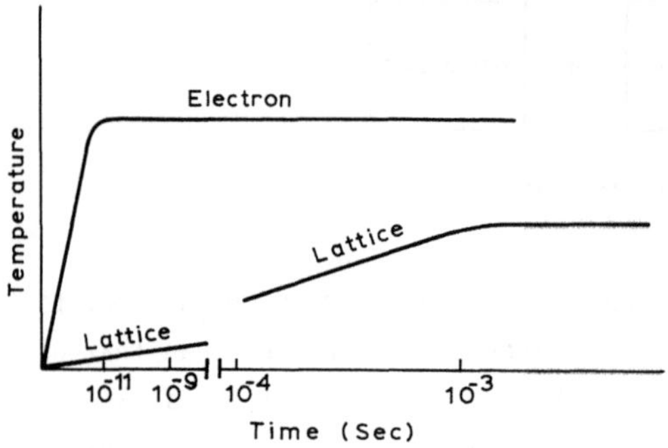

Fig. 11.2. Development of electron and lattice temperature in a semiconductor sample after the application of a high voltage

elapses before the arrival of the next pulse so that the small amount of energy transferred to the lattice in the duration of the pulse may be radiated or conducted away to the heat sink. This may be arranged by applying the pulses at a very low repetition rate of the order of a few cycles per second. A suitable pulse duration and a pulse repetition rate may be worked out knowing the conductivity and dimensions of the experimental sample, but in practice it is more convenient to adjust the two so that the current through the sample does not show any significant change within the pulse or from pulse to pulse.

The contact effects are also minimised when a dumbbell-shaped sample is used along with the pulse voltage. The thin part of the sample is made with a dimension such that it constitutes the major part of the sample resistance. Then the high field appears mostly in this region and the end parts have low fields. Hence, even if charge carriers are injected from the contacts, they travel with a low drift velocity and do not have sufficient time to reach the filamentary part within the duration of the pulse. The carrier injection thus affects only the resistance of the end parts of the sample, which in any case, is a small fraction of the total sample resistance.

First experiments were carried out on the elemental semiconductors. The current densities through dumbbell-shaped samples were measured using pulses

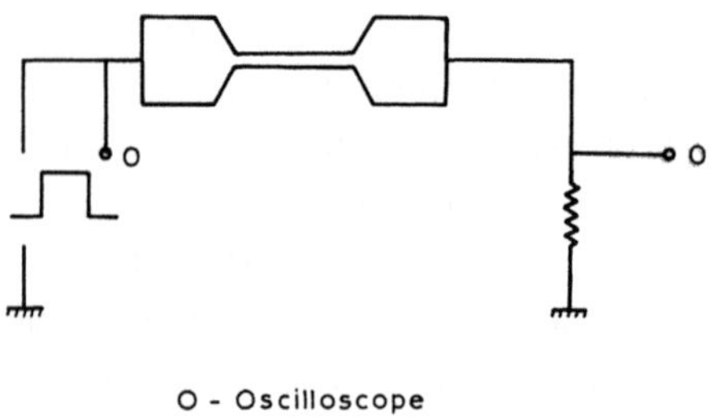

Fig. 11.3. Hot electron experimental set-up

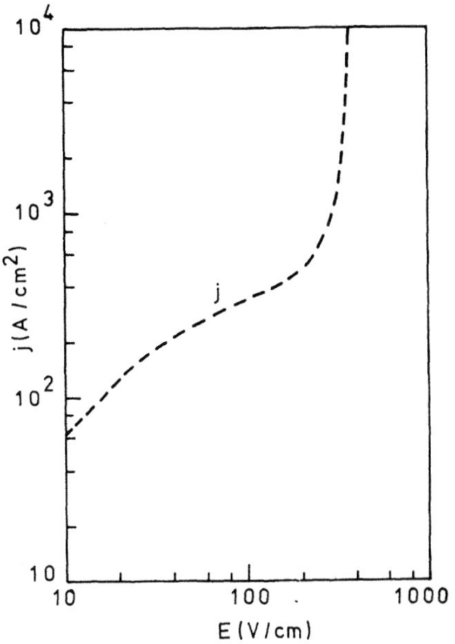

Fig. 11.4. Current-voltage characteristic of an indium antimonide sample [11.11]

of a duration of about 1 μs and repetition rate of about 1-10 cycles per second. The experimental arrangement is shown in Fig.11.3. The voltage across the sample was measured on an oscilloscope and the current through the sample was also obtained by measuring on the oscilloscope the voltage across a small resistance connected in series with the sample. The experiments gave results confirming the earlier theoretical predictions that the current will vary nonlinearly with the voltage at high fields. Many experiments were carried out in the following two decades to elucidate the transport characteristics of semiconductors in high electric fields.

Experiments were also carried out on the high-field transport characteristics of compound semiconductors. The first few experiments were done with indium antimonide essentially using the same technique [11.12]. Pulse voltages were applied, but rectangular samples were used. The lifetime of injected carriers and consequently the diffusion length being small (a few micrometres), the injection effects could be considered negligible. We have presented in Fig.11.4 the experimental current voltage characteristic of an indium antimonide sample at 77 K. The Hall coefficient was also monitored in the experiment and found to be constant till the field reached a value of 200 V/cm. The variation of current density with the field may thus be taken to represent the variation of electron drift velocity with the electric field. It is seen that the drift velocity at first increases linearly with the electric field,

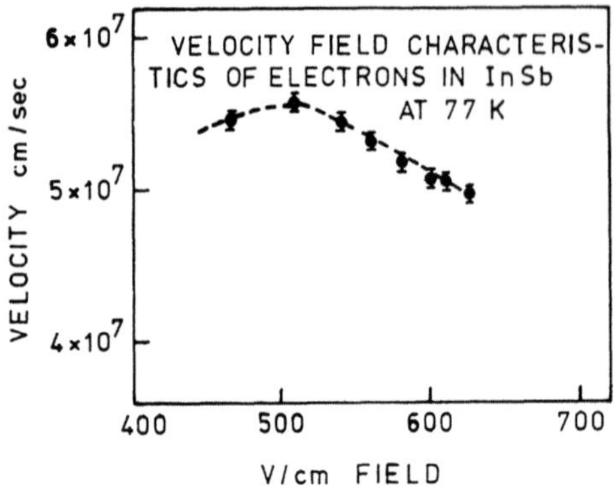

Fig. 11.5. Current-voltage characteristic of an indium antimonide sample in the high-field region for nanosecond pulses [11.13]

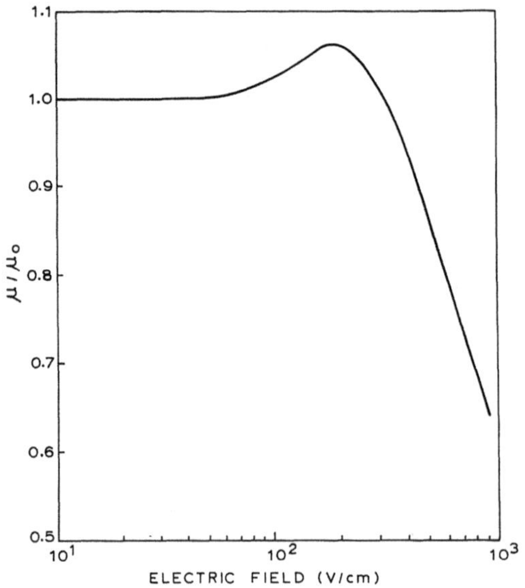

Fig. 11.6. Hot-electron conduction characteristic of indium arsenide [11.15c]

then increases at a slower rate and finally tends to a saturation value. But this region of the characteristic is masked by the increase in carrier concentration apparently due to avalanche breakdown. In a later experiment, carrier generation due to avalanching was avoided by using pulses of very short duration [11.13]. The build-up of carrier concentration takes a longer time than the establishment of the hot-electron condiction. The electron temperature reaches the steady-state value but significant change in carrier concentration does not occur in about 100 ns. The experimental characteristic for such pulses is shown in Fig.11.5. A significant feature of this characteristic is that the drift velocity passes through a maximum at a field of about 500 V/cm and then starts to decrease with increase in the field. The differ-

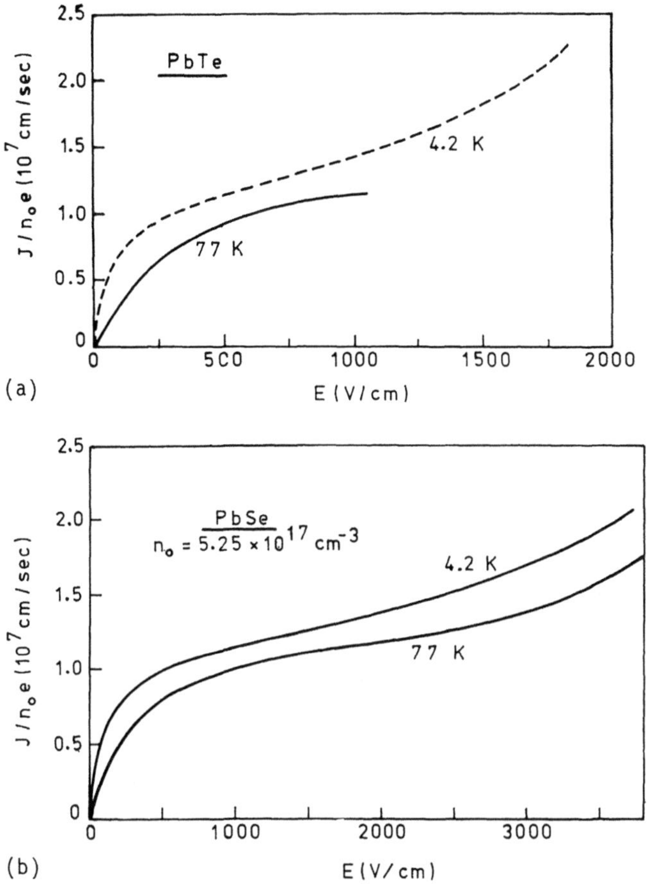

Fig. 11.7. (a) Hot-electron conduction characteristic of lead telluride [11.86]; (b) Hot-electron conduction characteristic of lead sellenide [11.86]

ential mobility thus becomes negative above a threshold field. Most of the hot-electron experiments on compound semiconductors have been carried out with indium antimonide. In fact various other transport coefficients under hot-electron conditions have been studied for indium antimonide [11.14]. However, as the discussion on the hot-electron transport will be confined in this book to the variation of drift velocity with field, we shall not present other results. Hot-electron experiments have also been carried out on a few other compound semiconductors. Specifically, characteristics of indium arsenide [11.15], lead chalcogenides [11.16], mercury cadmium telluride [11.17], and cadmium sulphide [11.18,19] have been studied by applying essentially the same technique. These characteristics are illustrated in Figs.11.6-9.

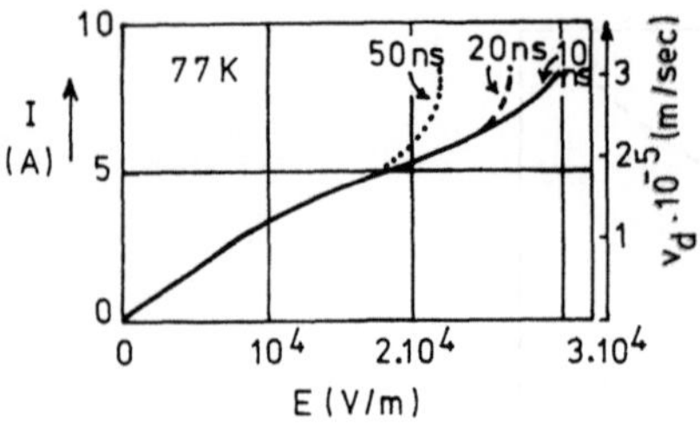

Fig. 11.8. Hot-electron conduction characteristic of $Hg_{1-x}Cd_xTe$ [11.17]

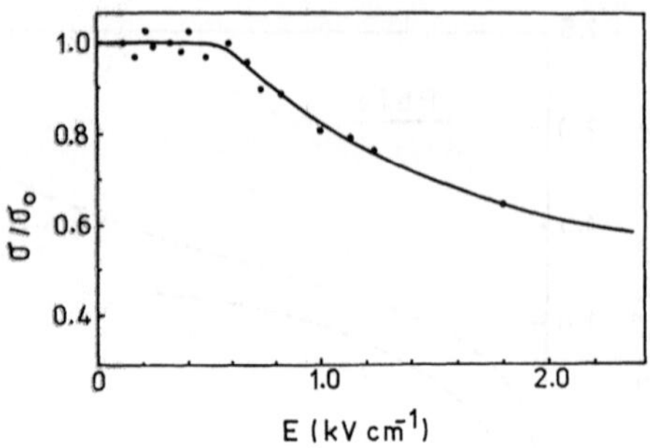

Fig. 11.9. Hot-electron conduction characteristic of cadmium sulphide [11.19b]

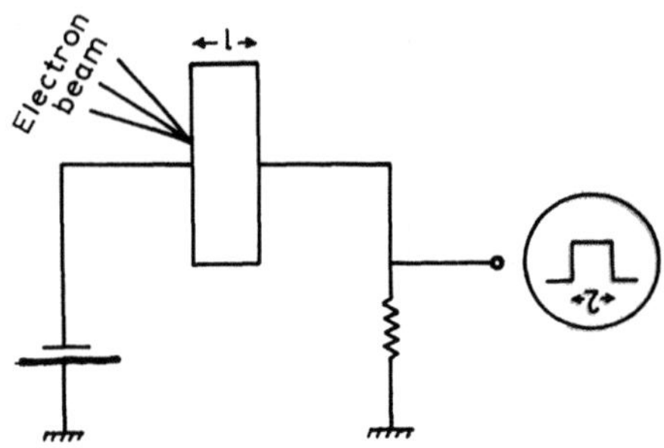

Fig. 11.10. Time-of-flight experimental set-up

More spectacular characteristics are obtained for gallium arsenide, indium phosphide, and cadmium telluride. In these materials, when the pulse voltage is applied, the current at first increases linearly and then sublinearly. But, as the voltage is made higher than a threshold value, the current becomes unstable and one observes regular oscillations in the current. These instabilities were first discovered by GUNN [11.4d] and the effect is often referred to as the Gunn effect. Although there were some exploratory discussions about the cause of this instability when the effect was first discovered, it is now well established that the instabilities are due to the appearance of negative differential conductivity at high fields in these materials. The negative differential conductivity leads to charge density fluctuations which appears as oscillations in the current. The hot-electron characteristics of these materials cannot therefore be obtained above the threshold voltage by applying the simple pulse voltage technique. A new method, the time-of-flight technique [11.20], was developed for the measurement of drift velocity in the negative-differential-conductivity region. In this method, a sample of high resistivity is taken, or the sample is depleted of carriers by producing a junction or Schottky barrier and applying a reverse bias. A short pulse of

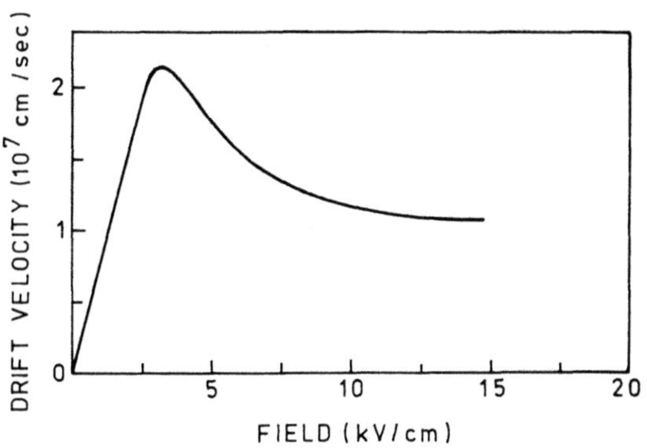

Fig. 11.11. Hot-electron conduction characteristic of gallium arsenide [11.20a]

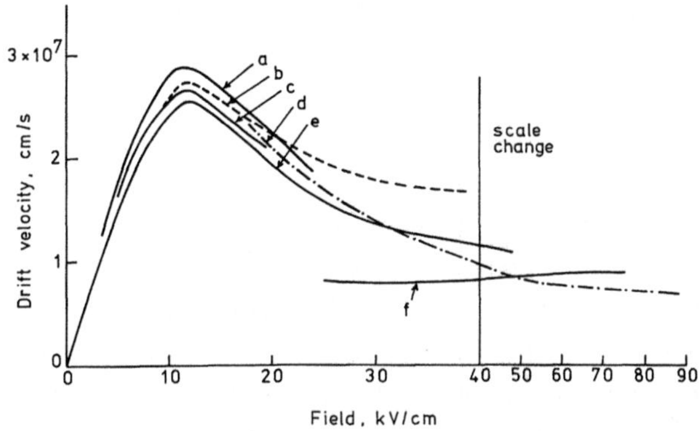

Fig. 11.12. Hot-electron conduction characteristics of indium phosphide [11.87]. a,b,c,d,e,f - Results of different experiments

carriers is then created at one end by shining an electron beam for a short duration on the sample. In the experimental arrangement shown in Fig.11.10, holes move out of the sample at the injection end but the electrons move towards the positive end with a velocity determined by the applied voltage and the consequential field. The current density is of magnitude $n|e|vL^{-1}$ and has a duration $\tau_D = L/v$. (L is the length of the sample, n is the concentration of electrons, v is the drift velocity.) The drift velocity may thus be determined by measuring the time-of-flight τ_D and the length of the sample. Experimental hot-electron characteristics of gallium arsenide, indium phosphide and cadmium telluride obtained by this technique are shown in Figs.11.11-13. In all these characteristics, it may be seen that the differential mobility becomes negative above a threshold voltage.

These characteristics have been determined experimentally by applying other techniques [11.21] also. Successful application of the time-of-flight

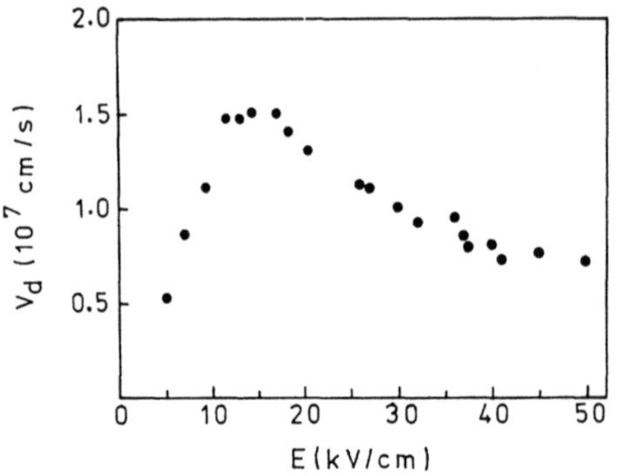

Fig. 11.13. Hot-electron conduction characteristic of cadmium telluride [11.88]

technique requires that the sample should be free of traps as otherwise the generated carriers are trapped in the process of flight and the current is terminated before the carriers reach the other end. A variation of the time-of-flight technique was developed which may be used for relatively poor quality crystals. In this method the electron beam is modulated by a microwave signal and the phase and amplitude of the microwave component of the output current signal is used to determine the time of flight. The technique has been successfully applied to indium phosphide and gallium arsenide.

The characteristics may also be determined by analysing the field distribution in the sample when the charge distribution becomes nonuniform at post-threshold fields [11.22]. The characteristics may be obtained also by using a high-power microwave signal to produce the hot-electron condition and then determining the conductivity with a small steady voltage [11.23]. Characteristics of gallium arsenide and indium phosphide have been determined by applying all these techniques. The important point about these characteristics is that they show the same broad features but the exact values of the drift velocity for an applied electric field are not the same but differ by large amounts. Apparently this difference is mostly due to the poor accuracy of most of these methods [11.24]. In fact, the discrepancy between the different experiments is such that it is very difficult to draw any conclusion about the validity of a particular theory by comparing the results of theoretical analysis with experiments. There has been a lot of confusion in the literature on this account particularly in relation to indium phosphide for which two alternative models of scattering were proposed [11.25]. It would appear that the experimental accuracy of hot-electron characteristics for materials with negative differential mobility is such that an exact comparison with theory is not justifiable at this stage.

Although hot-electron transport experiments were started about three decades ago, interest in such experiments has not yet completely died out. Experiments are being conducted on new materials with the expectation of discovering a larger negative differential mobility than so far observed. As a result of this search some of the mixed compound semiconductors have been found to have a negative differential conductivity in recent experiments [11.26].

Intensive work is also being carried out to determine directly the energy distribution of electrons in the steady state and the time evolution of the distribution by using laser probing signals [11.27]. These experiments reveal directly the establishment of the hot-electron condition and help us to understand the phenomenon. We shall, however, confine our attention in this chapter mostly to the theory of the conductivity characteristics and some related transport coefficients.

11.3 Negative Differential Mobility and Electron Transfer Effect

We find from the hot-electron characteristics of the previous section that in many compound semiconductors at some fields the differential conductivity is negative. This result, both theoretical and experimental, is of great significance. In fact, the discovery of this effect has been the greatest payoff from the study of the physics of electron transport. A new class of solid-state microwave sources and fast-switching circuits have been developed for commercial use utilising this effect [11.28]. The physics underlying this effect is also somewhat complex and to understand it required development of new methods for the evaluation of the hot-electron transport coefficients. One may say that recent developments in transport theory occurred through attempts to calculate these characteristics accurately. Before presenting these methods we discuss the physics of the phenomenon in this section.

The discussion will be confined to n-type materials, since only in this type of material has significant negative differential conductivity (NDC) been observed. The conduction band in most of the semiconductors is of the multivalley type (see Sect.3.5) and the average drift velocity of the electrons may, in general, be written as

$$\underline{v}_d = \sum_\nu n_\nu e < \tau_\nu(E)/m^*_\nu(E) > \underline{\mathscr{E}} \Big(\sum_\nu n_\nu\Big)^{-1} \quad , \tag{11.5}$$

where n_ν is the electron concentration in a particular valley, $\tau_\nu(E)$ is the momentum relaxation time for an electron of energy E and $m^*_\nu(E)$ is the effective mass component in the direction of the field $\underline{\mathscr{E}}$ for the same valley corresponding to an electron energy E. The multivalley character of the conduction band is of great significance for the understanding of the NDC that is observed experimentally. We shall, however, first confine our attention to materials having the lowest conduction band minimum at the Γ point and to conditions for which all the electrons remain confined to this minium. For this case, the expression for the average drift velocity simplifies to

$$\underline{v}_d = e < \tau(E)/m^*(E) > \underline{\mathscr{E}} \quad . \tag{11.6}$$

At low electric fields, the average energy of the electrons is taken to be unaffected by the electric field. Hence, although the relaxation time is a function of electron energy and the effective mass also varies with electron energy in materials with nonparabolic bands, the mobility is not affected by the electric field. On the other hand, under hot-electron conditions as explained earlier, the average energy of the electrons depends on the electric field and in most of the semiconductors the average energy increases with increase of the electric field. This increase in electron energy affects the mobility by altering the value of $\tau(E)$ or by altering $m^*(E)$ or both. In materials for which the band gap is large, the energy dependence of the effective mass is not significant and the observed variation of mobility is mostly due to the change in $\tau(E)$. This is the condition in elemental semiconductors and some of the large band-gap compound semiconductors. The analysis of hot-electron conductivity in such materials involves evaluation of the average energy of the electrons at the experimental field and determination of the corresponding value of τ. The reduction in mobility with electric field that may occur through the alteration of τ is not however faster than $\mathscr{E}^{-1}$. NDC is not therefore expected to occur in single-valley parabolic materials. On the other hand, if the material is highly nonparabolic (e.g. indium antimonide, mercury cadmium telluride), the increase in electron energy with the applied field causes a significant increase in the value of the average value of the effective mass. The change in mobility in these materials under hot-electron conditions is, therefore, due to changes in both $< \tau(E) >$ and $< m^*(E) >$. Since the overall value of $< \tau(E) >$ decreases with increase in $< E >$, and $< m^*(E) >$ increases with $< E >$, it is expected that the mobility will decrease faster with the electric field. Under favourable conditions, additional decrease in the mobility brought about by the increase of $< m^*(E) >$ may cause the mobility to decrease faster than $\mathscr{E}^{-1}$. In that case, the drift velocity decreases with increase in $\mathscr{E}$ and NDC may be observed. It has been predicted that an NDC may be produced in indium antimonide [11.29], in some mixed compounds [11.30], and in semiconductor superlattices [11.31] through this mechanism. However, there is yet no clear experimental evidence of NDC produced through this process.

In materials with the lowest conduction band minima along the X or L directions, an additional mechanism for the lowering of the mobility comes into play. The effective mass of the electrons is anisotropic for these valleys and the magnitude of the component of the effective mass in the direction of the field depends on the orientation of the field in relation to the principal axes of the constant-energy ellipsoids. For example, when the electric field is applied in the [100] direction, the effective mass for this direction has a value larger for the [100] and [$\bar{1}$00] valleys than for the [010], [001], [0$\bar{1}$0], and [00$\bar{1}$] valleys. The mobility of the electrons in the valleys with the lower values of the effective mass being larger, the energy gained from the electric field by the electrons in these valleys is also larger. The average energy of the electrons for the lower mass valleys is as a result larger. This difference in the average energy of the electrons in the hot and cold valleys, even when these are equivalent, destroys the symmetry of the valleys by making the average values of the relaxation times different in the different valleys. The scalar nature of the conductivity and Hall mobility of cubic semiconductors with anisotropic effective mass is thereby changed. These transport coefficients become tensor quantities under hot-electron conditions. The conductivity is dependent on the direction of the electric field and the current is not collinear with the electric field. These new effects were first observed in germanium and silicon [11.32]. A more significant effect occurs through the inequality of the average energy of the electrons in the hot and cold valleys. The electrons in these valleys being coupled by intervalley scattering, there is also a net transfer of electrons from the hot to the colder valleys. The electron concentrations in the different valleys, which are equal under low-field conditions, become different under hot-electron conditions. The colder valleys have more electrons than the hotter valleys and the transfer of electrons from the hotter to colder valleys increases with increase in the electric field. Since the mobility is lower for the colder valleys (it is why the valleys are colder), this transfer of electrons further lowers the value of mobility at high fields. This electron transfer was first suggested by SASAKI et al. [11.33] to explain the magnitude of the nondiagonal components of the conductivity tensor for hot electrons in germanium. It has been suggested and apparently proved experimentally that the reduction in mobility caused by the electron transfer from the [010], [0$\bar{1}$0], [001], [00$\bar{1}$] valleys to the [100] and [$\bar{1}$00] valleys for an electric field in the [100] direction may be large enough to produce NDC in silicon. The observed effect is, however, small [11.34]. There is also no experimental evidence of this kind of NDC in any of the compound semiconductors.

Electron transfer may, however, occur also in materials with a Γ-point minimum as the lowest valley. If the increase in the average energy of the electrons due to the electric field is larger than the separation between the Γ-point valley and the X or L valleys, electrons will be transferred from the Γ valley to the higher valleys through nonequivalent intervalley scattering. The effect of this transfer will be evident in experiments if the energy band gap is larger than the separation, as otherwise avalanche breakdown will set in before the electron transfer is effective. However, if electrons are transferred from the Γ valley to the X or L valleys, a significant NDC may be expected under favourable conditions. The effective mass of the electrons is smallest in the Γ valley and as a result the mobility of electrons in the valley is much higher than the value for the X or L valleys. Hence, when the electrons are transferred from the Γ valley to higher valleys, the average mobility of the electrons is reduced drastically. This reduction in mobility reduces the average mobility with increasing field and if the rate of decrease of average mobility is higher than $\mathscr{E}^{-1}$ we should get an NDC. The required condition may be made more evident by considering the expression for the differential conductivity for fields at which the electron transfer is taking place.

At low fields when all the electrons are in the Γ valley, the conductivity is

$$\sigma = ne^2 \langle \tau_\Gamma / m^*_\Gamma \rangle = n|e|\mu_\Gamma \quad . \tag{11.7}$$

The subscript Γ indicates values for the Γ valley. At high fields, when electron transfer takes place, the conductivity is

$$\sigma = |e|(n_\Gamma\mu_\Gamma + n_{X,L}\mu_{X,L}) \quad . \tag{11.8}$$

The subscript X,L indicates the values for X or L valleys. The differential conductivity at high fields

$$\sigma_d = \sigma + \mathscr{E}(d\sigma/d\mathscr{E}) = |e|\Big[(n_\Gamma\mu_\Gamma + n_{X,L}\mu_{X,L}) + (\mu_{X,L} - \mu_\Gamma)(dn_{X,L}/d\mathscr{E})\mathscr{E}$$
$$+ n_{X,L}(d\mu_{X,L}/d\mathscr{E})\mathscr{E} + n_\Gamma(d\mu_\Gamma/d\mathscr{E})\mathscr{E}\Big] \quad . \tag{11.9}$$

Particle conservation requires that

$$dn_{X,L}/d\mathscr{E} = -dn_\Gamma/d\mathscr{E} \quad . \tag{11.10}$$

At fields where electron transfer starts taking place $n_{X,L} \ll n_\Gamma$, and since $\mu_\Gamma \gg \mu_{X,L}$, we may simplify the expression for σ_d to

$$\sigma_d \simeq |e|(n_\Gamma\mu_\Gamma - \mu_\Gamma\mathscr{E}dn_{X,L}/d\mathscr{E} + n_\Gamma\mathscr{E}d\mu_\Gamma/d\mathscr{E}) \quad . \tag{11.11}$$

If the electrons are confined to the Γ valley, a negative σ_d is obtained if $d\mu_\Gamma/d\mathscr{E}$ is negative and $\mathscr{E}|d\mu_\Gamma/d\mathscr{E}| > \mu_\Gamma$. In the presence of the electron-transfer effect, on the other hand, we get an additional negative term and the possibility of NDC is much enhanced. It will be comparatively easily produced if $\mathscr{E}dn_{X,L}/d\mathscr{E} > n_\Gamma$. This requires that the effective mass in the X,L valleys be much larger than that in Γ valleys, so that all the electrons having energies larger than Δ are transferred from the Γ minimum to the X,L valleys. An NDC due to the nonequivalent intervalley electron transfer effect would be produced if the lowest valley is the Γ valley and the higher X or L valleys are separated by an energy Δ smaller than the band gap and at the same time if $m^*_{X,L}$ is much larger than m^*_Γ. These conditions are obeyed in most of the semiconductors in which NDC, evidenced in the hot-electron characteristics presented in Figs.11.11-13, has been established to be due to this electron-transfer effect. The theory of hot-electron transport in compound semiconductors is therefore mostly related to this effect.

11.4 Analytic Theories

It should be evident from the preceding discussion that hot-electron transport in compound semiconductors involves a number of complex phenomena. The average energy of the electrons is given by the symmetric part of the distribution function, which at low fields is taken as known. But, for hot-electron conditions, both symmetric and asymmetric parts of the distribution function have to be determined as a solution of the Boltzmann equation or by other means. Since electrons are also distributed over interacting equivalent or nonequivalent valleys, we have to determine distribution functions for electrons in different valleys, which are coupled through intervalley scattering. Further, the predominant scattering mechanism being polar optic phonon scattering, the symmetric part is not isotropic and the relaxation-time formalism is inapplicable. The problem is further complicated by the effects of nonparabolicity of the energy band which are very significant for the direct-gap materials.

The difficulties discussed above make it almost impossible to develop analytic methods for the analysis of hot-electron transport in compound semiconductors. Analytic theories may be worked out only by introducing very drastic assumptions, often not justified. However, it is also true that the basic features of the characteristics are brought out very clearly from these approximate theories, although no quantitative accuracy may be claimed. Also, for some of the phenomena, such as warm-electron mobility and diffusion, only analytic methods may be conveniently applied. Accurate calculations using numerical methods are also not often very meaningful as our knowledge of the physical parameters involved in the coupling constants for the various scattering mechanisms or of the energy band structure is yet rather inadequate. Analytic theories have therefore an important role to play for the understanding of hot-electron transport although a number of accurate numerical methods are available. We briefly review the available analytic methods in this section.

11.4.1 Differential Equation Method

The starting point for the analytic theories of hot-electron transport is the Boltzmann equation. The solution of the equation is worked out introducing suitable assumptions in one of three ways. In the differential equation method, the unknown distribution function is expanded as follows in a spherical harmonic series with the direction of the field as the polar axis [11.35]:

$$f(\underline{k}) = \sum_{n=0}^{\infty} f_n(k)P_n(\cos\theta) \quad , \tag{11.12}$$

where θ is the polar angle of $\underline{k}$. It is then substituted in the Boltzmann equation and the equation is separated into a series of coupled equations by multiplying successively by $P_n(\cos\theta)$ and integrating over θ from 0 to π. We give below first few of such equations.

$$(e\mathscr{E}/\hbar)\left[(2/3)(\partial f_1/\partial k)+(4/3)(f_1/k)\right] = \int_{-1}^{+1}(\partial f/\partial t)_c P_0(\cos\theta)d(\cos\theta) \quad ,$$

$$(e\mathscr{E}/\hbar)\left[(2/3)(\partial f_0/\partial k)+(4/15)(\partial f_2/\partial k)+(12/15)(f_2/k)+\dots\right]$$
$$= \int_{-1}^{+1}(\partial f/\partial t)_c P_1(\cos\theta)d(\cos\theta) \quad ,$$

$$(e\mathcal{E}/\hbar)\Big[(4/15)(\partial f_1/\partial k)-(4/15)(f_1/k)+(6/35)(\partial f_3/\partial k)+\ldots\Big]$$
$$= \int_{-1}^{+1} (\partial f/\partial t)_c P_2(\cos\theta)d(\cos\theta) \quad . \tag{11.13}$$

It is evident that an analytic solution is almost impossible unless some assumptions are introduced. In most of the analyses discussed in the literature, the series is terminated at the second term, so that the equations to be solved are

$$(e\mathcal{E}/\hbar)\Big[(2/3)(\partial f_1/\partial k)+(4/3)(f_1/k)\Big] = \int_{-1}^{+1}(\partial f/\partial t)_c P_0(\cos\theta)d(\cos\theta) \quad , \tag{11.14}$$

and

$$(e\mathcal{E}/\hbar)(2/3)(\partial f_0/\partial k) = \int_{-1}^{+1}(\partial f/\partial t)_c P_1(\cos\theta)d(\cos\theta) \quad . \tag{11.15}$$

The collision term is nonlinear if the material is degenerate.

Some attempts are reported in the literature for solving the equations for degenerate materials [11.36]. But, in degenerate materials the electron concentration would also be high and electron-electron scattering, which was neglected, should have a very significant effect. The analysis for degenerate materials is thus incomplete when electron-electron scattering is neglected. A complete analysis of hot-electron conduction in degenerate materials is still lacking in the literature.

On the other hand, the collision terms are linear for nondegenerate materials and an analytic solution may be developed with suitable approximations. We shall confine our discussion to nondegenerate materials only. We may then reduce (11.14) and (11.15) to

$$(1/3)(e\mathcal{E}/\hbar)\Big[(\partial f_1/\partial k)+2f_1/k\Big] = (1/2)\int_{-1}^{+1}(\partial f_0/\partial t)_c d(\cos\theta) \quad , \tag{11.16}$$

and

$$(e\mathcal{E}/\hbar)(\partial f_0/\partial k) = (3/2)\int_{-1}^{+1}\Big[\partial(f_1 \cos\theta)/\partial t\Big]_c \cos\theta\, d(\cos\theta) \quad . \tag{11.17}$$

The differential equation method may be applied when the collision term admits a relaxation time so that

$$f_1 = -(e\mathcal{E}/\hbar)\tau(\partial f_0/\partial k) \quad , \tag{11.18}$$

τ being given by

$$-f_1/\tau = (3/2)\int_{-1}^{+1}\left[\partial(f_1 \cos\theta/\partial t\right]_c \cos\theta\, d(\cos\theta) \quad . \qquad (11.19)$$

The function f_1 in (11.16) is replaced by the expression (11.18) and the collision term on the right-hand side is expressed in terms of f_0, $\partial f_0/\partial k$, $\partial^2 f_0/\partial k^2$, etc. To take an example, for isotropic parabolic bands, the equation obtained for acoustic phonon scattering is [Ref.11.37,p. 183]

$$k_BTE(d^2f_0/dE^2) + (E+2k_BT)(df_0/dE) + 2f_0 = -p(k_BT)^2E^{-1}\frac{d}{dE}(Edf_0/dE) \quad ,$$
$$p = 2e^2\mathcal{E}^2(3m^*)^{-1}\tau_{ac}^2(2m^*s^2)^{-1} \quad ; \quad \tau_{ac} = \pi\rho\hbar^4s^2/\left(\sqrt{2}m^{*3/2}k_B^{3/2}T^{3/2}E_1^2\right) \quad , \qquad (11.20)$$

where the symbols have the same meaning as defined earlier. This equation has the solution

$$f_0(E) = (n/I_n)\left[(E/k_BT)+p\right]^p \exp(-E/k_BT) \quad , \qquad (11.21)$$

I_n being the normalisation integral. The complete distribution function is now known since f_1 is expressed in terms of f_0. The conductivity may now be calculated using the expression for the mobility. The distribution function may be simplified for very high fields when $p \gg E/k_BT$ to

$$f_0 = (n/I_n)\exp\left[-E^2/2p(k_BT)^2\right] \quad . \qquad (11.22)$$

This distribution function was discovered through studies on electron transport in ionised gases and is often called the Druyvesteyn function [11.9]. On evaluating the conductivity we find that the conductivity at a field $\mathcal{E}$ is

$$\sigma(\mathcal{E}) = \sigma(0)\mathcal{E}^{-\frac{1}{2}} \quad . \qquad (11.23)$$

The conductivity thus decreases as the square root of the electric field. This result being available from the study of gases was used in the beginning to understand the hot-electron results when the first experiment of Ryder and Shockley was reported. However, as is evident from the discussion of earlier chapters, electrons are scattered in semiconductors through a number of scattering mechanisms. The above result based on acoustic phonon scattering alone does not therefore explain the conductivity characteristics of semiconductors even qualitatively. The differential equation method, however, is difficult to apply when we include other important scattering mechanisms. Some analyses have been reported by including nonpolar optic and intervalley phonon scattering. For these scattering mechanisms a relaxation time may be defined and the right-hand side of (11.16) may be written in terms of the differential coefficients if we assume that $E \gg \hbar\omega_0$. The resultung equation and the solutions are [Ref.11.37,p. 186]

$$\frac{d}{dE}\left\{E\left[(2n_0+1)(\hbar\omega_0/2)(df_0/dE)+f_0\right]\right\} = -p_1 k_B T \frac{d}{dE}(E df_0/dE) \quad , \qquad (11.24)$$

$$p_1 = 2e^2\mathscr{E}^2(3m^*)^{-1}2^{\frac{1}{2}}\pi\rho\hbar^2(k_BT)^{-\frac{1}{2}}(D_0^2 m^{*3/2})^{-1}\tau_{m0} \quad ;$$

$$\tau_{m0}^{-1} = 2^{\frac{1}{2}}m^{*3/2}(k_BT)^{\frac{1}{2}}(\pi\rho\hbar^3)^{-1}D_0^2(2n_0+1)(2\omega_0)^{-1} \quad ;$$

$$f_0 = (n/I_n)\exp\left[-E/(\gamma_2 k_B T)\right] \quad ,$$

$$\gamma_2 = p_1 + (2n_0+1)\hbar\omega_0/(2k_BT) \quad . \qquad (11.25)$$

An important result from the analysis is that

$$\sigma(\mathscr{E}) = \sigma(0)\mathscr{E}^{-1} \quad . \qquad (11.26)$$

In other words, the current density saturates at high fields when electrons are scattered predominantly by nonpolar optic phonons. This result should apply also when intervalley scattering predominates as its nature is identical to that of nonpolar optic phonon scattering.

An important characteristic of hot-electron conduction in elemental semiconductors, namely, velocity saturation, is thus explained from the above analysis. In fact for more than twenty years after the first hot-electron experiment, an attempt was made to explain the available experimental results on various transport coefficients for elemental semiconductors by combining the effects of acoustic phonons and nonpolar optic phonons for germanium, and those of the acoustic and intervalley phonons for silicon. The expression for the distribution function for combined scattering is [Ref.11.37,p. 190]

$$f_0 = (n/I_n)\left[\alpha_1 E/(k_BT)+\gamma_2\right]^{(\gamma_2-1)/\alpha_1}\exp(-E/k_BT) \quad , \qquad (11.27)$$

$$\alpha_1 = 4k_BTE_1^2 m^*/(D_0^2\hbar^2) \quad , \quad \gamma_2 = p_2 + (2n_0+1)\hbar\omega_0/(2k_BT) \quad ,$$

$$p_2 = p_1\tau_{m1}/\tau_{m0} \quad , \quad \tau_{m1}^{-1} = \tau_{m0}^{-1}\left[1+2E_1^2 k_BT\omega_0(\hbar s^2 D_0^2)^{-1}(2n_0+1)^{-1}\right] \quad .$$

It should be accepted that the success of this analytic theory was remarkable in that it explained fairly well the experimental results on conductivity anisotropy, the Sasaki effect, galvanomagnetic coefficients, and AC conductivity of both germanium and silicon [11.38-41].

The method was also applied to explain the results of compound semiconductors, notably of gallium arsenide. In this analysis the same assumption, $E \gg \hbar\omega_\ell$, is made, so that a relaxation time may be defined and the collision term may be reduced to differential form. We quote below the expression derived for polar optic phonon scattering to illustrate the nature of the equation:

$$\left[x^2/\gamma+(n_\ell+1/2)\log 4x\right]f' + (\log 4x)f = 0 \quad , \tag{11.28}$$

where $x = E/\hbar\omega_\ell$, $\gamma^{-1} = (4/3)(2n_\ell+1)^{-1}(\mathscr{E}/E_0)^2$, $f' = df/dx$, E_0 is given by (10.15).

An analytic expression for the distribution function cannot be obtained for all energies and numerical methods are required for the solution of the equation. Such solutions were obtained in the case of gallium arsenide [11.42]. Two equations were obtained involving the distribution functions for electrons in Γ and in L valleys. These equations were solved numerically and conductivity characteristics were evaluated which showed NDC. However, experimental results could not be fitted with these analytic results. It appears that, although the differential equation method allows us to understand the characteristics qualitatively, the required approximations introduce large errors. The method has not, therefore, been much used for the analysis of hot-electron characteristics of compound semiconductors.

11.4.2 Maxwellian Distribution Function Method

The difficulty of reducing the collision term corresponding to the symmetric part to a differential equation for inelastic scattering may be avoided if it is assumed that electron concentration is high enough [11.43] to dominate the energy relaxation of the hot electrons. The effect of electron-electron scattering is proportional to the electron concentration whereas the effect of lattice scattering is proportional to the phonon concentration.

We may therefore expect that for concentrations above a certain critical value, electrons will exchange energy at a faster rate through electron-electron collisions than they lose through electron-phonon collisions. When this happens, electrons share energy among themselves to establish an equilibrium distribution, which would be Maxwellian for nondegenerate materials. In the Maxwellian distribution method of hot-electron calculations, it is assumed that the symmetric part of the distribution function is Maxwellian, it being tacitly assumed that the electron concentration is high enough to ensure validity of this assumption. The distribution function is written as

$$f_0 = (n/I_n)\exp(-E/k_BT_e) \quad , \tag{11.29}$$

where I_n is the normalisation integral

$$I_n = (4\pi^3)^{-1} \int \exp(-E/k_BT_e)d\underline{k} \quad , \tag{11.30}$$

and n is the electron concentration. Although, the form of f_0 is assumed, the temperature parameter T_e is required to be treated as unknown, and it should be chosen so as to ensure that for its steady-state value, the energy gained by the electrons from the electric field should be balanced by that lost to the lattice. The value of T_e may be obtained by solving the energy balance equation which is obtained by multiplying the two sides of (11.16) by E and integrating over $\underline{k}$ space. It reads

$$\begin{aligned}&(4\pi^3)^{-1} \int E(e\mathscr{E}/\hbar)(1/3)(2f_1/k+\partial_k f_1)d\underline{k} \\ &= (1/2)(4\pi^3)^{-1} \int E\left[\int_{-1}^{+1} (\partial f_0/\partial t)_c d(\cos\theta)\right]d\underline{k} \quad .\end{aligned} \tag{11.31}$$

The left-hand side may be shown to reduce to

$$-e\,\mathscr{E}(4\pi^3)^{-1} \int (1/\hbar)(\partial E/\partial k_x)f_1 \cos\theta\, d\underline{k} = -n|e|\mu_d\mathscr{E}^2 \quad , \tag{11.32}$$

since

$$\mu_d = e|e|^{-1}(n\mathscr{E})^{-1}(4\pi^3)^{-1} \int (1/\hbar)(\partial E/\partial k_x)f_1 \cos\theta\, d\underline{k} \quad . \tag{11.33}$$

We thus get

$$\mu_d\mathscr{E}^2 = -(n|e|)^{-1}(8\pi^3)^{-1}\int E\left[\int_{-1}^{+1} (\partial f_0/\partial t)_c d(\cos\theta)\right]d\underline{k} \quad . \tag{11.34}$$

Now, μ_d may be obtained by solving (11.18), since the form of f_0 has been assumed. The expression for μ_d for collisions admitting a relaxation time is

$$\mu_d = -(|e|/n4\pi^3) \int (1/\hbar^2)\tau(E)(\partial E/\partial k_x)(\partial f_0/\partial k)\cos\theta\, d\underline{k} \quad , \tag{11.35}$$

which, when evaluated using the expression for $\tau(E)$, gives μ_d as a function of T_e. The integral on the right-hand side of (11.34), when worked out, gives a function $F(T_e)$ of T_e. Equation (11.34) thus reduces to

$$|e|\mu_d(T_e)\mathscr{E}^2 = F(T_e) \quad , \tag{11.36}$$

and it may be solved to obtain first T_e. The mobility $\mu_d(T_e)$ corresponding to the field $\mathscr{E}$ may also be evaluated from (11.35) using this values of T_e. The method has been used extensively to analyse hot-electron transport results for elemental semiconductors. The analytical results were also in very good agreement with experiments. The method has not, however, been used much for compound semiconductors. One apparent reason is that for compound semiconductors, polar optic phonon scattering being the most important component of scattering, analytic expressions for $\mu_d(T_e)$ cannot be obtained. This difficulty may now be removed. We may obtain accurate values of $\mu_d(T_e)$ for assumed values of T_e by solving (11.17) for polar optic phonon scattering by applying the numerical methods discussed in Chap.7 for the evaluation of low-field mobility. Equation (11.36) may then be solved for T_e and corresponding value of $\mu_d(T_e)$ be determined. Such results should have the same kind of accuracy as obtained for elemental semiconductors by assuming a relaxation time. The method was recently applied by NAG and CHATTOPADHYAY [11.44] for the evaluation of warm-carrier mobility in InSb and the theoretical results were in better agreement with experiments. The calculations may also be extended easily to the hot-electron regime.

11.4.3 Displaced Maxwellian Distribution Function Method

The argument justifying the assumption of a Maxwellian distribution for a high electron concentration, when extended to the case of still higher electron concentrations, leads us to a distribution function, called the displaced Maxwellian function [11.45]. In many semiconductors, the energy relaxation process is slower than momentum relaxation. Hence, for some electron concentration, the energy relaxation may be dominantly determined by electron-electron scattering, but the momentum relaxation may be determined by lattice scattering. However, even in these materials when the electron concentration has a still higher value, the momentum relaxation may also be established through electron-electron collisions. In compound semiconductors for which polar optic phonon scattering is the most dominant scattering mechanism, energy relaxation and momentum relaxation times are of the same order. Hence when energy relaxation is established by electron-electron collisions, momentum relaxation may also take place through the same process. In such a situation, when electrons share the momentum between themselves, all the electrons will have a common momentum or a common drift velocity in the direction of the electric force. The distribution function will therefore be Maxwellian, but displaced in the velocity space by the common velocity, and hence the name displaced Maxwellian. The form of this function for parabolic bands is

$$f(\underline{k}) = (n/I_n)\exp\left[-\hbar^2|\underline{k}-\underline{k}_0|^2/(2m^* k_B T_e)\right] \quad , \tag{11.37}$$

where $\hbar\underline{k}_0/m^*$ is the common drift velocity, and I_n is the normalisation integral, n being the carrier concentration. There has been some argument about the form of the function for nonparabolic bands. The form given by HAMMAR and WEISGLASS [11.46] appears most convenient:

$$f(\underline{k}) = (n/I_n)\exp\left[-(E-\hbar\underline{v}_d\cdot\underline{k})/(k_B T_e)\right] \quad . \tag{11.38}$$

In this form the distribution function is Maxwellian in a frame moving with velocity $\underline{v}_d$ relative to the lattice. Although the form of the distribution function may be chosen from physical arguments given above, T_e and $\underline{v}_d$ are unknown. They have to be so chosen that the momentum and energy gained by the electrons from the field should be balanced by those lost to the lattice, as is required for the establishment of a steady-state condition. To ensure the balance, $\underline{v}_d$ and T_e are determined from the following equations:

$$\int E(e\mathscr{E}/\hbar)(\partial f/\partial k_x)d\underline{k} = \int E(\partial f/\partial t)_c d\underline{k} \quad ,$$
$$\int \hbar\underline{k}(e\mathscr{E}/\hbar)(\partial f/\partial k_x)d\underline{k} = \int \hbar\underline{k}(\partial f/\partial t)_c d\underline{k} \quad . \tag{11.39}$$

The equations are easily reduced to

$$e\underline{v}_d\cdot\underline{\mathscr{E}} = -(1/4\pi^3)\int E(\partial f/\partial t)_c d\underline{k} \quad ,$$
$$e\underline{\mathscr{E}} = \underline{\mathscr{E}}(1/4\pi^3)\int \hbar\underline{k}(\partial f/\partial t)_c d\underline{k} \quad . \tag{11.40}$$

The right-hand sides of the equations may be reduced analytically since the form of f is assumed. The integrals have been evaluated for nonparabolic bands taking into account the polar optic, deformation potential acoustic, and ionised impurity scattering. The expressions are fairly complex and the derivation involves only tedious algebra. We refrain from quoting these expressions and the interested reader may obtain them from [11.47]. The result of the integrations are two functions $F_1(T_e, \underline{v}_d)$ and $F_2(T_e, \underline{v}_d)$, which often involve integrals of modified Bessel functions. We quote below the forms of F_1 and F_2 for polar optic phonon scattering only to give an idea of the nature of functions involved. We get for a parabolic band and polar optic phonon scattering [Ref.11.37,p. 196]

$$
\begin{aligned}
F_1(T_e, \underline{v}_d) &= (2k_B\theta_0/m^*\pi)^{\frac{1}{2}} eE_0 n_\ell \left[\exp(\theta_0/T_L - \theta_0/T_e) - 1\right](\theta_0/T_e)^{\frac{1}{2}} \\
&\quad \times \exp(\theta_0/2T_e) K_0(\theta_0/2T_e) \quad , \\
F_2(T_e, \underline{v}_d) &= (\sqrt{2}/3)(m^*\pi k_B\theta_0)^{-\frac{1}{2}} eE_0 n_\ell \hbar k_0 (\theta_0/T_e)^{3/2} \exp(\theta_0/2T_e) \\
&\quad \times \left\{\left[\exp(\theta_0/T_L - \theta_0/T_e) + 1\right] K_1(\theta_0/2T_e)\right. \\
&\quad \left. + \left[\exp(\theta_0/T_L - \theta_0/T_e) - 1\right] K_0(\theta_0/2T_e)\right\} \quad ,
\end{aligned}
\qquad (11.41)
$$

where K_0, K_1 are modified Bessel functions of the second kind (see Chap.9). θ_0, T_L, T_e are respectively the optic phonon, lattice and electron temperature and E_0 is the optic phonon field defined by (10.15). Values of $\underline{v}_d$ for a given value of $\underline{\mathcal{E}}$ may be obtained from the above two equations using tabulated values of K_1 and K_2.

It should be evident that the method is analytically complete and easy to apply. The method was developed in this form by FRÖHLICH [11.48] for the study of dielectric breakdown. It was studied later in detail by STRATTON [11.49]. Because of the simplicity (although it involves tedious mathematical steps), the method was very popular in the early studies of hot-electron transport in compound semiconductors. Several workers used the method to calculate the hot-electron-conductivity characteristics of InSb [11.50]. The possibility of producing NDC through the increase of electron mass with energy in nonparabolic bands was studied by the method [11.46,51]. The first calculations of the hot-electron characteristics of gallium arsenide was also carried out by using this method [11.52]. A large amount of detailed calculations using the method are also reported in the literature for the hot-electron characteristics of many other compound semiconductors. We show in Fig.11.14 a few calculated characteristics. However, the method has also been strongly criticised in the literature. It is stated that the displaced Maxwellian distribution function cannot be produced in an experimental sample. For the large electron concentration that is required for the validity of the function, impurity scattering would be as important as the electron-electron scattering since large electron concentration requires an equally large

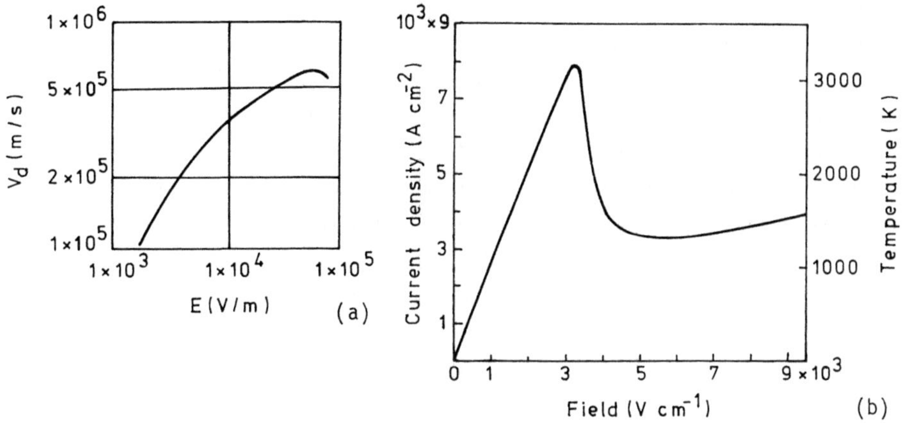

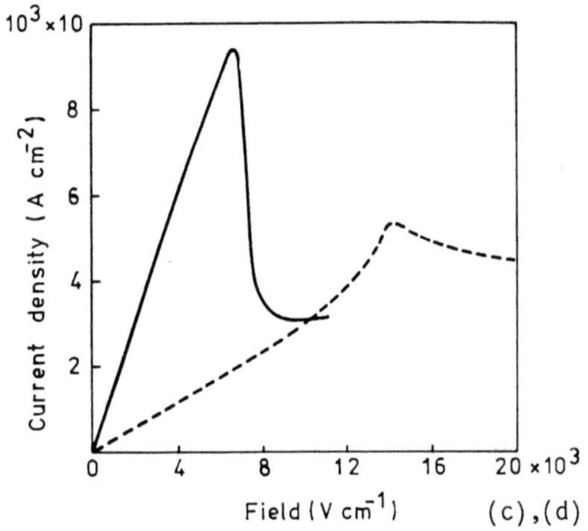

Fig. 11.14. Hot-electron conduction characteristic calculated by the displaced Maxwellian method. (a) Indium antimonide [11.46]; (b) Gallium arsenide; (c) Indium phosphide (solid line); (d) Cadmium telluride (dotted line) [11.52]

impurity concentration. Experimental evidence for the distribution function also shows that it is far from the displaced Maxwellian [11.27a,53]. The success of the method appears to be mostly because it is a kind of variational method.

11.5 Numerical Methods

The analytical methods reviewed in the last section played a major role in the understanding of the hot-electron transport phenomena. Most of the important features of the hot-electron conductivity characteristics could be explained by applying these methods. In many cases, analytically calculated values were also found to agree very well with experiments. However, it should be evident from the preceding discussion that the analytic methods are based on gross assumptions. The distribution function has been determined experimentally and is found to be different from the Maxwellian function. The agreement between theory and experiment is also obtained in most cases by adjusting the values of coupling constants, which cannot be determined by independent experiments. Attempts have therefore been made to evaluate the hot-electron characteristics without recourse to any assumption. As a result of these attempts, two numerical methods have been developed, which give accurately the distrbution function as well as the conductivity. We discuss these methods in this section.

11.5.1 Iteration Method

Let us consider the time-dependent Boltzmann equation

$$\partial_t f + (e\mathscr{E}/\hbar)(\partial f/\partial k_x) = (\partial f/\partial t)_c \quad , \tag{11.42}$$

for an electric field applied in the x direction. Using the variables,

$$\underline{k}_1 = \underline{k} - \hat{\underline{x}}(e\mathscr{E}/\hbar)t \quad , \quad t_1 = t \quad , \tag{11.43}$$

we may reduce the equation to

$$\partial_t f(\underline{k}_1+\hat{\underline{x}}e\,\mathscr{E}t_1/\hbar,t_1) = \left[\partial_t f(\underline{k}_1+\hat{\underline{x}}e\,\mathscr{E}t_1/\hbar,t_1)\right]_c \quad . \tag{11.44}$$

We recall that for nondegenerate semiconductors (assuming $V_c = 1$)

$$(\partial f/\partial t)_c = (8\pi^3)^{-1}\int_{\underline{k}'} [S(\underline{k}',\underline{k})f(\underline{k}')-S(\underline{k},\underline{k}')f(\underline{k})]d\underline{k}' \quad . \tag{11.45}$$

Defining

$$\lambda(\underline{k}) = (8\pi^3)^{-1}\int_{\underline{k}'} S(\underline{k},\underline{k}')d\underline{k}' \quad , \tag{11.46}$$

$$g(\underline{k},t) = (8\pi^3)^{-1}\int_{\underline{k}'} S(\underline{k}',\underline{k})f(\underline{k}',t)d\underline{k}' \quad , \tag{11.47}$$

we may write (11.44) as

$$\partial f(\underline{k}_1+\hat{\underline{x}}e\,\mathscr{E}t_1/\hbar,t_1)/\partial t_1 + \lambda(\underline{k}_1+\hat{\underline{x}}e\,\mathscr{E}t_1/\hbar)f(\underline{k}_1+\hat{\underline{x}}e\,\mathscr{E}t_1/\hbar,t_1)$$
$$= g(\underline{k}_1+\hat{\underline{x}}e\,\mathscr{E}t_1/\hbar,t_1) \quad . \tag{11.48}$$

Let us substitute for f

$$f(\underline{k}_1+\hat{\underline{x}}e\,\mathcal{E}t_1/\hbar,t_1) = \phi(\underline{k}_1+\hat{\underline{x}}e\,\mathcal{E}t_1/\hbar)\psi(\underline{k}_1+\hat{\underline{x}}e\,\mathcal{E}t_1/\hbar) \quad , \tag{11.49}$$

where ϕ is chosen such that

$$\frac{\partial}{\partial t_1}\phi(\underline{k}_1+\hat{\underline{x}}e\,\mathcal{E}t_1/\hbar) + \lambda(\underline{k}_1+\hat{\underline{x}}e\,\mathcal{E}t_1/\hbar)\phi(\underline{k}_1+\hat{\underline{x}}e\,\mathcal{E}t_1/\hbar) = 0 \quad . \tag{11.50}$$

The solution for $\phi(\underline{k}_1+\hat{\underline{x}}e\,\mathcal{E}t_1/\hbar)$ is

$$\phi(\underline{k}_1+\hat{\underline{x}}e\,\mathcal{E}t_1/\hbar) = C\exp\left(-\int_0^{t_1}\lambda(\underline{k}_1+\hat{\underline{x}}e\,\mathcal{E}t_1'/\hbar)dt_1'\right) \quad , \tag{11.51}$$

where C is the value of ϕ at $t_1=0$ and it may depend on $\underline{k}_1$. Substituting (11.49) in (11.48)

$$\frac{\partial}{\partial t_1}\psi(\underline{k}_1+\hat{\underline{x}}e\,\mathcal{E}t_1/\hbar,t_1) = \left[\phi(\underline{k}_1+\hat{\underline{x}}e\,\mathcal{E}t_1/\hbar)\right]^{-1}g(\underline{k}_1+\hat{\underline{x}}e\,\mathcal{E}t_1/\hbar) \quad . \tag{11.52}$$

Integrating

$$\psi(\underline{k}_1+\hat{\underline{x}}e\,\mathcal{E}t_1/\hbar,t_1) = \psi(\underline{k}_1,0) + \int_0^{t_1}(1/C)\exp\left(-\int_{t_1'}^{t_1}\lambda(\underline{k}_1+\hat{\underline{x}}e\,\mathcal{E}t_1''/\hbar)dt_1''\right) \times g(\underline{k}_1+\hat{\underline{x}}e\,\mathcal{E}t_1'/\hbar)dt_1' \quad . \tag{11.53}$$

The distribution function $f(\underline{k}_1,t_1)$ is thus given by

$$f(\underline{k}_1+\hat{\underline{x}}e\,\mathcal{E}t_1/\hbar,t_1) = \psi(\underline{k}_1,0)C\exp\left(-\int_0^{t_1}\lambda(\underline{k}_1+\hat{\underline{x}}e\,\mathcal{E}t_1'/\hbar)dt_1'\right) + \int_0^{t_1}\exp\left(-\int_{t_1'}^{t_1}\lambda(\underline{k}_1+\hat{\underline{x}}e\,\mathcal{E}t_1''/\hbar)dt_1''\right)g(\underline{k}_1+\hat{\underline{x}}e\,\mathcal{E}t_1'/\hbar)dt_1' \quad . \tag{11.54}$$

Reverting to the variables k and t by putting

$$\underline{k} = \underline{k}_1 + \hat{\underline{x}}e\,\mathcal{E}t/\hbar \quad , \quad t = t_1 \quad , \quad t - t_1' = t' \quad , \quad t - t_1'' = t'' \quad ,$$

we get

$$f(\underline{k},t) = f(\underline{k}-\hat{\underline{x}}e\,\mathcal{E}t/\hbar,0)\exp\left[-\int_0^t\lambda(\underline{k}-\hat{\underline{x}}e\,\mathcal{E}t'/\hbar)dt'\right] + \int_0^t\exp\left[-\int_0^{t'}\lambda(\underline{k}-\hat{\underline{x}}e\,\mathcal{E}t''/\hbar)dt''\right]g(\underline{k}-\hat{\underline{x}}e\,\mathcal{E}t'/\hbar)dt' \quad . \tag{11.55}$$

For $t\to\infty$,

$$f(\underline{k},\infty) = \int_0^\infty \exp\left[-\int_0^{t'} \lambda(\underline{k}-\hat{\underline{x}}e\mathscr{E}t''/\hbar)dt''\right]g(\underline{k}-\hat{\underline{x}}e\mathscr{E}t'/\hbar)dt' \quad . \tag{11.56}$$

The distribution function $f(\underline{k})$ may thus be obtained by solving the above equation.

The equation may be given a simple physical interpretation. The second term in the integral on the right-hand side, i.e. $g(\underline{k}-\hat{\underline{x}}e\mathscr{E}t/\hbar)$, represents the electrons which were scattered to the $\underline{k}'$ state at an earlier time t and the first term represents the probability of the electron not suffering a collision within the time t. Hence, the product of the two terms gives the number of electrons which were in the $\underline{k}' = (\underline{k}-\hat{\underline{x}}e\mathscr{E}t/\hbar)$ state at an earlier time t and travelled to the $\underline{k}$ state without being scattered. The number of electrons in the $\underline{k}$ state is evidently given by the sum of these electrons for t extending to infinity as is written in the equation.

We note that when $f(\underline{k})$ is determined by solving (11.56), the conductivity may be obtained by putting this $f(\underline{k})$ in (8.4).

The equation to be solved is, however, an integral equation as the function $g(\underline{k})$ involves the unknown distribution function $f(\underline{k})$. It is a Fredholm equation [11.54] of second kind with the kernel

$$\int_0^\infty S(\underline{k}',\underline{k}-\hat{\underline{x}}e\mathscr{E}t/\hbar)\exp\left[-\int_0^t\lambda(\underline{k}-\hat{\underline{x}}e\mathscr{E}t'/\hbar)dt'\right]dt \quad . \tag{11.57}$$

It may be solved by Kellog's iteration method [11.55] in which one starts with a known function for $f(\underline{k})$ and proceeds with the iteration using the following scheme:

$$f^n(\underline{k}) = \int_0^\infty \exp\left[-\int_0^{t'} \lambda(\underline{k}-\hat{\underline{x}}e\mathscr{E}t''/\hbar)dt''\right]\left[\int_{\underline{k}'} S(\underline{k}',\underline{k}-\hat{\underline{x}}e\mathscr{E}t'/\hbar)f^{n-1}(\underline{k}')dk'\right]dt' \quad . \tag{11.58}$$

The above method for obtaining $f(\underline{k})$ was first suggested and applied to hot electron problems by BUDD [11.56] as a generalisation of Chamber's method and was called the path variable method. A direct application of the method to compound semiconductors, however, is rather difficult as the integral in the exponential term assumes a complex form when energy band nonparabolicity and inelastic scattering processes are to be taken into account. A simple way of getting around this difficulty was suggested and implemented by REES [11.57]. In the modified method, a fictitious scattering, called self-scattering, is added to the scattering processes so that the scattering probability is

$$S_M(\underline{k},\underline{k}') = S(\underline{k},\underline{k}') + W(\underline{k})\delta(\underline{k}-\underline{k}') \quad . \tag{11.59}$$

The fictitious scattering, being defined with the Dirac $\delta(\underline{k}-\underline{k}')$ function, does not cause a change in $\underline{k}$ and, in effect, means that when this scattering occurs there is no change in electron wave vector. It may be looked upon as a mathematical artifice introduced for the simplification of computations. The Boltzmann equation with the inclusion of the additional scattering term reads

$$\begin{aligned} &\partial_t f + \hat{\underline{x}}(e\mathscr{E}/\hbar)\cdot\underline{\nabla}_{\underline{k}} f + (8\pi^3)^{-1}\int_{\underline{k}'} S_M(\underline{k},\underline{k}')f(\underline{k})d\underline{k}' \\ &= (8\pi^3)^{-1}\left[\int_{\underline{k}'} S(\underline{k}',\underline{k})f(\underline{k}')d\underline{k}' + \int_{\underline{k}'} W(\underline{k})\delta(\underline{k}-\underline{k}')f(\underline{k})d\underline{k}'\right] \quad . \end{aligned} \tag{11.60}$$

Proceeding as before, the above equation may be transformed to

$$f(\underline{k}) = \int_0^\infty \exp\left\{-\int_0^{t'}\left[\int_{\underline{k}'} S_M(\underline{k}-\hat{\underline{x}}t''e\mathscr{E}/\hbar,\underline{k}')d\underline{k}'\right]dt''\right\}g'(\underline{k}-\hat{\underline{x}}e\mathscr{E}t'/\hbar)dt' \quad , \tag{11.61}$$

where now

$$g'(\underline{k}) = (8\pi^3)^{-1}\int_{\underline{k}'} S(\underline{k}',\underline{k})f(\underline{k}')d\underline{k}' + (8\pi^3)^{-1}\int_{\underline{k}'} W(\underline{k})\delta(\underline{k}-\underline{k}')f(\underline{k}')d\underline{k}' \quad . \tag{11.62}$$

It should be noted that $W(\underline{k})$ is an arbitrary function and it may be chosen such that $(8\pi^3)^{-1}\int S_M(\underline{k},\underline{k}')d\underline{k}'$ is independent of $\underline{k}$ and $\underline{k}'$. Writing this as a constant, Γ, we get

$$\begin{aligned} f(\underline{k}) &= \int_0^\infty [\exp(-\Gamma t')]g'(\underline{k}-\hat{\underline{x}}e\mathscr{E}t'/\hbar)dt' \quad , \\ g'(\underline{k}-\hat{\underline{x}}e\mathscr{E}t/\hbar) &= (8\pi^3)^{-1}\int_{\underline{k}'}\left[S(\underline{k}',\underline{k}-\hat{\underline{x}}e\mathscr{E}t/\hbar)+[\Gamma-\lambda(\underline{k}')]\delta(\underline{k}'-\underline{k}+\hat{\underline{x}}e\mathscr{E}t/\hbar)\right]f(\underline{k}')d\underline{k}', \end{aligned} \tag{11.63}$$

and

$$(8\pi^3)^{-1}W(\underline{k}-\hat{\underline{x}}e\mathscr{E}t/\hbar) = \Gamma - \lambda(\underline{k}-\hat{\underline{x}}e\mathscr{E}t/\hbar) \quad . \tag{11.64}$$

The introduction of the self-scattering term thus simplifies the integrations such that it becomes computationally simpler to carry out the integrations even when all the complexities of band structure and scattering mechanisms are included.

The computation is started with a chosen function for $f(\underline{k})$ and its values are evaluated at a chosen number of grid points in the $\underline{k}$ space. Using these values, the function $g(\underline{k})$ is first evaluated for the same values of $\underline{k}$. We recall that the scattering probability $S(\underline{k}',\underline{k})$ is, in general, of the form

$$S(\underline{k}',\underline{k}) = CA(|\underline{k}-\underline{k}'|)\delta(E_{\underline{k}'}-E_{\underline{k}}\pm\Delta E)G(\underline{k},\underline{k}') \quad , \tag{11.65}$$

where C is a coupling constant determined by the physical parameters of the sample, $A(|\underline{k}-\underline{k}'|)$ is a function of the magnitude of the difference wave vector $\underline{k}-\underline{k}'$, $G(\underline{k},\underline{k}')$ is the overlap integral factor, ΔE is the energy lost or gained in the particular kind of scattering. Let $\underline{k}$ and $\underline{k}'$ be written in polar coordinates with the direction of the field chosen as the polar axis. Then,

$$\begin{aligned} g(k,\theta,\phi) &= (8\pi^3)^{-1}\int S(k',\theta',\phi';k,\theta,\phi)f(k',\theta',\phi')d\underline{k}' \\ &= (8\pi^3)^{-1}\int CA\left(\left\{k^2+k'^2-2kk'[\cos\theta\ \cos\theta'+\sin\theta\ \sin\theta'\ \cos(\phi-\phi')]\right\}^{\frac{1}{2}}\right) \\ &\quad \times G(\underline{k},\underline{k}')f(k',\theta',\phi')\delta(E_{\underline{k}}-E_{\underline{k}'}\pm\Delta E)d\underline{k}' \\ &= (8\pi^3)^{-1}\int_{\theta',\phi'} CA\left((2m^*/\hbar^2)^{\frac{1}{2}}\left\{\gamma(E_{\underline{k}})+\gamma(E_{\underline{k}}\pm\Delta E)-2\sqrt{\gamma(E_{\underline{k}})}\sqrt{\gamma(E_{\underline{k}}\pm\Delta E)}\right.\right. \\ &\quad \left.\left.\times[\cos\theta\ \cos\theta'+\sin\theta\ \sin\theta'\ \cos(\phi-\phi')]\right\}^{\frac{1}{2}}\right) \\ &\quad \times G(\gamma(E_{\underline{k}}),\gamma(E_{\underline{k}}\pm\Delta E),\theta,\theta',\phi,\phi')f((E_{\underline{k}}\pm\Delta E),\theta',\phi') \\ &\quad \times(1/2)(2m^*/\hbar^2)^{3/2}\left[\gamma(E_{\underline{k}}\pm\Delta E)\right]^{\frac{1}{2}}\gamma'(E_{\underline{k}}\pm\Delta E)\sin\theta'd\theta'd\phi' \quad . \end{aligned} \tag{11.66}$$

The double integration over θ' and ϕ' is to be worked out numerically. It is usual to choose for this purpose the Gaussian quadrature method which has been explained earlier in Sect.8.1.3. We note that for materials with isotropic effective mass, the problem has ϕ symmetry and we may, therefore, choose ϕ equal to zero and evaluate the function $g(k,\theta,\phi)$ only for chosen points in the k-θ plane.

Once $g(\underline{k})$ is known we may evaluate new iterated values of $f(\underline{k})$ using (11.64) with the values of $g(\underline{k})$ and $[\Gamma-S(\underline{k})]f^0(\underline{k})$. Note that expressions for $S(\underline{k})$ have been given in Sect.6.6.

We give below the numerical formulae to be used in computation for carrying out the above integration:

$$\begin{aligned} f^n(k_i,\theta_j) &= \sum_{\ell} B_\ell \exp(-\Gamma t_\ell)\left[g(k_i\ \cos\theta_j-t_\ell e\mathcal{E}/\hbar,k_i\ \sin\theta_j)\right] + \left[\Gamma\right. \\ &\quad \left.-S\left(\left[(k_i\ \cos\theta_j-et_\ell\mathcal{E}/\hbar)^2+(k_i\ \sin\theta_j)^2\right]^{\frac{1}{2}}\right)\right]f^{n-1}(k_i,\theta_j) \quad , \end{aligned} \tag{11.67}$$

$$g(k_i \cos\theta_j - t_\ell e\mathcal{E}/\hbar, k_i \sin\theta_j) = \sum_s \Big\{ \sum_m B_m \sum_n B_n C_s A_s \Big((2m^*/\hbar^2)^{\frac{1}{2}} \Big[\gamma(E_{k_i}) + \gamma(E_{k_i} \pm \Delta E_s) - 2\sqrt{\gamma(E_{k_i})\gamma(E_{k_i} \pm \Delta E_s)} (\cos\theta_j \cos\theta_n + \sin\theta_j \sin\theta_n \cos\phi_m) \Big]^{\frac{1}{2}} \Big) \times G\Big(\gamma(E_{k_i} \pm \Delta E_s), \gamma(E_{k_i}), \theta_i, \theta_n, 0, \phi_m\Big)(1/2)(2m^*/\hbar^2)^{3/2} \times \Big[\gamma(E_{k_i} \pm \Delta E_s)\Big]^{\frac{1}{2}} \gamma'(E_{k_i} \pm \Delta E_s) \Big\} f^{n-1}(E_{k_i} \pm \Delta E_s, \theta_n, \phi_m) \quad (11.68)$$

where B_ℓ are the Gaussian quadrature multipliers for the chosen values of t, t_ℓ; the integral for t from 0 to ∞ is normalised to unity. The largest value of t, t_∞, is chosen so that $\exp(-\Gamma t_\infty)$ is less than the desired fractional accuracy. B_m, B_n are the Gaussian quadrature multipliers for the θ' and ϕ' integrations. Note that θ' extends from 0 to π and ϕ' from 0 to 2π. The Gaussian quadrature points for θ' are distributed for $\cos\theta$ between -1 and +1. The multiplying constants are normalised accordingly. The summation over s is for the various scattering mechanisms.

The scheme outlined above when used for the iteration is found to give convergent values within a few iterations. Introduction of the self-scattering term simplifies the computations but increases the required number of iterations. It may be seen that when Γ is made large the new iterated value of $g(\underline{k})$ is obtained by including only a few values of $g(\underline{k})$ and for the inclusion of all the $g(\underline{k})$ values corresponding to real scattering, the iteration has to be repeated a larger number of times. The introduction of self-scattering in effect makes available the distribution function at intermediate steps of development and may be used to study the time dependent behaviour. The value of Γ suggested by REES, as required for convergence, should be larger than $S(\underline{k})$. More detailed consideration indicates that Γ may be given lower value and convergence may be assured provided the kernel remains always positive [11.58].

The iterative method has been applied to obtain the hot-electron conductivity and distribution function for indium antimonide and gallium arsenide [11.59]. We present in Fig.11.15 the distribution obtained for indium antimonide in the directions of the field for different magnitudes of the field. It is seen from these plots that the distribution function $f(\underline{k})$ is quite different from the Maxwellian or the displaced Maxwellian function. In fact, it shows a sharp bend at the energy corresponding to the polar optic phonon energy. The number of electrons is much lowered below the Maxwellian value due to the emission of optic phonons, as may be expected physically.

The method, in principle, should yield any desired accuracy. But in practice, the accuracy is limited by the finite number of grid points that may

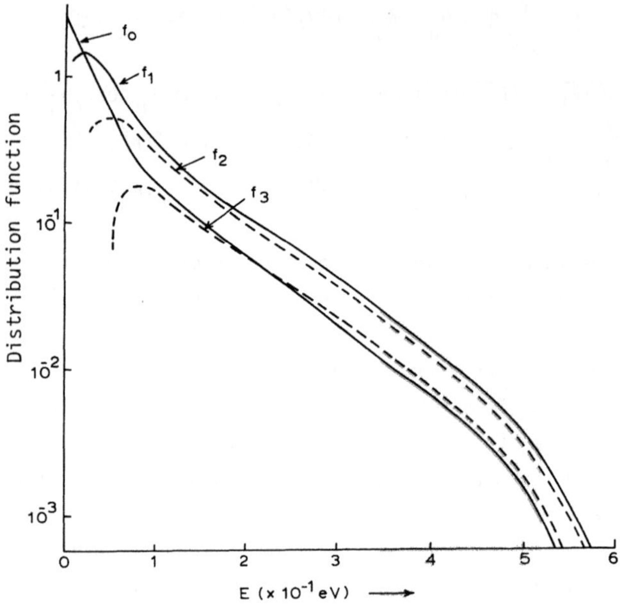

Fig. 11.15. Electron distribution under hot-electron conditions in indium antimonide [11.59b]. f_0, f_1, f_2, f_3 — Harmonic components of the distribution function

be chosen for actual computation. The computer memory capacity required for the execution of the method is also very large. The time required for carrying out the iterations is also very large when self-scattering is introduced and when $S(\underline{k})$ is strongly dependent function of $\underline{k}$. The method may in principle be used to degenerate materials and also to include the effects of electron-electron collisions. Time-dependent and space-dependent or magnetic transport coefficients may also be obtained by applying the method. We shall not discuss the details for such applications any further.

11.5.2 Monte Carlo Method

The second numerical method for the computation of hot-electron characteristics is the Monte Carlo method [11.60]. We discussed the principle of this method in Sect.7.6.4 and shall not repeat it here, but indicate the modification to be applied for high fields. The first step is the determination of the next collision time after a collision. The formula used for this purpose is

$$r_1 = \left\{\int_0^t \exp\left[-\int_0^{t'} S(\underline{k})dt''\right]dt'\right\}\left\{\int_0^\infty \exp\left[-\int_0^{t'} S(\underline{k})dt''\right]dt'\right\}^{-1} \quad , \tag{11.69}$$

r_1 being a random number uniformly distributed between 0 and 1.

For low fields, the gain in electron momentum from the applied field could be considered negligible in comparison to the electron wave vector $\underline{k}$ and the integration could be easily carried out to obtain the collision time t. At high fields, however, the gain in electron momentum $(e\underline{\mathcal{E}}/\hbar)t$ is comparable to $\underline{k}$ and the time has to be evaluated using the formula

$$r_1 = \left\{\int_0^t \exp\left[-\int_0^{t'} S(\underline{k}+\hat{\underline{x}}e\mathcal{E}t''/\hbar)dt''\right]dt'\right\} \times\left\{\int_0^\infty \exp\left[-\int_0^{t'} S(\underline{k}+\hat{\underline{x}}e\mathcal{E}t''/\hbar)dt''\right]dt'\right\}^{-1} \quad . \tag{11.70}$$

Depending on the formula for $S(\underline{k})$ the integration may be quite complex. This complexity is avoided by including a self-scattering term [11.61], as in the iterative method. The total scattering rate $S(\underline{k})$ is now defined so that

$$S_T(\underline{k}) = S(\underline{k}) + W(\underline{k}) \tag{11.71}$$

is a constant equal to Γ. The equation then giving the collision time is

$$r_1 = 1 - \exp(-\Gamma t) \quad . \tag{11.72}$$

It has been shown by REES and others that the steady-state values obtained by including the self-scattering term is indeed the values corresponding to the real scatterings. The inclusion of the term, in effect, results in computations in steps between two real scatterings. The computation time is thereby increased but the integrals are rendered manageable. The rest of the computations is as described earlier. Having determined the collision time, the kind of collision is determined by using a second random number r_2. Now, however, the self-scattering term being included, $\sum_{i=n} S_i(\underline{k})$ may be less than r_2. In such case the scattering is to be identified as a self-scattering. The electron wave vector does not change as a result of this scattering and the initial state of the electron after such a scattering is

$$\underline{k} = \underline{k}_0 + (e\underline{\mathcal{E}}/\hbar)t_c \quad , \tag{11.73}$$

where $\underline{k}_0$ was the initial state after the previous scattering and t_c is the present collision time. In the case, however, that

$$\sum_{n-1} S_i(\underline{k}) < r_2 < \sum S_i(\underline{k}) \tag{11.74}$$

the $(n-1)^{th}$ kind of real scattering occurs, and the state of the electron after collision should be determined by generating two more random numbers r_3 and r_4 as discussed.

In contrast to the low-field case, the drift velocity for high fields may be computed straightaway, as it is comparable to the random component of the velocity. It is obtained by using the formula

$$v_d = \hbar^{-1} \sum (E_f - E_i) / \sum (k_{f\mathscr{E}} - k_{i\mathscr{E}}) \quad , \tag{11.75}$$

where E_f and E_i denote respectively the energy after a collision and before the next collision. $k_{f\mathscr{E}}$ and $k_{i\mathscr{E}}$ are the corresponding components of the wave vector in the field direction. It should be recalled that such straight calculation was not possible at low fields as the unbalanced part of the random velocities could be comparable to the drift velocity. To avoid this difficulty low-field mobility is evaluated by Monte Carlo method indirectly through the diffusion constant. This is not, however, necessary at high fields.

In fact, the Monte Carlo method was first applied for the calculation of electron transport coefficients under hot-electron conditions. Particular attention was paid to the conductivity characteristics of gallium arsenide. The method was later used very successfully to calculate high-field conductivity of many other compound semiconductors [11.62]. Convergence is usually obtained within 15-150 thousand scattering events depending on the nature of collisions. Convergence is usually rapid for inelastic collisions but it is comparatively slow for elastic collisions, particularly for ionised impurity scattering.

We have presented in Fig.11.16 hot-electron conductivity characteristics obtained by the Monte Carlo method for gallium arsenide, indium antimonide, cadmium telluride, cadmium sulphide, and indium phosphide. Corresponding experimental characteristics are also presented for comparison for some of the materials. A computer program for Monte Carlo calculation of conductivity including deformation potential acoustic, polar optic, and ionised impurity scattering is presented in Appendix D.

The method has been very popular and it has been extensively used. It is very convenient as the program is rather simple and does not require much

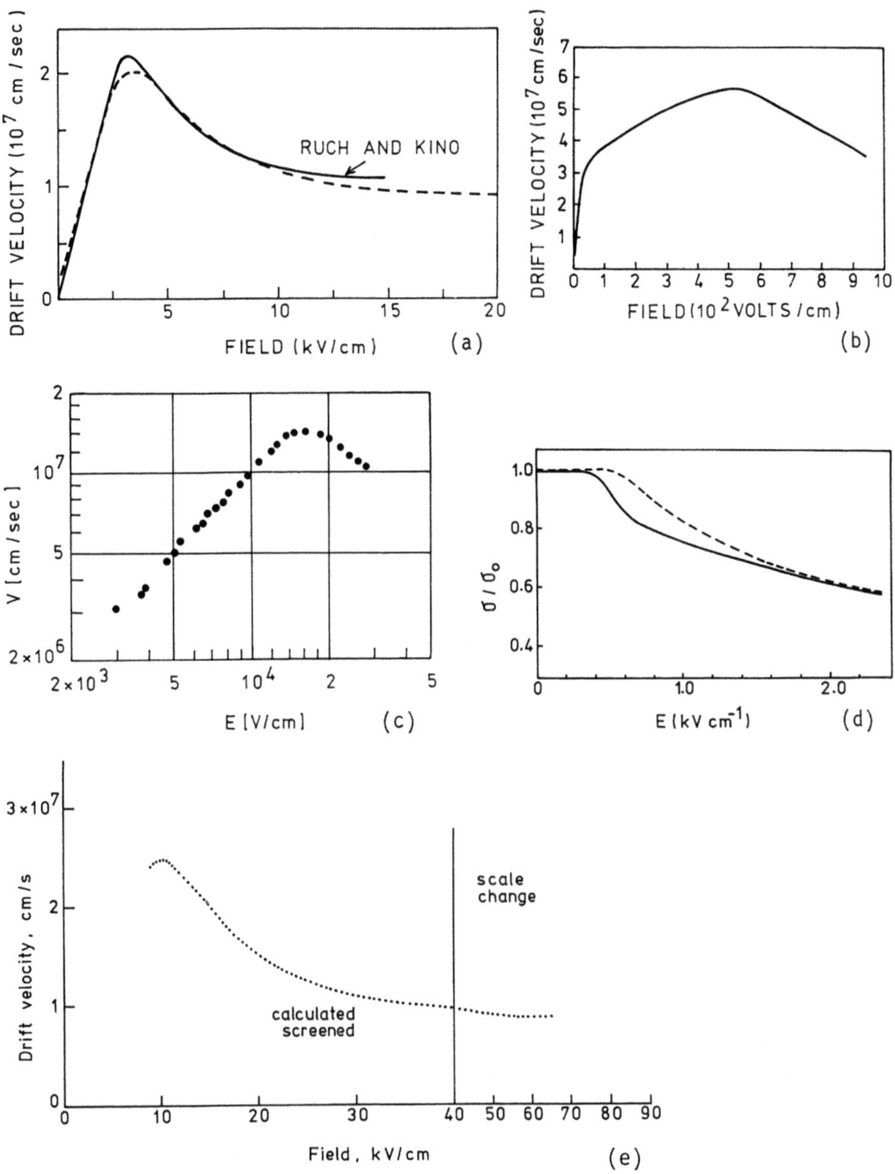

Fig. 11.16. Hot-electron conductivity characteristics calculated by the Monte Carlo method. (a) Gallium arsenide [11.60], solid line – experiment, dotted line – calculated; (b) Indium antimonide [11.62g]; (c) Cadmium telluride [11.89]; (d) Cadmium sulphide [11.90]; (e) Indium phosphide [11.87]

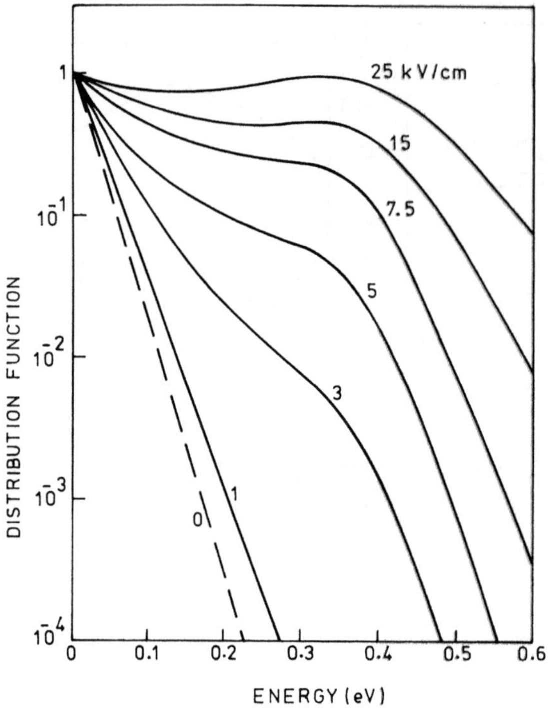

Fig. 11.17. Electron distribution under hot-electron conditions in gallium arsenide [11.60]

computer memory space. Variations of the method are also possible. For example, the scattering probabilities may be precalculated for some chosen energies and that required for a particular electron energy may be obtained by interpolation. The value of Γ may also be chosen to be different in different ranges of energy to reduce the number of self-scatterings and enhance the rate of convergence in a given computation time [11.63].

The distribution function for the electrons may also be obtained from the results of the computation. For this purpose, the time the electron spends in a particular cell of $\underline{k}$ space is logged and the value normalised by the total computation time gives the probability of the electron being in that cell, and hence the distribution function. Distribution functions obtained this way for gallium arsenide in the field direction are shown in Fig.11.17.

The two numerical methods discussed above yield the conductivity characteristics to any desired accuracy. It is evident from the preceding discussion that no simplifying assumptions, as required for the analytic methods, are made. If one is interested in accurately calculating the hot-electron characteristics, either of the two methods may be employed. The Monte Carlo method would be preferred if one is interested in calculating only the drift velocity as the program is simpler and the computation time may not be very large. On the other hand, if one is interested in determining the distribution function or in calculating the characteristics of a degenerate material including electron-electron scattering the iterative method may be preferred. Although methods have been suggested for adapting the Monte Carlo calculations to degenerate materials [11.64] or for including the effects of electron-electron scattering [11.65], the calculations including these effects are not convenient.

It may be pointed out before closing this discussion that the numerical methods appear to be the answer for the accurate calculation of hot-electron distribution function and the corresponding transport coefficients. However, it may not be necessary to apply these methods for all kinds of hot-electron studies. Particularly, when the physical parameters are not accurately known, application of these methods may be hardly justified in view of the expensive computations involved. It is now fairly well established that the numerical methods do not bring to light strikingly new features in the hot-electron characteristics which were not revealed by the analytic methods reviewed earlier. These numerical methods become prohibitively expensive when the interest lies in determining the qualitative nature of the variation of the characteristics with temperature [11.65] or impurity concentration [11.66]. The analytic methods may be preferred for the study of such variations unless a very high accuracy is desired and constants involved in the scattering mechanisms are well established.

There are, however, some transport problems where the characteristics obtained by analytic methods may become suspect in view of the gross approximations that may be required. One such problem is the NDC through electron transfer effect, which is the most important characteristic of hot-electron conduction in compound semiconductors. Although analytic solutions may be obtained which appear to be in close agreement with experiment, it is required to assume that the intravalley scatterings are more frequent than the equivalent or nonequivalent intervalley scatterings so that electrons in different valleys may be treated as different systems. This assumption was made by REIK and RISKEN [11.67] in their treatment of intervalley scattering in germanium, by BUTCHER and FAWCETT [11.52] in their displaced Maxwellian analysis of transport in gallium arsenide, and also by CONWELL and VASSELL [11.42] in their differential equation method. The actual conditions are, however, such that intervalley scattering may be as frequent as or more frequent than intravalley scattering. Electrons in different valleys may not therefore be considered as different systems but should be treated as belonging to one system. Analytic solutions are not obtainable by treating electrons distributed in nonidentical valleys as belonging to one system. Proper evaluation of the characteristics may then be carried out only by applying either the Monte Carlo or the iterative method. It may become important to carry out such calculations particularly to judge if the nature of the phenomenon is altered when the above assumption, which has to be made to obtain the analytic solution, is given up. This apparently provided the impetus for the Monte Carlo calculation of the gallium arsenide characteristics. It was very much necessary to examine in the initial stages of the discovery of the Gunn effect whether the NDC may be definitely ascribed to electron transfer. But it now appears that the qualitative nature of the characteristics are not significantly affected by the invalidity of the simplifying assumption involved in the analytic theories. The uncertainties in the scattering parameters and alterations in the exact form of the energy band structure emerging from later work [11.68] have thrown doubt on the accuracy of Monte Carlo calculations of gallium arsenide. Our knowledge about the various parameters involved in the transport calculations is yet far from complete for many materials and application of the numerical methods may not be justified. However, these uncertainties are being removed gradually and we may foresee in not too remote future more intensive use of the numerical methods for the calculation of hot-electron characteristics. This may be particularly relevant in connection with the search for a suitable material for extending the frequency of operation of FETs by increasing the value of the saturation velocity.

11.6 Hot-Electron AC Conductivity

The conductivity of semiconductors for microwave signals in the presence of steady high electric fields was studied in early days to obtain hot-electron conductivity avoiding the effects of carrier injection. It was believed that the experiment would yield the differential conductivity if the frequency of the microwave signal was low in comparison to the time constants for momentum relaxation, and that knowing the differential conductivity for different values of the steady field, the current-voltage characteristic could be constructed [11.69].

Microwave experiments were conducted later with the purpose of obtaining the hot-electron characteristics in the NDC region avoiding the effects of charge carrier instabilities. In these later experiments, high-power microwave signals were used to produce the high hot-electron fields and the conductivity was measured by applying a small steady voltage [11.23]. It was expected that the time constant of charge instabilities would be high in comparison to the period of the microwave signals and hence the charge redistribution would not occur as a consequence of NDC and the conductivity obtained in the experiment would represent that for uniform charge distribution. Experimental results obtained in both these experiments were, however, found to be very different from the expectations [11.70]. The interpretation of the results required a more detailed insight into the physical phenomena involved in hot-electron microwave conduction.

Interest in microwave conductivity in the presence of steady electric fields also stems from the fact that important solid-state microwave devices (e.g. impatt, Gunn diodes, FETs) operate under hot-electron conditions. An acquaintance with the characteristics of hot-electron microwave conduction is useful for the understanding of the operating characteristics and frequency limitations of these devices.

We first present a phenomenological theory and review briefly the accurate methods available for the calculation of the hot-electron microwave conductivity. Experimental results for some important compound semiconductors together with calculated results are then presented.

11.6.1 Phenomenological Theory for Single-Valley Materials

The theory of low-field microwave conductivity was presented in detail in Chap.9. It may be recalled that the conductivity may be evaluated by using the phenomenological momentum balance equation repeated below:

$$d(m\langle v\rangle)/dt = e\mathscr{E} - m\langle v\rangle\left[\tau_m(\langle E\rangle)\right]^{-1} \quad , \tag{11.76}$$

where $\langle v\rangle$ is the drift velocity, $\langle E\rangle$ is the average electron energy, the first term on the right-hand side is the rate of gain of momentum from the electric field and the second term is the rate of loss of momentum through collisions, $\tau_m(\langle E\rangle)$ being the momentum relaxation time corresponding to the average energy $\langle E\rangle$. The electric field $\mathscr{E}$ is the total field $\mathscr{E}_0+\mathscr{E}_1\exp(i\omega t)$, $\mathscr{E}_0$ being the steady hot-electron field and $\mathscr{E}_1\exp(i\omega t)$ the microwave field of frequency ω. We note that the drift velocity and the average energy would be modulated about the average values corresponding to $\mathscr{E}_0$ by the microwave field and these two terms may be expressed as

$$\langle v \rangle = v_0 + v_1 \exp(i\omega t) \quad , \tag{11.77}$$

$$\langle E \rangle = E_0 + E_1 \exp(i\omega t) \quad , \tag{11.78}$$

where the subscripts 0 and 1 indicate the values corresponding to the steady and alternating fields.

We should also note that the momentum relaxation time τ_m, being a function of energy, may also be expanded as

$$\tau_m(\langle E \rangle) = \tau_m(E_0) + (d\tau_m/dE)E_1 \quad . \tag{11.79}$$

We neglect the energy dependence of m^*. Substituting (11.77-79) in (11.76) and separating the steady and alternating components, we get for the conductivities

$$\sigma_{dc} = ne^2\tau_m(E_0)/m^* \quad , \tag{11.80}$$

and

$$\sigma_{ac} = ne^2\tau_m(E_0)m^{*-1}\left[1+(E_1\mathscr{E}_0/\mathscr{E}_1)(d\tau_m/dE)\tau_m^{-1}\right]\left[1+i\omega\tau_m(E_0)\right]^{-1} \quad , \tag{11.81}$$

where σ_{dc} and σ_{ac} are respectively the conductivities for steady and alternating fields, n being the electron concentration. The microwave conductivity at low fields is given by

$$\sigma_{ac} = ne^2\tau_m(E_L)m^{*-1}\left[1+i\omega\tau_m(E_L)\right]^{-1} \quad , \tag{11.82}$$

where E_L is the electron energy corresponding to the lattice temperature. The expressions for the hot-electron microwave conductivity differs from the low-field by the extra term $(E_1\mathscr{E}_0/\mathscr{E}_1)(d\tau_m/dE)$, which arises from the energy dependence of the momentum relaxation time and from the fact that the electron energy is determined by the electric field under hot-electron conditions.

The extra component is required to be evaluated by applying the energy balance equation which may be expressed phenomenologically as

$$d\langle E \rangle/dt = e\langle v \rangle\mathscr{E} - (\langle E \rangle - E_L)/\tau_E(\langle E \rangle) \quad , \tag{11.83}$$

where $\tau_E(\langle E \rangle)$ is the phenomenological energy relaxation time, the first term on the right-hand side is the rate of gain of energy from the electric field and the second term is the rate of loss of energy through collisions. Expanding $\tau_E(\langle E\rangle)$ as

$$\tau_E(\langle E\rangle) = \tau_E(E_0) + (d\tau_E/dE)E_1 \quad , \tag{11.84}$$

and substituting (11.77,78), and (11.84) in (11.83) and separating the steady and alternating components, we get

$$ev_0\mathcal{E}_0 = (E_0-E_L)/\tau_E(E_0) \quad , \tag{11.85}$$

$$i\omega E_1 = ev_1\mathcal{E}_0 + ev_0\mathcal{E}_1 - E_1/\tau_E + E_1(E_0-E_L)(d\tau_E/dE)\tau_E^{-2} \quad . \tag{11.86}$$

Using expression (11.86) we may eliminate E_1 from (11.81). The AC conductivity is then found to be [11.71]

$$\begin{aligned}\sigma_{ac} = (ne^2\tau_m)m^{*-1}&\Big\{1+i\omega\tau_E-|e|\mu_d(\mathcal{E}_0)\mathcal{E}_0^2\Big(d\tau_E/dE-\tau_E\tau_m^{-1}d\tau_m/d\tau_E\Big)\\ &\times\Big[(1+i\omega\tau_m)\Big(1+i\omega\tau_E-|e|\mu_d(\mathcal{E}_0)\mathcal{E}_0^2 d\tau_E/dE\Big)\\ &-|e|\mu_d(\mathcal{E}_0)\mathcal{E}_0^2\tau_E\tau_m^{-1}d\tau_m/dE\Big]^{-1}\Big\} \quad .\end{aligned} \tag{11.87}$$

The study of hot-electron microwave conductivity in a semiconductor requires the evaluation of τ_m and τ_E for the predominant scattering mechanisms in the material. We have seen that the scattering mechanisms are often such that τ_m and τ_E can only be regarded as phenomenological time constants and cannot be evaluated analytically for all scattering mechanisms. Orders of magnitude of these quantities may, however, be obtained if some simplifying assumptions are made. We may obtain expressions for τ_m and τ_E if we assume that the distribution function is displaced Maxwellian. We get for the case of polar optic phonon scattering

$$\begin{aligned}\tau_m^{-1} &= (2/3)n_0C_0(\theta_0/T_e)^{\frac{1}{2}}(\theta_0/2T_e)\exp(\theta_0/2T_e)\Big\{\Big[\exp(\theta_0/T_L-\theta_0/T_e)+1\Big]\\ &\quad\times K_1(\theta_0/2T_e)+\Big[\exp(\theta_0/T_L-\theta_0/T_e)-1\Big]K_0(\theta_0/2T_e)\Big\} \quad ,\\ \tau_E^{-1} &= n_0C_0(\theta_0/T_e)^{3/2}\Big[\exp(\theta_0/T_L-\theta_0/T_e)-1\Big]\exp(\theta_0/2T_e)K_0(\theta_0/2T_e)(4/3)(1-T_L/T_e)^{-1},\end{aligned} \tag{11.88}$$

where

$$C_0 = 2m^*e^2k_B\theta_0(2\pi m^*k_B\theta_0)^{-\frac{1}{2}}(4\pi\hbar^2)^{-1}(\varepsilon_\infty^{-1}-\varepsilon_S^{-1}) \quad .$$

Further analysis of AC conductivity may be carried out by using the above expressions for τ_m and τ_E. However, the exact expressions are fairly complex when all scattering mechanisms are taken into account and instead of presenting here a quantitative discussion for any particular material we refer to [11.72], where the detailed expressions are dealt with. We shall only discuss the qualitative characteristics of σ_{ac}.

We may discuss separately some special cases:

a) The signal frequency ω *is so low that* $\omega\tau_E$ *and* $\omega\tau_m$ *may be neglected in comparison to 1.*

The conductivity is

$$\sigma_{ac} = \sigma_{ac}\left[1-|e|\mu_d(\mathscr{E}_0)\mathscr{E}_0^2\left(d\tau_E/dE)-\tau_E\tau_m^{-1}(d\tau_m/dE)\right)\right]$$
$$\times\left[1-|e|\mu_d(\mathscr{E}_0)\mathscr{E}_0^2\left(d\tau_E/dE+\tau_E\tau_m^{-1}d\tau_m/dE\right)\right]^{-1} \quad . \qquad (11.89)$$

It is the same as the low-frequency differential conductivity and it is significantly different from the DC conductivity $\sigma(\mathscr{E}_0)$ if τ_E/τ_m is larger. The difference between σ_{ac} and $\sigma(\mathscr{E}_0)$ would decrease with decrease in the value of τ_E/τ_m. This result is as expected from simple arguments. A small value of τ_E would mean that the electron energy E_0 is not significantly different from E_L and hence the conductivity is not much altered by the field. The AC conductivity σ_{ac} would be different from $\sigma(\mathscr{E}_0)$ only if τ_E is much larger than τ_m and it would be smaller if τ_m decreases with electron energy.

b) The value of τ_E *is such that* $\omega\tau_E \ll 1$ *but* $\omega\tau_m$ *is comparable to* 1.

The AC conductivity is then

$$\sigma_{ac} = \sigma_{df}(1+i\omega\tau_m)^{-1} \quad , \qquad (11.90)$$

where σ_{df} is the differential conductivity. The microwave conductivity in this case differs from the differential conductivity by the factor $(1+i\omega\tau_m)^{-1}$ as at low fields.

c) The relaxation times are such that $\omega\tau_E \gg 1$ *and* $\omega\tau_m \gg 1$.

The AC conductivity in this case is identical to the DC conductivity. In this case the signal changes in magnitude before the electron energy adjusts to the altering field. The AC field cannot thus alter the electron energy and the conductivity is the same as that corresponding to the DC field. If, however, $\omega\tau_m$ is comparable to 1, the conductivity in this case would be $\sigma(\mathscr{E}_0)$ $(1+i\omega\tau_m)^{-1}$, i.e. the DC conductivity multiplied by the momentum relaxation time factor.

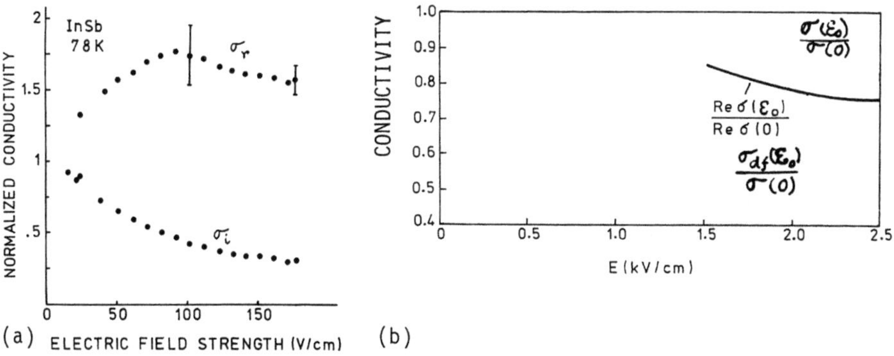

Fig. 11.18. Hot-electron microwave conductivity characteristics. (a) Indium antimonide [11.91] for 83.7 GHz; (b) Gallium arsenide [11.92]

d) In the general case, when $\omega\tau_E$ *is comparable to 1, we may hence expect that* σ_{ac} *would be intermediate between* σ_{df} *and* $\sigma(\mathcal{E}_0)$.

We illustrate in Fig.11.18 some experimental characteristics of hot-electron microwave conductivity in indium antimonide and in gallium arsenide. It may be seen that σ_{ac} is intermediate between σ_{df} and $\sigma(\mathcal{E}_0)$ for the experimental frequency.

It may be mentioned in this connection that σ_{ac} is in general complex, as evident from the expression (11.87). It has, hence, both a real part and an imaginary part. The imaginary part causes effectively a change in the dielectric constant. At low fields this contribution is negative, being given by

$$\Delta K = \sigma_i(\omega\varepsilon_0)^{-1} = -(\sigma_0\tau_m/\varepsilon_0)(1+\omega^2\tau_m^2)^{-1} \quad . \tag{11.91}$$

Under hot-electron conditions, however, it is found from a detailed plotting of the expression that the electron contribution to the permittivity may be positive depending on the relative values of τ_E and τ_m. This interesting new physical phenomenon was first brought to light through microwave hot-electron experiments in silicon and germanium [11.70] and it was for understanding this result that the significant effect of the energy relaxation time constant had to be taken into account.

11.6.2 Phenomenological Theory for Two-Valley Materials

The phenomenological theory presented in the preceding section is related to materials for which the electrons are confined to one valley in the energy band. The theory may also be extended to materials for which the electrons are distributed in two valleys and the distribution depends on the hot-electron electric field. In a two-valley material, as explained earlier, electrons are transferred from one valley to another by intervalley scattering, when the field is high enough to raise the average energy of the electrons in the low valley above the energy separation between the two valleys. Electrons in the two valleys get mixed up due to this transfer. If, however, the intravalley scattering is much stronger than intervalley scattering, we may still consider electrons in the two valleys as separate systems. This is usually the case and we shall assume that the electrons in the two valleys may be considered to have different average velocities and energies, which are determined by separate momentum and energy balance equations:

$$dJ_i/dt = n_i e^2 \mathscr{E}_i/m_i - J_i/\tau_{mi} \quad , \tag{11.92}$$

$$d(n_i E_i)/dt = J_i \mathscr{E} - n_i(E_i - E_L)/\tau_{Ei} + n_j E_j/\tau_{ji} \quad , \tag{11.93}$$

where J_i, n_i, E_i, m_i, τ_{mi}, τ_{Ei} represent respectively the current density, electron concentration, average energy, effective mass, momentum relaxation time, energy relaxation time for the ith valley. The subscript j indicates corresponding values for the jth valley, τ_{ji} represents the intervalley scattering time constant for energy transfer. The equations given above are for the ith valley and we note that these are essentially the same as for a single-valley material except for an additional term in the energy balance equation, which accounts for the energy brought by electrons transferred from the jth valley to the ith valley. We also note that the electron concentration in each valley, being determined by the magnitude of the electric field, would vary with time and the instantaneous values are to be determined by using the particle number conservation equation

$$dn_i/dt = n_i/\tau_{ni} + n_j/\tau_{nj} \quad , \tag{11.94}$$

where τ_{ni}, τ_{nj} represent the intervalley scattering time constants for the transfer of electrons. An expression for the AC conductivity may now be derived by separating as before J, E, and n into a steady and time-dependent part like the field, i.e. writing

$$\mathscr{E} = \mathscr{E}_0 + \mathscr{E}_1 \exp(i\omega t) \quad , \tag{11.95}$$

$$J_i = J_{i0} + J_{i1} \exp(i\omega t) \quad , \tag{11.96}$$

$$E_i = E_{i0} + E_{i1} \exp(i\omega t) \quad , \tag{11.97}$$

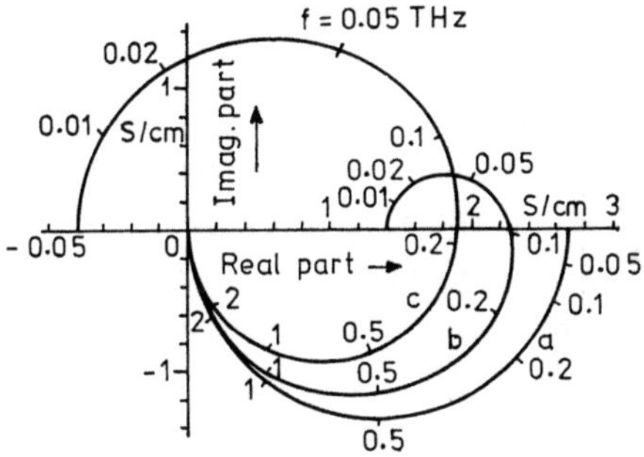

Fig. 11.19. Calculated microwave conductivity for hot-electron conditions in gallium arsenide [11.72]. a) $\mathscr{E}_0 = 0$; b) $\mathscr{E}_0 = 2.98$; c) $\mathscr{E}_0 = 3.46$, $\mathscr{E}_0$ in kV/cm

$$n_i = n_{i0} + n_{i1}\exp(i\omega t) \quad . \tag{11.98}$$

Note that $n_{i1} = -n_{j1}$. Substituting into (11.92-94) and separating the steady and time-dependent terms, we obtain

$$J_{i0} = n_{i0}e^2(\tau_{mi}/m_i)\ {}_0 \quad , \tag{11.99}$$

$$J_{i0}\ {}_0 = n_{i0}(E_{i0}-E_L)/\tau_{Ei} - n_{j0}E_{j0}/\tau_{ji} \quad , \tag{11.100}$$

$$n_{i0}/\tau_{ni} = n_{j0}/\tau_{nj} \quad , \tag{11.101}$$

$$(J_{i1}/J_{i0})(1+i\omega\tau_{mi}) - (\tau'_{mi}/\tau_{mi})E_{i1} - (n_{i1}/n_{i0}) = \mathscr{E}_1/\mathscr{E}_0 \quad , \tag{11.102}$$

$$\begin{aligned}
&-(J_{i1}/J_{i0})n_{i0}^{-1} + \left(\tau_{Ei}^{-1}-i\omega-E_{i0}\tau'_{Ei}\tau_{Ei}^{-2}\right)E_{i1}J_{i0}^{-1}\mathscr{E}_0^{-1}\\
&\quad - n_{j0}n_{i0}^{-1}\left(\tau_{ij}^{-1}-E_{j0}\tau'_{ij}\tau_{ij}^{-2}\right)E_{j1}J_{i0}^{-1}\mathscr{E}_0^{-1}\\
&\quad + \left(i\omega+\tau_{Ei}^{-1}+E_{j0}E_{i0}^{-1}\tau_{ij}^{-1}\right)E_{i0}J_{i0}^{-1}\mathscr{E}_0^{-1}n_{i1}n_{i0}^{-1} = \mathscr{E}_1/\mathscr{E}_0 \quad ,\\
&-\tau_{ni}^{-2}\tau'_{ni}E_{i1} + n_{j0}n_{i0}^{-1}\tau_{nj}^{-2}\tau'_{nj}E_{ji} + \left(i\omega+\tau_{ni}^{-1}+\tau_{nj}^{-1}\right)n_{i1}n_{i0}^{-1} = 0 \quad .
\end{aligned} \tag{11.103}$$

The prime on the relaxation time constants indicate the derivative for the energy.

We may solve the above equations to obtain J_{11} and J_{21} and hence the AC conductivity $(J_{11}+J_{21})/\mathscr{E}_1$. The relaxation time constants may be analytically evaluated as in the single-valley case only by assuming a displaced Maxwellian distribution function for electrons in the two valleys. The expressions are fairly involved and we refrain from quoting them (see [11.72,73]). In [11.72] detailed results on the frequency dependence of the conductivity for gallium arsenide are also given. We reproduce in Fig.11.19 some of the calculated results. The important result of technological consequence is the fact that the NDC disappears above a critical frequency. This occurs as a result of the finite value of the energy relaxation time τ_{E1} and τ_{E2} and of the intervalley-particle-transfer time constants τ_{n1} and τ_{n2}. Due to their finite values en-energy of the electrons cannot change in step with the field and the electrons are not also transferred from the lower valley to the higher valley. The limiting frequency is therefore of the order of the inverse of these time constants. Early estimate of the limiting frequency for gallium arsenide on the basis of these considerations was about 200 GHz [11.73]. This value has been revised later in the light of more detailed analysis. However, no experimental result is available for direct comparison with these theoretical values. It would appear that a qualitative knowledge of the phenomena involved in frequency limitation is sufficient for the understanding of the frequency behaviour of the devices using NDC in compound semiconductors.

11.6.3 Large-Signal AC Conductivity

It should be evident from the preceding discussion that an accurate analytic theory of hot-electron AC conductivity is difficult to work out and it has not been attempted so far. The actual operating conditions in most of the hot-electron microwave devices are also such that the microwave field cannot be considered small in comparison to the steady field. The small-signal AC conductivity is therefore not of much practical significance either.

The iterative method of calculation of hot-electron conductivity may, however, be applied to determine the frequency dependence of the differential conductivity for all magnitudes of the AC field. It is seen that if the self-scattering rate is chosen to have a large value, the distribution function develops in steps and this development represents the development of the distribution function in time due to a step field input. The AC conductivity may therefore be computed by first determining the distribution function for a steady field and then determining the time-evolving distribution function due to a step increase in the steady field. The time development of the change in current may be computed from this distribution function and the AC conductivity may be obtained by analysing the shape of the current change. This method was developed by REES [11.74] and applied to study the frequency dependence of AC conductivity of gallium arsenide. We reproduce in Fig.11.20 the differential mobility obtained by him. The results are quantitatively different from those obtained by the analytic theories. The limiting frequency is found to be about 140 GHz, which is significantly lower than that obtained by analytic theories. An important conclusion about the importance of various scattering mechanisms was also arrived at. It was found that the

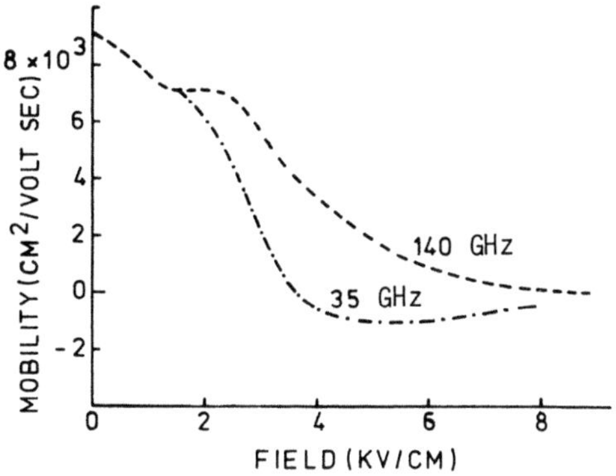

Fig. 11.20. Hot-electron microwave conductivity in gallium arsenide calculated by the iterative technique [11.74]

intravalley scattering time constant for the lower valley being much larger than other time constants the frequency limitation is introduced predominantly by this process and not by the intervalley scattering rates.

11.7 Hot-Electron Diffusion

In solid-state devices operated under hot-electron conditions the charge distribution is usually nonuniform. The nonuniformity results from the NDC in Gunn diodes, from nonuniform fields in FETs and from avalanche injection in impatts. Variation in the electron concentration produces a diffusion current which significantly affects the device performance. It is also found that the thermal noise in these devices is related to the value of the diffusion coefficient. There is therefore great interest in the values of the diffusion constant for hot-electron fields. We present in this section the methods that may be applied for the evaluation of the constant.

11.7.1 Einstein Relation for Hot-Electron Diffusivity

It was shown in Sect.8.4.1 that the diffusion constant is related to the drift mobility by the Einstein relation which for nondegenerate materials has the simple form

$$D = \mu_d k_B T/|e| \quad . \tag{11.104}$$

For degenerate materials, it has the form

$$D = \mu_d n|e|^{-1}(\partial n/\partial E_F)^{-1} \quad . \tag{11.105}$$

One may expect that the same relation will be applicable also for hot-electron conditions if the distribution function is Maxwellian. We find, however, as shown below, that this is not generally true.

Let us assume that there is a small concentration gradient $\partial n/\partial x$ in the x direction. To obtain the diffusion coefficient, we must solve the Boltzmann equation by including the space-dependent term. We assume that the electric field is also in the x direction and the energy band is isotropic and parabolic. The Boltzmann equation for these assumptions is

$$\partial_t f = -(\hbar k_x/m^*)\partial_x f - (e\mathcal{E}/\hbar)(\partial f/\partial k_x) + (\partial f/\partial t)_c \quad . \tag{11.106}$$

Expanding the distribution function f as

$$f(\underline{k},x,t) = f_0(E,x,t) + kf_1(E,x,t)\cos\theta \quad , \tag{11.107}$$

and separating (11.108) into the symmetric and asymmetric parts, we get

$$\partial f_0/\partial t = -(2/3)(E/\hbar)(\partial f_1/\partial x) - (2/3)(e\mathcal{E}/\hbar)E^{-\frac{1}{2}}\partial(E^{3/2}f_1)/\partial E + (\partial f_0/\partial t)_c \quad , \tag{11.108}$$

$$\partial f_1/\partial t = -(\hbar/m^*)(\partial f_0/\partial x) - (e\hbar\mathcal{E}/m^*)(\partial f_0/\partial E) + (\partial f_1/\partial t)_c \quad . \tag{11.109}$$

We assume further that the concentration gradient introduces a local perturbation in the distribution function so that we may separate (11.108,109) into two sets of equations:

$$(\partial f_{0e}/\partial t)_c = (2/3)(e\mathcal{E}/\hbar)E^{-\frac{1}{2}}\partial(E^{3/2}f_{1e})/\partial E \quad , \tag{11.110}$$

$$(\partial f_{1e}/\partial t)_c = (e\mathcal{E}\hbar/m^*)(\partial f_{0e}/\partial E) \quad ; \tag{11.111}$$

and

$$\partial f_{1p}/\partial t = -(\hbar/m^*)(\partial f_{0p}/\partial x) - (e\mathcal{E}\hbar/m^*)(\partial f_{0p}/\partial E) + (\partial f_{1p}/\partial t)_c \quad ,$$

$$\partial f_{0p}/\partial t = -(2/3)(e\mathcal{E}/\hbar)E^{-\frac{1}{2}}\partial(E^{3/2}f_{1p}) - (2/3)(E/\hbar)(\partial f_{1p}/\partial x) + (\partial f_{0p}/\partial t)_c \quad , \tag{11.112}$$

where the perturbation and the steady-state components are defined as

$$f_0 = f_{0e}(E) + f_{0p}(E,x,t) \quad , \tag{11.113}$$

$$f_1 = f_{1e}(E) + f_{1p}(E,x,t) \quad . \tag{11.114}$$

The perturbation component of f_0, f_{0p} is related to the excess carrier concentration, $n_1(x,t)$, by the equation

$$n_1(x,t) = \int f_{0p}(E,x,t)d\underline{k} \quad , \tag{11.115}$$

whereas the uniform part of n, n_0 is given by

$$n_0 = \int f_{0e} d\underline{k} \quad . \tag{11.116}$$

Any further progress in the analysis may be made only if we assume that the collision mechanisms admit a relaxation time for both energy and momentum. Equations (11.112,113) may be solved by the moment method by expanding f_{0p} as a polynomial in E. This method has been applied to study hot-electron diffusion in silicon [11.75]. We present here a simpler analysis, assuming that the symmetric part of the distribution function is Maxwellian with a space- and time-dependent electron temperature. Substituting

$$(\partial f_1/\partial t)_c = -f_1/\tau_m \quad , \quad f_{0e} = nI_n^{-1} \exp(-E/k_B T_e) \quad , \tag{11.117}$$

we get for f_{0e} and f_{0p}

$$f_{0e} = n_0 \hbar^3 (2\pi m^* k_B T_0)^{-3/2} \exp(-E/k_B T_0) \quad , \tag{11.118}$$

$$f_{0p} = \left[n_1/n_0 + (E/k_B T_0 - 3/2)(T_1/T_0)\right] f_{0e} \quad , \tag{11.119}$$

and for f_{10} and f_{1p}

$$f_{10} = (e\hbar\tau_m \mathscr{E}/m^*)\partial_E f_0 \quad , \tag{11.120}$$

$$f_{1p} = -(\hbar\tau_m/m^*)\partial_x f_{0p} - (e\hbar\tau_m \mathscr{E}/m^*)\partial_E f_{0p} \quad , \tag{11.121}$$

where T_0 and T_1 are the equilibrium and perturbation components of the electron temperature and $\partial f_{1p}/\partial t$ has been neglected in comparison to f_{1p}/τ_m.

It we now apply the energy balance condition, we get a fairly involved equation for T_1, which does not admit a simple solution. However, we may arrive at a simple final equation [11.76] for n_1 by neglecting high-order derivatives and expressing the energy relaxation time in terms of the differential mobility:

$$\partial n_1/\partial t = D_{11}(\partial^2 n_1/\partial x^2) - \mu_1 \mathscr{E}(\partial n_1/\partial x) - (3/2)a\mathscr{E}(\partial^2 n_1/\partial t \partial x) \quad , \tag{11.122}$$

where

$$D_{11} = D_1\left[1-(\mu_1-\mu_{df})(\mu_1+\mu_2)/(2\mu_1^2)\right] \quad ,$$

$$D_1 = \mu_1 k_B T_0/|e| \quad , \quad a = (\mu_1-\mu_d)k_B T_0/(2|e|\mu_1\mathscr{E}^2) \quad ,$$

$$\mu_1 = (2|e|/3n_0 m^*)\int (E/k_B T_0)^n \tau_m f_{0e} d\underline{k} \quad .$$

μ_{df} is the differential mobility. The equation appears different from the low-field equation and the diffusion behaviour of electrons under hot-electron conditions may, therefore, be very different from that at low fields. However, if the diffusion current is negligible in comparison to the drift current we may reduce the equation to the usual form. We may then replace $\partial_t n_1$ by $-\mu_1\mathscr{E}\partial_x n_1$ in the last term on the right-hand side and simplify (11.122) to

$$\partial_t n_1 = \left(D_{11}+3\mu_1 a\mathscr{E}^2/2\right)\partial^2 n_1/\partial x^2 - \mu_1\mathscr{E}\partial_x n_1 \quad , \tag{11.123}$$

which is of the same form as the low-field diffusion equation. The diffusion constant is then given by

$$D = D_1\left[1+(\mu_1-\mu_d)(\mu_1-2\mu_2)/4\mu_1^2\right] \quad . \tag{11.124}$$

If we assume further that the relaxation time varies as E^{-s}, we may simplify (11.124) to

$$D = D_1\left[\mu_{df}/\mu_1+s(\mu_1-\mu_{df})/2\mu_1\right] \quad . \tag{11.125}$$

Had the conventional Einstein relation been applicable then the diffusion constant would have been given by D_1. We find instead that it is given by a more involved relation. The difference in the two expressions originates from the effect of the diffusion current on the hot-electron distribution. It may be shown from detailed considerations that the diffusion current causes a first-order perturbation in the electron temperature and this in turn produces an extra alteration in the current, proportional to the concentration gradient. It is essentially the same kind of phenomenon which causes the differential mobility to be different from the chord mobility.

The diffusion constant for electrons diffusing perpendicular to the electric field may also be derived following the procedure outlined above. In this case, however, the diffusion current produces a second-order perturbation, and it may be neglected. The perpendicular diffusion current is hence given by D_1, i.e. the expression obtained by applying the Einstein relation.

No detailed study of hot-electron diffusion in compound semiconductors starting from the Boltzmann equation has yet been reported. But, as the knowledge of the diffusion constant is necessary for the understanding of hot-electron devices, some preliminary results were obtained by using intuitive relations for gallium arsenide. We present briefly in the following section these results.

11.7.2 Electron Diffusivity in Gallium Arsenide

The problem of calculation of the electron diffusivity in gallium arsenide is complicated by the transfer of electrons from the lower to higher valleys in the energy band and vice versa. In the initial results reported in the literature these complications were ignored and the diffusion constant was taken to be given by [11.77]

$$D = (n_1D_1+n_2D_2)/(n_1+n_2) \quad , \quad D_i = \mu_i k_B T_i/|e| \quad , \tag{11.126}$$

i.e. the average of the values for the two valleys, the value for the individual valleys being given by the Einstein relation with chord mobility.

As mentioned already, the expressions for D_i is modified from the Einstein relation, but if the distribution function is displaced Maxwellian we may simplify (11.125), since s is equal to zero for this distribution function. Then, we are only required to replace μ_i by μ_{dfi}, i.e. replace chord mobilities by the corresponding differential mobilities. However, we find that even with this modification (11.126) is not complete. Let us assume that the total electron concentration is n and let n_1 and n_2 be the electron concentrations in the two valleys. The particle conservation equations for n_1 and n_2 are

$$\partial_t n_1 = D_1(\partial^2 n_1/\partial x^2) - v_1\partial_x n_1 + n_2/\tau_2 - n_1/\tau_1 \quad , \tag{11.127}$$

$$\partial_t n_2 = D_2(\partial^2 n_2/\partial x^2) - v_2\partial_x n_2 + n_1/\tau_1 - n_2/\tau_2 \quad , \tag{11.128}$$

where D_1 and D_2 are the diffusion constants; v_1 and v_2 are the drift velocities and τ_1 and τ_2 are the time constants for intervalley scattering. The subscripts 1 and 2 refer to electrons in the two valleys. We get the last two additional terms to account for particle transfer from one valley to another by intervalley scattering. We note that

$$n_1 + n_2 = n \quad . \tag{11.129}$$

Let us also express n_1 and n_2 as

$$n_1 = p_1 n + \Delta n \quad , \quad n_2 = p_2 n - \Delta n \quad , \tag{11.130}$$

where

$$p_1 = \tau_1(\tau_1+\tau_2)^{-1} \quad , \quad p_2 = \tau_2(\tau_1+\tau_2)^{-1} \quad .$$

In the absence of any concentration gradient Δn would be zero, since then n_1 and n_2 are determined by intervalley scattering alone. Substituting (11.130) in (11.127,128), and neglecting the derivatives of p_1 and p_2 we get

$$\begin{aligned} p_1 \partial_t n = {} & p_1 D_1 \partial^2 n/\partial x^2 + D_1 \partial^2 \Delta n/\partial x^2 - p_1 v_1 \partial_x n - v_1 \partial_x \Delta n \\ & - (\tau_1+\tau_2)(\tau_1\tau_2)^{-1} \Delta n \quad , \end{aligned} \tag{11.131}$$

$$\begin{aligned} p_2 \partial_t n = {} & p_2 D_2 \partial^2 n/\partial x^2 - D_2 \partial^2 \Delta n/\partial x^2 - p_2 v_2 \partial_x n + v_2 \partial_x \Delta n \\ & + (\tau_1+\tau_2)(\tau_1\tau_2)^{-1} \Delta n \quad . \end{aligned} \tag{11.132}$$

Adding the two equations,

$$\begin{aligned} \partial_t n = {} & -(p_1 v_1 + p_2 v_2)\partial_x n + (p_1 D_1 + p_2 D_2)\partial^2 n/\partial x^2 \\ & + (v_2 - v_1)\partial_x \Delta n + (D_1 - D_2)\partial^2 \Delta n/\partial x^2 \quad . \end{aligned} \tag{11.133}$$

We may eliminate Δn by using (11.131). We note that of all the different terms the drift current term is of the largest magnitude and in the first approximation we may neglect other terms on the right-hand side in comparison. We then obtain

$$\begin{aligned} \Delta n & = -\tau_1\tau_2(\tau_1+\tau_2)^{-1}(p_1 v_1 \partial_x n + p_1 \partial_t n) \\ & = -\tau_1\tau_2(\tau_1+\tau_2)^{-1} p_1(v_1 - p_1 v_1 - p_2 v_2)\partial_x n \\ & = -\tau_1\tau_2(\tau_1+\tau_2)^{-1} p_1 p_2 (v_1 - v_2)\partial_x n \quad . \end{aligned} \tag{11.134}$$

Replacing Δn and neglecting the third-order derivative, we may reduce (11.133) to the usual diffusion equation,

$$\partial_t n = D_{eq}\partial^2 n/\partial x^2 - v_{av}\partial_x n \quad , \tag{11.135}$$

where

$$D_{eq} = p_1 D_1 + p_2 D_2 + \tau_1\tau_2(\tau_1+\tau_2)^{-1} p_1 p_2 (v_1 - v_2)^2 \quad ,$$

and

$$v_{av} = p_1 v_1 + p_2 v_2 \quad .$$

It is evident from the above equation that in a two-valley material the diffusion of electrons is controlled by the average drift velocity v_{av} and an equivalent diffusion constant D_{eq}, which contains in addition to the term representing the average of the diffusion constants for the two valleys, an extra term arising from the inequality of the drift velocities in the valleys, through intervalley scattering. We have given here a simple derivation using the continuity equation to show how the term gets included. However, a more sophisticated derivation using statistical arguments is given by SHOCKLEY et al. [11.77b].

Diffusion constants for gallium arsenide have been calculated by applying (11.135) and it appears to give good agreement with the experiments. However, it should be evident that the formula for the diffusion constant may be regarded only as an approximate one as it involves a number of simplifying assumptions. Attempts to obtain more accurate theoretical values were therefore based on the Monte Carlo method.

11.7.3 Monte Carlo Calculation of the Diffusion Constant

The diffusion constant may be calculated from the variance of the electron displacement as shown below. Let us consider the continuity equation in three dimensions,

$$\partial_t n = D_{\parallel}\partial^2 n/\partial x^2 + D_{\perp}(\partial^2 n/\partial y^2 + \partial^2 n/\partial z^2) - v\partial_x n \quad , \tag{11.136}$$

where we have assumed that the diffusion constant may be anisotropic as expected under hot-electron conditions; $D_{\parallel}$ and $D_{\perp}$ represent respectively the values of the constant in the direction and transverse to the direction of the electric field, v is the drift velocity of the electrons. Solution of the equation for a Dirac delta function input of carriers is

$$n(x,y,z,t) = At^{-3/2} \exp\left[-(x-vt)^2/4D_{\parallel}t-(y^2+z^2)/4D_{\perp}t\right] \quad , \tag{11.137}$$

where A is a normalising constant. The displacements have the variance,

$$(x-<x>)^2 = <x^2> - <x>^2 = 2D_{\parallel}t \quad , \tag{11.138}$$

$$<y^2> = <z^2> = 2D_{\perp}t \quad . \tag{11.139}$$

It is therefore expected that if the motion of a single electron from an initial position (0,0,0) is observed over a sampling time T and a large number of samples are taken then the displacements for the different samples should have the same variance.

The diffusion coefficients may therefore be calculated from the Monte Carlo simulation of trajectories, details of which have been given in Sect. 11.5.2. In practice it is more convenient to calculate $D_{\parallel}$ and $D_{\perp}$ by using the values of the velocities and the formulae

$$D_{\parallel} = 0.5(<v_x^2>-<v_x>^2)T \quad , \tag{11.140}$$

$$D_{\perp} = 0.5<v_y^2>T = 0.5<v_z^2>T \quad , \tag{11.141}$$

where T is the samplig time, v_x, v_y, v_z are the average velocities for a sample over this time. The coefficients may also be obtained by evaluating the higher order moments, since it may be shown that

$$(x-<x>)^4 = 12D_{\parallel}^2t^2 \quad , \tag{11.142}$$

and so on.

This method of evaluation of the diffusion constants has been applied for gallium arsenide [11.78], indium phosphide [11.79], and cadmium telluride [11.80]. We present in Fig.11.21 the curves obtained from such calculations.

It should be mentioned that the method involves the basic assumption that the motion of a single electron over a long time can be taken to represent the behaviour of an ensemble over a shorter time. A rigorous statistical proof of (11.138) has been given by KATZ [11.81] for small departures from equilibrium. It is intuitively assumed that the same formula is valid also at high fields. However, under hot-electron conditions, the electrons are far from equilibrium and the diffusion current associated with the motion of the ensemble of the electrons affects the electron temperature significantly.

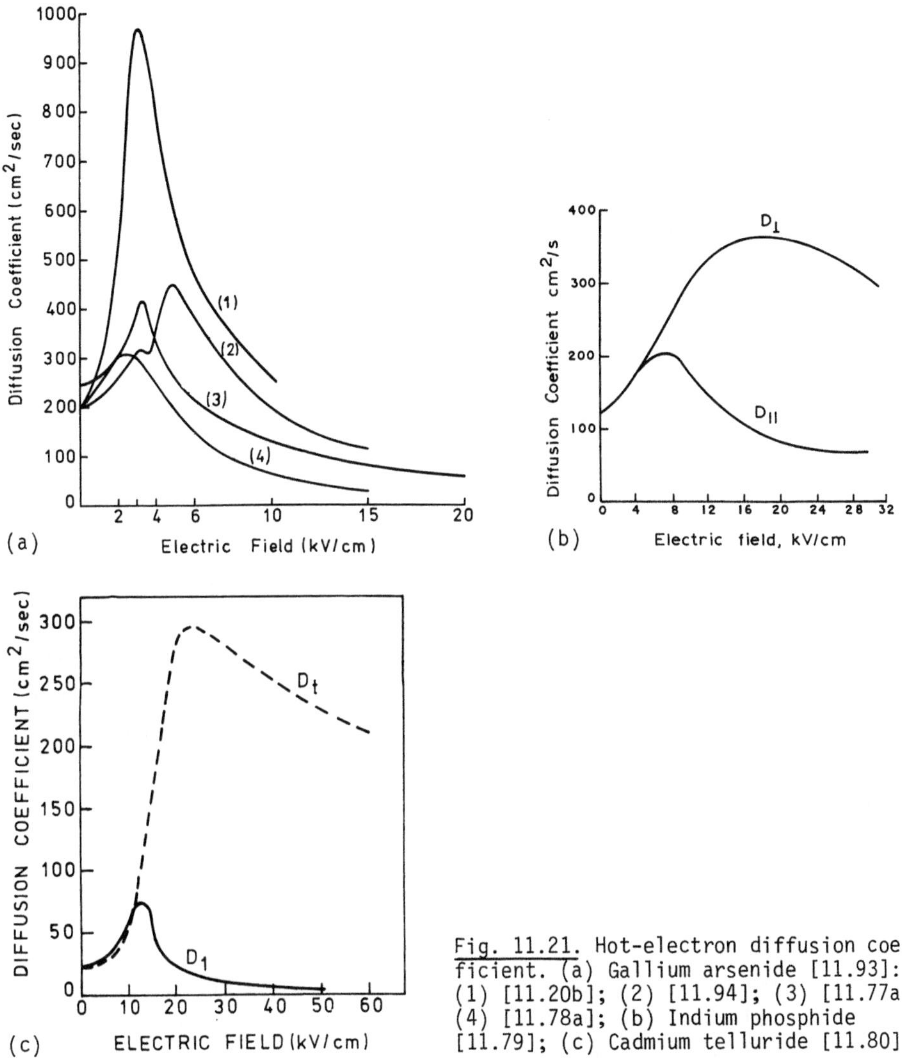

Fig. 11.21. Hot-electron diffusion coefficient. (a) Gallium arsenide [11.93]: (1) [11.20b]; (2) [11.94]; (3) [11.77a]; (4) [11.78a]; (b) Indium phosphide [11.79]; (c) Cadmium telluride [11.80]

This current is not simulated in the single electron trajectories and this raises some doubt about the validity of the single-particle approach of solving the diffusion problem under hot-electron conditions [11.76]. The results for electron diffusion in gases under high fields as obtained by solving the Boltzmann equation are found to be different in some cases from the Monte Carlo results [11.82]. In view of this disagreement it appears that a more rigorous proof than given above for (11.138) is required for the satisfactory removal of the doubt.

11.8 Conclusion

We have discussed in this chapter the phenomenon of hot-electron conduction and reviewed the experimental and theoretical results for compound semiconductors. We find that the studies have been mostly confined to materials of device interest. A general study of hot-electron conduction in various compounds is still lacking.

The broad features of the hot-electron characteristics may be obtained by the analytic theories. The displaced Maxwellian distribution function method is particularly suitable for this purpose and it has been extensively applied for studying the hot-electron properties of compound semiconductors. However, for obtaining numerically reliable results we can apply either the iteration or the Monte Carlo method. These methods are good enough for obtaining accurate results, and hot-electron characteristics may be accurately calculated by these methods for any material, provided the physical constants are known. It is the lack of our knowledge about the physical constants, particularly those related to intervalley scattering and to the energy-wave-vector dispersion relation of higher valleys, that limits the theoretical computation of hot-electron characteristics. However, the upper frequency limit of operation of FETs is determined by the value of the saturation velocity. FETs are the most promising devices for the amplification of GHz signals and for fast switching. A material, for which the saturation velocity is higher than that in the materials of current use, would therefore be of great technical interest. A search may be made for such a material by making both experimental and theoretical study of compound semiconductors. Some work along this line has been carried out in recent years [11.30a,c].

A subject of great interest in relevance to high-frequency FETs is the time development of hot-electron currents. The dimensions of the devices are usually so small that the electrons are swept out before they reach the steady state. The current in such devices cannot be calculated by applying the steady-state theory presented here. We may study such currents by straightforward application of the Monte Carlo method [11.83]. It is however found that the response may also be calculated by applying phenomenological equations involving energy-dependent momentum and energy relaxation times [11.84]. Such analyses may be made by using the values of these constants revealed by recent hot-electron optical interaction experiments [11.85]. Work along these lines could be useful in the search for a suitable material for high-frequency FETs.

Before concluding this chapter it should be pointed out that the study of hot-electron transport started from an interest in classical physics. However, through these studies a commercially useful device has been invented. At the same time, the theory of electrical transport in semiconductors, particularly in compound semiconductors, has been perfected more and more in the way of understanding the behaviour of this device. The interest in the hot-electron transport phenomenon, however, seems to have diminished in recent years. This is mainly because the important features of the phenomenon are now well understood. But hot-electron transport in mixed compound semiconductors, in superlattice structures, and in inversion layers of compound semiconductors have not yet been fully explained and it may be expected that these aspects will receive greater attention in the future.

12. Review of Experimental Results

The theory of electron transport has been presented in the earlier chapters in all its complexity. The characteristics of important transport phenomena have also been described. We have not, however, considered any detailed experimental results or discussed how far the theory explains the available results. The total number of compound semiconductors is very large. There are about twelve important III-V compounds, six II-VI compounds, three IV-VI compounds and innumerable kinds of mixed and chalcopyrite compounds. It is impossible to discuss the transport properties of all these materials, nor would such coverage be meaningful at this stage, since reproducible data for well-characterised samples are generally not available. The data reported in the literature are mostly obtained by using available samples, which are not often in perfectly crystalline form and may be very impure, with unknown impurity concentration. The samples have been characterised by stating the electron concentration determined from the value of the Hall coefficient by assuming that the Hall factor is unity. This assumption may introduce an error of about 10-20% in the values of the electron concentration. Even if this error is ignored there remains a factor of greater uncertainty. The experimental samples are mostly compensated and the impurity concentration may be radically different from the electron concentration. We find on plotting the transport parameters against electron concentration that the data points are

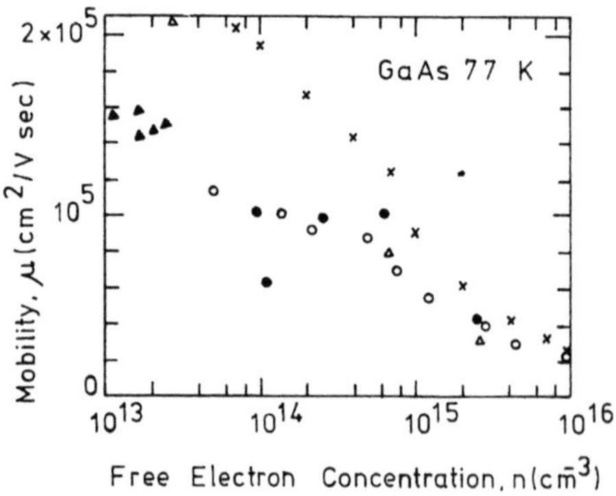

Fig. 12.1. Electron mobility in n-GaAs for different electron concentrations [12.13]

scattered within ranges differing by orders of magnitude. As an illustration, we have plotted in Fig.12.1 experimental results on electron mobility in gallium arsenide for a lattice temperature of 77 K. Agreement of theory with such data may be accepted only for a qualitative verification. In some cases even such verification may be important, for a confirmation of a scattering model of a material, the physical constants of which are completely unknown. However, transport theory with all its refinements as presented in this book should be useful for predicting the transport coefficients with greater significance. Encouragingly, in recent years, some experiments have been carried out in which the Hall coefficient of the samples was also studied at very low temperatures, where the impurity atoms are partially ionised. It is possible to obtain the acceptor as well as the donor concentration from the analysis of such data by assuming a single donor ionisation energy level as is usually the case. The samples for such experiments are characterised by stating both the donor and the acceptor concentration and the total impurity concentration for samples of such experiments are also known. A detailed comparison with theory is meaningful for such experiments only.

From the above considerations, we have chosen to present in this chapter results on the transport coefficients of compound semiconductors for which experiments were conducted with samples of known impurity content or of very pure form with negligible contribution from impurity scattering. Our discussion will be confined mostly to the Hall mobility and the low-field galvanomagnetic coefficients. We shall also present data on free-carrier absorption, as such data have been very useful in identifying the scattering mechanisms. The thermal transport coefficients are discussed for materials for which data are available. A discussion of the coefficients for quantising magnetic fields is omitted as it has been partially covered in Chap.10 and also because no elegant method has yet been developed for the accurate calculation of the coefficients for practical samples. Experimental results on hot-electron transport, wherever available, have already been presented and we shall not discuss these results any further.

We shall present calculated values of the coefficients along with the experimental results to indicate the applicability of the theory. The theoretical evaluation of the coefficients requires knowledge of a large number of physical constants, namely, mass density, lattice dispersion relations, dielectric constants, elastic constants, energy band diagrams, and deformation potentials. All the constants except the deformation potentials may be determined by physical measurements (see [12.1]). There is as yet no unambiguous method of determination of the deformation potentials. Some theoretical me-

Table 12.1. Physical constants of compound semiconductors

Material	Static dielectric constant K_s	High-frequency dielectric constant K_∞	Optic phonon temperature θ_ℓ [K]	Mass density ρ[gm cm^{-3}]	Longitudinal elastic constant (average value) $\langle \rho s^2 \rangle$ [Nm$^{-2}\times 10^{-10}$]	Intravalley deformation potential constant E_1 [eV]	Piezoelectric coupling constant $[\langle h_{pz}^2/\rho s^2 \varepsilon \rangle]^{1/2}$	Ionic charge in units of $\lvert e \rvert$	Highest observed mobility at 77 K [m^{2}V^{-1}s^{-1}]	300 K
InSb	17.54	15.68	278	5.8	8.1	20	0.027	0.16	110	-
InAs	14.54	11.74	337	5.61	10.3	5.8	0.0168	0.22	10	2
InP	12.38	9.55	497	4.79	12.7	18	0.0131	0.26	8	0.38
GaSb	15.69	14.44	346	5.66	10.7	8.3	-	0.13	-	0.54
GaAs	12.91	10.91	419	5.31	14.5	12	0.052	0.20	10	0.9
GaP	11.10	9.11	580	4.13	16.6	13	0.036	0.24	0.22	0.012
AlSb	11.21	9.88	480	4.22	5.8	-	-	0.19	-	-
ZnO	8.12	3.72	837	S 5.66 W 5.21	20.5	3.8	⊥ 0.21 ∥ 0.36	0.51	0.22	0.022
ZnS	8.32	5.13	506	S 4.08 W 6.26	12.8	4.9	0.076	0.39	-	0.014
ZnSe	8.33	5.9	360	S 5.42 W 6.62	10.3	4.2	0.024	0.34	8	0.053
ZnTe	9.67	7.28	297	S 5.72 W 6.99	8.4	3.5	0.0146	0.28	-	0.034
CdS	E⊥C 8.58 E∥C 9.1	E⊥C 5.26 E∥C 5.31	428	S 4.82 W 6.75	8.5	3.3	⊥ 0.143 ∥ 0.192	0.34	2	0.035
CdSe	9.40	6.10	303	S 5.81 W 7.01	7.4	3.7	⊥ 0.104 ∥ 0.148	-	0.34	0.065
CdTe	9.65	7.21	246	S 6.06 W 7.47	7.0	4.0	0.022	0.32	1.0	0.105
HgSc	25.6	12.0	268	-	8	4	0.022	-	6.0	1.0
HgTe	20.0	14.0	199	-	6.1	4	0.022	-	6.0	2.5
PbS	175	17	300	7.6	-	20	-	0.26	1.1	0.06
PbTe	400	33	160	8.2	7.1	25	-	0.18	3.16	0.17
PbSe	250	24	190	8.3	-	24	-	0.21	1.65	0.1

The quoted values are collected from Ref.12.31,p.1; Ref.2.8,p.566; Refs.12.7,20; Ref.12.32; Ref.12.1,p.120; Ref.12.44; Ref.12.42,p.352.

thods [12.2] have been recently developed and applied to obtain the values of the constants for some materials. The uncertainty in the values of the deformation potentials introduces an ambiguity in the values of the transport coefficients when acoustic phonon scattering or intervalley and nonpolar optic phonon scattering dominate the scattering processes. In compound semiconductors, however, polar optic phonon scattering is usually the most important scattering mechanism and the uncertainty in the values of the deformation potential constant may not have very important consequences. Further, in many cases only the acoustic phonon deformation potential scattering may have to be considered in addition to the optic phonon and impurity scattering and its value may be obtained empirically from the analysis of the transport data for certain conditions and then used to explain the results over a larger range of ambient conditions. We shall apply this procedure in most cases.

The values of the physical constants required for transport calculations are listed in Table 12.I for the various compound semiconductors. The energy-band parameters have been omitted as these are already listed in Tables 3.1-3. It should be pointed out that the constants are mostly temperature dependent. However, the temperature coefficients not being accurately known and being often compensatory in their effects on the transport coefficients, we shall ignore the variation of the constants with temperature. Only the temperature dependence of the effective mass is included for some materials.

12.1 Transport Coefficients of III-V Compounds

It has been mentioned in the preceding section that some compound semiconductors occur freely in nature, while others have been made in the laboratory. The freely occurring materials are rarely pure and their characteristic properties have been recorded only with crystals carefully grown in the laboratory in recent years. On the other hand, the III-V compounds were studied by using crystals grown in the laboratory almost from the beginning. Among the III-V compounds, indium antimonide has been studied the most extensively, and we shall start our account with this material.

12.1.1 Indium Antimonide

Experimental results for mobility are available over the temperature range 30-700 K for electron concentrations ranging between 10^{13} and 10^{16} cm^{-3}. It is evident from these results that the mobility is significantly controlled

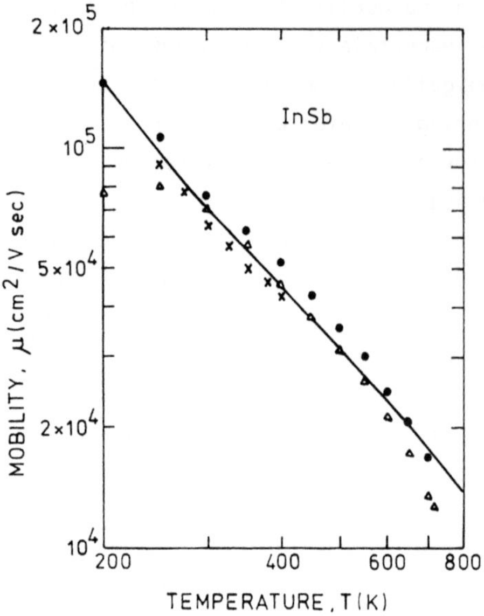

Fig. 12.2. Electron mobility in n-InSb [12.19]

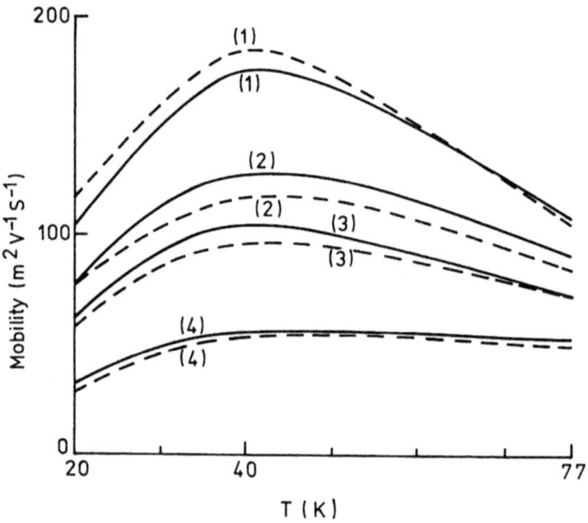

Fig. 12.3. Hall mobility in n-InSb. Solid line – Calculated values. (1) $N_i = 1.0 \times 10^{14}$; $n = 1.9 \times 10^{13}$; (2) $N_i = 2.0 \times 10^{14}$; $n = 6.5 \times 10^{13}$; (3) $N_i = 2.8 \times 10^{14}$; $n = 6.2 \times 10^{13}$; (4) $N_i = 5.4 \times 10^{14}$; $n = 5.0 \times 10^{13}$ in units of cm^{-3}

by impurity scattering below 200 K, but above this temperature the data may be considered to be characteristic of pure indium antimonide as the samples become intrinsic. Experimental results covering this temperature range are presented in Fig.12.2. Theoretical values of drift mobility for the same conditions were calculated by RODE [12.3] using the parameters given in Table 12.1 but taking $E_1 = 7.2$ eV. It is seen that the theoretical curve agrees very well with the experimental results. Similar agreement was also found by EHRENREICH [12.4]. Electron mobility for this temperature range is, however,

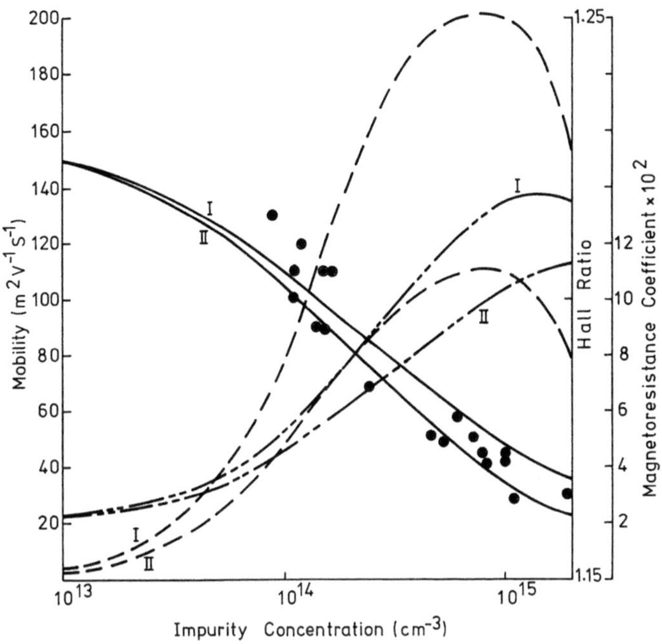

Fig. 12.4. Galvanomagnetic coefficients of n-InSb [12.7]. I — Compensation ratio = 0. II — Compensation ratio = 0.9. Solid line — Mobility. Dotted line — Hall ratio. Chain line — Magnetoresistance coefficient

mostly controlled by electron-hole scattering. The hole mass in indium antimonide being about 30 times the electron mass, the effect of this kind of scattering was calculated by the formula given for ionised impurity scattering.

The data for temperatures below 200 K, however, show a large amount of scattering when plotted against the electron concentration, indicating thereby that the impurity concentration is very different from the electron concentration. A few years ago experiments were carried out with purer samples and the impurity concentration in the samples was also characterised by using very low temperature Hall coefficient data [12.5].

We show in Fig.12.3 the experimental results on Hall mobility for different lattice temperatures for three samples as obtained by GERSHENZON et al. [12.6]. Mobility values were calculated by NAG and DUTTA [12.7] using the parameter values given in Table 12.1 and impurity concentration quoted in the paper. The program used for the calculation is given in Appendix B. It was found that the calculated values could be fitted well with the experimental data if the acoustic phonon deformation potential was taken to be 20 eV. The

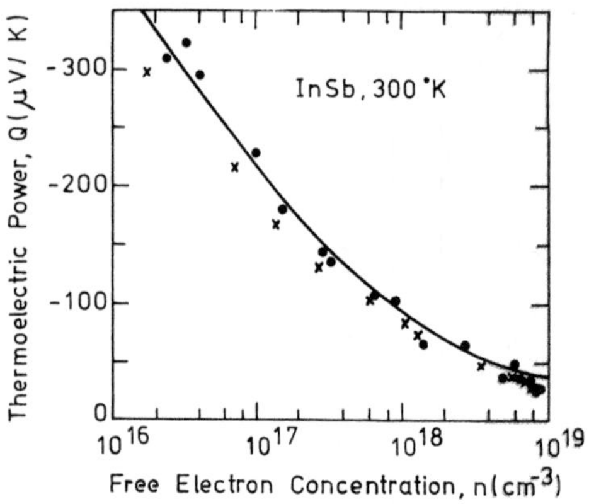

Fig. 12.5. Thermoelectric power of InSb for different electron concentrations [12.19]

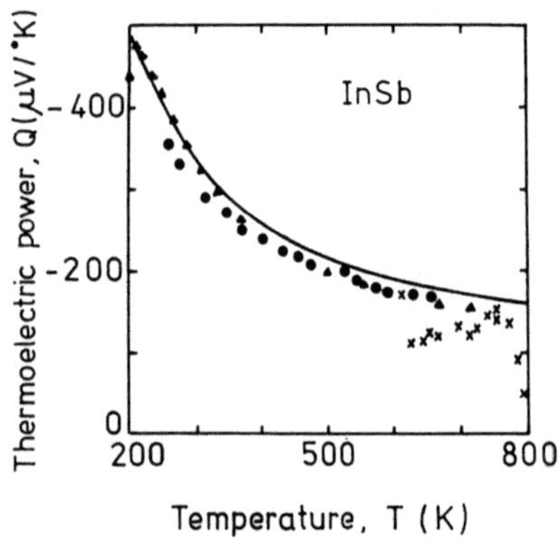

Fig. 12.6. Thermoelectric power of InSb at different temperatures [12.19]

calculated values are also given in Fig.12.3. It thus appears from these results that transport coefficients of indium antimonide may be calculated in agreement with the experiments with the parameter values of Table 12.1 and a value of E_1 equal to 20 eV.

There has been a long-standing controversy in the literature about the value of E_1. The present value is in agreement with theoretical calculation [12.8] and is close to the value required to give agreement with the available AC conductivity and magnetoresistance values [12.9].

We present in Fig.12.4 the calculated values of Hall mobility, Hall factor, and low-field magnetoresistance for different impurity concentrations together with the available experimental results. These curves should be useful as standard values for indium antimonide on the basis of currently available information about the scattering mechanisms.

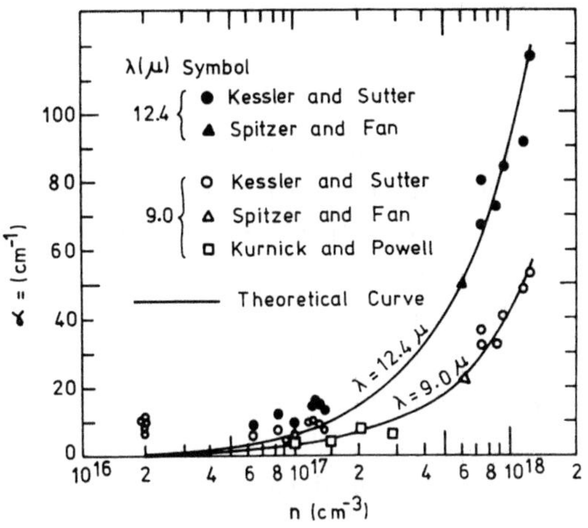

Fig. 12.7. Free carrier absorption in InSb [12.12]

Thermoelectric power for InSb has been studied experimentally by FILIPCHENKO and NASLEDOV, and by BARRIE and EDMOND [12.10]. Theoretical values were given by EHRENREICH [12.4] and RODE [12.3] and good agreement was obtained between theory and experiment. We show in Figs.12.5 and 12.6 the variation of thermoelectric power with electron concentration and temperature.

Infrared free-carrier-absorption data for InSb were obtained by KESSLER and SUTTER, and by SPITZER and FAN [12.11]. These results are shown in Fig. 12.7. HAGA and KIMURA [12.12] analysed these results treating E_1 as an unknown and assuming that the impurity concentration is the same as the electron concentration. The effective ionic charge for e* polar scattering was taken as $0.13|e|$, while the other constants were essentially as given in Table 12.1. It was found that experimental results could be explained by theory if E_1 is taken as 30 eV. The calculated curves of HAGA and KIMURA are also shown in Fig.12.7. This result was in conflict with the finding of EHRENREICH that E_1 should be taken as 7.2 eV and started the controversy about the value of E_1 in InSb which continued in the literature for the next fifteen years. However, the value of e* has been revised and a value of $0.16|e|$ appears to be more appropriate than $0.13|e|$. Also, since the experiments were conducted with samples having large impurity concentration which may have been compensated, the impurity concentration was likely to be larger than the electron concentration. The contribution of polar optic phonon scattering and impurity scattering, therefore, were underestimated in HAGA and KIMURA's work. We may expect that agreement with experiment may be obtained by using a value of 20 eV for E_1 as found from the analysis of recent mobility data. This point may be settled only by further analysis.

12.1.2 Gallium Arsenide

Gallium arsenide is the most important III-V compound semiconductor, being a strong commercial competitor to silicon. It is the material for a large number of microwave and optical devices and intensive work has been done to grow good quality crystals of it. Its properties have been studied extensively in the literature. It is, however, interesting to find that standardised values of the transport parameters of this material have been available only recently. A detailed analysis of the results available up to 1971 was made by RODE and KNIGHT [12.13]. But up to that time, most samples were characterised by stating the electron concentration only. We show in Fig.12.1 the reported values of electron mobility at 77 K available prior to 1971. It should be evident from the scatter in the results that no meaningful conclusion may be made by comparing theory with these data.

Experiments were, however, conducted in recent years with samples for which the impurity concentrations were also estimated from the low-temperature Hall coefficient data [12.14]. We present in Fig.12.8 results of these experiments. The mobility values obtained from different experiments above 100 K are found to be identical and may therefore be taken to represent the lattice-scattering-limited mobility. We also present in Figs.12.9,10 the mobility values at 100 K and 50 K for samples with different impurity concentrations. The results do not show much scatter and agree within the experimental error.

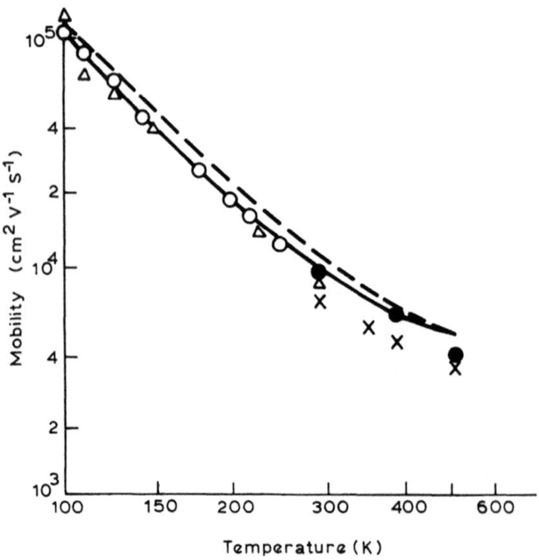

Fig. 12.8. Electron mobility in GaAs at temperatures of 100-500 K. Different symbols for different experiments. Solid line — Drift mobility; Dotted line — Hall mobility

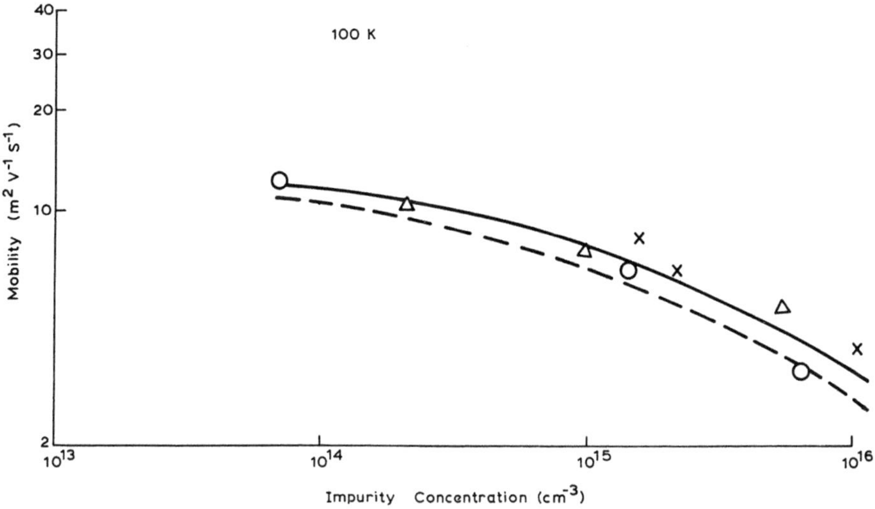

Fig. 12.9. Electron mobility in GaAs at 100 K for impurity concentrations of 10^{14}-10^{16} cm^{-3}. Different symbols for different experiments. Solid line – Hall mobility; dotted line – drift mobility

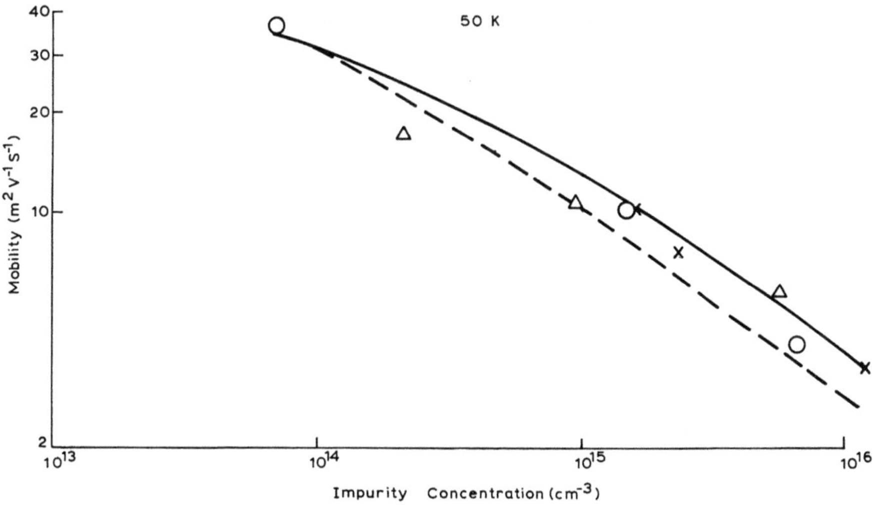

Fig. 12.10. Electron mobility in GaAs at 50 K for impurity concentrations of 10^{14}-10^{16} cm^{-3}. Different symbols for different experiments. Solid line – Hall mobility; dotted line – drift mobility

Theoretical values of Hall mobility, drift mobility, and Hall factor have been given by RODE and KNIGHT [12.13]. These results were found to agree very well with the experiments. However, the Hall mobility was apparently evaluated

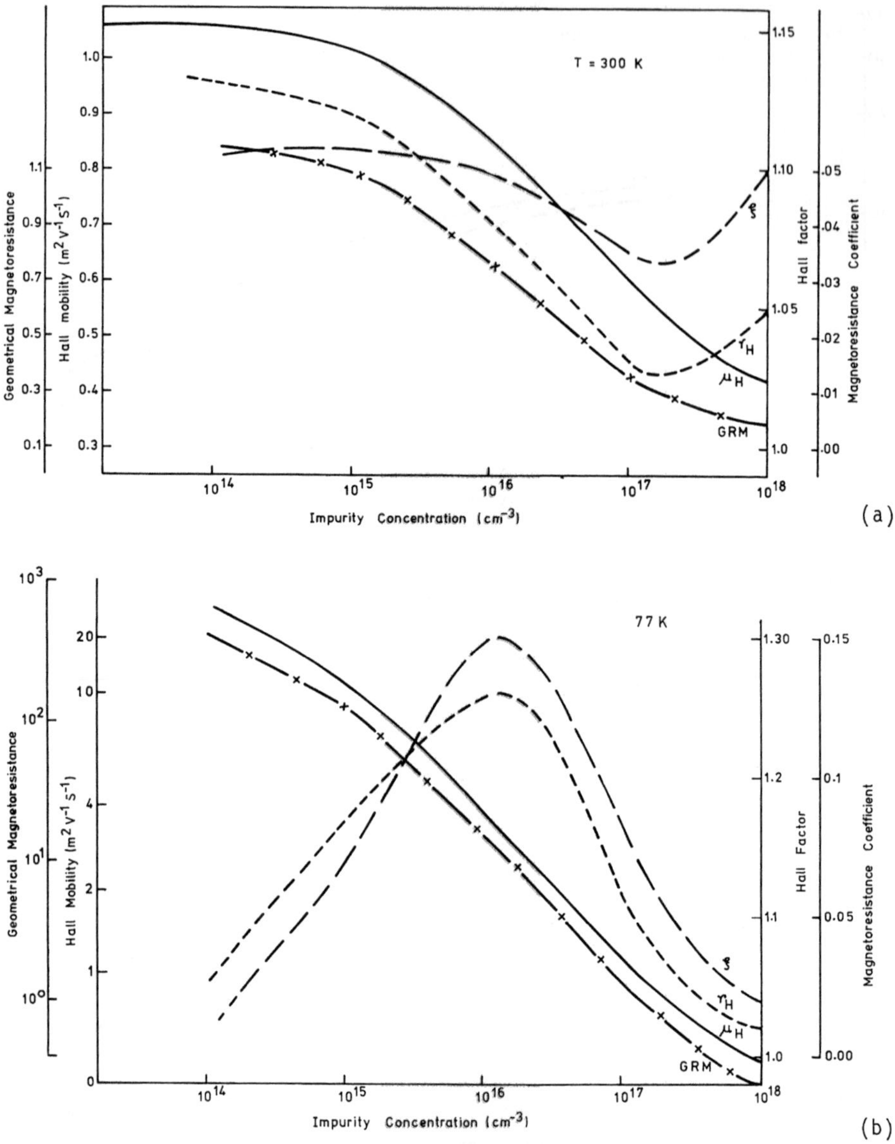

Fig. 12.11. Galvanomagnetic coefficients of GaAs (a) 300 K, (b) 77 K. μ_H — Hall mobility; r_H = Hall factor; ξ — magnetoresistance coefficient; GRM — geometrical magnetoresistance

by using a Hall factor obtained by using Matheissen's rule. The author calculated the coefficients by using the method discussed in Sect.8.3.5 and these values are presented in the same figures. The value of the acoustic phonon deformation potential constant E_1 was treated as an adjustable parameter and

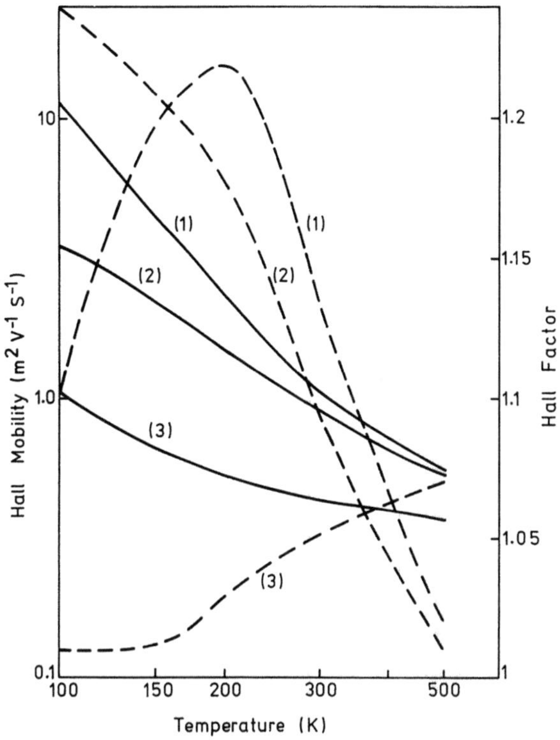

Fig. 12.12. Hall mobility and Hall factor in GaAs at temperatures of 100-500 K. Solid line – Hall mobility; dotted line – Hall factor. (1) N_i - 10^{18} cm^{-3} (2) N_i - 10^{16} cm^{-3}; (3) N_i - 0

a value of 12 eV was found to give the best fit. We have presented results for this value of E_1 only. As different values of magnetic field were used in the different experiments the experimental values really gave a value intermediate between the drift mobility and the zero-field Hall mobility. We have hence given results of both drift and Hall mobility. It may be concluded from the results of Figs.12.8-10 that the values calculated with the constants of Table 12.1 explain the experimental results within the expected errors. Results were also calculated for the drift mobility, Hall mobility, Hall factor, magnetoresistance coefficient, and geometrical magnetoresistance for different impurity concentrations assuming that the samples are uncompensated. These are given in Figs.12.11,12 and may be taken to represent the characteristic values for gallium arsenide. It should be mentioned that some results on magnetoresistance have been reported in the literature [12.15]. The theoretical results presented here are only in qualitative agreement with these results.

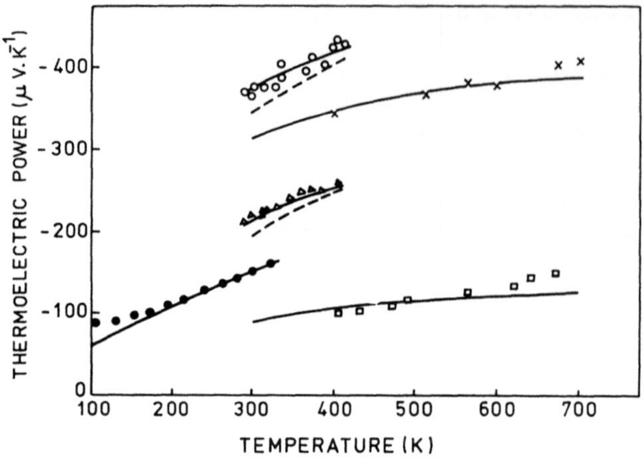

Fig. 12.13. Thermoelectric power of GaAs [12.16d]

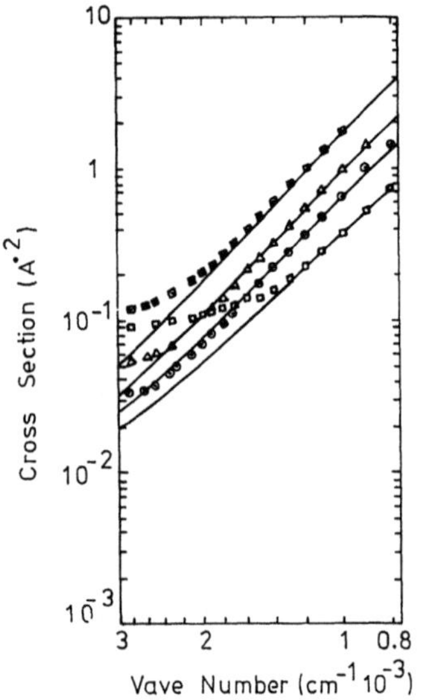

Fig. 12.14. Infrared free-carrier absorption in GaAs [12.17]. Different curves are for different carrier concentrations

Thermoelectric powers in GaAs were measured by EDMOND et al., AMITH et al., and more recently by EMELYANENKO et al. [12.16]. These are presented in Fig. 12.13. Theoretical values have been calculated by CHATTOPADHYAY and SUTRADHAR [12.16] using the method discussed in Sect.8.4.3. The theoretical values are shown in the same figure by solid lines and we find that there is very good agreement between experiment and theory. It may be mentioned in this connection that the Fermi energy term is of comparable magnitude to the scattering term and hence the agreement proves indirectly the validity of the assumed scattering constants as well as of the band structure.

Infrared free carrier absorption in GaAs has been studied by a number of workers and a thorough study was made by OSAMURA and MURAKANI [12.17]. We show in Fig.12.14 the experimental absorption cross sections. Theoretical values were obtained by the authors by assuming a parabolic band and values of K_s, K_α, and E_1 as 12.53, 11.10, and 6.3 eV. The impurity concentration being unknown the authors adjusted the values so as to fit the experimental values of the absorption constant at a wave number of 1000 cm^{-1} at room temperature. Theoretical values obtained by using the same impurity concentration for different wave numbers are also shown in Fig.12.14. The agreement is found to be good. However, in spite of this agreement it would be useful to examine the data by using the more recent values of K_s, K_α, and E_1 and a nonparabolic band structure, as nonparabolicity has been shown to affect the FCA signifi-

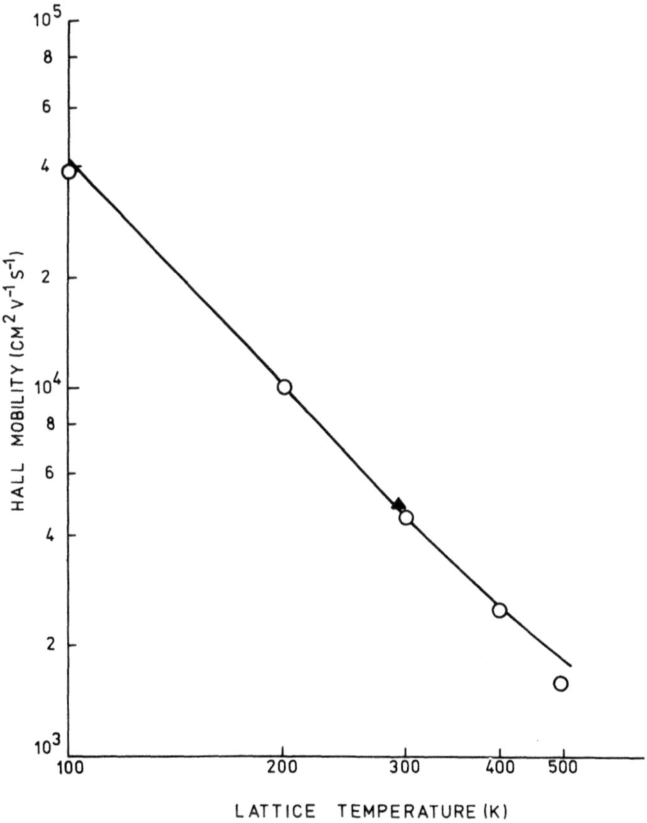

Fig. 12.15. Electron mobility in indium phosphide for temperatures of 100-500 K [12.21]

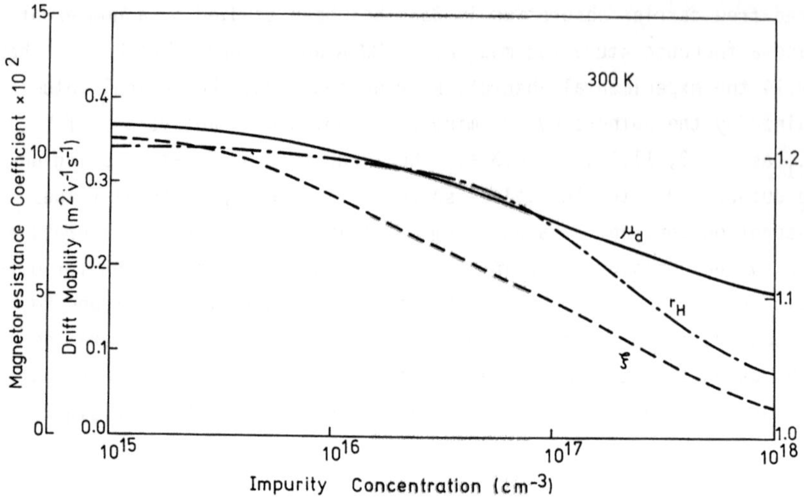

(a)

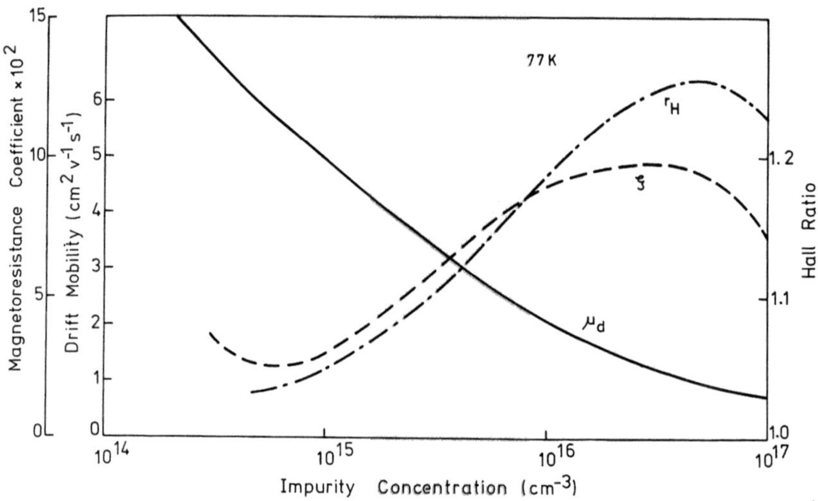

(b)

Fig. 12.16. Hall mobility, Hall factor, and magnetoresistance coefficient of indium phosphide at (a) 300 K (b) 77 K

cantly even in large band gap materials. It is also important to obtain FCA data for samples with known impurity concentration for a closer examination of the theory.

12.1.3 Indium Phosphide

Indium phosphide has assumed technological importance in recent years. It was suggested [12.18] that this material may provide Gunn diodes for frequencies higher than that obtained by using gallium arsenide. The peak velocity was also predicted to be higher so that it was expected to be useful for microwave FETs. Although these initial predictions have not been fully realised, the spate of research activities that followed has yielded good quality crystals and values of transport coefficients for pure materials. Experimental results available up to 1970 have been reviewed by RODE [12.19]. Theoretical values calculated by using the parameter values of Table 12.1 were found to give order of magnitude agreements with experiments. However, as results for very pure samples or for samples with known impurity concentration were not available no quantitative comparison was possible. In a more recent paper NAG and DUTTA [12.20] analysed the data obtained by POTTER [12.21] on relatively pure samples. The mobility for a number of samples studied by POTTER were found to give the same mobility above about 100 K, so that the mobility values could be taken to represent the characteristic values of pure samples. The results could be fitted very well by using the parameter values of Table 12.1 and a value of 18 eV for E_1. The calculated values together with the experimental results are reproduced in Fig.12.15. It was also observed that the same parameter values explain the recent experimental results of ARCHER and BARRERA [12.22] for lattice temperatures of 77 K and 300 K with reasonable values of the compensation ratio.

Some experimental results on the Hall factor and magnetoresistance [12.21] in InP are also available. But these are not sufficiently reliable for quantitative comparison with theory. We present in Fig.12.16 the theoretical values of the galvanomagnetic coefficients which may be of interest when more reliable experimental results are available.

Thermoelectric power in InP was studied by KESAMANLY et al. [12.23] and the results were found to be in good agreement with theory by RODE [12.19].

Free carrier absorption in InP was studied by DUMKE et al. [12.24] for a temperature of 77 K. We present the experimental results in Fig.12.17. The samples were characterised by stating the electron concentration. We find on calculating the coefficients that the experimental results are much higher than the theoretical values if the samples are assumed to be uncompensated. If, however, we assume that the samples are compensated then agreement between experiment and theory may be obtained for each sample by suitably choosing the values of compensation ratio. In Fig.12.17, we show the calculated values with impurity concentrations chosen to fit the experimental result for a particular wavelength. The agreement observed for other wavelengths with the same value of impurity concentration leads us to conclude that FCA in InP can also be explained by the same scattering model which explains the DC transport coefficients.

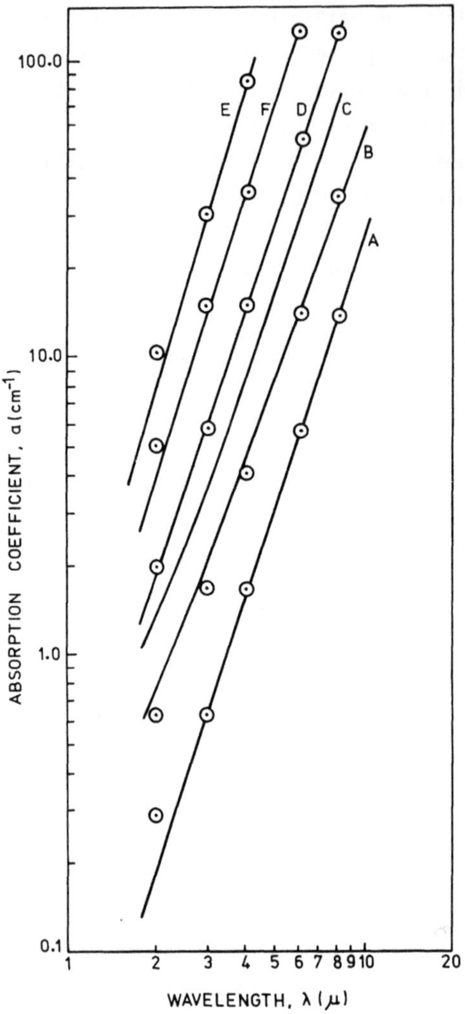

Fig. 12.17. FCA in indium phosphide [12.24]. Solid line — experiment; points — theory. Carrier concentration ($\times 10^{-18}$ cm^{-3}) - A - 0.35, B - 0.71, D - 2.1, E - 5.2, F - 9.6. Assumed impurity concentration ($\times 10^{-18}$ cm^{-3}) - A - 2.6, B - 4.1, D - 6.9, E - 7.8, F - 10.5

12.1.4 Indium Arsenide

Indium arsenide is a direct-gap compound like the three compounds the properties of which have been discussed in the preceding sections. It has been used in galvanomagnetic devices, but otherwise the material is not of much commercial importance. In spite of this the material has received attention in the literature, being similar in characteristics to the other direct-gap III-V compounds.

Electron mobility in indium arsenide has been studied by several workers over the temperature range between 50 K and 900 K [12.25]. The scatter in the values of mobility below 400 K indicate that the samples used by the different workers had different impurity contents. Since the impurity concentrations were not determined, a direct comparison between theory and experiment is not possible. The data above 400 K apparently represent the characteristic values for the intrinsic material. These results were found to be in good agreement with the theoretical values of RODE obtained by using the parameter values of Table 12.1. The thermoelectric power data of [12.26] may also be explained by the same model. Free carrier absorption in InAs was measured by CULPEPPER and DIXON [12.27]. These results were first examined by HAGA and KIMURA [12.28] and general agreement was obtained for most of the samples (assumed to be uncompensated) using nearly the same values of parameters as given in Table 12.1, but a value of 10 eV for E_1. Similar agreement was also obtained by JENSEN [12.29].

12.1.5 Indirect-Band-Gap III-V Compounds

The III-V compounds other than those discussed in the preceding sections have the conduction band minima at the X or L point. The constant-energy surfaces are therefore ellipsoidal and we are required to consider intervalley phonon scattering in addition to the scattering mechanisms taken into account for the direct-band-gap materials. Although experimental values are available for the conductivity effective mass, values of the components may only be guessed by assuming values of K from the knowledge of its value for germanium or silicon. The intervalley scattering being also important we are required to know its deformation potential D_i in addition to that for the intravalley scattering E_1. For the analysis of the transport coefficients of direct-band-gap compounds, E_1 is the only unknown which may be estimated (as in the preceding sections) from the analysis of data for pure samples. For the indirect-band-gap materials we find, on the other hand, that we have three unknowns E_1, K, D_i and no unique solution may be found from the analysis of transport coefficients. Good quality crystals of these materials being also relatively scarce, very few detailed studies of the transport coefficients of indirect-gap materials are reported in the literature. Gallium phosphide is the only material which has been studied in some details, apparently because it is an important commercial material for application to electroluminescent devices. We discuss here briefly the results obtained from this analysis. Data for GaP have been collected by several workers [12.30]. In the experiments of CASEY Jr. et al., the electron and acceptor concentrations were also determined from Hall measurements. RODE [12.31] obtained a fit with these data by choosing the value of E_1 as 13 eV, and that of D_i as 1.2×10^9 eV/cm. The mobility values calculated by using the same constant for pure materials were also found to be in good agreement with the experimental results for pure samples. Electron mobility in GaP has also been analysed by FLETCHER and BUTCHER [12.32]. They considered the experimental results of TAYLOR et al. [12.30] and assumed for K, E_1, and D_i the values of 5, 7 eV, and 1×10^9 eV/cm. The theoretical values were found to be larger than those obtained in experiments. However, fairly close agreement could be obtained by including the effect of space-charge scattering.

In spite of these attempts to explain the experimental results, it appears that an explanation of the observations requires further experiments with purer and good quality crystals. Experiments for obtaining independently the values of K, E_1, and D_i must also be carried out.

Thermoelectric power or Seeback coefficients have been measured for GaSb [12.33] and AlSb [12.34] but no detailed theoretical analysis are reported for these data. Experimental results on thermoelectric power for other compounds are scarcely reported in the literature.

Free-carrier-absorption coefficients have been studied for GaP [12.35] and AlSb [12.36]. In both the cases it is found that the experimental results may be fitted by taking E_1 around 10 eV and D_i around 1×10^9 eV/cm. It was also shown by DAS and NAG [12.37] that even though the band gap in these materials is large, the effects of band nonparabolicity have, in general, an important effect on the absolute values of the absorption coefficients and the index of wavelength dependence.

12.2 II-VI Compounds

The combinations of the group II elements Zn, Cd, and Hg with the group VI elements O, S, Se, and Te give twelve compounds which occur freely in nature. These compounds have been studied extensively and some of them have been used as phosphors and electroluminescent materials. Good quality single crystals have also been produced after the development of the techniques of artificial crystal growth and new applications have been found for these materials. Zinc oxide and cadmium sulphide, being strongly piezoelectric, are widely used in acousto-electric devices. The compounds have often very large band gap in the range between 1.8 and 4 eV, and have been used to construct lasers for different optical wavelengths. The mercury compounds are, on the other hand, semimetals and are therefore used to grow mixed compounds with controllable band gap, which are useful for infrared detection. Because of these various applications there has always been considerable interest in the II-VI compounds. We find, however, that the interest has been mostly confined to the study of Hall mobility and that also in as-grown materials with unknown impurity content. A comparison between theory and experiment for the various transport coefficients is not therefore meaningful for these materials. A few detailed studies are, however, reported in the recent literature and it is worthwhile to note the main conclusions of these studies. We discuss them below, classifying the compounds into three groups on the basis of the similarity of the band structure.

12.2.1 Cubic Compounds of Zinc and Cadmium

The II-VI compounds may have either the sphalerite or the wurtzite crystal structure or allotropic modifications. ZnS, ZnSe, ZnTe, and CdTe are obtainable in the zinc-blende form at room temperature. Electron mobility has been studied in all these materials. We note that the band gaps (see Tables 3.1,2) for these compounds are large, and nonparabolic effects are not therefore significant. The Γ-point minimum being lowest and the crystal structure being cubic, the effective mass and as a result the electron mobility is isotropic. The theory presented in Sect.8.1.3 has been used by RODE [12.38] to calculate

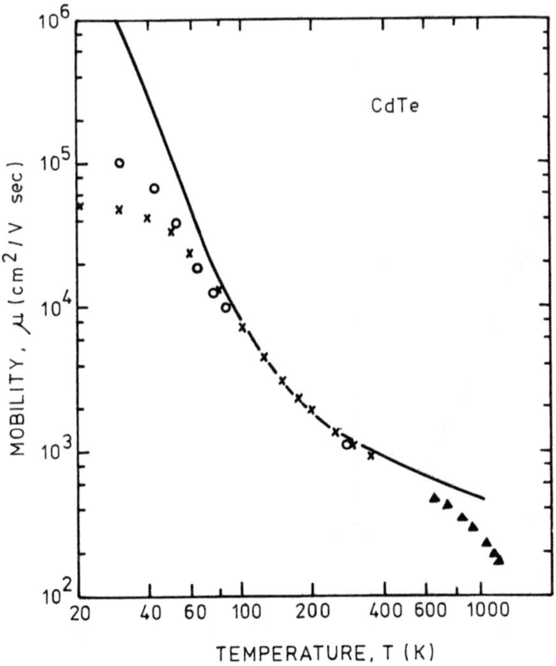

Fig. 12.18. Electron mobility in cadmium telluride. Solid curve—Theory. Points — Experiment. [12.31]

the electron mobility in all these materials and fairly good agreement was observed for the temperatures between 100 and 600 K. As an illustration we reproduce in Fig.12.18 the results for CdTe.

There was no agreement at lower or higher temperatures. The discrepancies were attributed for low temperatures to ionised impurity scattering and for higher temperatures, to multivalley effects. Evidently for making a definite conclusion more detailed calculations and experiments on samples with known impurity content are required.

12.2.2 Wurtzite Compounds of Zinc and Cadmium

Zinc oxide, cadmium sulphide, and cadmium selenide are commonly obtained in wurtzite form. These materials also have a large direct band gap so that the effects of band nonparabolicity may be neglected. However, even the Γ-point valley is anisotropic due to asymmetry in the distribution of atomic potentials. The anisotropy in the effective mass and in some of the coupling constants for the scattering mechanisms (e.g. the piezoelectric tensor) is expected to make the electron mobility anisotropic, which is a significant deviation from the characteristics of cubic materials. The anisotropy in the mobility has been found theoretically as well as experimentally. Detailed comparison between theory and experiment has also been made by RODE [12.38],

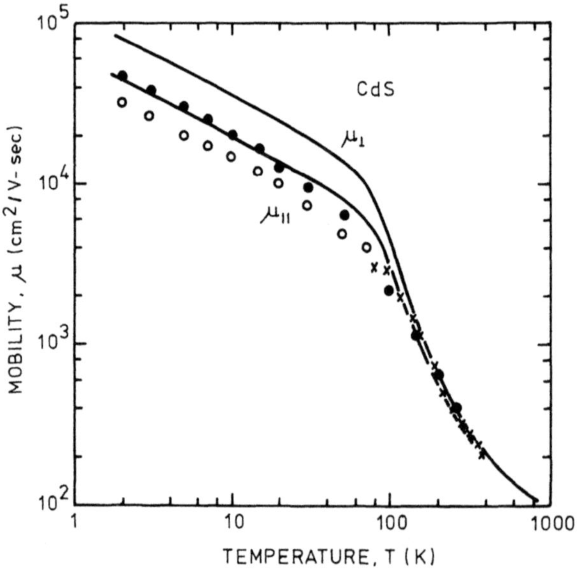

Fig. 12.19. Electron mobility in cadmium sulphide. Solid curve – Theory. Points – Experiment. [12.31]

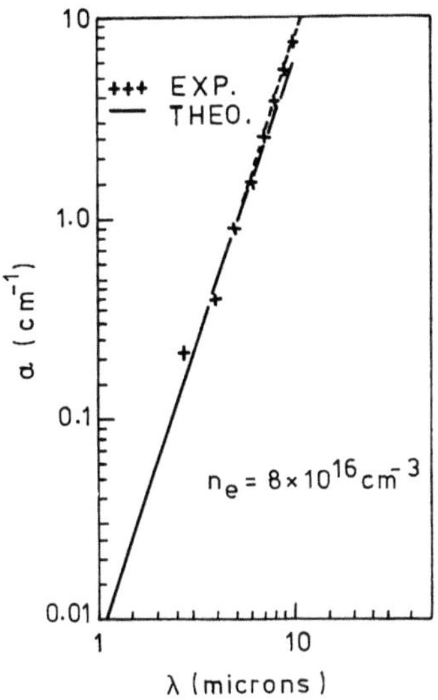

Fig. 12.20. Free-carrier absorption in cadmium telluride [12.40]

and close agreement was observed in the intermediate range of temperatures. Ionised impurity and multivalley scattering apparently account for the discrepancies at other temperatures. We present in Fig.12.19 the results for CdS to illustrate the nature of agreement.

12.2.3 Mercury Compounds

Mercury compounds are all semimetals and the band structures are inverted, i.e. the electrons in the conduction band minimum have p-like wave functions. The overlap integrals are therefore altered and are similar to those given by (6.45). The Kane dispersion relation is supposed to be applicable so that the coefficients for these materials may also be calculated by the theory of Sect.8.3. Such calculations have been carried out by RODE [12.38] and good agreement was found for HgTe at all temperatures, and for HgSe, below 200 K. The mobility is experimentally found to fall very rapidly above 200 K, apparently due to nonequivalent intervalley scattering which was not taken into account in the theory.

As mentioned earlier, other transport coefficients of II-VI compounds have not yet been studied in details either theoretically or experimentally. Only free carrier absorption in CdTe [12.39] has received some attention. Although some conflicting conclusions have been drawn from the results, the recent study of JENSEN [12.40] indicates that the absorption may be explained as being predominantly due to scattering by polar optic phonons. The results for an electron concentration of 8×10^{16} cm^{-3} are shown in Fig.12.20, indicating the nature of agreement observed between theory and experiment.

12.3 IV-VI Compounds

The three IV-VI compounds, lead sulphide, lead selenide, and lead telluride are available freely in nature. These materials have been used extensively for different purposes. Lead sulphide in the naturally occurring form of galena crystals was used in detectors of early radios. In later years the materials have been used in thermoelectric generators and infrared imaging systems. The lead chalcogenides have also some distinctive properties: the static permittivity is very large; the valence band maxima are almost the mirror images of the conduction band minima, and lie at the L points; the forbidden bandwidth has a positive temperature coefficient and the effective mass of the carriers increases with temperature. Due to the high value of static permittivity, ionised impurity scattering is weak in these materials and the impurity levels are almost merged with the conduction band so that the free carrier concentration does not vary much with the lattice temperature. These distinctive properties make the study of these materials interesting, and all the various transport properties have been studied in details.

Detailed study of electron mobility in all the three compounds was made by ALLGAIER and SCANLON [12.41]. We present in Fig.12.21 the experimental characteristics obtained by these authors. The most important feature of these characteristics is that at temperatures above about 100 K the electron mobilities for different samples agree to within a few per cent and vary with lattice temperature as T^{-n}, and n is larger than 1.5. As the mobility values for different samples are nearly the same, the contribution of impurity scattering is, as expected, negligible. In the initial theoretical studies, acoustic phonon scattering was considered to be the only important scattering mechanism. The value of the index n for this scattering for temperature-independent effective mass and parabolic bands is 1.5 (see Sect.8.2). The deviation in the value of the index from 1.5 has been explained as being due to the variation of the effective mass with temperature assuming that the temperature coefficient for the effective mass is equal to that for the band gap. The required value of E_1 for a quantitative fit with experiments lies in the range of 25-30 eV [12.42]. However, the chalcogenides being polar materials, polar optic phonon scattering is expected to be important. It has been pointed out by RAVICH et al. [12.43] that at a temperature of 77 K, the

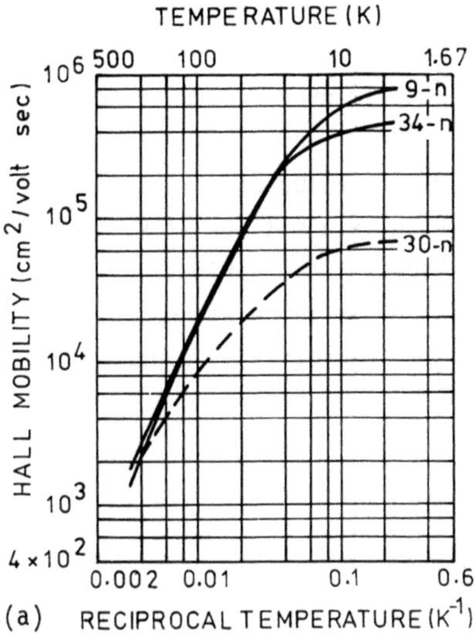

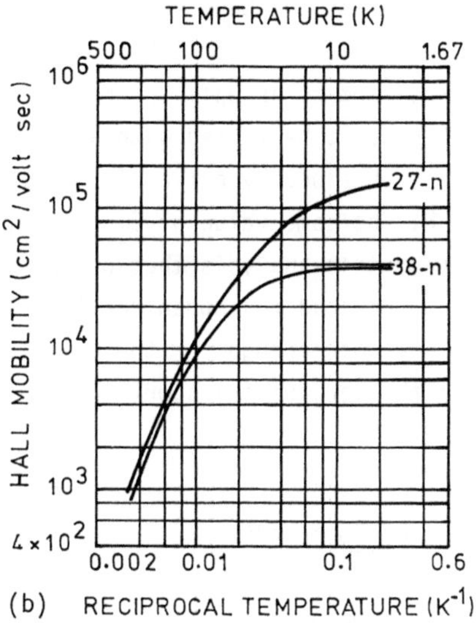

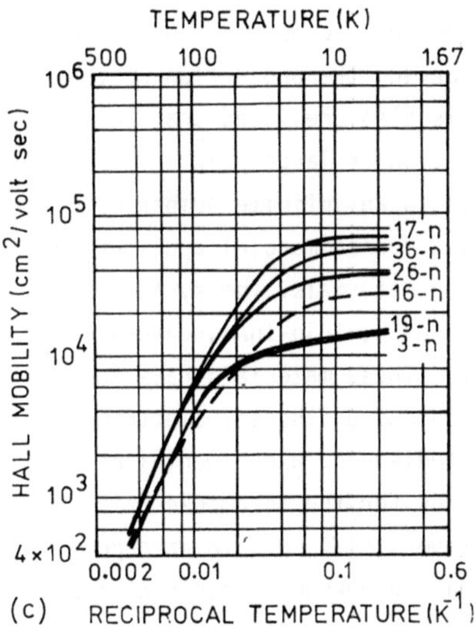

Fig. 12.21. Electron mobility in chalcogenides [12.41]. (a) PbTe; (b) PbSe; (c) PbS. Different curves are for different samples

contribution of polar optic phonon scattering may be comparable to that of acoustic phonon scattering. Detailed quantitative analysis of electron mobility in all the lead chalcogenides taking into account the effects of the non-parabolic anisotropic band structure, the complex wave functions corresponding to the L-band minima, and polar optic phonon scattering are not yet reported. A detailed analysis of electron mobility in lead telluride only has been recently carried out by DAS and NAG [12.44]. In this analysis the complexities of the band structure were taken into account and the iterative method presented in Sect.8.2 was used to take into account the effect of mass anisotropy and polar optic phonon scattering. Experimental values of mobility for 300 K could be fitted with a value of 15 eV for E_1 and a value for effective mass obtained by extrapolating the experimental value of effective mass for 77 K. The same value of E_1 did not, however, give a fit for 77 K. It appears from this analysis that our present knowledge of the various physical constants of lead chalcogenides (likely to be strongly temperature dependent) is not accurate enough for the calculation of the theoretically expected values of the various transport coefficients.

12.4 Mixed Compounds

Two III-V compounds or II-VI compounds may be mixed while growing a crystal by the zone melting technique or the vapour phase technique. These mixed compounds behave as semiconductors like the component compounds. One important property of these materials is that by varying the proportion of the components, the band gap may be adjusted. It is also predicted that an optimum composition may be found for the realisation of large peak drift velocity and a large NDR. In principle, we may obtain an infinite number of such compounds, but only some of them are attractive for commercial applications and studies have been concentrated only on such materials. The systems studied extensively are InAs-GaAs, InAs-InP, InSb-GaSb, InSb-InAs, GaAs-GaP, GaSb-GaAs, GaSb-AlSb, HgTe-CdTe, and PbTe-SuTe. These materials have found applications in lasers, infrared detectors, and photocathodes. Electron transport in these mixed compounds has been studied and experimental results are available for the mobility and in some cases for the thermoelectric coefficients. The experimental results are, however, difficult to interpret. The samples had microinhomogeneities and contained unknown amounts of impurity. In spite of these difficulties attempts have been made to explain the observed mobilities by theory. Extensive calculations for III-V systems have been reported by MAKOWSKI and GLICKSMAN [12.45], and by HARRISON and HAUSER [12.46]. In these calculations, the relaxation-time formalism was used to calculate the effects of all scattering mechanisms including even polar optic phonon scattering. The values of the different physical constants were obtained by interpolation using the known values for the constituent compounds. The impurity concentration was assumed to be the same as the electron concentration. An additional scattering mechanism, the so-called alloy scattering,

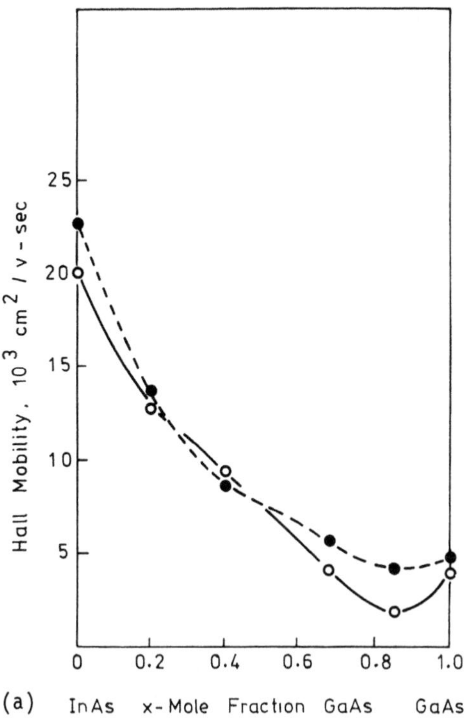

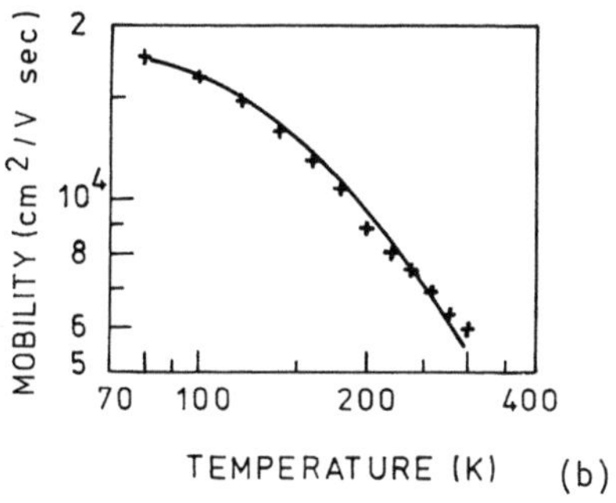

Fig. 12.22. Electron mobility in $In_xGa_{1-x}As$. (a) [12.46] Dotted line — theory. (b) [12.48] Solid line — theory

was included and its effect was evaluated by using the formula of [12.47] given in Table 6.2. We present in Fig.12.22 the results for InAs-GaAs for which the agreement between theory and experiment appears to be most satisfactory. For most of the alloys, however, there are large discrepancies between the theoretical values and the experimentally observed mobilities. The difference cannot always be accounted for by alloy scattering. It appears at this stage that the variation of mobility with composition is understood for the InAs-InP and InAs-GaAs system [12.48]. But, for other systems, apparently materials are not yet available with sufficient purity and homogeneity to make the comparison between theory and experiment meaningful. There have been some attempts to find whether alloy scattering is important for a particular compound. The conclusions were mostly drawn by calculating the alloy-scattering-limited mobility and comparing it with the observed mobility [12.45]. A comparatively high value of the former was taken to indicate its insignificance. However, a good theoretical estimate of the alloy-scattering-limited mobility is difficult to make. There is some uncertainty about the scattering potential to be associated with this scattering. In some analyses it was taken to be given by the difference of the band gaps [12.45], while in others it was taken as the difference of electron affinities [12.46]. Further, it has recently been suggested [12.49] that this potential should be obtained by applying the electronegativity theory of PHILLIPS [12.50]. The value of the potential obtained from these different approaches do not agree and as a result the estimated magnitude of the alloy-scattering-limited mobility turns out to be quite different. A more detailed analysis of the electron mobility in mixed III-V compounds would therefore be possible only when the role of alloy scattering is properly ascertained.

Electron transport in $Hg_{1-x}Cd_xTe$ has also been analysed in some details. The results are reviewed by DORNHAUS and NIMITZ [12.51]. The experimental values of mobility [12.52] cannot be explained, particularly for low values of x, by considering only lattice scattering and ionised impurity scattering. The inclusion of alloy scattering improves agreement only for high values of x. $Hg_{1-x}Cd_xTe$ shows two-mode behaviour [12.53] and in the estimation of mobility, the two-phonon characteristics of HgTe and CdTe have to be included. A detailed analysis, using an iteration method to take account of these two-mode phonons, was carried out by CHATTOPADHYAY and NAG [12.54]. It was found that the theoretical results could not generally be fitted with theory. It was, however, shown that inclusion of a transverse phonon, interaction with which came to light in some experiments [12.55], may explain the discrepancy between theory and experiment. However, the values of the constants used in the calculation were obtained by interpolation, as directly measured values were not available. Since the applicability of these values are not fully established, a definite conclusion cannot be made at this stage.

Experimental data are available for a few other mixed compounds, but no detailed analysis is reported in the literature.

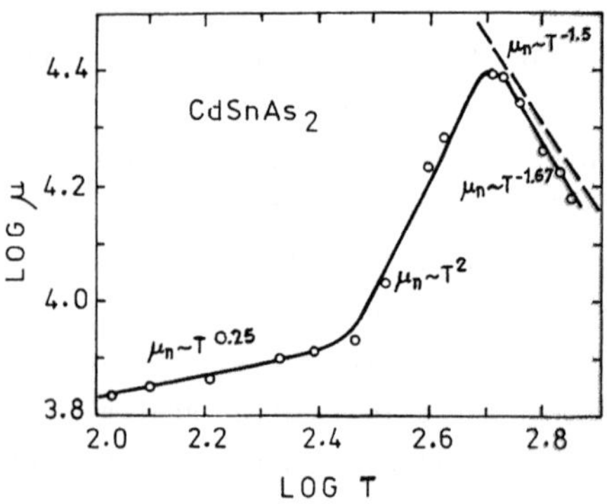

Fig. 12.23. Electron mobility in $CdSnAs_2$ [12.57]

12.5 Chalcopyrites

The chalcopyrites are birefringent and optically nonlinear with large values of the nonlinearity coefficients. Useful applications of these materials for mixing and harmonic generation at optical wavelengths were therefore envisaged and intensive work was carried out in the last decade for growing good quality crystals in two groups, II-IV-V_2 and I-III-VI_2, of chalcopyrites. As crystals became available, electron transport properties were also studied. These studies were undertaken to characterise the materials, and also to explore the possibility of using them for making devices. However, the technology of growth of these crystals is not yet perfected and good quality single crystals are not easily produced. Transport studies have therefore been mostly confined to the measurement of resistivity and thermoelectric power. The available results are reviewed in detail by SHAY and WERNICK [12.56]. Many of the chalcopyrites may be grown only with p-type conductivity. Electron mobility data are, however, available for some of these materials. We present in Fig.12.23 the observed electron mobility [12.57] in $CdSnAs_2$ to illustrate the characteristic features. It may be seen that the variation of mobility with temperature in these materials is qualitatively different from that in other semiconductors. The number of defects is usually so large that the characteristic behaviour of semiconductors is not exhibited even at temperatures of 500 K. The observed values of electron mobility in most of the compounds is between 10 and 100 $cm^2V^{-1}s^{-1}$, and the low value is likely to be due to crystal defects. Application of the theory to explain the observed results has therefore been rarely attempted. We may expect that with further improvement in the techniques of growth of these crystals, experimental results will be available for meaningful comparison with theory. The characteristics of electron transport in these materials may, however, be expected to show new interesting features as the crystal structure is very different from the other well-studied semiconductors.

12.6 Conclusion

We have reviewed in this chapter the available results on the transport coefficients of some of the compound semiconductors. We find that experimental results are not available on all the transport coefficients for many compounds. The absence of data is partly due to the difficulties in the growth of good quality crystals, but mostly due to lack of commercial interest in the materials. A close comparison between theory and experiment is rendered difficult even for compounds for which experimental data are available by our lack of knowledge about the impurity content of the samples. It is also found that in many of the compounds the contribution of deformation potential acoustic phonon and intervalley phonon scattering is comparable to that of other scattering mechanisms. Since the deformation potential constants related to these scattering mechanisms are not accurately known, it is difficult to arrive at a definite conclusion about the applicability of a particular scattering model. However, for some compounds, e.g. InSb, GaAs, and InP, it is possible to choose a model which explains all the available results. In such cases, the theory may be usefully employed to predict the transport coefficients for all sorts of experimental conditions. For other materials, on the other hand, our understanding is only qualitative at this stage and further studies are required before quantitative predictions can be made.

13. Conclusions

The phenomena associated with the transport of electrons in compound semiconductors have been described in this book. The theory currently used for explaining these phenomena has also been given. Electrons are taken in this theory as particles with an energy-dependent effective mass, determined by the atomic potentials in the crystal. The low-energy value of the effective mass is treated as an experimentally determined parameter. It is obtained from measurements on transport phenomena exhibiting resonance characteristics: cyclotron resonance and magnetophonon resonance. The energy dependence of the effective mass is incorporated in the theory by using the Kane dispersion relation which is derived from the $\underline{k}\cdot\underline{p}$ perturbation analysis. According to this relation, the variation is linear in energy and the proportionality constant is approximately the inverse of the energy band gap, which is also an experimentally determined constant. The effective mass, in general, varies with the lattice temperature. The temperature coefficient is taken to be the same as that of the band gap but for some materials experimental values, determined from magnetophonon resonance experiments, are also available [13.1].

Electron assembly in the crystal is assumed to behave statistically like the gas particles and the statistical nature of electron transport is taken into account in the theory by using the Boltzmann transport equation for the evaluation of the coefficients of transport. Electron-electron collisions are usually neglected so that the equation may be reduced to a linear integral equation. Expressions for scattering probabilities of the various kinds of scattering are derived by using perturbation theory and these involve the physical constants: mass density, elasticity tensor or the sound velocity, dielectric constant, piezoelectric coefficient, phonon frequencies, deformation potential constants, and impurity density. All these constants except the deformation potential constants and impurity density may be obtained from non-transport experiments. The values of deformation potential constants have been estimated from the pressure dependence of the band gap [13.2]. These may also be obtained from theoretical band structure calculations [13.3]. But, for most of the materials the constants, particularly the intervalley deformation potential constants, are determined from the analysis of transport results. The impurity concentration is also determined from the analysis of the variation of Hall coefficient with temperature at very low temperatures.

The physical constants involved in the expressions for scattering probabilities are more or less temperature dependent [13.4]. The temperature coefficients are not, however, strong enough to alter the values of the transport coefficients significantly and the constants are assumed to be temperature independent in the analyses reported in the literature.

The theory has been applied to calculate the values of the various transport coefficients for many semiconductors. For some materials, calculations may be done by using analytic solutions of the Boltzmann equation, but for most compound semiconductors, analytic solutions are not obtainable. Numerical methods have been presented for obtaining the solutions for such materials. Calculated results have been compared with the available experimental results in the preceding chapter. It was seen that the distinctive features of the various transport coefficients are explained very well by the theory. The nature of variation of the coefficients with temperature and with the magnetic induction are given correctly by theory for all compounds. A quantitative comparison between theory and experiment is, however, not possible as experimental data for well-characterised samples are not available for many materials. But, wherever such data are available, theoretical results are found to agree with experiment within a few per cent. Only the acoustic phonon deformation potential constant, which cannot yet be determined independently, has to be adjusted. But it may be chosen so as to give an all round fit to the available results on the different transport coefficients for the important compounds. We may conclude from this general agreement between theory and experiment that electron transport in compound semiconductors is well explained by the current theory. The confidence in theory is at present such that expected values of the coefficients are being predicted from theory. These values are often used to estimate the impurity density and to assess the purity of as-grown crystals [13.5]. The theory is also being applied to judge the suitability of a material for the construction of devices. New aspects of transport are being explored and interpreted by this theory. Data are also being collected for new experimental conditions for checking the theory. We discuss some of these current studies in the following section.

13.1 Problems of Current Interest

Studies on electron transport were confined in the past mainly to homogeneous and lightly doped bulk materials. Currently, on the other hand, electron transport is being studied under conditions for which the material cannot be treated as homogeneous or in some cases as three dimensional. In this section some of these current topics are reviewed and the related processes are indicated.

13.1.1 Heavily Doped Materials

Semiconducting compounds used for the construction of lasers, FETs and similar devices are usually doped in the range of 10^{19} cm^{-3}. Transport properties of materials with such doping are therefore of great technical interest.

It was shown in Sect.4.5.1 that heavy doping distorts the band structure radically and causes the so-called band tailing. The transport coefficients of heavily doped materials may be expected to exhibit some effects of the band distortion. But, not effect of band tailing can be definitely isolated for the transport coefficients. Electron mobility has been measured for impurity concentration up to 10^{19} cm^{-3} and the results may be explained by the Brooks-Herring theory of impurity scattering provided the effects of degeneracy are properly taken into account [13.6]. A closer fit between theory and experiment, however, appears to require some modifications in the theory to account for central cell corrections [13.7], impurity charge correlations [13.8], electron-electron scattering, etc. It is also evident that the impurity atoms cannot be considered completely isolated. For large impurity densities, the distance between the impurity atoms may be so small that electrons will be under the influence of the field of more than one impurity atom when it is being scattered [13.6]. Refinements are required in the expressions for scattering probabilities from all these considerations. But the essentials of the theory of transport presented in this volume remain valid, and band distortions need not be considered for impurity concentrations even up to 10^{19} cm^{-3}. For larger impurity concentrations also the above statement apparently remains valid. The conduction becomes metallic and the coefficients are then weakly temperature dependent. The relative insensitivity of the transport coefficients to the band distortion appears to be due to the fact that the contribution of the electrons occupying the distorted part of the band is very small compared to the contribution of electrons occupying the higher undistorted regions of the conduction band [13.9].

The effect of band distortion, however, becomes important in heavily doped but compensated materials. The electron concentration being small for compensated conditions only the conduction band edge and the impurity levels are populated by the electrons and the transport coefficients are as a result significantly determined by the properties of this region of the band. The effects are manifested significantly at low temperatures when the electrons are confined to these levels. The resistivity for these conditions is found to decrease exponentially with the inverse of the temperature. The activation energy and the exact law of temperature dependence are intimately related to the nature of band distortion.

A successful theory for explaining the observed characteristics of electron transport in heavily compensated materials is based on the assumption that the electrons are localised at the impurity centres and that the electrons hop — from one impurity state to another [13.10]. The characteristics of this so-called hopping conduction have been studied in the recent past in elemental semiconductors. The studies are being extended to compound semiconductors. The phenomenon is not yet well understood and results are not available for transport coefficients other than conductivity. This aspect of conduction in heavily doped materials is likely to be of interest in the near future.

13.1.2 Alloy Semiconductors

The need for the development of semiconductors with controllable energy band gaps arose from the requirement of LEDs giving out light of a desired colour. Alloy semiconductors, the solid solutions of similar compounds, were developed to meet this need. Initial studies on these materials were, therefore, mostly limited to the optical properties and detailed experimental results are available about the energy-band-structure parameters of these materials.

In recent years applications of these materials have been suggested in Gunn diodes, FETs, and superlattices. Practical realisation of the ideas requires a thorough knowledge of transport properties, and attempts are being made to explain the observed electron mobilities in alloy semiconductors. A review of the available results was given in Sect.12.4. The important conclusion of the work reviewed is that for many mixed compounds an additional scattering mechanism, the so-called alloy scattering [13.11], plays a very significant role. The nature of this scattering, the perturbing potentials to be associated, and the importance of the distribution of disorder are not yet well understood. The characteristics of the lattice vibrations are also not very clear. The lattice vibrations may be characterised by single phonons determined by the alloy composition and localised, or they may be a combination of the phonons characteristic of the two compounds forming the alloy [13.12]. Studies are in progress for the clarification of these new aspects of scattering.

Ternary compounds, i.e. alloys of two compounds with a common element, were the materials of interest in the past. But, in recent years, interest has grown in quaternary compounds, e.g. $Ga_x\, In_{1-x}\, As_y\, P_{1-y}$ [13.13], as for these materials, the band gap as well as the lattice spacing may be individually controlled by controlling the composition. It is therefore possible to grow unstrained heterostructures with a desired band gap. These materials are being studied for the development of laser diodes with a radiation output in the 1 μm wavelength range so that the window of low attenuation in the optical fibres may be utilised for optical communication. Transport properties of such alloys are being currently studied. Electron scattering mechanisms, the physical constants, the deformation potentials, and effective mass are not accurately known for these materials. Theoretical calculations reported in the literature are based on good guesses. Work is in progress to obtain accurate information about all these parameters. This line of work is likely to be further intensified with the availability of new materials.

13.1.3 Transport Under Magnetically Quantised Conditions

A strong magnetic field quantises the component of electron energy associated with its transverse motion. We reviewed in Chap.10 the phenomena resulting from this quantisation. It was shown that oscillations may be observed in all the transport coefficients due to the periodicity in the density of states. Information may be obtained about the effective mass and phonon spectrum by studying these oscillations. The phenomena have been intensively studied in compound semiconductors as the value of effective mass being low, it is easy to produce the quantised conditions in these materials. Although the phenomena are generally understood, a detailed quantitative theory is not yet available. Work for the development of such theory is in progress.

More current interest in transport in quantising fields is related to the effects of a high electric field. New techniques have recently been developed, e.g. Fourier transform spectroscopy [13.14], field-modulation spectroscopy [13.15], and third-harmonic detection [13.16], which enable very accurate identification of the resonance peaks. Experimental results are becoming available through the application of these techniques about the phonon spectrum including the very weak phonons, the electron temperature, electron distribution function, impurity levels, and the effective mass of electrons in high-lying Landau levels. Current work on quantised magnetotransport is mostly concentrated on these aspects. A major part of the present efforts is aimed at correlating the spectrum with our knowledge of the phonon behaviour obtainable from other studies. Attempts are also being made to develop quantitative theory to explain the observed effects. It is a field of intense current activity and there are many controversies about the interpretations [13.17] of the observed peaks in the literature. New experimental results and new interpretations are being reported in almost all the international conferences on semiconductor physics. The subject is not only of great theoretical interest but also of practical importance. Infrared sources tunable by varying the magnetic field have been proposed [13.18]. Results are also of importance for the understanding of the behaviour of some Raman lasers. Studies have so far been confined to the important compounds like InSb, GaAs, InP, and CdTe, but we may expect the studies to be extended to other compounds with the availability of good quality crystals.

13.1.4 Inversion Layers

The band structure of a semiconductor is distorted when an electric field is applied perpendicular to its surface. This distortion occurs due to the redistribution of the charge carriers under the influence of the field. In a p-type material, electrons accumulate near the surface when the applied field is directed towards the surface. The conduction band is bent downwards as a result of this accumulation. If the field is strong, the surface region becomes n type and thus the type becomes inverted. These layers are called the inversion layers. Evidently, in an n-type material, a strong field directed away from the surface would cause the inversion. The fields are usually such that the extent of the inversion layer is a few angstroms and comparable to the electron wavelength. The component of electron energy corresponding to motion perpendicular to the surface is quantised as in a potential well. However, in this case, the shape of the well is to be determined from a solution of the Schrödinger equation and Poisson equation combined with the Fermi function giving the electron concentration for the distorted band. The lower part of the well is often approximated as a triangle and the energy levels are worked out by assuming a triangular well. The separation between the levels is of the order of few millivolts.

The inversion layer is essentially two dimensional as the electron motion perpendicular to the surface is restricted. At very low temperatures, the electrons are more concentrated in the lower levels and the effects of this quantisation appear as oscillatory variations with electric field in many transport coefficients. At higher temperatures the electrons are distributed over the different levels, but due to the two-dimensional character of the layers the transport coefficients assume values very different from that in the bulk. Such inversion layers are also produced in the MOSFETs under operating conditions and since the fields parallel to the surface of such devices is often very high electrons are transported in hot-electron condition. Characteristics of the hot electron transport in inversion layers are also very different from that in the bulk [13.19]. The inversion layers hence pose many interesting problems and are therefore of great current interest.

The theory of electron transport in inversion layers has so far been developed by a modification of the theory for the bulk transport [13.20], only by taking into account the two dimensional nature of the motion. The effective mass approximation, the Kane nonparabolicity relation, and the perturbation theory of scattering based on the bulk phonon dispersion relations have been used. Although the theory so developed explains the observed phenomena, the validity of this bulk picture is in doubt. The effective mass approximation is known to give unexplainable discrepancies with the observed energy levels. There is no direct confirmation of the variation of effective mass with energy in accordance with the Kane relation in the inversion layers. The phonon spectrum may also be altered by the termination of the lattice at the surface. Electron-electron scattering and charge correlation effects may also be very important due to the very large values of electron concentration. Work is currently in progress for the development of theory taking all these factors into account.

Studies of inversion were mostly confined in the past to silicon as it is easy to grow an oxide layer on silicon and apply the perpendicular field by placing an electrode on top of the oxide layer. As it was difficult to grow an insulating layer on compound semiconductors, there were very few attempts to study the characteristics of inversion layers in compounds. In the studies reported [13.21], a dielectric sheet was used as an insulator but as the surface was not good enough and the surface state densities were large the results were difficult to correlate with theory. Attempts were made in the past to grow oxides by the anodic oxidation method [13.22] or by a thermal method [13.23], but were not very successful in producing low surface-state densities in the inter-oxide-semiconductor surface. Recently some progress has been made in this direction. Using a new technique [13.24], oxides have been produced on gallium arsenide surfaces with low state densities. Good quality MOSFET devices have been produced by this technique. We may expect that this success in producing good quality oxides will lead to a spurt in the studies of electron transport in inversion layers of compound semiconductors. We may expect to observe new phenomena from such studies due to the strong nonparabolicity and strong polar interaction in these materials.

13.1.5 Superlattices and Heterostructures

It was suggested in 1970 that NDR may be conveniently produced in superlattice structures constituted by growing a series of layers alternately of two different compounds [13.25]. The difference in the energy band gap of the compounds produces a periodic potential in the structure with a period equal to the sum of the thicknesses of two successive layers. The electron energy band corresponding to motion in the direction of the axis of the structure breaks into minibands (see Sect.4.4) and due to the bending of the minibands near the edges it should be possible to realise NDR as discussed for nonparabolic bands in Sect.11.3. Although this prediction has not yet been confirmed beyond doubt, superlattice structures have been studied with great expectations all these years. The structures have been grown by molecular beam epitaxy, gallium arsenide and $Ga_xAl_{1-x}As$ with values of x ranging between 0.15 and 1 being used for the alternate layers. Studies have been mostly concerned with the exact nature of the modification in the electron energy dispersion relation by the superlattice potential [13.26]. Experimental findings are found to be in fair agreement with theory for some structures but some discrepancies are not yet explained.

Electron transport was studied in the superlattice structures with the objective of discovering a good source of NDR. Some early experimental results seemed to exhibit NDR, but its origin was not clear. Theoretical studies [13.27] were also aimed at predicting the values of NDR that could be expected and mostly simple band structures as discussed in Sect.4.4 were used in these calculations. However, in spite of the intense efforts given to the realisation of the predicted effect it appears that the expectations are yet to be fulfilled. The difficulty is mostly due to the incompleteness of the techniques of growth of superlattice structures.

Although the expectation of a good source of NDR in superlattice structures may or may not be successful, these structures are of great interest as media of many interesting physical phenomena. The technique of molecular beam epitaxy is improving day by day and when perfected it is likely to give us structures of greater perfection than available so far. All the transport coefficients of these structures would be different from those in the bulk materials due to the quasi-two-dimensional nature of the structures. We may, therefore, expect that when the superlattice materials are more freely available, electron transport in these structures will receive more attention.

13.2 Scope of Further Studies

Work on electron transport in compound semiconductors was started mainly from an academic point of view. The main objective appears to have been to confirm the prediction that compounds of elements of group III and group V should behave similarly to the then known semiconductors. However, after the initial work confirming this prediction was done, the transport properties were studied and the high electron mobility found in these materials made them practically useful. Since those days the interest has been more in exploring the possible uses of the materials, and studies were mostly concentrated on materials for which some practical use was found. In the early days indium antimonide was the most intensely studied material as it was found useful for galvanomagnetic devices. Studies were then concentrated more on gallium arsenide as it became useful for lasers and later for Gunn diodes. The studies were extended in time to other compounds that were found useful. It appears that in this development, utility was the main criterion. The search was for a material which had useful properties not possessed by other materials. The studies were concentrated on only the useful properties and work on theory was mostly carried out to understand the available experimental results. It is because of this attitude that developments in theory occurred very rapidly after the discovery of the Gunn effect due to the need for understanding the effect so that Gunn devices could be commercially perfected.

Our understanding of electron transport has, however, reached such a stage that academic study of the various materials would again be very fruitful. The theory has been developed to give accurate values of the various transport coefficients in terms of the physical constants. But experimental results are not widely available for many compound semiconductors. A systematic experimental study of the different transport coefficients of all the compounds with well-characterised samples would be very useful. When detailed results are available it may become possible to discover interrelations for the prediction of the properties of one compound knowing the properties of another. Some attempts have been made in this direction [13.28], but to make these attempts successful, more data than are available at present are needed.

The compound semiconductors have served as useful media for the verification of the one-electron theory of transport. In most cases, however, when comparing theory with experiment we are required to adjust certain unknown parameters to obtain a fit. This approach has often resulted in different values being obtained for the same constant by different authors. The weaknesses of the theory are also often hidden in this method of analysis and the importance of mechanisms that have been left out of consideration may not come to light. The two-phonon scattering is a mechanism which has been mostly neglected in the theoretical calculations reported in the literature. It has, however, been recently pointed out that it may be important [13.29]. A conclusion cannot be reached because the constants related to this scattering mechanism are not known. Until methods are developed for the accurate determination of all the physical constants involved in the expressions for the transport coefficients, we have to accept a particular scattering model on the basis of its suitability for explaining the available data. More and more experimental data are, therefore, needed so that a model may be judged in the light of all the coefficients. It is only through detailed comparison of the theory and experiment with more extensive data than we have at present that the limitations of the current theory will be brought to light, and efforts can be directed to improving it. Techniques of material growth are improving from day to day and we may expect that various kinds of compounds will soon be available freely and that such detailed studies will be undertaken.

Appendix A

Table of Fermi integrals

η	$F_{1/2}(\eta)$	$F_{3/2}(\eta)$	η	$F_{1/2}(\eta)$	$F_{3/2}(\eta)$
-4.0	0.0182	0.0183	2.0	2.82	4.16
-3.5	0.0299	0.030	2.5	3.61	5.77
-3.0	0.0489	0.0493	3.0	4.49	7.79
-2.5	0.0798	0.0809	3.5	5.46	10.27
-2.0	0.129	0.132	4.0	6.51	13.26
-1.5	0.207	0.215	5.0	8.84	20.91
-1.0	0.328	0.347	6.0	11.45	31.04
-0.5	0.507	0.553	7.0	14.29	43.89
0.0	0.765	0.867	8.0	17.35	59.69
0.5	1.12	1.33	9.0	20.62	78.67
1.0	1.58	2.00	10.0	24.09	101.0
1.5	2.14	2.93			

Appendix B

Computer Program for the Evaluation of the Transport Coefficients by the Iteration Method

EPS - $\varepsilon_0 \times 10^{12}$[Fm^{-1}]; HL - $\hbar \times 10^{34}$[Js]; EC - $|e| \times 10^{19}$[C]; BK - $k_B \times 10^{23}$[J K^{-1}]

TOP - θ_ℓ[K] ; EMO - $m^* \times 10^{31}$[kg]; DCI - K_α; DCS - K_s; BG - E_g[eV];

E1 - E_1[eV]; E14 - h_{pz}[C m^{-2}]; RO - $\rho \times 10^{-3}$[kg m^{-3}]; AV - $s \times 10^{-3}$[m s^{-1}];

PV - $u_p \times 10^{-3}$[m s^{-1}]; EMT - dm^*/dT; WWS - $B \times 10^4$[T]; TTL - T[K]

LLL - Integration up to LLL $\times k\theta_\ell$; NNYT - Subintervals of integration;

EFF - E_F[eV]; CNE - n[m^{-3}]; CNI - $N_i \times 10^{-20}$[m^{-3}]; RHNE - R_H;

HMB - μ_H[m^{2}V^{-1}s^{-1}]; RM - $\Delta\rho/\rho$; GAI - ξ;

GRM - Geometrical magnetoresistance

Program (Appropriate for IBM 1130 in Fortran)

```
// FOR
*IOCS(CARD,DISK,1403PRINTER)
* LIST SOURCE PROGRAM
*ONE WORD INTEGERS
*EXTENDED PRECISION
      DIMENSION A(16),B(16),XX(15),SX(15),CX(15),TO(15),SP(15),SM(15),
     1TOR(15),TOI(15),SPR(15),SMR(15),SPI(15),SMI(15),FAR(15),FAI(15),
     2FBR(15),FBI(15),GA(15),GI(15),GB(15),HA(15),PA(15),EA(15),EB(15),
     3CB(15),AB(15),F1(16),F2(16),F3(16),F4(16),F5(16),F6(16),F7(16),
     4F8(16),F9(16),F10(16),F11(16),F12(16),AALE(15),WWS(10),HM(10),
     5DM(10),RHNE(10),HMB(10),RM(10),GAI(10),GRM(10),TTL(6),NNYT(6)
      DIMENSION LLL (6),FO(15),EFF(5),AD(15)
      A(1)=0.095012510
      A(2)=0.281603551
      A(3)=0.458016778
      A(4)=0.617876244
      A(5)=0.755404408
      A(6)=0.865631202
      A(7)=0.944575023
      A(8)=0.989400935
      B(1)=0.189450610
      B(2)=0.182603415
      B(3)=0.169156519
      B(4)=0.149595989
      B(5)=0.124628971
      B(6)=0.095158512
      B(7)=0.062253524
      B(8)=0.027152459
      DO 10 II=1,8
      A(II+8)=-A(II)
      B(II+8)=B(II)
   10 CONTINUE
      EPS=8.8542
      HL=1.05444
      EC=1.602
      BK=1.3805
      PAI=3.14159
       READ(2,86) (WWS(I),I=1,5)
       READ(2,87) (TTL(I),        I=1,5)
       READ(2,87)  (EFF(I),I=1,5)
       READ(2,88) (LLL(I),        I=1,5)
       READ(2,88)  (NNYT(I),I=1,5)
      READ(2,89)MM1,MM2,MM3,MM4,IP1,IP2
      READ(2,85)CNE,EMT
       READ(2,90) TOP,EMO,DCI,DCS,BG
       READ(2,90) E1,E14,RO,AV,PV
 85   FORMAT(2E8.1)
 86    FORMAT(5E8.1)
 87    FORMAT(5F8.1)
 88    FORMAT(5I8)
 89   FORMAT(6I4)
 90    FORMAT(5F10.5)
        WRITE(5,76) (WWS(K),K=1,5)
```

```
 76     FORMAT(5X,'WWS='5E16.1)
       WRITE(5,77)(TTL(K),K=1,5)
 77    FORMAT(5X,'TTL='5F16.1)
       WRITE(5,177)(EFF(K),K=1,5)
177    FORMAT(5X,'EFF='5F16.1)
       WRITE(5,78)(LLL(K),K=1,5)
 78    FORMAT(5X,'LLL='5I8)
       WRITE(5,178)(NNYT(K),K=1,5)
178    FORMAT(5X,'NNYT='5I8)
       WRITE(5,79)CNE,EMT
 79    FORMAT (5X,'CNE,EMT='2E16.1)
      WRITE(5,179)MM1,MM2,MM3,MM4,IP1,IP2
179    FORMAT(5X,'MM1,MM2,MM3,MM4,IP1,IP2='6I4)
       WRITE(5,80)TOP,EMO,DCI,DCS,BG
 80    FORMAT(5X,'TOP,EMO,DCI,DCS,BG='5F16.5)
       WRITE(5,180)E1,E14,RO,AV,PV
180    FORMAT(5X,'E1,E14,RO,AV,PV='5F16.5)
       AL= 1.E-04*TOP*BK/(BG*EC)
      DO 301ITL=MM1,MM2
       TL = TTL(ITL)
       NYT = NNYT(ITL)
      LL=LLL(ITL)
      I1=LL+9
      I2=LL+8
      I3=LL+7
       I4 = LL+6
      EM=EMO+EMT*TL
      Y=TOP/TL
      DO 302 ICN=MM3,MM4
      EF=EFF(ICN)
      B1NT=0
      ECIT=0
      DO 1207 IJK=1,NYT
      DO 1207 IJ=1,16
      X=0.5*(1.+A(IJ))
      X=(X+(IJK-1))/NYT
      BC=0.5*B(IJ)/NYT
      DO 1207 IK=1,I1
      XX(IK)=X+(I<-1)
      SX(IK)=SQRT(XX(IK))
      AALE(IK)=AL*XX(IK)
      ALE=AALE(IK)
      GA(IK)=1.+2.*ALE
      EA(IK)=SX(IK)*SQRT(1.+ALE)
      FO(IK)=1./(EXP(XX(IK)*Y-EF)+1.)
      AD(IK)=EA(IK)*GA(IK)*BC*FO(IK)
      ECIT=ECIT+AD(IK)
      B1NT=B1NT+AD(IK)*(1.-FO(IK))
 1207 CONTINUE
      CNE=SQRT(2.*EM*BK*TOP)*EM*BK*TOP/(PAI*PAI*HL*HL*HL)*1.0E+21*ECIT
      CNI= CNE*1.0E-20
      CIMI=8.*EPS*DCS*BK*TL*PAI*PAI*HL/(EC*EC*SQRT(2.*EM*BK*TOP)*B1NT)
     1*1.0E-4
       WRITE(5,444) TL,CNI,CNE
      B1=EXP(Y)
      B2=1./B1
      CA=EC*EC*SQRT(BK*TOP*EM/2.)*(1./DCI-1./DCS)/(4.*PAI*EPS*HL*HL*(B1-
     11.))*1.E+15
      CAC=SQRT(BK*TOP*2.*EM)*EM*EC*E1*E1*BK*TL/(PAI*HL*HL*HL*RO*AV*AV
     1*CA)*1.E+08*EC
      CPE=SQRT(EM/(2.*BK*TOP))*EC*EC*E14*E14*BK*TL/(2.*PAI*EPS*EPS*DCS*D
     1CS*HL*HL*RO*PV*PV*CA)*1.E+18
      CIM=EC*EC*EC*EC*CNI/(16.*PAI*EPS*EPS*DCS*DCS*SQRT(2.*EM*BK*TOP)*BK
     1*TOP*CA)*1.E+18
      WRITE(5,230)CA,CAC,CPE,CIM,CIMI
```

```
 230 FORMAT(5E18.9)
     EBM=-EC/(CA*EM)*1.E+12
     HMR=0
     DO 15 MA=IP1,IP2
     DM(MA)=0
     HM(MA)=0
15   CONTINUE
     CON= 0
     SF=2./CIMI
     DO 201 I=1,16
     F1(I)=A(I)
     F2(I)=F1(I)*F1(I)
      F3(I)=F1(I)*F2(I)
     F4(I)=1-F1(I)
     F5(I)=F4(I)*F4(I)
 201 CONTINUE
     DO 207 JK=1,NYT
     DO 200 J=1,16
     X=0.5*(1.+A(J))
     X=(X+(JK-1))/NYT
     EX=1./EXP(X*Y)
     BC=0.5*B(J)/NYT
     DO 20 K=1,I1
     XX(K)=X+(K-1)
     SX(K)=SQRT(XX(K))
     CX(K)=XX(K)*SX(K)
     AALE(K)=AL*XX(K)
     ALE=AALE(K)
     GA(K)=1.+ALE
     EB(K)=XX(K)*GA(K)
     HA(K)=SQRT(GA(K))
     EA(K)=SX(K)*HA(K)
     GB(K)=GA(K)+ALE
     PA(K)=SQRT(GB(K))
     QA=SQRT(ALE)
     CB(K)=QA/PA(K)
     AB(K)=HA(K)/PA(K)
     GI(K)=1./GB(K)
     FO(K)=1./(EXP(XX(K)*Y-EF)+1.)
  20 CONTINUE
     DO 30 KA=1,I2
      AK= GA(KA)/GB(KA)
     CK=AALE(KA)/GB(KA)
     AA=AK*AK
     BB=2.*AK*CK
     CC=CK*CK
     SE=SF/EB(KA)
     DO 701 I=1,16
      DN=F4(I)+SE
     F6(I)=(AA+BB*F1(I)+CC*F2(I))
     F7(I)=F6(I)*B(I)*F4(I)/(DN*DN)
701   CONTINUE
     GNT=0
     DO '01 I=1,16
     GNT=GNT+F7(I)
 801 CONTINUE
     SIM=CIM*GB(KA)/(EA(KA)*EB(KA))*GNT
      GNT=0
      DO 901 I = 1,16
     GNT=GNT+F7(I)*F4(I)
 901 CONTINUE
     SOPE=CPE*GB(KA)*0.5*GNT/EA(KA)
     GNT=0
     DO 1001 I=1,16
     GNT=GNT+F7(I)*F5(I)
```

```
1001 CONTINUE
     SOA=CAC*GB(KA)*EA(KA)*0.5*GNT
     GNT=0
     GTM=EB(KA+1)/EB(KA)
     BTM=SQRT(GTM)
     DO 1501 I=1,16
     F8(I)=1.-2.*BTM*F1(I)+GTM
     DN=F8(I)+SE*2.
     F9(I)=F8(I)*B(I)/(DN*DN)
      AKD=GA(KA+1)/GB(KA+1)
     CKD=AALE(KA+1)/GB(KA+1)
     AA=AK*AKD
      BB=SQRT(AK*AKD*CK*CKD)*2.
     CC=CK*CKD
     F10(I)=(AA+BB*F1(I)+CC*F2(I))
     F11(I)=F10(I)*F9(I)
     F12(I)=0
1501 CONTINUE
     GNT=0
     DO 1601 I=1,16
     GNT=GNT+F11(I)
1601  CONTINUE
     SOP=GB(KA+1)*EA(KA+1)*GNT/EB(KA)
      SOP=SOP*(1.-FO(KA+1))/(1.-FO(KA))
     GNT=0
     DO 1701 I=1,16
     GNT=GNT+F11(I)*F1(I)
1701 CONTINUE
     SP(KA)=GB(KA+1)*EA(KA+1)*GNT/EB(KA)*BTM
      SP(KA)=SP(KA)*(1.-Fo(KA+1))/(1.-FO(KA))
     SO=SOP+SOA+SOPE+SIM
     IF(KA-1) 31,31,32
 31  TO(KA)=1./SO
     GO TO 30
 32  GTM=EB(KA-1)/EA(KA)
     BTM=SQRT(GTM)
     DO 1201 I=1,16
     F8(I)=1.-2.*BTM*F1(I)+GTM
      DN=F8(I)+SE*2.
     F9(I)=F8(I)*B(I)/(DN*DN)
     AKD=GA(KA-1)/GB(KA-1)
     CKD-AALE(KA-1)/GB(KA-1)
     AA=AK*AKD
     BB=SQRT(AK*AKD*CK*CKD)*2.
     CC=CK*CKD
     F10(I)=(AA+BB*F1(I)+CC*F2(I))
     F11(I)=F10(I)*F9(I)
1201 CONTINUE
     GNT=0
     DO 1301I=1,16
     GNT=GNT+F11(I)
1301 CONTINUE
      SOO=B1*GB(KA-1)*EA(KA-1)*GNT/EB(KA)
      SOO=SOO*(1.-FO(KA-1))/(1.-FO(KA))
     SO=SO+SOO
      GNT= 0
     DO 1401I=1,16
     GNT=GNT+F11(I)*F1(I)
1401 CONTINUE
     SM(KA)=B1*GB(KA-1)*FA(KA-1)*GNT/EB(KA)*BTM
      SM(KA)=SM(KA)*(1.-FO(KA-1))/(1.-FO(KA))
     TO(KA)=1./SO
  30 CONTINUE
     DO 156JA=IP1,IP2
     WS=EBM*EBM*WWS(JA)*WWS(JA)*1.E-08
```

```
      TOR(1)=TO(1)/(1.+WS*GI(1)*GI(1) *TO(1)*TO(1))
       TOI(1)=-TO(1)*TOR(1)
      SPR(1)=SP(1)*TO(1)
       SPI(1)=SP(1)/TO(1)
      DO 35 KA=2,I2
      TOR(KA)=TO(KA)/(1.+WS*GI(KA)*GI(KA)*TO(KA)*TO(KA)
      TOI(KA)=-TO(KA)*TOR(KA)
      SPR(KA)=SP(KA)*TO(KA)
      SPI(KA)=SP(KA)/TO(KA)
      SMR(KA)=SM(KA)*TO(KA)
      SMI(KA)=SM(KA)/TO(KA)
   35 CONTINUE
      FAR(1)=TOR(1)*(GI(1)+WS*GI(1)*SPR(1)*TOI(2)+SP(1)*TOR(2))
      FAI(1)=TOI(1)*(GI(1)*(GI(1)+SP(1)*TOr(2))-SPI(1)*TOI(2))
      DO 100 IB=2,I3
      FAR(IB)=TOR(IB)*(GI(IB)+WS*GI(IB)*(SPR(IB)*TOI(IB+1)+SMR(IB)*TOI(I
     1B-1))+SP(IB)*TOR(IB+1)+SM(IB)*TOR(IB-1)
       FAI(IB)=TOI(IB)*(GI(IB)*(GI(IB)+SP(IB)*TOR(IB+1)+SM(IB)*TOR(IB-1)
     1)-(SPI(IB)*TOI(IB+1)+SMI(IB)*TOI(IB-1)))
  100  CONTINUE
       ID=I4
       DO 155 N=1,7
      FBR(1)=TOR(1)*(GI(1)+WS GI(1)*SPR(1)*FAI(2)+SP(1)*FAR(2))
      FBI(1)=TOI(1)*(GI(1)*(G1(1)+SP(1)*FAR(2))-SPI(1)*FAI(2))
      DO 110 IC=2,ID
      FBR(IC)=TOR(IC)*(GI(IC)+WS*GI(IC)*(SPR(IC)*FAI(IC+1)+SMR(IC)*FAI
     (1IC-1))+SP(IC)*FAR(IC+1)+SM(IC)*FAR(IC-1))
      FBI(IC)=TOI(IC)*(G1(IC)*(GI(IC)+SP(IC)*FAR(IC+1)+SM(IC)*FAR(IC-1))
     1-(SPI(IC)*FAI(IC+1)+SMI(IC)*FAI(IC-1)))
  110  CONTINUE
       DO 150 M=1,ID
       FAR(M)=FBR(M)
       FAI(M)=FBI(M)
  150 CONTINUE
       ID=ID-1
  155 CONTINUE
        EX1=EX
       DO 120 L=1,I4
       ECX=CX(L)*GA(L)*HA(L)*BC
        ECX=ECX*FO(L)*(1.-FO(L))
       DM(JA)=DM(JA)+ECX*FBR(L)
       HM(JA)=HM(JA)+ECX*FBI(L)
       IF(JA-1) 119,118,119
  118 HMR=HMR+FBR(L)*FBI(L)*ECX
  119 EX1=Ex1*B2
  120 CONTINUE
  156 CONTINUE
      EX1=EX
      DO 158 LA=1,I4
      CNC=SX(LA)*GB(LA)*HA(LA)*BC
       CNC=CNC*FO(LA)
      CON=CON+CNC
      EX1=EX1*B2
  158 CONTINUE
  200 CONTINUE
  207  CONTINUE
      CON=2.*EC*TOP/(3.*EM*TL*CA*CON)*1.E+12
      HMR=HMR*CON
      DO159 LB=IP1,IP2
      DM(LB)=DM(LB)*CON
      HM(LB)=HM(LB)*CON
  159 CONTINUE
      DMO=DM(1)
      DO 300 JB=IP1,IP2
      WS=EBM*EBM*WWS(JB)*WWS(JB)*1.E-08
```

```
      RHNE(JB)=EBM*HM(JB)/(DM(JB)*DM(JB)+WS*HM(JB)*HM(JB))
      HMB(JB)=RHNE(JB)*DMO
      IF(JB-1)291,292,291
  292 RM(JB)=-(HMR/DM(JB)+HM(JB)*HM(JB)/(DM(JB)*DM(JB)))*EBM*EBM
      GAI(JB)=-1.-HMR*DM(JB)/(HM(JB)*HM(JB))
      GRM(JB)=-HMR/DM(JB)*EBM*EBM
      GO TO 295
  291 RM(JB)=1.-EBM*HM(JB)/(RHNE(JB)*DM(JB)*DMO)
      RM(JB)=RM(JB)/(1.-RM(JB))
      GAI(JB)=RM(JB)*EBM*EBM/(WS*HMB(JB)*HMB(JB))
      GRM(JB)=DMO/DM(JB)-1.
  295 WRITE(5,240)WWS(JB),RHNE(JB), HMB(JB),RM(JB),GAI(JB),GRM(JB),
     1DM(JB),HM(JB)
444    FORMAT(//5E20.9)
  240 FORMAT(5E20.9)
  300 CONTINUE
302    CONTINUE
  301 CONTINUE
        CALL EXIT
       END
```

The Program given above may be used to compute μ_H, R_H, ξ, and geometrical magnetoresistance for degenerate materials with given Fermi levels. For nondegenerate materials with given electron concentrations, the following changes are to be made:

Replace 1207 + 001, + 002, + 003 by

CNI = CNE

CIMI = 8.*EPS * DCS * EM * BK * TL * BK * TOP/
(EC * EC * HL * HL * CNE)* 1. E - 03

Omit 1701 + 002, 1401 + 002, 1601 + 002, 1301 + 002,

Replace 260 + 004 by

ECX = ECX * EX1

156 + 004 by

CNC = CNC * EX1

Appendix C

Values of a_i and b_i for Gaussian Quadrature Integration (Interval -1 to +1)

a_i	b_i
n = 5	
0.000000000	0.568888889
0.538469310	0.478628670
0.906179846	0.236926885
n = 10	
0.148874339	0.295524225
0.433395394	0.269266719
0.679409568	0.219086362
0.865063367	0.149451349
0.973906528	0.066671344
n = 16	
0.095012510	0.189450610
0.281603551	0.182603415
0.458016778	0.169156519
0.617876244	0.149595989
0.755404408	0.124628971
0.865631202	0.095158512
0.944575023	0.062253524
0.989400935	0.027152459

Appendix D

Computer Program for the Monte Carlo Calculation of Hot-Electron Conductivity and Diffusivity

Meanings of Symbols

DL2 $= 8m^*k_BT_LK_S\varepsilon_0(n\hbar^2|e|)^{-1}$; A $= |e|(2m^*)^{-1}(\mathcal{E}/\Gamma)^2$ B $= \hbar\omega_\ell/|e|$;

D $= \mathcal{E}/\Gamma$; AL $= 1/E_g$;

CN1 $= m^{*1/2}|e|^{3/2}\omega_\ell\ (4\ \sqrt{2}\pi\hbar\varepsilon_0\Gamma)^{-1}(K_\alpha^{-1} - K_S^{-1})[\exp(\hbar\omega_\ell/k_BT_L)-1]^{-1}$;

CN2 $=$ CN1.$\exp(\hbar\omega_\ell/k_BT_L)$; CNA $= m^{*3/2}|e|^{5/2}k_BT_LE_1^2(2^{1/2}\pi\hbar^4\langle\rho s^2\rangle\Gamma)^{-1}$;

CNI $= N_ik_BT_Lm^{*1/2}|e|^{3/2}\ (4\ \sqrt{2}n\pi\hbar^2\varepsilon_0k_S\Gamma)^{-1}$;

CNA1 $= 2m^{*1/2}s|e|^{-1/2}$ CNA2 $= 4\ \sqrt{2}m^{*1/2}|e|^{1/2}s\ (3k_BT_L)^{-1}$; CNA3 $= 2^{1/2}$.CNA1;

Γ - a constant larger than the maximum value of scattering probability.

EL1, EL2, EL3, and EL4 are selected values of energy. If the energy is below EL1 the electron is assumed not to be scattered and allowed to continue its journey. The maximum value of scattering probability occurs for EL1 < E < EL2. SF1, SF2, SF3 are constants larger than the maximum values of scattering probability for the intervals EL2 < E < EL3, EL3 < E < EL4 and E > EL4. FS, EH, SO are limiting constants introduced to prevent the program going amuck. EL is the value of E for which $E \to 0$ formulae apply. ERO is the error limiting constant.

TL - Lattice temperature;
F - $\mathcal{E}$; TT - Time interval for a sample normalised by Γ;
JJ - Input for random number program;
SSX - $\sum_i x_i$; SSY - $\sum_i y_i$; SSZ - $\sum_i z_i$, SXQ $= \sum_i x_i^2$;
SYQ $= \sum_i y_i^2$; SZQ $= \sum_i z_i^2$; KPR - Normalized magnitude of k;

CTA - $\cos\theta$; STA-$\sin\theta$; CPA - $\cos\phi$; SPA - $\sin\phi$; θ and ϕ are the polar and azimuthal angles of k. JA - Initial sample number (i.e. value of i at start). JB - Sample number up to which computation should be done.

JC - Sample numbers for which results should be printed out.

VD - Drift velocity [m s^{-1}]

DFX, DFY, DFZ are the diffusion constants [m^2 s^{-1}] for the x, y, and z (direction of the field) directions respectively. QSX, QSY and QSZ are the total displacements [m] in the x, y, and z directions.

Program (Appropriate for Burroughs 6700 in Fortran)

```
      FUNCTION ROT A (X)
      COMMON E,ED,AL,RE,RED,B,FS,ERD,ALE
      REAL X
      C=ALE;IF(ALE-ERD)20,20,10
   10 A=1.0/3.0
      Z=((1+(1-X)*C*(6+C*(12+C*8)))**A-(1+C))/C
      GO TO 30
20    Z=1-2*X
30    ROT A =Z
      RETURN
      END
      FUNCTION ROTOP(X)
      COMMON E,ED,AL,RE,RED,B,FS,ERO,ALE
      REAL X
      F=(RE*RED*2)/(RED-RE)**2;IF(F-FS)10,10,20
10    Z1=1-2*X;GO TO 30
20    Z1=(1+F-(1+2*F)**X)/F
30    IF(ALE-ERO)60,60,40
40    G=E*(1+ALE);GD=ED*(1+AL*ED);C=G+GD;B1=ALE*ED
      D=2*SQRT(G*GD);D2=D**2;AO=B1*(1+B1*C/D2);A1=D2/4+C*AO
      A 3 =B1**2/(2*A1)
      A 2 =D*AO/A1;CD=ALOG((C-D)/(C+D));CD1=C/D;CD2=CD1-1
50    Y=EXP(A3*(1-Z1**2)+A2*(1-Z1-2*X)-X*CD)
      Y1=CD2*Y;Z2=Z1-(Z1+Y1-CD1)/(1+Y1*(A2+2*A3*Z1))
      ER1=ABS(Z2-Z1);Z1=Z2
      IF(ER1-ERO)60,60,50
60    ROTOP=Z1
      RETURN
      END
      FUNCTION SOP(X,Y)
      COMMON E,ED,AL,RE,RED,B,FS,ERO,ALE
      REAL X,Y
      GA=1+AL*X;GAD=1+AL*Y;GB=1+2*AL*X;GBD=1+2*AL*Y
      G=X*GA;GD=Y*GAD;GS=SQRT(G);GSD=SQRT(GD)
      A1=(2*GA*GAD+AL*(G+GD))**2
      A2=-(2*AL*GS*GSD*(4*GA*GAD+AL*(G+GD)));A3=4*GA*GAD*GB
      X2=ABS((GS+GSD)/(GS-GSD));SOP=(A1*ALOG(X2)+A2)/(A3*GS)
      RETURN
      END
      REAL LM1,LM2,LM3,KPR,LMA,LM4,LMO,KZ1,KZO
      REAL KR,KRQ,KX,KY,KAQ,KBQ
      COMMON E,ED,AL,RE,RED,B,FS,ERO,ALE
      DATA DL2,EL1,EL2,EL3,EL4,SF1,SF2,SF3,FS,ERO,EL,EH,SO/1.887E5,
     +.0016,.016,.16,1.6,.55,.35,.25,.001,.0001,.0001,1000.,1.E-6/
      A=1.413E-5;B=2.5E-2;D=6.667 E8;AL=4.364;CN1=3.87E-4;CN2=1.647E-2
      CNA=4.695E-2;CNI=3.81E-2;CNA1=2.186E-3;CNA2=3.294E-1;CNA3=3.092E-3
```

```
          READ (5.500)JA,JB,JC,JJ,TT,F
          READ (5,502)SSY,SSY,SSZ,SXG,SYQ,SZQ,KPR,CTA,STA,CPA,SPA
          WRITE(6,500)JA,JB,JC,JJ,TT,F
          WRITE(6,502)SSX,SSY,SSZ,SXG,SYQ,SZQ,KPR,CTA,STA,CPA,SPA
          FA=SQRT(A/AL);FB=.5 FA;FC=1./F;FX=FA*FC;FW=.5/AL
          RT=1./TT;FZ=.5*D*F*RT
          C=1/SQRT(A);FY=;/(4*A*AL);I=1;J=JA
  115     KZO=KPR*CTA;KR=KPR*STA;KRQ=KR**2;KX=KR*CPA;KY=KR*SPA
  100     S1=RANDOM(JJ);IF(S1-SO)100,100,60
    60    T=-ALOG(S1)
  401     SIGT=SIGT+T;IF(SIGT-TT)101,102,103
  101     II=1;GO TO 2
  102     II=2;GO TO 2
  103     II=3;DT=SIGT-TT;T=T-DT
  2       KZ1=KZO+T;KBQ=KZ1**2;E1=A*(KRQ+KBQ)
          KAQ=KZO**2
  125     ALE1=AL*E1;IF(ALE1-ERO)21,21,22
    21    E=E1;GO TO 23
    22    E=(SQRT(1+4*ALE1)-1)/(2*A L)
  23      EI=A*(KRQ+KAQ)
          ALE2=AL*EI;IF(ALE2-ERO)25,25,24
    24    E1=(SQRT(1+4*ALE2)-1)/(2*AL)
  25      RE=SQRT(E)
  8       DEN=SQRT(KBQ+KRQ);ALE=AL E
          CTA=KZ1/DEN;STA=KR /DEN
          WA=FY+KRQ;WB=SQRT(WA+KBQ);WC=SQRT(WA+KAQ);WD=KZ1+WB;WE=KZO+WC
          WL=ALOG(WD/WE);WF=FX*WL
          DZ=(E-EI)*FC;DX=WF*KX;DY=WF*KY
          SX=SX+DX;SY=SY+DY;SZ=SZ+DZ;IF(II-2)106,105,105
  105     SSX=SSX+SX;SSY=SSY+SY;SSZ=SSZ+SZ
          SXQ=SXQ+SX**2;SYQ=SY**2;SZQ=SZQ+SZ**2
  142     SIGT=0;SX=0;SY=0;SZ=0
          IF(II.GT.2)GO TO 126;II=0;GO TO 106
  126     KZO=KZ1;T=DT;GO TO 127
  106     IF(E-EH)3,13,13
    13    ED=E-8;RED=SQRT(ED);GO TO 4
    3     S2=RANDOM(JJ);IF(E-EL1)14,14,302
    302   IF(E-EL2)310,310,304
    304   IF(E-EL3)305,305,306
    305   IF(S2-Sf1)310,310,14
    306   IF(E-EL4)307,307,308
    307   IF(S2-Sf2)310,310,14
    308   IF(S2-SF3)310,310,14
    310   LMO=CNI*(1+2*ALE)/SQRT(E*( 1+ALE));IF(S2-LMO)224,224,301
  224     KPR=DEN;S3=RANDOM(JJ)
          Qt=1-(2*S3)/(1+DL2*E*(1+ALE)*(1-S3));AQ=ABS(QT)
          IF(AQ,LE.1)GO TO 7
          QT=SIGN(1.,QT);GO TO 7
  301     ED=E+B;RED=SQRT(ED)
    70    IF(E-EL)20,20,30
  20      LM1=2* CN1/SQRT(B)+LMO;GO TO 40
    30    LM1=LMO+CN1*SOP(E,ED)
    40    IF(S2-LM1)4,11,11
    11    If(E-B)200,200,12
    12    ED=E-B;RED=SQRT(ED)
          LM2=LM1+CN2*SOP(E,ED);IF(S2-LM2)4,211,211
    200   LM2=LM1
```

```
211 GD=SQRT(E*(1+ALE))
    LMA=CNA*(1+ALE)**2+(ALE**2)/3)/(1+2*ALE)
    LM3=LM2+LMA*(GD+CNA3);IF(S2-LM3)221,221,231
221 S3=RANDOM(JJ);QT=ROT A (S3);AQ=ABS(QT)
    IF(AQ.LE.1.)  GO TO 33;QT=SIGN(1.,QT)
33  ED=E+CNA1*GD*SQRT( 1-QT)
    KPR=C*SQRT(ED*(1+AL*ED));GO TO 7
231 LM4=LM3+LMA*(GD*(1+CNA2*GD)-CNA3)
    IF(S2.GT.LM4)GO TO 14;S3=RANDOM(JJ);QT=ROTA(S3)
    AQ=ABS(QT);IF(AQ.LE.1)GO TO 53
    QT=SIGN(1.,QT)
53  ED=E-CNA1*GD*SQRT(1-QT)
    KPR=C*SQRT(ED*(1+AL*ED));GO TO 7
4   KPR=C*RED*SQRT(1+AL*ED)
6   S3=RANDOM(JJ);QT=ROTOP(S3)
    AQ=ABS(QT);IF(AQ.LE.1.)GO TO 7;QT=SIGN(1.,QT)
7   S4=RANDOM(JJ);BETA=6.28318*S4
    CP8=COS(BETA);SPB=SIN(BETA);STB=SQRT(1-QT**2)
    CX=CTA*CPA*STB*CPB-SPA*STB*SPB+STA*CPA*QT
    CY=CTA*SPA*STB*CPB+CPA*STB*SPB+STA*SPA*QT
113 CTA=-STA*STB*CPB+CTA*QT;AQ=ABS(CTA)
    IF(AQ.LT.1)GO TO 114;CTA=SIGN(.999999,CTA)
114 STA=SQRT(1-CTA**2) ;CPA=CX/STA;IA=0
    AQ=ABS(CPA);IF(AQ.LT.1)GO TO 112
    CPA=SIGN(.999999,CPA)
112 SPA=SQRT(1-CPA**2) ;SPA=SIGN(SPA,CY)
410 IF(II-1)127,115,115
14  KZ0=KZ1 ;IA=1
411 IF(II-1)127,100,100
127 I=I+1;J=J+1;IF(I-64)108,108,107
107 IF(J.LT.JC)GO TO 108
    RJ=1./J;RTJ=RT*RJ;FZJ=FZ*RJ
    QSX=SSX*RJ;QSY=SSY*RJ;QSZ=SSZ*RJ;DQZ=FZ*QSZ**2;I=1
    DFX=SXQ*FZJ;DFY=SYQ*FZJ;DFZ=SZQ*FZJ-DQZ
    VD=GSZ*RT*D*F
    WRITE(6.136)J,JJ ,VD
    WRITE(6.63)DFX,DFY,DFZ,QSX,QSY,QSZ
    IF(J.GE.JB)GO TO 999
108 IF(II.GE.3)GO TO 401
    IF(IA-1)115,100,100
999 WRITE(6.502)SSX,SSY,SSZ,SXQ,SYQ,SZQ,KPR,CTA,STA,CPA,SPA
63  FORMAT(6E15.5)
136 FORMAT(//I10,I15,E15.5)
500 FORMAT(3I10,I15,2F10.3)
502 FORMAT(6E12.5)
    STOP
    END
```

The values of constants as given in the program apply to indium antimonide at 77 K for $\mathscr{E} = 100$ V/cm.

In the first run of the program, the variables SSX, SSY, SSZ, KPR, CTA, STA, CPA, SPA may be given values of zero. In subsequent runs the values obtained in the previous run should be given as inputs. TT, F, JA, JB, JC, JJ should be given values in accordance with the definitions.

List of Symbols

A	Normalisation constant; Area; Magnitude of principal axis of an ellipse
$\underline{A}$	Vector potential
$A(q)$	Fourier coefficient
a	Lattice constant; width
a_o	Bohr radius (5.29177×10^{-11}m)
$\underline{a}$	Unit vector
a_k, a_k^+	Annihilation, creation operators
$\underline{a}_1, \underline{a}_2, \underline{a}_3$	Primitive translation vectors of a crystal
$\underline{B}$	Magnetic induction [T]
B_i, b_i	Weighting factors for Gaussian quadrature integration
B_r, B_o	Resonance magnetic induction
b_o	Width
$\underline{b}_1, \underline{b}_2, \underline{b}_3$	Reciprocal lattice vectors
C	Coupling constant; normalising constant
$C_{ij}, C_{11}, C_{12}, C_{44}$	Components of elasticity tensor
c	Lattice constant of a wurtzite crystal along the c axis; velocity of light in free space (2.99793×10^8 m s^{-1})
$\underline{c}$	Primitive translation vector of a wurtzite crystal
c_o, c_n	Constants
$D, D_{\parallel}, D_{\perp}$	Diffusion constant [m^2 s^{-1}]
D_i	Intervalley phonon deformation potential constant [eV m^{-1}]
D_N	Probability of a particle being at a point in phase space
D_o	Non-polar optic phonon deformation potential constant [eV m^{-1}]
$\mathscr{D}$	Displacement
d	Thickness; period of semiconductor superlattice
$\underline{d}$	Reciprocal lattice vector for a wurtzite crystal; Primitive translation vector of a lattice
E	Energy [eV]
$E(\underline{k})$	Energy for wave vector $\underline{k}$
E_c	Conduction band minimum energy

E_d	Donor energy level; Deformation potential constant for dilation [eV]
E_F	Fermi energy
E_{F0}	Fermi energy in the absence of a magnetic field
E_g	Band gap energy
E_i	Energy of ith electron
E_ℓ	Energy eigen value for a quantising magnetic field
E_n	Energy at the band edge
E_u	Deformation potential constant for shear [eV]
E_v	Valence band maximum energy
E_{zs}	Energy at the centre of the sth miniband
E_o	Band tailing parameter; Average of the Gaussian function; Polar optic phonon coupling constant; Steady component of energy
E_{oe}	Difference between the atomic energy levels and electron levels in a solid
E_1	Acoustic phonon deformation potential [eV]; Magnitude of time-dependent component of energy
E'	Energy of electron after collision
e	Charge of an electron (-1.60219×10^{-19} C)
$\mathscr{E}$	Electric field [V m^{-1}]
$\mathscr{E}_x, \mathscr{E}_y$	x, y components of electric field
$\mathscr{E}_o$	Magnitude of electric field; DC field
$\mathscr{E}_1$	Magnitude of time-dependent component of electric field
$\underline{F}$	External force
$F_j(\eta)$	Fermi integral
f, f(E), f(k)	Distribution function
$f_{\ell\pm}$	Distribution function for the ℓth level for the two spins
$f_o, f_o(E)$	Fermi function; even part of the distribution function
G_{12}	Overlap integral for scattering out of the heavy hole band
G_{21}	Overlap integral for scattering into the heavy hole band
g	Reciprocal lattice vector; spectroscopic splilling factor
g(E)	Density of states for energy E
g_s	Number of levels at energy E_s
H	Hamiltonian operator
$\underline{H}$	Magnetic field intensity [A m^{-1}]
H_n	Hermite polynomial of order n; Hamiltonian for n particles
$H(x)$	Heaviside unit function
h	Planck's constant
$\hbar$	Dirac's constant, $h/2\pi$ (1.0546×10^{-34} J s)

h_{ij}	Components of piezoelectric tensor
h_{pz}	Piezoelectric constant [C m^{-2}]
I_n	Normalisation integral
$I(\underline{k},\underline{k}')$	Overlap integral
i	$\sqrt{-1}$
J	Current density
J_x,J_y,J_z	x, y, z component of current density
J_{xo}	Current density in the absence of a magnetic field
J_{xB}	Current density in the presence of a magnetic induction B
K	Anisotropy ratio; Dielectric constant
K_s	Static dielectric constant
K_T	Energy flux per unit temperature gradient
K_∞	High-frequency dielectric constant
$\underline{k}$	Wave vector
k_x,k_y,k_z	x, y, z components of $\underline{k}$
k_{xi}	Initial value of k_x
$\underline{k}_o$	Wave vector of the electron at the band minimum
$k_\pm$	Wave vector corresponding to $(E \pm \hbar\omega_\ell)$
$k_{\parallel}$	Longitudinal component of $\underline{k}$
$k_\perp$	Transverse component of $\underline{k}$
k_B	Boltzmann constant $(1.38062 \times 10^{-23}$ J $K^{-1})$
L_c	Collision operator
L_y,L_z	Length of a sample in the y and z direction
L_1	Collision operator
ℓ	Electron mean free path; an integer; length of a sample
ℓ,m,n	Miller indices; direction cosines of magnetic induction
$M(\underline{k},\underline{k}')$	Matrix element for scattering from $\underline{k}$ state to $\underline{k}'$ state
$M^\pm_{pt},M^\pm_{pn}$	Matrix elements for transitions by interaction with photons, phonons
m_d	Density of states effective mass
m_i	Component of effective mass in the ith direction
m_ℓ,m_t	Longitudinal and transverse effective mass
m_p	Polaron mass
m_x,m_y,m_z	x, y, z components of effective mass
m_o	Free electron mass $(9.10956 \times 10^{-31}$ kg)
m_1,m_2,m_3	Components of effective mass tensor

$m_{\parallel}$	Longitudinal component of effective mass
$m_{\perp}$	Transverse component of effective mass
m^*	Effective mass
$m^*(E)$	Effective mass for energy E
$m^*(\underline{k})$	Effective mass for wave vector $\underline{k}$
N	Number of unit cells
N_A	Acceptor concentration
N_A^-	Ionised acceptor concentration
N_D	Donor concentration
N_D^+	Ionised donor concentration
N_{di}	Concentration of ionised donors
N_{dn}	Concentration of neutral donors
N_i	Impurity concentration
N_n	Concentration of neutral impurity atoms
N_o	Number of atoms in a crystal
$n_o, n_\ell, n_{\underline{q}}$	Phonon occupation number
n, n_v, n_i	Electron concentration; an integer
n_{po}	Photon number
n_s	Electron concentration of energy E_s
P	Absorbed power
P_ℓ, P_n	Legendre polynomials of order ℓ, n
P(t)	Probability of collision at time t
P_o	Incident power
$\underline{p}$	Momentum
Q	A function of temperature
Q_1	Q for semiconductor
Q_2	Q for metals
$\underline{q}$	Phonon wave vector
R_H	Hall coefficient $[m^3\ C^{-1}]$
R_m	Magnetoresistance
r	Hall ratio
$\underline{r}$	Position coordinate
$\underline{r}_1$	Coordinate of atom nearest to origin
r_1, r_2	Random numbers
$S(\underline{k},\underline{k}')$	Probability of scattering from $\underline{k}$ state to $\underline{k}'$ state
$S_i^{e\ell}$	Scattering probability for the ith kind of elastic scattering

S_{pa}, S_{pe}	Scattering probabilities for the absorption and emission of polar optic phonons
s	Sound velocity
T_e	Electron temperature
T, T_L	Lattice temperature
$T_{\underline{k}\underline{k}'}$	Transition rate from $E_{\underline{k}}$ to $E_{\underline{k}'}$
t	Time
t_c	Collision time
t_i	Time of flight
U_a	Overlap of the atomic functions
$U_c, U_{v1}, U_{v2}, U_{v3}$	Cell periodic parts of the Bloch functions for the conduction band and the three valence bands
U_f, U_i	Wave functions of the radiation oscillator
$U_{\underline{k}}(\underline{r})$	Cell periodic part of the Bloch function
U_o	Height of potential barrier; atomic wave function
V	Potential energy
V_a	Atomic potential
V_c	Crystal volume
V_n, V_n^*	Fourier coefficient and its conjugate for the crystal potential
v	Velocity
v_t	Component of electron velocity perpendicular to dislocation lines
W	Heat flux
W_i	Normalised energy
$\hat{\underline{x}}, \hat{\underline{y}}, \hat{\underline{z}}$	Unit vectors in the cartesian system
Z_e	Charge on the impurity atom
α	Nonparabolicity parameter of Kane relation; Polaron coupling constant; Attenuation constant
α_+, α_-	Attenuation constant for circularly polarised waves
$\alpha_1, \alpha_2, \alpha_3$	Direction cosines of $\underline{B}$
β	Donor level degeneracy; Angle; Phase constant; Warm-electron transport coefficient
Γ	Centre (0,0,0) of Brillouin zone
Γ'	Dingle temperature
$\Gamma(n)$	Gamma function

γ	Propagation constant
$\gamma, \gamma(E)$	Function of energy in Kane relation
Δ	Spin-orbit splitting energy, Intervalley separation energy
Δ_s	Width of forbidden band in a superlattice
$\delta(x)$	Dirac delta function
$\varepsilon, \varepsilon_d$	Permittivity [F m^{-1}]
ε_s	Static permittivity
ε_∞	High-frequency permittivity
ε_0	Free-space permittivity (8.854×10^{-12} F m^{-1})
ε_1	Band width of first miniband
η	Mean deviation of a Gaussian function
Θ	Absolute thermoelectric power
θ	Polar angle; Angle of Faraday rotation
θ_0	Optic phonon temperature
κ	Wave vector of a photon
λ	Decay coefficient; Wavelength
λ_s	Screening wavelength
μ	Thompson coefficient
μ_B	Bohr magneton (9.2741×10^{-24} J T^{-1})
$\mu_i, \mu_d, \mu_\Gamma, \mu_{X,L}$	Drift mobility [m^2 V^{-1} s^{-1}]
μ_H	Hall mobility [m^2 V^{-1} s^{-1}]
μ_0	Free-space permeability ($4\pi \times 10^{-7}$ H m^{-1})
Ξ_u	Deformation potential constant for uniaxial shear
Ξ_d	Deformation potential constant for dilation
ξ	Magnetoresistance coefficient
π_{12}	Peltier coefficient
ρ	Mass density
ρ_0	Density of states; Resistivity in the absence of a magnetic field
ρ_B	Resistivity in the presence of a magnetic field
$\sum_{+,-}$	Summation for the two signs + and -
σ	Conductivity [S]
$\underline{\underline{\sigma}}$	Conductivity tensor
$[\sigma]$	Spin matrix

σ_{ij}	Components of conductivity tensor
$\sigma_{ij\ell}$	Third-rank tensor associated with Hall effect
$\sigma_{ij\ell m}$	Fourth-rank tensor associated with magnetoconductivity
σ_z	AC conductivity in the absence of a magnetic field
$\sigma(B)$	Conductivity in a magnetic field B
$\tau, \tau(E)$	Relaxation time
τ_D	Dielectric relaxation time
τ_E	Energy relaxation time
τ_m	Momentum relaxation time
ϕ, ϕ'	Azimuthal angle
ϕ_ℓ	Wave function for harmonic oscillator
$\phi_{n,\underline{q}}$	Wave function for lattice vibration
ϕ_r, ϕ_i	Real and imaginary parts of ϕ
ϕ_{xx}, ϕ_{xy}	Components of ϕ
$\phi, \phi(E)$	Perturbation in distribution function
$\phi(x)$	Potential
$\chi_{\underline{k},n}$	Wave function of electron-phonon system
$\psi, \psi_{\underline{k}}$	Electron wave function
Ω	Crystal volume
ω	Angular frequency
ω_B	Cyclotron resonance frequency
ω_ℓ	Polar optic phonon frequency
$\omega_{\underline{q}}$	Angular frequency of lattice vibration for wave vector $\underline{q}$
ω_t	Angular frequency for transverse vibration
ω_o	Optic phonon frequency
∇	Gradient operator
∇^2	Laplacian operator

References

1.1 J.C. Bose: U.S. Patent 755840 (1904)
1.2 J. Bardeen, W.H. Brattain: Phys. Rev. *75*, 1208 (1949)
1.3 C. Hilsum, A.C. Rose-Innes: *Semiconducting III-V Compounds* Pergamon, Oxford 1961) p. 1
1.4 M.L. Huggins: Phys. Rev. *27*, 286 (1926)
1.5 V.M. Goldschmidt: Trans. Faraday Soc. *25*, 253 (1929)
1.6 H. Welker: Z. Naturforsch. *7a*, 744 (1952)
1.7 J.C. Woolley: In *Compound Semiconductors*, Vol.1, ed. by R.K. Willardson, H.L. Goering (Reinhold, New York 1962) pp. 3-20
E.K. Muller, J.L. Richards: J. Appl. Phys. *35*, 1233 (1964)
D. Long: In *Semiconductors and Semimetals*, Vol.1, ed. by R.K. Willardson, A.C. Beer (Academic Press, New York 1966) pp. 143-6
1.8 L. Esaki, R. Tsu: IBM J. Res. Dev. *14*, 61 (1970)
E. Kasper, H.J. Herzog, H. Kibbel: Appl. Phys. *8*, 199 (1975)
1.9 W.M. Sharpless: Bell Syst. Tech. J. *38*, 259 (1959)
1.10 D.A. Jenny: Proc. IRE *46*, 959 (1958)
A.G. Foyt, W.T. Lindley, J.P. Donnelly: Appl. Phys. Lett. *16*, 335 (1970)
C.J. Nuese, A.G. Sigai, M. Ettenberg, J.J. Gannon, S.L. Gilbert: Appl. Phys. Lett. *17*, 90 (1970)
1.11 J.B. Gunn: Solid State Commun. *1*, 88 (1963)
1.12 H. Weiss: *Structure and Applications of Galvanomagnetic Devices* (Pergamon, Oxford 1969) Ref. 1.3, p. 204
1.13 J. Tauc: *Photo and Thermoelectric Effects in Semiconductors* (Pergamon, Oxford 1962)
1.14 E.J. Ryder, W. Shockley: Phys. Rev. *81*, 139 (1951)
E.M. Conwell: "High Field Transport in Semiconductors", in *Solid State Physics*, ed. by F. Seitz, D. Turnbull, H. Ehrenreich (Academic Press, New York 1967) Suppl. 9
G. Bauer: "Determination of Electron Temperatures and of Hot Electron Distribution Functions in Semiconductors", in *Springer Tracts in Modern Physics*, Vol. 74 (Springer, Berlin, Heidelberg, New York 1974)
1.15 G.W. Ludwig: IEEE Trans. *ED-14*, 547 (1967)
J.C. McGroddy, M.R. Lorenz, T.S. Plaskett: Solid State Commun. *7*, 901 (1969)
B.J. Elliot, J.B. Gunn, J.C. McGroddy: Appl. Phys. Lett. *11*, 253 (1967)
J.C. McGroddy, M.I. Nathan, J.E. Smith, Jr.: IBM J. Res. Dev. *13*, 543 (1969)
M.H. Jorgensen, N.O. Gram, N.I. Meyer: Solid State Commun. *10*, 337 (1972)
1.16 A.R. Hutson, J.H. McFee, D.L. White: Phys. Rev. Lett. *7*, 237 (1961)
A.R. Hutson, D.L. White: J. Appl. Phys. *33*, 40 (1962)
D.L. White: J. Appl. Phys. *33*, 2547 (1962)
J.H. McFee: J. Appl. Phys. *34*, 1548 (1963)
G. Weinreich, T.M. Sanders, Jr., H.G. White: Phys. Rev. *114*, 33 (1959)

A.A. Oliner (ed.): *Acoustic Surface Waves*, Topics in Applid Physics, Vol. 24 (Springer, Berlin, Heidelberg, New York 1978)
1.17 C.K.N. Patel, E.D. Shaw: Phys. Rev. Lett. *24*, 451 (1970)
1.18 R.H. Bube: *Photoconductivity of Solids* (John Wiley, New York 1957)
Ref. 1.13
W. Shockley: *Electrons and Holes in Semiconductors* (McGraw-Hill, New York 1958)
M. Cardona, L. Ley (eds.): *Photoemission in Solids, Part I and Part II*, Topics in Applied Physics, Vols. 26,27 (Springer, Berlin, Heidelberg, New York 1978,1979)
J.I. Pankove (ed.): *Electroluminescence*, Topics in Applied Physics, Vol. 17 (Springer, Berlin, Heidelberg, New York 1977)
1.19 E.O. Kane: J. Phys. Chem. Sol. *1*, 249 (1957)
1.20 R.T. Delves: Proc. Phys. Soc. *73*, 572 (1959)
1.21 N.N. Grigor'ev, I.M. Dykman, P.M. Tomchuk: Sov. Phys. - Solid State, *10*, 837 (1968)
1.22 D.L. Rode: Phys. Rev. B*2*, 1012 (1970)
1.23 B.R. Nag: J. Phys. C*7*, 3541 (1974)
B.R. Nag: Phys. Status Solidi (b) *66*, 719 (1974)
B.R. Nag: J. Appl. Phys. *46*, 4819 (1975)
B.R. Nag, G.M. Dutta: Phys. Status Solidi (b) *71*, 401 (1975)
1.24 D.L. Rode: Phys. Rev. B*3*, 3287 (1971)
D. Chattopadhyay, B.R. Nag: Phys. Rev. B*12*, 5676 (1975)
J. Schnakenberg: "Electron Phonon Interaction and Boltzmann Equation in Narrow-Band Semiconductors", in Springer Tracts in Modern Physics, Vol. 51 (Springer, Berlin, Heidelberg, New York 1969)
2.1 J.L. Shay, J.H. Wernick: *Ternary Chalcopyrite Semiconductors: Growth, Electronic Properties and Applications* (Pergamon, Oxford 1975)
2.2 C. Kittel: *Introduction to Solid State Physics*, 4th ed. (John Wiley and Sons, New York 1971) p. 5
2.3 H. Jones: *The Theory of Brillouin Zones and Electronic States in Crystals*, 2nd ed. (North-Holland, Oxford 1975) p. 32
2.4 J.F. Cornwell: *Group Theory and Electronic Energy Bands in Solids* (North-Holland, London 1969) p. 77
2.5 F.J. Blatt: *Physics of Electronic Conduction in Solids* (McGraw-Hill, New York 1968) p. 19
2.6 F. Herman: J. Electron. *1*, 103 (1955)
2.7 C. Hilsum, A.C. Rose-Innes: *Semiconducting III-V Compounds* (Pergamon, Oxford 1961) p. 10
R.W. Keys: In *Solid State Physics*, Vol. 11, ed. by F. Seitz, D. Turnbull (Academic Press, New York 1960) p. 152
2.8 W.L. Roth: In *Physics and Chemistry of II-VI Compounds*, ed. by M. Aven, J.S. Prener (North-Holland, Amsterdam 1967) p. 123
2.9 D.L. Rode: Phys. Rev. B*2*, 4036 (1970)
2.10 C. Kittel: *Introduction to Solid State Physics*, 2nd ed. (Wiley, New York 1956) p. 35
3.1 G.H. Wannier: Phys. Rev. *52*, 191 (1937)
J.C. Slater: Phys. Rev. *76*, 1592 (1949)
J.M. Ziman: *Electrons and Phonons* (Oxford University Press, London 1960)
J.M. Luttinger: Phys. Rev. *84*, 814 (1951)
J.M. Luttinger, W. Kohn: Phys. Rev. *97*, 869 (1955)
W. Kohn: Phys. Rev. *115*, 1460 (1959)
A.C. Smith, J.F. Janak, P.B. Adler: *Electronic Conduction in Solids* (McGraw-Hill, New York 1967) p. 121
3.2 B.R. Nag: *Theory of Electrical Transport in Semiconductors* (Pergamon, Oxford 1972) p. 7

3.3 H. Jones: *The Theory of Brillouin Zones and Electronic States in Crystals* (North-Holland, Amsterdam 1975, 2nd Revised ed.) pp. 93-131, pp. 141-166
3.4 J.L. Birman: Phys. Rev. *115*, 1493 (1959)
3.5 M.L. Cohen, T.K. Bergstresser: Phys. Rev. *141*, 789 (1966)
F.H. Pollak, C.W. Higginbotham, M. Cardona: J. Phys. Soc. Jpn. *21*, Suppl. 20 (1966)
D.J. Stukel, R.N. Euwema: Phys. Rev. *188*, 1193 (1969)
R. Braunstein, E.O. Kane: J. Phys. Chem. Sol. *23*, 1423 (1962)
T.C. Collins, D.J. Stukel, R.N. Euwema: Phys. Rev. B*1*, 724 (1970)
D.J. Stukel, R.N. Euwema, T.C. Collins, F. Herman, R.L. Kortum: Phys. Rev. *179*, 740 (1969)
T.K. Bergstresser, M.L. Cohen: Phys. Rev. *164*, 1069 (1967)
P. Eckelt, O. Madelung, J. Treusch: Phys. Rev. Lett. *18*, 656 (1967)
J.L. Shay, W.E. Spicer, F. Herman: Phys. Rev. Lett. *18*, 649 (1967)
R.N. Euwema, T.C. Collins, D.G. Shankland, J.S. DeWitt: Phys. Rev. *162*, 710 (1967)
Y.W. Tung, M.L. Cohen: Phys. Rev. *180*, 823 (1969)
H.I. Zhang, J. Callaway: Phys. Rev. *181*, 1163 (1969)
F. Herman, R.L. Kortum, C.D. Kuglin, J.P. VanDyke, S. Skillman: In *Methods in Computational Physics*, Vol. 8, ed. by B. Adler, S. Fernbach, M. Rotenberg (Academic Press, New York 1968)
J.P. Walter, M.L. Cohen: Phys. Rev. *183*, 763 (1969)
U. Rössler: Phys. Rev. *184*, 733 (1969)
D.J. Stukel, R.N. Euwema: Phys. Rev. *186*, 754 (1969)
M. Cardona: J. Phys. Chem. Sol. *24*, 1543 (1963)
Y.W. Tung, M.L. Cohen: Phys. Rev. B2, 1216 (1970)
S.E. Kohn, P.Y. Yu, Y. Petroff, Y.R. Shen, Y. Tsang, M.L. Cohen: Phys. Rev. B*8*, 1477 (1975)
F. Herman, R.L. Kortum, I.B. Ortenburger, J.P. VanDyke: J. Phys. (Paris) *29*, 4 (1962)
3.6 G. Dresselhaus: Phys. Rev. *100*, 580 (1955)
R.H. Parmenter: Phys. Rev. *100*, 573 (1955)
E.O. Kane: J. Phys. Chem. Sol. *1*, 249 (1957)
J.J. Hopfield: J. Appl. Phys. *32*, Suppl. 2277 (1961)
G.D. Mahan, J.J. Hopfield: Phys. Rev. *135*, A428 (1964)
M.L.A. Robinson: Phys. Rev. Lett. *17*, 963 (1966)
F. Matossi, F. Stern: Phys. Rev.*111*, 472 (1958)
3.7 F. Bassani: In *Semiconductors and Semimetals*, Vol. 1, ed. by R.K. Willardson, A.C. Beer (Academic Press, New York 1966) p. 70
3.8 A.S. Poplavnoi: Sov. Phys. - Solid State *8*, 1779 (1967)
3.9 E.O. Kane: In *Semiconductors and Semimetals*, Vol. 1, ed. by R.K. Willardson, A.C. Beer (Academic Press, New York 1966) p. 75
3.10 G. Dresselhaus, A.F. Kip, C. Kittel: Phys. Rev. *98*, 368 (1955)
3.11 J.G. Mavroides, B. Lax: Phys. Rev. *107*, 1530 (1957); *108*, 1648 (1957)
3.12 B. Segall: In *Physics and Chemistry of II-VI Compounds*, ed. by M. Aven, J.S. Prener (North-Holland, Amsterdam 1967) p. 52
3.13 R.G. Wheeler, J.O. Dimmock: Phys. Rev. *125*, 1805 (1962)
J.J. Hopfield, D.G. Thomas: Phys. Rev. *122*, 35 (1961)
3.14 H. Fröhlich: Adv. Phys. *3*, 325 (1954)
W.E. Tefft: Phys. Rev. *164*, 1032 (1967)
C.G. Kuper, G.D. Whitfield (eds.): *Polarons and Phonons* (Plenum, New York 1963)
I.G. Austin, N.F. Mott: Adv. Phys. *18*, 41 (1969)
J. Appel: In *Solid State Physics*, Vol. 21, ed. by F. Seitz, D. Turnbull, H. Ehrenreich (Academic Press, New York 1968) p. 193

3.15 D.C. Langreth: Phys. Rev. *159*, 717 (1967)
3.16 U.J. Hansen, J.H. Gardner, K.F. Cuff: Bull. Amer. Phys. Soc. Ser II, *11*, 755 (1966)
S. Bermon: Phys. Rev. *158*, 723 (1967)
P.J. Lin, J. Kleinman: Phys. Rev. *142*, 478 (1966)
3.17 J.N. Walpole, A.L. McWhorter: Phys. Rev. *158*, 708 (1967)
3.18 R. Dalven: In *Solid State Physics*, Vol. 28, ed. by H. Ehrenreich, F. Seitz, D. Turnbull (Academic Press, New York 1973) p. 204
3.19 J.A. Van Vechten, T.K. Bergstresser: Phys. Rev. B*1*, 3351 (1970)
3.20 A.G. Thompson, M. Cardona, K.L. Shaklee, J.C. Woolley: Phys. Rev. *146* 601 (1966)
A.G. Thompson, J.E. Rowe, M. Ruben: J. Appl. Phys. *40*, 3280 (1969)
J.C. Woolley, J.A. Evans: Proc. Phys. Soc. London *78*, 354 (1961)
3.21 J.L. Shay, J.H. Wernick: *Ternary Chalcopyrite Semiconductors: Growth, Electronic Properties and Applications* (Pergamon, Oxford 1974)
3.22 A. Shileika: Surf. Sci. *37*, 730 (1973)
3.23 J.L. Shay: In *Proc. 11th Int. Conf. Physics of Semiconductors*, ed. by M. Miasek (PWN-Polish Scientific Publishers, Warsaw 1972) p. 787
R.A. Bendorius, G.Z. Krivaiti, A.Yu. Shileika, G.F. Karavaev, V.A. Chaldyshev: In *Proc. 11th Int. Conf. Physics of Semiconductors*, ed. by M. Miasek (PWN-Polish Scientific Publishers, Warsaw 1972) p. 834
3.24 H. Kildal: Ph.D. Thesis, Stanford University (1972)
4.1 H. Jones: *The Theory of Brillouin Zones and Electronic States in Crystals*, 2nd ed. (North-Holland, Amsterdam 1975)
J.F. Cornwell: *Group Theory and Electronic Energy Bands in Solids* (North-Holland, Amsterdam 1969)
4.2 J. Callaway: *Energy Band Theory* (Academic Press, New York 1964)
J. Reitz: In *Solid State Physics*, Vol. 1, ed. by F. Seitz, D. Turnbull (Academic Press, New York 1955) p. 1
F. Bassani: In *Semiconductors and Semimetals*, Vol. 1, ed. by R.K. Willardson, A.C. Beer (Academic Press, New York 1966) p. 21
4.3 R. de L. Kronig, W.G. Penney: Proc. Roy. Soc. *130*, 499 (1931)
4.4 J.P. McKelvey: *Solid State and Semiconductor Physics* (Harper and Row, New York 1966) p. 227
4.5 Zh.I. Alferow, Yu.V. Zhilyaev, Yu.V.Shmartsev: Sov. Phys. Semicond. *5*, 174 (1971)
L. Esaki, L.L. Chang, W.E. Howard, V.L. Rideout: In *Proc. 11th Int. Conf. Physics of Semiconductors*, ed. by M. Miasek (PWN-Polish Scientific Publishers, Warsaw 1972) p. 431
4.6 R. Dingle, W. Wiegmann, C.H. Henry: Phys. Rev. Lett. *33*, 827 (1974)
4.7 B.R. Nag: *Theory of Electrical Transport in Semiconductors* (Pergamon, Oxford 1972) p. 4
4.8 L. Esaki, L.L. Chang: Phys. Rev. Lett. *33*, 495 (1974)
4.9 R.A. Smith: *Wave Mechanics of Crystalline Solids* (Chapman and Hall, London 1961) p. 150
4.10 D. Mukherji, B.R. Nag: Phys. Rev. B *12*, 4338 (1975)
J.N. Schulman, T.C. McGill: Phys. Rev. Lett. *39*, 1680 (1977)
4.11 E.O. Kane: In *Semiconductors and Semimetals*, Vol. 1, ed. by R.K. Willardson, A.C. Beer (Academic Press, New York 1966) p. 75
W.C. Wang: *Solid State Electronics* (McGraw-Hill, New York 1966) p. 659
4.12 F. Herman, C.D. Kuglin, K.F. Cuff, R.L. Kortum: Phys. Rev. Lett. *11*, 541 (1963)
4.13 E. Haga, H. Kimura: J. Phys. Soc. Jpn. *19*, 471 (1964)
4.14 B.R. Nag, A.N. Chakravarti: Phys. Status Solidi (b) *71*, K45 (1975)
4.15 W.P. Dumke: Phys. Rev. B *2*, 987 (1970)
A.K. Das, B.R. Nag: Phys. Status Solidi (b) *69*, 329 (1975)

A.K. Das, B.R. Nag: Phys. Rev. *13*, 1857 (1976)
4.16 M.H. Cohen: Phys. Rev. *121*, 387 (1961)
4.17 R. Dalven: In *Solid State Physics*, Vol. 28, ed. by H. Ehrenreich, F. Seitz, D. Turnbull (Academic Press, New York 1973) p. 179
4.18 T.N. Morgan: Phys. Rev. *139*, A343 (1965)
4.19 N.F. Mott: Philos. Mag. *19*, 835 (1969)
B.I. Shklovskii: Sov. Phys. Semicond. *6*, 77 (1973)
4.20 I.I. Eldumiati, G.I. Haddad: J. Appl. Phys. *44*, 395 (1973)
4.21 R.A. Logan, A.G. Chynoweth: Phys. Rev. *131*, 89 (1963)
G. Lucovsky: Appl. Phys. Lett. *5*, 37 (1964)
J.I. Pankove: Phys. Rev. *140*, A2059 (1965)
J.I. Pankove: J. Appl. Phys. *35*, 1890 (1964)
4.22 E.O. Kane: Phys. Rev. *131*, 79 (1963)
4.23 B.I. Halperin, M. Lax: Phys. Rev. *148*, 722 (1966)
4.24 V.I. Bonch-Bruevich: In *Semiconductors and Semimetals*, Vol. 1, ed. by R. K. Willardson, A.C. Beer (Academic Press, New York 1966) p. 132
4.25 D.D. Kleppinger, F.A. Lindhom: Solid-State Electron *14*, 199 (1971)
4.26 R.J. Van Overstraeten, H.J. DeMan, R.P. Mertens: IEEE Trans. *ED-20* 290 (1973)
4.27 W. Franz: Z. Naturforsch. *13a*, 484 (1958)
L.V. Keldysh: Sov. Phys.-JETP *6*, 763 (1958)
T.S. Moss: *Optical Properties of Semiconductors* (Butterworths, London 1961) p. 47
4.28 P. Guetin, D. Boccon - Gibod: Appl. Phys. Lett. *13*, 161 (1968)
4.29 L. Landau: Z. Phys. *64*, 629 (1930)
4.30 R. Peierls: Z. Phys. *80*, 763 (1933)
J.M. Luttinger: Phys. Rev. *84*, 814 (1951)
4.31 J.M. Luttinger, W. Kohn: Phys. Rev. *97*, 869 (1955)
4.32 B. Lax: *Proceedings of the International School of Physics, 'Enrico Fermi' Course XXII,* ed. by R.A. Smith (Academic Press, New York 1963)
4.33 E.J. Johnson, D.H. Dickey: Phys. Rev. B *1*, 2686 (1970)
p. 263
5.1 E. Fermi: Z. Phys. *36*, 902 (1926)
P.A.M. Dirac: Proc. Roy. Soc. *112*, 661 (1926)
5.2 A. Sommerfeld: Z. Phys. *47*, 1 (1928)
J. McDougall, E.C. Stoner: Roy. Soc. Philos. Trans. *237*A, 67 (1938)
5.3 C. Herring, E. Vogt: Phys. Rev. *101*, 944 (1956)
5.4 B. Lax, J.G. Mavroides: Phys. Rev. *100*, 1650 (1955)
J.G. Mavroides, B. Lax: Phys. Rev. *107*, 1530 (1957)
5.5 E.G.S. Paige: In *Progress in Semiconductors*, Vol. 8, ed. by A.F. Gibson, R.E. Burgess (Heywood, London 1964) p. 29
5.6 T.S. Moss: Proc. Phys. Soc. London B *67*, 775 (1954)
5.7 J.P. McKelvey: *Solid State and Semiconductor Physics* (Harper and Row, New York 1966) p. 95
5.8 L.M. Roth, P.N. Argyress: In *Semiconductors and Semimetals*, Vol. 1, ed. by R.K. Willardson, A.C. Beer (Academic Press, New York 1966) p. 167
5.9 J.S. Blakemore: *Semiconductor Statistics* (Pergamon, Oxford 1962) p. 89
5.10 B.R. Nag, A.N. Chakrabarti: Int. J. Electron. *36*, 275 (1974)
5.11 Y. Yafet, R.W. Keyes, E.N. Adams: J. Phys. Chem. Sol. *1*, 137 (1956)
5.12 G.E. Stillman, C.M. Wolfe, J.O. Dimmock: J. Phys. Chem. Sol. *31*, 1199 (1970)
C.M. Wolfe, G.E. Stillman, J.O. Dimmock: J. Appl. Phys. *41*, 504 (1970)
H. Kahlert, J. Geiger, G. Landwehr: In *Proc. 12th Int. Conf. Physics of Semiconductors*, ed. by M.A. Pilkhun (Teubner, Stuttgart 1974) p. 269
P. Norton, T. Braggins, H. Levinstein: Phys. Rev. B *8*, 5632 (1973)
R.J. Sladek: J. Phys. Chem. Sol. *5*, 157 (1958)

V.F. Bannaya, E.M. Gershenzon, L.B. Litvak-Gorskaya: Sov. Phys. Semicond. *2*, 812 (1969); *4*, 159 (1970)
R.S. Crandall: Phys. Rev. B *1*, 730 (1970)
E.M. Gershenzon, V.A. Il'in, L.B. Litvak-Gorskaya, R.I. Rabinovich, E.Z. Shapiro: Sov. Phys. Semicond. *6*, 1639 (1973)
E.M. Gershenzon, V.A. Il'in, I.N. Kurilenko, L.B. Litvak-Gorskaya: Sov. Phys. Semicond. *6*, 1612 (1975)
Sh.M. Gasanli, O.V. Emel'yanenko, V.K. Ergakov, F.P. Kesamanly, T.S. Lagunova, D.N. Nasledov: Sov. Phys. Semicond. *5*, 1641 (1972)
K. Somogyi: Phys. Status Solidi (a) *20*, K127 (1973)

5.13 P. Norton, H. Levinstein: Phys. Rev. B *6*, 470 (1972)
E.M. Gershenzon, V.A. Il'in, I.N. Kurilenko, L.B. Litvak-Gorskaya: Sov. Phys. Semicond. *6*, 1457 (1973)
E. Gornik, W. Muller, F. Kohl: In *Proc. 12th Int. Conf. Physics of Semiconductors*, ed. by M.A. Pilkhun (Teubner, Stuttgart 1974) p. 839
D. Kranzer: Thesis, T.H. Vienna (1972)
B.R. Nag, G.M. Dutta: J. Appl. Phys. *48*, 3621 (1977)
B.R. Nag, G.M. Dutta: J. Phys. C *11*, 119 (1978)

6.1 B.R. Nag: *Theory of Electrical Transport in Semiconductors* (Pergamon, Oxford 1972)

6.2 F.J. Blatt: *Physics of Electronic Conduction in Solids* (McGraw-Hill, New York 1968)
J.M. Ziman: *Electrons and Phonons* (Clarendon Press, Oxford 1960)
A.H. Wilson: *Theory of Metals* (Cambridge University Press, Cambridge 1936)
K. Seeger: *Semiconductor Physics* (Springer, Wien, New York 1973)
O. Madelung: *Introduction to Solid-State Theory*, Springer Series in Solid-State Sciences, Vol. 2 (Springer, Berlin, Heidelberg, New York 1978)

6.3 D.L. Dexter, F. Seitz: Phys. Rev. *86*, 964 (1952)
W.T. Read, Jr.: Philos. Mag. *45*, 775 (1954); *46*, 111 (1955)
G.L. Pearson, W.T. Read, Jr., F.J. Morin: Phys. Rev. *93*, 666 (1954)
R.S. Allgaier, W.W. Scanlon: Phys. Rev. *111*, 1029 (1958)

6.4 C. Erginsoy: Phys. Rev. *79*, 1013 (1950)
N. Sclar: Phys. Rev. *104*, 1559 (1956)

6.5 S.N. Bose: Z. Phys. *26*, 178 (1924); *27*, 384 (1924)
A. Einstein: Preuss. Akad. Wiss. Berlin, Ber. *22*, 261 (1924); Ber. *1*, 3 (1925)

6.6 W.A. Harrison: Phys. Rev. *104*, 1281 (1956)

6.7 L.I. Schiff: *Quantum Mechanics* (McGraw-Hill, New York 1949) p. 114

6.8 E. Fermi: *Nuclear Physics* (University of Chicago Press, Chicago 1960) p. 142

6.9 J.P. McKelvey: *Solid State and Semiconductor Physics* (Harper and Row, London 1966) p. 100

6.10 D. Pines: *Elementary Excitations in Solids* (Benjamin, New York 1963) pp. 17, 25

6.11 H. Ehrenreich: J. Phys. Chem. Sol. *8*, 130 (1959)

6.12 P.A. Lindsay: *Introduction to Quantum Mechanics for Electrical Engineers* (McGraw-Hill, New York 1967) p. 182

6.13 J.D. Wiley: Phys. Rev. B *4*, 2485 (1971)

6.14 S. Wang: *Solid State Electronics* (McGraw-Hill, New York 1966) p. 659

6.15 W.P. Dumke: Phys. Rev. B *2*, 987 (1970)

6.16 A. Honig: Phys. Rev. Lett. *17*, 186 (1966)
I. Solomon: In *Proc. 11th Int. Conf. Physics of Semiconductors*, ed. by M. Miasek (PWN-Polish Scientific Publishers, Warsaw 1972) p. 27

6.17 J. Bardeen, W. Shockley: Phys. Rev. *80*, 72 (1950)

6.18 H.J.G. Meijer, D. Polder: Physica *19*, 255 (1953)
E.M. Conwell: *High-field Transport in Semiconductors*, Vol. 9, Solid State Physics, ed. by F. Seitz, D. Turnbull, H. Ehrenreich (Academic Press, New York 1967) Suppl. p. 111
6.19 W.A. Harrison: Phys. Rev. *104*, 1281 (1956)
6.20 H.B. Callen: Phys. Rev. *76*, 1394 (1949)
6.21 E.G.S. Paige: In *Progress in Semiconductors*, Vol. 8, ed. by A.F. Gibson, R.E. Burger (Heywood, London 1964) p. 76
6.22 D.L. Rode, S. Knight: Phys. Rev. B *3*, 2534 (1971)
M. Cuevas: Phys. Rev. *164*, 1021 (1967)
L.M. Falicov, M. Cuevas: Phys. Rev. *164*, 1025 (1967)
H.I. Ralph, G. Simpson, R.J. Elliott: Phys. Rev. B *11*, 2948 (1975)
6.23 B. Pödör: Phys. Status Solidi *16*, K167 (1966)
6.24 D.L. Dexter, F. Seitz: Phys. Rev. *86*, 964 (1952)
6.25 L. Makowski, M. Glicksman: J. Phys. Chem. Sol. *34*, 487 (1973)
6.26 C. Herring, E. Vogt: Phys. Rev. *101*, 944 (1956)
6.27 P. Norton, T. Braggins, H. Levinstein: Phys. Rev. B *8*, 5632 (1973)
6.28 J.W. Harrison, J.R. Hauser: J. Appl. Phys. *47*, 292 (1976)
6.29 D. Kranzer: Phys. Status Solidi (a) *26*, 11 (1974)
7.1 J.H. Ferziger, H.G. Kaper: *Mathematical Theory of Transport Processes in Gases* (North-Holland, Amsterdam 1972)
7.2 R.B. Thomas: Phys. Rev. *152*, 138 (1966)
M. Dresden: Rev. Mod. Phys. *33*, 265 (1961)
7.3 G. Goudet, C. Meuleau: *Semiconductors* (McDonald and Evans, London 1957) p. 27
7.4 A. Katz: *Principles of Statistical Mechanics* (W.H. Freeman, San Francisco 1967) p. 158
7.5 H.J. Stocker, H. Kaplan: Phys. Rev.*150* , 619 (1966)
7.6 E.M. Conwell: *High-field Transport in Semiconductors*, Solid State Physics, Vol. 9, ed. by F. Seitz, D. Turnbull, H. Ehrenreich (Academic Press, New York 1967) Suppl. p. 127
7.7 A. Honig: Phys. Rev. Lett. *17*, 186 (1966)
Ref. 7.4, p. 162
P.T. Landsberg (ed.): *Solid State Theory* (Wiley Interscience, New York 1969) p. 284
7.8 A.H. Wilson: *Theory of Metals* (Cambridge University Press, Cambridge 1936) p. 300
7.9 M. Kohler: Z. Phys. *124*, 772 (1948)
7.10 D.J. Howarth, E.H. Sondheimer: Proc. Roy. Soc. A *219*, 53 (1953)
B.F. Lewis, E.H. Sondheimer: Proc. Roy. Soc. A *227*, 241 (1955)
7.11 H. Ehrenreich: J. Phys. Chem. Sol. *2*, 131 (1957)
7.12 S.S. Devlin: In *Physics and Chemistry of II-VI Compounds*, ed. by M. Aven, J.S. Prener (North-Holland, Amsterdam 1967) p. 551
7.13 N.N. Grigor'ev, I.M. Dykman, P.M. Tomchuk: Sov. Phys. Solid State *10*, 837 (1968)
7.14 D. Kranzer: Phys. Status Solidi *46*, 591 (1971)
7.15 R.T. Delves: Proc. Phys. Soc. London *73*, 572 (1959)
7.16 K. Fletcher, P.N. Butcher: J. Phys. C *5*, 212 (1972)
K. Fletcher, P.N. Butcher: J. Phys. C *6*, 976 (1973)
7.17 D.L. Rode: Phys. Rev. B *2*, 1012 (1970)
7.18 P.N. Dr. K. Rektorys: *Survey of Applicable Mathematics* (Illife Books Ltd., London 1969) p. 1152
7.19 B. Adler, S. Fernbach, M. Rotenberg (eds.): *Methods of Computational Physics*, Vol. 1 (Academic Press, New York 1961)
Yu.A. Shreider (ed.): *The Monte Carlo Method* (Pergamon, Oxford 1966)
J. Spannier, E.M. Gellbard: *Monte Carlo Principles and Neutron Transport Problems* (Addison Wesley, London 1969)

J.M. Hammersley, D.C. Handscomb: *Monte Carlo Methods* (Methuen, London 1964)
R. Stuart, F. Wooten, W.E. Spicer: Phys. Rev. *135*, A495 (1964)
7.20 W. Fawcett, H.D. Rees: Phys. Lett. *28a*, 731 (1969)
7.21 C. Canali, C. Jacoboni, F. Nava, G. Ottaviani, A. Alberigi-Quaranta: Phys. Rev. B *12*, 2265 (1975)
G. Ottaviani, L. Reggiani, C. Canali, F. Nava, A. Alberigi-Quaranta: Phys. Rev. B *12*, 3318 (1975)
7.22 D.H. Lehmer: In *Proc. 2nd Symp. Large Scale Digital Calculating Machinery* (Harvard University Press, Harvard 1949) p. 141
T.E. Hull, A.R. Dobell: SIAM Rev. *4*, 230 (1962)
7.23 H. Jeffreys: *Theory of Probability* (Oxford University Press, Oxford 1948) p. 371
7.24 C. Herring, E. Vogt: Phys. Rev. *101*, 944 (1956)
7.25 C. Kittel: *Introduction to Solid State Physics*, 3rd ed. (John Wiley and Sons, New York 1966) p. 139
7.26 J.D. Zook: Phys. Rev. *136*, A 869 (1964)
D.L. Rode: Phys. Rev. B *2*, 4036 (1970)
7.27 P. Lawaetz: Phys. Rev. *183*, 730 (1969)
7.28 A.G. Samoilovich, I.Ya, Korenblit, I.V. Dakhovskii: Sov. Phys. Doklady *6*, 606 (1962)
A.G. Samoilovich, I.Ya, Korenblit, I.V. Dakhovskii, V.D. Iskra: Sov. Phys. Solid State *3*, 2148 (1962)
A.G. Samoilovich, I.Ya. Korenblit, I.V. Dakhovskii, V.D. Iskra: Sov. Phys. Solid State *3*, 2385 (1962)
P. Norton, T. Braggins, H. Levinstein: Phys. Rev. B *8*, 5632 (1973)
7.29 A.K. Das, B.R. Nag: J. Phys. Chem. Sol. *39*, 259 (1978)
8.1 A. Matthiessen: Ann. Phys. Chem. *7*, 761, 892 (1864)
8.2 J.P. McKelvey: *Solid State and Semiconductor Physics* (Harper and Row, New York 1966) p. 316
8.3 J.M. Ziman: *Electrons and Phonons* (Oxford University Press, London 1960) p. 434
D.L. Rode, S. Knight: Phys. Rev. B *3*, 2534 (1971)
8.4 D.J. Howarth, E.H. Sondheimer: Proc. Roy. Soc. London A *219*, 53 (1953)
8.5 C. Canali, C. Jacoboni, F. Nava, G. Ottaviani, A. Alberigi-Quaranta: Phys. Rev. B *12*, 2265 (1975)
8.6 L. Onsager: Phys. Rev. *37*, 405 (1931); *38*, 2265 (1931)
8.7 P.J. Price: Phys. Rev. B *6*, 4882 (1972)
8.8 P.J. Price: IBM J. Res. Dev. *1*, 239 (1957)
8.9 B.F. Lewis, E.H. Sondheimer: Proc. Roy. Soc. London A *227*, 241 (1955)
S.S. Devlin: In *Physics and Chemistry of II-VI Compounds*, ed. by M. Aven, J. Prener (North-Holland, Amsterdam 1967) p. 561
8.10 N.N. Grigor'ev, I.M. Dykman, P.M. Tomchuk: Sov. Phys. Solid State *10*, 837 (1968)
8.11 A. Fortini, D. Diguet, J. Lugand: J. Appl. Phys. *41*, 3121 (1970)
8.12 R.T. Delves: Proc. Phys. Soc. London *73*, 572 (1959)
8.13 K. Fletcher, P.N. Butcher: J. Phys. C *5*, 212 (1972)
8.14 D.L. Rode: Phys. Rev. B*2*, 1012, 4036 (1970)
8.15 B.R. Nag: J. Phys. C*7*, 3541 (1974)
B.R. Nag, G.M. Dutta: Phys. Status Solidi (b) *71*, 401 (1975)
B.R. Nag: J. Appl. Phys. *46*, 4819 (1975)
8.16 K. Fletcher, P.N. Butcher: J. Phys. C*6*, 976 (1973)
8.17 R.J. Stirn: In *Semiconductors and Semimetals*, Vol. 8, ed. by R.K. Willardson, A.C. Beer (Academic Press, New York 1973)
8.18 R.A. Smith: *Semiconductors* (Cambridge University Press, Cambridge 1961) p. 240-260

8.19 R. Stratton: IEEE Trans. *ED-19*, 1288 (1972)
B.R. Nag, A.N. Chakravarti: Solid-State Electron. *18*, 109 (1975)
8.20 W. Nernst: Z. Phys. Chem. Leipzig *2*, 613 (1888)
J.S. Townsend: Trans. Roy. Soc. (London) A *193*, 129 (1900)
A. Einstein: Ann. Phys. Leipzig *17*, 549 (1905)
8.21 P.T. Landsberg: Proc. Roy. Soc. London A *213*, 226 (1952)
8.22 A.N. Chakravarti, D.P. Parui: Phys. Status Solidi (a) *10*, K 141 (1972); *14*, K 55 (1972)
A.N. Chakravarti, B.R. Nag: Phys. Status Solidi (a) *22*, K 153 (1974)
H. Kroemer: IEEE Trans. *ED-25*, 850 (1978)
8.23 A.N. Chakravarti, B.R. Nag: Int. J. Electron. *37*, 281 (1974)
8.24 D. Mukherji, A.N. Chakravarti, B.R. Nag: Phys. Status Solidi (a) *26*, K 27 (1974)
8.25 P.S. Kireev: *Semiconductor Physics* (Mir Publishers, Moscow 1974) p. 315
8.26 H.J. Goldsmid: Proc. Phys. Soc. London *72*, 17 (1958)
9.1 J.B. Gunn: Solid State Commun. *1*, 88 (1963)
9.2 H. Jacobs, F.A. Brand, J.D. Meindl, S. Weitz, R. Benjamin: IRE Int. Conv. Rec. *10*, part 3, 30 (1962)
G.E. Hambleton, W.W. Gärtner: J. Phys. Chem. Sol. *8*, 329 (1959)
A.P. Ramsa, H. Jacobs, F.A. Brand: J. Appl. Phys. *30*, 1054 (1959)
R.D. Larrabee: RCA Rev. *21*, 124 (1960)
H. Jacobs, F.A. Brand, J.D. Meindl, M. Benanti, R. Benjamin: Proc. IRE *49*, 928 (1961)
A.N. Datta, B.R. Nag: IEEE Trans. *MTT-18*, 162 (1970)
9.3 R. Pethig, R.B. South: IEEE Trans. *IM-23*, 460 (1974)
M.M. Sayed, C.R. Westgate: Rev. Sci. Instrum. *46*, 1074, 1080 (1975)
9.4 G. Dresselhaus, A.F. Kip, C. Kittel: Phys. Rev. *92*, 827 (1953)
B. Lax, H.J. Zeiger, R.N. Dexter, E.S. Rosenblum: Phys. Rev. *93*, 1418 (1954)
R.N. Dexter, H.J. Zeiger, B. Lax: Phys. Rev. *95*, 557 (1954)
9.5 W.G. Spitzer, H.Y.Fan: Phys. Rev. *106*, 882 (1957)
B. Lax, G.B. Wright: Phys. Rev. Lett. *4*, 16 (1960)
9.6 T.S. Benedict, W. Shockley: Phys. Rev. *89*, 1152 (1953)
T.S. Benedict: Phys. Rev. *91*, 1565 (1953)
9.7 T. Stubb: State Inst. for Tech. Res. Finland, Publication 42
9.8 K.S. Champlin, J.D. Holm, G.H. Glover: J. Appl. Phys. *38*, 96 (1967)
9.9 G. Dresselhaus, A.F. Kip, C. Kittel: Phys. Rev. *100*, 618 (1955)
9.10 R. Pethig: J. Biolog. Phys. *1*, 193 (1973)
9.11 B.R. Nag: J. Appl. Phys. *46*, 4819 (1975)
9.12 I.I. Eldumiati, G.I. Haddad: J. Appl. Phys. *44*, 395 (1973)
9.13 J.K. Furdyna, S. Broersma: Phys. Rev. *120*, 1995 (1960)
J.K. Furdyna, M.E. Brodwin: Phys. Rev. *124*, 740 (1961)
B. Donovan, J. Webster: Proc. Phys. Soc. London *78*, 120 (1961)
9.14 H. Kawamura, M. Fukai, K. Sekido, I.Imai: J. Phys. Soc. Jpn. *18*, 322 (1963)
M. Fukai, H. Kawamura, I. Imai, K. Tomishima: J. Phys. Soc. Jpn. *17*, 1191 (1962)
9.15 R.N. Dexter, H.J. Zeiger, B. Lax: Phys. Rev. *104*, 637 (1956)
9.16 B. Lax: In *Proc. International School of Physics, 'Enrico Fermi' Course XXII*, ed. by R.A. Smith (Academic Press, New York 1963) p. 252
9.17 R.A. Stradling: Solid State Commun. *6*, 665 (1968)
9.18 P.N. Argyres: In *Proc. 12th Int. Conf. Physics of Semiconductors*, ed. by M.H. Pilkhun (Teubner, Stuttgart 1974) p. 517
9.19 H. Piller: In *Semiconductors and Semimetals*, Vol. 8, ed. by R.K. Willardson, A.C. Beer (Academic Press, New York 1972) p. 105
9.20 S.D. Smith, T.S. Moss, K.W. Taylor: J. Phys. Chem. Sol. *11*, 131 (1959)

S.D. Smith, C.R. Pidgeon, V. Prosser: In *Proc. 6th Int. Conf. Physics of Semiconductors*, ed. by A.C. Stickland (Institute of Physics and Physical Society, London 1962) p. 301
J. Kolodziejczak, S. Zukotynski, H. Stramska: Phys. Status Solidi *14*, 471 (1966)
9.21 A.I. Kosinskaya, S.I. Uritsky: Phys. Status Solidi *23*, 57 (1967)
B. Lax, L.M. Roth: Phys. Rev. *98*, 548 (1955)
B. Donovan, J. Webster: Proc. Phys. Soc. London *81*, 90 (1963)
9.22 F.R. Kessler, E. Sutter: Z. Naturforsch. *16a*, 1173 (1961)
J.R. Dixon: In *Proc. 5th Int. Conf. Physics of Semiconductors* (Publishing House of the Czechoslovak Academy of Sciences, Prague 1961) p. 366
W.P. Dumke, M.R. Lorenz, G.D. Pettit: Phys. Rev. B *1*, 4668 (1970)
A.J. Strauss: J. Nonmetals *1*, 133 (1973)
K. Osamura, Y. Murakami: Jpn. J. Appl. Phys. *11*, 365 (1972)
9.23 S.N. Bose: Z. Phys. *26*, 178 (1924); *27*, 384 (1924)
9.24 W. Heitler: *The Quantum Theory of Radiation*, 3rd. ed. (Clarendon Press, Oxford 1954) Sect. 14
9.25 G.N. Watson: *A Treatise on the Theory of Bessel Functions* (Cambridge University Press, Cambridge 1944)
M. Abramowitz, J.A. Stegun: *Handbook of Mathematical Functions* (Dover Publications, New York 1968)
9.26 W.P. Dumke: Phys. Rev. *124*, 1813 (1961)
9.27 E. Haga, H. Kimura: J. Phys. Soc. Jpn. *18*, 777 (1963)
9.28 E. Haga, H. Kimura: J. Phys. Soc. Jpn. *19*, 658 (1964)
9.29 H.J.G. Meyer: Phys. Rev. *112*, 298 (1958)
R. Rosenberg, M. Lax: Phys. Rev. *112*, 843 (1958)
9.30 A.K. Das, B.R. Nag: J. Phys. Chem. Sol. *36*, 945 (1975)
9.31 A.K. Das, B.R. Nag: Phys. Status Solidi (b) *69*, 329 (1975)
9.32 B. Jensen: Ann. Phys. (N.Y.) *80*, 284 (1973)
9.33 A.K. Ganguly, K.L. Ngai: Phys. Rev. B *8*, 5654 (1973)
K.L. Ngai, E.J. Johnson: Phys. Rev. Lett. *29*, 1607 (1972)
J. Szymanski: Phys. Status Solidi (b) *72*, 667 (1975)
9.34 J.D. Wiley, M. DiDomenico, Jr.: Phys. Rev. B *1*, 1655 (1970)
A.K. Das, B.R. Nag: Phys. Rev. B *13*, 1857 (1976)
9.35 A.K. Das, B.R. Nag: J. Phys. Chem. Sol. *39*, 259 (1978)
9.36 B. Lax, H.J. Zeiger, R.N. Dexter: Physica *20*, 818 (1954)
10.1 L. Schubnikow, W.J. de Haas: Leiden Commun. 207a, 207c, 207d, 210a (1930)
10.2 V.L. Gurevich, Yu.A. Firsov: Sov. Phys. - JETP *13*, 137 (1961)
10.3 S.M. Puri, T.H. Geballe: Bull. Amer. Phys. Soc. *8*, 309 (1963)
10.4 S. Zwerdling, B. Lax, L.M. Roth, K.J. Button: Phys. Rev. *114*, 80 (1959)
10.5 R. Kubo, H. Hasegawa, N. Hashitsume: Phys. Rev. Lett. *1*, 279 (1958)
10.6 P.M. Morse, H. Feshbach: *Methods of Theoretical Physics* (McGraw-Hill, New York 1953) p. 784
10.7 V.S. Teteica: Ann. d. Phys. (Paris) *22*, 129 (1935)
10.8 E.N. Adams, T.D. Holstein: J. Phys. Chem. Sol. *10*, 254 (1959)
10.9 R. Kubo, S.J. Miyake, N. Hashitsume: In *Solid State Physics*, Vol. 17, ed. by F. Seitz, D. Turnbull (Academic Press, New York 1965) p. 270
10.10 B. Magnusson: Phys. Stat. Sol. *52*, 361 (1972)
10.11 P.K. Basu, B.R. Nag: Phys. Status Solidi (b) *73*, K15 (1976)
10.12 L.M. Roth, P.N. Argyress: In *Semiconductors and Semimetals*, Vol. 1, ed. by R.K. Willardson, A.C. Beer (Academic Press, New York 1966) p. 184

10.13 P.G. Harper, J.W. Hodby, R.A. Stradling: Rep. Prog. Phys. *36*, 1 (1973)
10.14 G.I. Kharus, I.M. Tsidil'kovskii: Sov. Phys. Semicond. *5*, 534 (1971)
10.15 B. Magnusson: Phys. Status Solidi (b) *56*, 269 (1973)
10.16 V.L. Gurevich, Yu.A. Firsov: Sov. Phys. JETP *20*, 489 (1965)
10.17 J.R. Barker: Phys. Lett. *33*A, 516 (1970)
J.R. Barker: J. Phys. C *5*, 1657 (1972)
10.18 S.S. Razkhov, P.M. Tomchuk, A.A. Chumak: Uker Fiz. Zh. (USSR) *20*, 7542 (1975)
L. Eaves, R.A. Stradling, R.A. Wood: In *Proc. 10th Int. Conf. Physics of Semiconductors*, ed. by S.P. Kelter, J.C. Hensel, F. Stern (US Atomic Energy Commission, Oak Ridge 1970) p. 816
J.C. Portal, L. Eaves, S. Askenazy, R.A. Stradling: Solid State Commun. *14*, 1241 (1974)
10.19 R.V. Pomortsev, A.L. Ponomarev, G.I. Kharus, J.M. Tsidil'kovskii: In *Proc. 9th Int. Conf. Physics of Semiconductors*, ed. by S. Ryvkin (Nauka, Leningrad 1968) p. 720
10.20 H.P.R. Frederikse, W.R. Hosler: Phys. Rev. *108*, 1136 (1957)
10.21 R.J. Sladek: Phys. Rev. *110*, 817 (1958)
10.22 W.M. Becker, H.Y. Fan: In *Proc. 7th Int. Conf Physics of Semiconductors* (Academic Press, New York 1964) p. 663
11.1 J.B. Gunn: In *Progress in Semiconductors*, Vol. 2, ed. by A.F. Gibson, R.E. Burgess (Temple Press, London 1957) p. 213
R.J. Sladek: Phys. Rev. *120*, 1589 (1960)
K. Seeger: Z. Phys. *172*, 68 (1963)
P. Kastner, E.P. Roth, K. Seeger: Z. Phys. *187*, 359 (1965)
Y. Kanai: J. Phys. Soc. Jpn. *15*, 830 (1960)
G. Bauer, F. Kuchar, D. Kranzer, E. Bonek: Acta Phys. Austrica *33*, 8 (1971)
T. Shirakawa, C. Hamaguchi: Phys. Lett. A *49*A, 231 (1974)
C.T. Elliott, I.L. Spain: J. Phys. C *7*, 727 (1974)
F.D. Hughes, R.J. Tree: J. Phys. C *3*, 1943 (1970)
C. Hamaguchi, Y. Inuishi: J. Phys. Soc. Jpn. *18*, 1755 (1963)
F. Kuchar, A. Philipp, K. Seeger: Solid State Commun. *11*, 965 (1972)
11.2 W. Sasaki, M. Shibuya, K. Mizuguchi: J. Phys. Soc. Jpn. *13*, 456 (1958)
W. Sasaki, M. Shibuya, K. Mizuguchi, G.M. Hatoyama: J. Phys. Chem. Sol. *8*, 250 (1959)
M. Shibuya: Phys. Rev. *99*, 1189 (1955)
M.I. Nathan: Phys. Rev. *130*, 2201 (1963)
B.R. Nag, H. Paria, S. Guha: Phys. Lett. A *26*A, 172 (1968)
B.R. Nag, H. Paria, P.K. Basu: Phys. Lett. A *28*A, 202 (1968)
11.3 E.J. Ryder, W. Shockley: Phys. Rev. *81*, 139 (1951)
E.J. Ryder: Phys. Rev. *90*, 766 (1953)
11.4 B.K. Ridley, T.B. Watkins: Proc. Phys. Soc. London *78*, 293 (1961)
B.K. Ridley: Proc. Phys. Soc. London *82*, 954 (1963)
C. Hilsum: Proc. IRE *50*, 185 (1962)
J.B. Gunn: Solid State Commun. *1*, 88 (1963)
11.5 V.V. Paranjape, T.P. Ambrose: Phys. Lett. *8*, 223 (1964)
V.V. Paranjape, E. De Alba: Proc. Phys. Soc. London *85*, 945 (1965)
A.C. Baynham, P.N. Butcher, W. Fawcett, J.M. Loveluck: Proc. Phys. Soc. London *92*, 783 (1967)
E.J. Aas, K. Bløtekjaer: J. Phys. Chem. Sol. *35*, 1053 (1974)
11.6 R. Stratton: Proc. Roy. Soc. London A *246*, 406 (1958)
11.7 S.V. Gantsevich, V.L. Gurevich, R. Katilius: Phys. Condens. Matter *18*, 165 (1974)
B.R. Nag, D. Chattopadhyay: Solid-State Electron *21*, 303 (1978)
11.8 E. Erlbach: Phys. Rev. *132*, 1976 (1963)

11.9 S. Chapman, T.G. Cowling: *The Mathematical Theory of Non-uniform Gases* (Cambridge University Press, Cambridge 1960) p. 346
11.10 M.J. Druyvesteyn: Physica *10*, 61 (1930)
11.11 M. Glicksman, W.A. Hicinbothem, Jr.: Phys. Rev. *129*, 1572 (1963)
11.12 A.C. Prior: J. Electr. Control *4*, 165 (1958)
H. Morisaki, Y. Inuishi: J. Phys. Soc. Jpn. *20*, 1814 (1965)
M.C. Steele, S. Tosima: Jpn. J. Appl. Phys. *2*, 381 (1963)
M.C. Steele, M. Glicksman: J. Phys. Chem. Sol. *8*, 242 (1959)
11.13 A. Neukermans, G.S. Kino: Appl. Phys. Lett. *17*, 102 (1970)
11.14 F. Okamoto, K. Ando: Jpn. J. Appl. Phys. *3*, 605 (1964)
L.E. Vorobev, V.S. Komissarov, V.I. Stafeev: Phys. Status Solidi *52*, 25 (1972)
E.M. Gershenzon, E.I. Kulevich, N.A. Serebryakova: Sov. Phys. Semicond. *5*, 137 (1971)
K.L.I. Kobayashi, E. Otsuka: Solid State Commun. *11*, 815 (1972)
V.T. Potapov, A.V. Sokolovskii, V.I. Trifonov, N.G. Yaremenko: Sov. Phys. Semicond *6*, 1081 (1973)
11.15 R.C. Curby, D.K. Ferry: Phys. Lett. A *32*A, 236 (1970)
A.G. Foyt, A.L. McWhorter: IEEE Trans. *ED-13*, 79 (1966)
R.C. Curby, D.K. Ferry: Phys. Rev. B *3*, 3379 (1971)
11.16 H.St. Onge, J.N. Walpole: Phys. Rev. B *6*, 2331 (1972)
11.17 R. Dornhaus, H. Happ, K.H. Muller, G. Nimitz, W. Schlabitz, P. Zabinski, G. Bauer: In *Proc. 12th Int. Conf Physics of Semiconductors*, ed. by M.H. Pilkhun (Teubner, Stuttgart 1974) p. 115
11.18 K.W. Böer, K. Bogus: Phys. Rev. *176*, 899 (1968)
M. Onuki, K. Shiga: J. Phys. Soc. Jpn. *21*, Suppl. 427 (1966)
11.19 R.S. Crandall: Phys. Rev. *169*, 577 (1968)
K. Hess, H. Vana: J. Phs. C *6*, L150 (1973)
11.20 J.G. Ruch, G.S. Kino: Appl. Phys. Lett. *10*, 40 (1967)
J.G. Ruch, G.S. Kino: Phys. Rev. *174*, 921 (1968)
11.21 A.G.R. Evans, P.N. Robson, M.G. Stubbs: Electron Lett. *8*, 195 (1972)
11.22 E.M. Bastida, G. Fabri, V. Svelto, F. Vaghi: Appl. Phys. Lett. *18*, 28 (1971)
B.A. Prew: Electron. Lett. *8*, 592 (1972)
P.M. Boers: Phys. Lett. A *34*A, 329 (1971)
P.M. Boers: Electron. Lett. *7*, 625 (1971)
11.23 N. Braslau, P.S. Hauge: IEEE Trans. *ED-17*, 616 (1970)
G.H. Glover: J. Appl. Phys. *42*, 4025 (1971)
H.T. Lam, G.A. Acket: Electron. Lett. *7*, 722 (1971)
G.H. Glover: Appl. Phys. Lett. *20*, 224 (1972)
M. Kawashima, K. Ohta, S. Kataoka: Electron. Lett. *12*, 71 (1976)
11.24 C. Hilsum: In *Proc. 11th Int. Conf Physics of Semiconductors*, ed. by M. Miąsek (PWN-Polish Scientific Publishers, Warsaw 1972) p. 595
11.25 L.W. James, J.P. Van Dyke, F. Herman, D.M. Chang: Phys. Rev. B *1*, 3998 (1970)
C. Hilsum, H.D. Rees: Electron. Lett. *6*, 277 (1970)
11.26 K. Segawa, H. Miki, M. Otsubo, K. Shirahata, K. Fujibayashi: Electron. Lett. *12*, 124 (1976)
A. Hojo, I. Kuru: Electron. Lett. *10*, 61 (1974)
R.E. Hayes, R.M. Raymond: Appl. Phys. Lett. *31*, 300 (1977)
11.27 G. Bauer: "Determination of Electron Temperatures and of Hot Electron Distribution Functions in Semiconductors", in Springer Tracts in Modern Physics, Vol. 74 (Springer, Berlin, Heidelberg, New York 1974)
J. Shah: Solid-State Electron. *21*, 43 (1978)
11.28 P.J. Bulman, G.S. Hobson, B.C. Taylor: *Transferred Electron Effects* (Academic Press, New York 1972)
G.S. Hobson: *The Gunn Effect* (Clarendon Press, Oxford 1974)

H.L. Hartnagel: Solid-State Electron *12*, 19 (1969)
O. Wada, S. Yanagisawa, H. Takanashi: Electron. Lett. *12*, 215 (1976)
K. Kurumada, T. Mizutani, M. Fujimoto: Electron. Lett. *10*, 161 (1974)
11.29 J.G. Ruch, W. Fawcett: J. Appl. Phys. *41*, 3843 (1970)
11.30 J.R. Hauser, M.A. Littlejohn, T.H. Glisson: Appl. Phys. Lett. *28*, 458 (1976)
M.A. Littlejohn, J.R. Hauser, T.H. Glisson: Appl. Phys. Lett. *30*, 242 (1977)
M.A. Littlejohn, J.R. Hauser, T.H. Glisson, D.K. Ferry, J.W. Harrison: Solid-State Electron *21*, 107 (1978)
11.31 L. Esaki, R. Tsu: IBM J. Res. Dev. *14*, 61 (1970)
11.32 M. Shibuya: Phys. Rev. *99*, 1189 (1955)
W. Sasaki, M. Shibuya: J. Phys. Soc. Jpn. *11*, 1202 (1956)
W. Sasaki, M. Shibuya, K. Mizuguchi: J. Phys. Soc. Jpn. *13*, 456 (1958)
M. Asche, B.L. Boitschenko, O.G. Sarbej: Phys. Status Solidi *9*, 323 (1965)
M.I. Nathan: Phys. Rev. *130*, 2201 (1963)
R. Barrie, R.R. Burgess: Canad. J. Phys. *40*, 1056 (1962)
D. Schweitzer, K. Seeger: Z. Phys. *183*, 207 (1965)
M. Asche, O.G. Sarbej: Phys. Status Solidi *7*, 339 (1964)
M.H. Jørgensen, N.I. Meyer, K.J. Schmidt-Tiedeman: In *Proc. 7th Int. Conf. Physics of Semiconductors*, ed. by M. Hulin (Dunod, Paris 1964) p. 457
W.E.K. Gibbs: J. Phys. Chem. Sol. *25*, 247 (1964)
11.33 W. Sasaki, M. Shibuya, K. Mizuguchi: J. Phys. Soc. Jpn. *13*, 456 (1958)
11.34 M.H. Jørgensen, N.O. Gram, N.I. Meyer: Solid State Commun. *10*, 337 (1972)
C. Canali, A. Loria, F. Nava, G. Ottaviani: Solid State Commun. *12*, 1017 (1973)
11.35 J. Yamashita, M. Watanabe: Progr. Theor. Phys. *12*, 443 (1954)
J. Yamashita: Phys. Rev. *111*, 1529 (1958)
J. Yamashita, K. Inoue: J. Phys. Chem. Sol. *12*, 1 (1959)
11.36 M.S. Sodha, B.M. Gupta: Phys. Rev. B *2*, 3249 (1970)
M.S. Sodha, U.P. Phadke, A.K. Chakravarti: Phys. Rev. B *3*, 2619 (1971)
11.37 B.R. Nag: *Theory of Electrical Transport in Semiconductors* (Pergamon, Oxford 1972) p. 183
11.38 B.R. Nag, S. Guha: Brit. J. Appl. Phys. *16*, 1461 (1965)
11.39 B.R. Nag, H. Paria: Phys. Rev. *150*, 632 (1966)
11.40 B.R. Nag: Solid-State Electron. *10*, 385 (1967); *11*, 209 (1968)
11.41 B.R. Nag, P.K. Basu: J. Phys. C *2*, 2396 (1969)
B.R. Nag, P.K. Basu: Phys. Rev. B *1*, 627 (1971)
B.R. Nag, D. Chattopadhyay: J. Phys. Chem. Sol. *32*, 2749 (1971)
B.R. Nag, D. Chattopadhyay: Phys. Rev. *4*, 1220 (1971)
B.R. Nag, P.K. Basu: Phys. Rev. B *5*, 1633 (1972)
E.M. Conwell: *High Field Transport in Semiconductors*, Solid State Physics, ed. by F. Seitz, D. Turnbull, H. Ehrenreich (Academic Press, New York 1967) Suppl. 9
E.A. Movchan: Sov. Phys. Semicond. *6*, 54 (1972)
11.42 E.M. Conwell, M.O. Vassell: Phys. Rev. *166*, 797 (1968)
11.43 R. Stratton: In *Proc. Int. Conf. Solid State Physics in Electronics and Telecommunications*; Brussels (1958), Vol. 1, ed. by M. Desirant, J.L. Michiels (Academic Press, New York 1960) p. 343
W. Shockley: Bell. Syst. Tech. J. *30*, 990 (1951)
11.44 D. Chattopadhyay, B.R. Nag: Commun. Phys. *2*, 163 (1977)
11.45 H. Fröhlich, B.V. Paranjape: Proc. Phys. Soc. London B *69*, 21 (1956)
11.46 C. Hammar, P. Weissglas: Phys. Status Solidi *24*, 531 (1967)

11.47 T.Y. Stokoe, J.F. Cornwell: Phys. Status Solidi *49*, 209 (1972)
11.48 H. Fröhlich: Proc. Roy. Soc. London A *188*, 521, 532 (1947)
11.49 R. Stratton: Proc. Roy. Soc. London A *242*, 355 (1957)
11.50 J. Bok, C. Guthmann: Phys. Status Solidi *6*, 853 (1964)
E. De Alba, J. Warman: J. Phys. Chem. Sol. *29*, 69 (1968)
11.51 I. Licea: Phys. Status Solidi *37*, 795 (1970)
11.52 P.N. Butcher, W. Fawcett: Proc. Phys. Soc. London *86*, 1205 (1965)
11.53 G. Bauer: Solid-State Electron *21*, 17 (1978)
W.E. Pinson, R. Bray: Phys. Rev. *136*, A1449 (1964)
A. Mooradian, A.G. Foyt: Bull. Amer. Phys. Soc. *15*, 303 (1970)
P.D. Southgate, D.S. Hall, A.B. Dreeben: J. Appl. Phys. *42*, 2868 (1971)
J. Shah, C. Lin, R.F. Leheny, A.E. Di Giovanni: Soldi State Commun. *18*, 487 (1976)
11.54 P.M. Morese, H. Feshbach: *Methods of Theoretical Physics* (McGraw-Hill, New York 1963) p. 949
11.55 M.O. Vassell: J. Math. Phys. *11*, 408 (1970)
11.56 H.F. Budd: Phys. Rev. *158*, 798 (1967)
11.57 H.D. Rees: J. Phys. Chem. Sol. *30*, 643 (1969)
11.58 E.J. Aas: "An Investigation of Numerical Methods for Hot Carrier Problems"; University of Trondheim, Tech. Rep. TE-192 (1972)
11.59 H.D. Rees: J. Phys. Chem. Sol. *30*, 643 (1969)
T.Y. Stokoe, J.F. Cornwell: Phys. Status Solidi (b) *58*, 493 (1973)
11.60 W. Fawcett, A.D. Boardman, S. Swain: J. Phys. Chem. Sol. *31*, 1963 (1970)
11.61 H.D. Rees: J. Phys. Chem. Sol. *30*, 643 (1969)
11.62 W. Fawcett, D.C. Herbert: J. Phys. C *7*, 1641 (1974)
N.I. Beletskiy, E.D. Prokhorov: Radio Eng. Electron Phys. *19*, 88 (1974)
V. Borsari, C. Jacoboni: Phys. Status Solidi (b) *54*, 649 (1972)
B.R. Nag: Solid State Commun. *11*, 987 (1972)
D. Chattopadhyay: Solid State Commun. *15*, 325 (1974)
M.A. Littlejohn, J.R. Hauser, T.H. Glisson: Appl. Phys. Lett. *26*, 625 (1975)
W. Fawcett, J.G. Ruch: Appl. Phys. Lett. *15*, 368 (1969)
B.R. Nag: J. Appl. Phys. *44*, 1888 (1973)
11.63 J.R. Barker, C.J. Hearn: Proc. Phys. Soc. (Solid State Phys.) *2*, 2128 (1969)
11.64 S. Bosi, C. Jacoboni: J. Phys. C *9*, 315 (1976)
11.65 L.B. Montefusco, C. Jacoboni: Solid State Commun. *10*, 71 (1972)
A. Matulionis, J. Pozela, A. Reklaitis: Solid State Commun. *16*, 1133 (1975)
J.G. Ruch, W. Fawcett: J. Appl. Phys. *41*, 3843 (1970)
W. Fawcett, G. Hill: Electron. Lett. *11*, 80 (1975)
11.66 D. Chattopadhyay, B.R. Nag: J. Phys. C *11*, 2055 (1978)
11.67 H.G. Reik, H. Risken: Phys. Rev. *124*, 777 (1961)
11.68 D.E. Aspnes: Phys. Rev. B *14*, 5331 (1976)
A.R. Adams, P.J. Vinson, C. Pickering, G.D. Pitt, W. Fawcett: Electron. Lett. *13*, 46 (1977)
11.69 A.F. Gibson, J.W. Granville: J. Electron. *2*, 259 (1956)
11.70 A.F. Gibson, J.W. Granville, E.G.S. Paige: J. Phys. Chem. Sol. *19*, 198 (1961)
11.71 M.T. Vlaardingerbroek, W. Kuypers, G.A. Acket: Phys. Lett. *28a*, 155 (1968)
11.72 L.M. Grasl, O.F. Zimmerl: Arch. Elektrischen Übertragung *23*, 165 (1969)
11.73 P. Das, R. Bharat: Appl. Phys. Lett. *11*, 386 (1967)
11.74 H.D. Rees: Solid State Commun. *7*, 267 (1969)
11.75 D. Chattopadhyay, B.R. Nag: Proc. Roy. Soc. London A*354*, 367 (1977); Solid State Commun. *22*, 569 (1977)

11.76 B.R. Nag: Phys. Rev. B *11*, 3031 (1975)
J.P. Nougier: Appl. Phys. Lett. *32*, 671 (1978)
11.77 P.N. Butcher, W. Fawcett, N.R. Ogg: Brit. J. Appl. Phys. *18*, 755 (1967)
W. Shockley, J.A. Copeland, R.P. James: In *Quantum Theory of Atoms, Molecules, Solid State* (Academic Press, New York 1967) p. 537
P.S. Cheung, C.J. Hearn: J. Phys. C *5*, 1563 (1972)
11.78 W. Fawcett, H.D. Rees: Phys. Lett. *29a*, 578 (1969)
A. Aishima, K. Yokoo, S. Ono: Jpn. J. Appl. Phys. *17*, 965 (1978)
M.S. Gupta: J. Appl. Phys. *49*, 2837 (1978)
11.79 P.E. Bauhahn, G.I. Haddad, N.A. Masnari: Electron. Lett. *9*, 460 (1973)
11.80 A. Alberigi-Quaranta, V. Borsari, J. Jacoboni, G. Zanarini: Appl. Phys. Lett. *22*, 103 (1973)
11.81 A. Katz: *Principles of Statistical Mechanics* (Freeman, London 1967) p. 138
11.82 H.R. Skullerud: J. Phys. B *6*, 728 (1973)
11.83 J.G. Ruch: IEEE Trans. *ED-19*, 652 (1972)
T.J. Maloney, J. Frey: J. Appl. Phys. *48*, 781 (1977)
S. Kratzer, J. Frey: J. Appl. Phys. *49*, 4064 (1978)
11.84 P.N. Robson: Private communication
11.85 G.H. Glover: J. Appl. Phys. *44*, 1295 (1973)
J.R. Sandercook: Proc. Phys. Soc. London *86*, 1221 (1965)
J.P. Maneval, A. Zylbersztejn, H.F. Budd: Phys. Rev. Lett. *23*, 848 (1969)
11.86 H.St. Onge, J.N. Walpole: Phys. Rev. B*6*, 2331 (1972)
11.87 W. Fawcett, D.C. Herbert: Electron. Lett. *9*, 308 (1973)
11.88 C. Canali, M. Martini, G. Ottoviani, K. Zanio: Phys. Lett. *33*A, 241 (1970)
11.89 C. Jacoboni, L. Reggiani: Phys. Lett. *33*A, 333 (1970)
11.90 K. Hess, H. Vana: J. Phys. C*6*, L150 (1973)
11.91 E. Bonek: J. Appl. Phys. *43*, 510 (1972)
11.92 M.T. Vlaardingerbroek, P.M. Boers, G.A. Acket: Philips Res. Rep. *24*, 705 (1969)
11.93 B.R. Nag: Indian J. Phys. *Commemoration*, Vol. 83 (1977)
11.94 P.S. Cheung, C.J. Hearn: J. Phys. C*5*, 1563 (1972)
12.1 B.R. Nag: *Theory of Electrical Transport in Semiconductors* (Pergamon, Oxford 1972)
12.2 G.T. Surratt, T.C. Collins: In *Proc. 11th Int. Conf. Physics of Semiconductors*, ed. by M. Miasek (PWN-Polish Scientific Publishers, Warsaw 1972) p. 959
12.3 D.L. Rode: Phys. Rev. B*3*, 3287 (1971)
12.4 H. Ehrenreich: J. Phys. Chem. Sol. *2*, 131 (1957)
12.5 E.M. Gershenzon, V.A. Il'in, I.N. Kurilenko, L.B. Litvak-Gorskaya: Sov. Phys. Semicond. *6*, 1457 (1970); *6*, 1612 (1973)
E.M. Gershenzon, V.A. Il'in, L.B. Litvak-Gorskaya, R.I. Rabinovich, E.Z. Shapiro: Sov. Phys. Semicond. *6*, 1626, 1639 (1973)
P. Blood: Rep. Prog. Phys. *41*, 157 (1978)
12.6 E.M. Gershenzon, I.N. Kurilenko, L.B. Litvak-Gorskaya, R.I. Rabinovich: Sov. Phys. Semicond. *7*, 1005 (1974)
12.7 B.R. Nag, G.M. Dutta: J. Appl. Phys. *48*, 3621 (1977)
12.8 J.L. Ivey: Phys. Rev. B *9*, 4281 (1974)
12.9 B.R. Nag, G.M. Dutta: Phys. Status Solidi (b) *71*, 401 (1975)
12.10 A.S. Filipchenko, D.N. Nasledov: Phys. Status Solidi *19*, 435 (1967)
R. Barrie, J.T. Edmond: J. Electron. *1*, 161 (1955)
A.I. Blum, G.P. Ryabtsova: Sov. Phys. Solid State *1*, 692 (1959)
H. Weiss: Z. Naturforsch. *11a*, 131 (1956)

G. Busch, E. Steigmeier: Helv. Phys. Acta *34*, 1 (1961)
12.11 F.R. Kessler, E. Sutter: Z. Naturforsch. *16a*, 1173 (1961)
W.G. Spitzer, H.Y. Fan: Phys. Rev. *106*, 882 (1957)
12.12 E. Haga, H. Kimura: J. Phys. Soc. Jpn *18*, 777 (1963)
12.13 D.L. Rode, S. Knight: Phys. Rev. B*3*, 2534 (1971)
12.14 G.E. Stillman, C.M. Wolfe, J.O. Dimmock: J. Phys. Chem. Sol. *31*, 1199 (1970)
H.G.B. Hicks, D.F. Manley: Solid State Commun. *7*, 1463 (1969)
P. Blood: Phys. Rev. B*6*, 2257 (1972)
V.F. Dvoryankin, O.V. Emel'yanenko, D.N. Nasledov, D.D. Nedeoglo, A.A. Telegin: Sov. Phys. Semicond. *5*, 1636 (1972)
12.15 T.I. Voronina, O.V. Emel'yanenko, D.N. Nasledov, D.D. Nedeoglo: Sov. Phys. Semicond. *7*, 925 (1974)
12.16 J.T. Edmond, R.F. Brooks, F.A. Cunnel: In *Report of the Meeting on Semiconductors, Rugby 1956* (The Physical Society, London 1957)
A. Amith, I. Kudman, E.F. Steigmeier: Phys. Rev. *138*, A 1270 (1965)
O.V. Emel'yanenko, V.A. Skripkin, V.A. Popova: Sov. Phys. Semicond. *7*, 667 (1973)
S.K. Sutradhar, D. Chattopadhyay: J. Phys. C *12*, 1693 (1979)
12.17 K. Osamura, Y. Murakami: Jpn. J. Appl. Phys. *11*, 365 (1972)
12.18 C. Hilsum, H.D. Rees: Electron. Lett. *6*, 277 (1970)
12.19 D.L. Rode: Phys. Rev. B*3*, 3287 (1971)
12.20 B.R. Nag, G.M. Dutta: J. Phys. C *11*, 119 (1978)
12.21 K.E. Potter: Ph.D. Thesis, Sheffield University (1975)
12.22 J.S. Barrera, R.J. Archer: IEEE Trans. *ED-22*, 1023 (1975)
12.23 F.P. Kesamanly, D.N. Nasledov, A.Ya. Nashel'skii, V.A. Skripkin: Sov. Phys. Semicond. *2*, 1221 (1969)
12.24 W.P. Dumke, M.R. Lorenz, G.D. Pettit: Phys. Rev. B*1*, 4668 (1970)
12.25 J.P. McCarthy: Solid-State Electron. *10*, 649 (1967)
G.R. Cronin, S.R. Borello: J. Electrochem. Soc. *114*, 1078 (1967)
G.R. Cronin, R.W. Conrad, S.R. Borrello: J. Electrochem. Soc. *113*, 1336 (1966)
J.P. McCarthy: Solid State Commun. *5*, 5 (1967)
12.26 F.P. Kesamanly, Yu.V. Mal'tsev, D.N. Nasledov, L.A. Nikolaeva, M.N. Pivovarov, V.A. Skripkin, Yu.I. Ukhanov: Sov. Phys. Semicond. *3*, 993 (1970)
12.27 J.R. Dixon: In *Proc. 5th Int. Conf. Physics of Semiconductors* (Publishing House of the Czechoslovak Academy of Sciences, Prague 1961) p. 366
R.M. Culpepper, J.R. Dixon: J. Opt. Soc. Am. *58*, 96 (1968)
12.28 E. Haga, H. Kimura: J. Phys. Soc. Jpn. *19*, 471 (1964)
12.29 B. Jensen: Ann. Phys. (N.Y.) *80*, 284 (1973)
12.30 M.G. Craford, W.O. Groves, A.H. Herzog, D.E. Hill: J. Appl. Phys. *42*, 2751 (1971)
R.C. Taylor, J.F. Woods, M.R. Lorenz: J. Appl. Phys. *39*, 5404 (1968)
A.S. Epstein: J. Phys. Chem. Sol. *27*, 1611 (1966)
H.C. Casey, Jr., F. Ermanis, J.C. Luther, L.R. Dawson, H.W. Verleur: J. Appl. Phys. *42*, 2130 (1971)
M. Toyama, M. Naito, A. Kasami: Jpn. J. Appl. Phys. *8*, 358 (1969)
12.31 D.L. Rode: In *Semiconductors and Semimetals*, Vol. 10, ed. by R.K. Willardson, A.C. Beer (Academic Press, New York 1975) p. 1
12.32 K. Fletcher, P.N. Butcher: J. Phys. C*6*, 976 (1973)
12.33 A. Sagar, R.C. Miller: J. Appl. Phys. *32*, 2073 (1961)
S.J. Silverman, R.O. Carlson, H. Ehrenreich: J. Appl. Phys. *34*, 456 (1963)
E.F. Steigmeier, I. Kudman: Phys. Rev. *141*, 767 (1966)
12.34 R.K. Willardson, A.C. Beer, A.E. Midleton: J. Electrochem. Soc. *101*, 354 (1954)

D.N. Nasledov, S.V. Slobodchikov: Sov. Phys. Solid State *1*, 681 (1959)
12.35 J.D. Wiley, M.DiDomenico, Jr.: Phys. Rev. B*1*, 1655 (1970)
12.36 W.J. Turner, W.E. Reese: Phys. Rev. *117*, 1003 (1960)
12.37 A.K. Das, B.R. Nag: Phys. Rev. B *13*, 1857 (1976)
12.38 D.L. Rode: Phys. Rev. B*2*, 4036 (1970)
12.39 A.J. Strauss, G.W. Iseler: Bull Amer. Phys. Soc. *17*, 326 (1972)
S. Yamada, Y. Kawasaki, O. Nishida: Phys. Status Solidi *26*, 77 (1968)
M.P. Lisitsa, V.N. Malinko, E.S. Nokonyuk, N.E. Novoseletskii, G.G. Tsebulya: Sov. Phys. Semicond *3*, 1321 (1970)
12.40 B. Jensen: J. Phys. Chem. Sol. *34*, 2235 (1973)
12.41 R.S. Allgaier, W.W. Scanlon: Phys. Rev. *111*, 1029 (1958)
12.42 Yu.I. Ravich, B.A. Efimova, I.A. Smirnov: *Semiconducting Lead Chalcogenides* (Plenum Press, New York 1970) p. 96
12.43 Yu.I. Ravich, L.Ya. Morgovskii: Sov. Phys. Semicond *3*, 1278 (1970)
Yu.I. Ravich, B.A. Efimova, V.I. Tamarchenko: Phys. Status Solidi *43*, 11, 453 (1971)
12.44 A.K. Das, B.R. Nag: J. Phys. Chem. Sol. *39*, 259 (1978)
12.45 L. Makowski, M. Glicksman: J. Phys. Chem. Sol. *34*, 487 (1973)
12.46 J.W. Harrison, J.R. Hauser: J. Appl. Phys. *47*, 292 (1976)
12.47 H. Brooks: In *Advances in Electronics and Electron Physics*, Vol. 7, ed. by L. Marton (Academic Press, New York 1955) p. 85
12.48 M. Glicksman, R.E. Endstrom, S.A. Mittleman, J.R. Appert: Phys. Rev. B*9*, 1621 (1974)
12.49 M.A. Littlejohn, J.R. Hauser, T.H. Glisson, D.K. Ferry, J.W. Harrison: Solid-State Electron. *21*, 107 (1978)
12.50 J.C. Phillips: Rev. Mod. Phys. *42*, 317 (1970)
12.51 R. Dornhaus, G. Nimtz: "The Properties and Applications of the $Hg_{1-x}Cd_xTe$ Alloy System", in *Springer Tracts in Modern Physics*, Vol. 78 (Springer, Berlin, Heidelberg, New York 1976)
12.52 W. Scott: J. Appl. Phys. *43*, 1055 (1972)
12.53 I.F. Chang, S.S. Mitra: Phys. Rev. *172*, 924 (1968)
12.54 D. Chattopadhyay, B.R. Nag: J. Appl. Phys. *45*, 1463 (1974)
12.55 E.I. Georgitse, V.I. Ivanov-Omskii, B.T. Kolomiets, A.A. Mal'Kova, K.P. Smekalova: Sov. Phys. Semicond. *6*, 1122 (1973)
12.56 J.L. Shay, J.H. Wernick: *Ternary Chalcopyrite Semiconductors: Growth Electronic Properties and Applications* (Pergamon, Oxford 1975)
12.57 M. Matyáš, P. Höschl: Czech. J. Phys. B *12*, 788 (1962)
13.1 R.A. Stradling, R.A. Wood: Proc. Phys. Soc. (Solid State Phys.) *1*, 1711 (1968); J. Phys. C*3*, L94 (1970)
P.G. Harper, J.W. Hodby, R.A. Stradling: Rep. Prog. Phys. *36*, 1 (1973)
13.2 H. Ehrenreich: J. Phys. Chem. Sol. *2*, 131 (1957)
13.3 G.T. Surratt, T.C. Collins: In *Proc. 11th Int. Conf. Physics of Semiconductors*, ed. by M. Miasek (PWN-Polish Scientific Publishers, Warsaw 1972) p. 959
J.L. Ivey: Phys. Rev. B*9*, 4281 (1974)
13.4 I. Strzalkowski, S. Joshi, C.R. Crowell: Appl. Phys. Lett. *28*, 350 (1976)
13.5 S.J. Bass: J. Cryst. Growth *31*, 172 (1975)
P. Blood, J.W. Orton: J. Phys. C*7*, 893 (1974)
V.T. Potapov, V.I. Trifonov, I.I. Chusov, N.G. Yaremenko: Sov. Phys. Semicond. *6*, 1076 (1973)
N.G. Yaremenko, V.T. Potapov, V.S. Ivleva: Sov. Phys. Semicond. *6*, 1084 (1973)
M.G. Milvidskii, N.S. Rytova, V.I. Fistul: Sov. Phys. Semicond. *11*, 1009 (1977)
13.6 D.L. Rode, S. Knight: Phys. Rev. *3*, 2534 (1971)

13.7 H.I. Ralph, G. Simpson, R.J. Elliott: Phys. Rev. B*11*, 2948 (1975)
13.8 L.M. Falicov, M. Cuevas: Phys. Rev. *164*, 1025 (1967)
13.9 V.L. Bonch-Bruevich: In *Semiconductors and Semimetals*, Vol. 1, ed. by R.K. Willardson, A.C. Beer (Academic Press, New York 1966) p. 141
13.10 N. Mikoshiba: Phys. Rev. *127*, 1962 (1962)
B.I. Shklovskii: Sov. Phys. Semicond. *6*, 1053 (1973)
L. Halbo, R.J. Sladek: J. Non-Cryst. Solids *4*, 192 (1970)
T.O. Poehler: Phys. Rev. B*4*, 1223 (1971)
V.M. Guslikov, O.V. Emel'yanenko, D.N. Nasledov, D.D. Nedeoglo, I.N. Timchenko: Sov. Phys. Semicond. *7*, 1191 (1974)
D. Lemoine, C. Pelletier, S. Rolland, R. Granger: Phys. Lett. A *56*A, 493 (1976)
13.11 L. Makowski, M. Glicksman: J. Phys. Chem. Sol. *34*, 487 (1975)
J.W. Harrison, J.R. Hauser: Phys. Rev. B *13*, 5347 (1976)
13.12 I.F. Chang, S.S. Mitra: Phys. Rev. *172*, 924 (1968)
G. Lucovsky, M.H. Brodsky, E. Burstein: Phys. Rev. B*2*, 3295 (1970)
M.H. Brodsky, G. Lucovsky: Phys. Rev. Lett. *21*, 990 (1968)
G. Lucovsky, M.F. Chen: Solid State Commun. *8*, 1397 (1970)
J. Baars, F. Sorger: Solid State Commun. *10*, 875 (1972)
13.13 M.A. Littlejohn, J.R. Hauser, T.H. Glisson: Appl. Phys. Lett. *30*, 242 (1977)
J.S. Escher, G.A. Antypas, J. Edgecumbe: Appl. Phys. Lett. *29*, 153 (1976)
J.J. Coleman, N. Holonyak, Jr., M.J. Ludowise, P.D. Wright, R. Chin, W.O. Groves, D.L. Keune: Appl. Phys. Lett. *29*, 167 (1976)
J.J. Hsieh, J.A. Rossi, J.P. Donnelly: Appl. Phys. Lett. *28*, 709 (1976)
T.H. Glisson, J.R. Hauser, M.A. Littlejohn, C.K. Williams: J. Electron. Mater. *7*, 1 (1978)
J.B. Restorff, Bland Houston, J.R. Burke, R.E. Hayes: Appl. Phys. Lett. *32*, 189 (1978)
13.14 L. Eaves, R.A. Hoult, R.A. Stradling, R.J. Tidey, J.C. Portal, S. Askenazy: J. Phys. C*8*, 1034 (1975)
13.15 K. Kasai, T. Shirakowa, C. Hamaguchi: In *Proc. 13th Int. Conf. Physics of Semiconductors*, ed. by F.G. Fumi (Tipographia Marves, Rome 1976) p. 1141
K. Kasai, C. Hamaguchi: J. Phys. Soc. Jpn. *44*, 1749 (1978)
13.16 C. Hamaguchi, T. Shirakawa, Y. Yamashita, J. Nakai: Phys. Rev. Lett. *28*, 1129 (1972)
R.J. Nicholas, R.A. Stradling: J. Phys. C*9*, 1253 (1976)
13.17 J.R. Barker: Solid-State Electron. *21*, 197 (1978)
13.18 K. Hess, C.T. Sah: Phys. Rev. B*10*, 3375 (1974)
E. Gornik: Opt. Laser Technol. *7*, 121 (1975)
13.18a J.R. Burke, G.P. Carver: Phys. Rev. B*17*, 2719 (1978)
E.A. Mozhaev, V.I. Ivanov-Omski, V.A. Mal'tseva, D.V. Mashovets, R.V. Parfen'ev: Sov. Phys. Semicond. *11*, 260 (1977)
13.19 K. Hess: Solid-State Electron. *21*, 123 (1978)
R.K. Kalia, J.J. Quinn: Phys. Rev. B*17*, 1383 (1978)
13.20 F. Stern: Crit. Rev. Solid State Sci. *4*, 499 (1974)
G. Dorda: *Advances in Solid State Physics*, ed. by H.J. Queisser (Pergamon, Oxford 1973) p. 215
H. Ezawa: Surf. Sci. *58*, 25 (1976)
13.21 A. Daerr, J.P. Kotthaus, J.F. Koch: Solid State Commun. *17*, 455 (1975)
13.22 H. Hasegawa, K.E. Forward, H.L. Hartnagel: Electron. Lett. *11*, 53 (1975)
D.L. Lile, A.R. Clawson, D.A. Collins: Appl. Phys. Lett. *29*, 207 (1976)
B. Bayraktaraglu, E. Kohn, H.L. Hartnagel: Electron. Lett. *12*, 53 (1976)

T. Mimura, N. Yokoyama, Y. Nakayama, M. Fukuta: Jpn. J. Appl. Phys. *17*, Suppl. 17-1, 153 (1977)
13.23 S.P. Murarka: Appl. Phys. Lett. *26*, 180 (1975)
H. Takagi, G. Kano, I. Teramoto: IEEE Trans. *ED-25*, 551 (1978)
13.24 N. Yokoyama, T. Mimura, K. Odani, M. Fukuta: Appl. Phys. Lett. *32*, 58 (1978)
R.P.H. Chang: J. Vac. Sci. Technol. *14*, 278 (1977)
13.25 L. Esaki, R. Tsu: IBM J. Res. Dev. *14*, 61 (1970)
13.26 L.L. Chang, L. Esaki, R. Tsu: Appl. Phys. Lett. *24*, 593 (1974)
R. Dingle, W. Wiegmann, C.H. Henry: Phys. Rev. Lett. *33*, 827 (1974)
D. Mukherji, B.R. Nag: Phys. Rev. B*12*, 4338 (1975)
L. Esaki, L.L. Chang: Phys. Rev. Lett. *33*, 495 (1974)
R. Dingle, A.C. Gossard, W. Wiegmann: Phys. Rev. Lett. *34*, 1327 (1975)
R. Tsu, L. Esaki: Appl. Phys. Lett. *22*, 562 (1973)
G. Abstreiter, E. Bauser, A. Fischer, K. Ploog: Appl. Phys. *16*, 345 (1978)
13.27 P.A. Lebwohl, R. Tsu: J. Appl. Phys. *41*, 2664 (1970)
P.J. Price: IBM J. Res. Dev. *17*, 39 (1973)
D.L. Andersen, E.J. Aas: J. Appl. Phys. *44*, 3721 (1973)
13.28 C. Hilsum: Electron. Lett. *10*, 259 (1974)
13.29 R.A. Stradling, R.A. Wood: J. Phys. C*3*, 2425 (1970)
H. Kahlert, G. Bauer: Phys. Status Solidi *46*, 535 (1971)
A.K. Ganguly, K.L. Ngai: Phys. Rev. B*8*, 5654 (1973)

Subject Index

Acoustic Surface Waves

Editor: A. A. Oliner
1978. 198 figures, 16 tables. XI, 331 pages
(Topics in Applied Physics, Volume 24)
ISBN 3-540-08575-0

Contents:
A. A. Oliner: Introduction. – *G. W. Farnell:* Types and Properties of Surface Waves. – *H. M. Gerard:* Principles of Surface Wave Filter Design. – *E. A. Ash:* Fundamentals of Signal Processing Devices. – *A. A. Oliner:* Waveguides for Surface Waves. – *A. J. Slobodnik, Jr.:* Materials and Their Influence on Performance. – *H. I. Smith:* Fabrication Techniques for Surface Wave Devices.

Light Scattering in Solids

Editor: M. Cardona
1975. 111 figures, 3 tables. XIII, 339 pages
(Topics in Applied Physics, Volume 8)
ISBN 3-540-07354-X

Contents:
M. Cardona: Introduction. – *A. Pinczuk, E. Burstein:* Fundamentals of Inelastic Light Scattering in Semiconductors and Insulators. – *R. M. Martin, L. M. Falicov:* Resonant Raman Scattering. – *M. V. Klein:* Electronic Raman Scattering. – *M. H. Brodsky:* Raman Scattering in Amorphous Semiconductors. – *A. S. Pine:* Brillouin Scattering in Semiconductors. – *Y.-R. Shen:* Stimulated Raman Scattering.

Nonlinear Infrared Generation

Editor: Y.-R. Shen
1977. 134 figures. XI, 279 pages
(Topics in Applied Physics, Volume 16)
ISBN 3-540-07945-9

Contents:
Y.-R. Shen: Introduction. – *R. L. Aggarwal, B. Lax:* Optical Mixing of CO_2 Lasers in the Far-Infrared. – *R. L. Byer, R. L. Herbst:* Parametric Oscillation and Mixing. – *V. T. Nguyen, T. J. Bridges:* Difference Frequency Mixing via Spin Nonlinearities in the Far-Infrared. – *J. J. Wynne, P. P. Sorokin:* Optical Mixing in Atomic Vapors. – *T. Y. Chang:* Optical Pumping in Gases.

Photoemission in Solids I

General Principles

Editors: M. Cardona, L. Ley
1978. 90 figures, 17 tables. XI, 290 pages
(Topics in Applied Physics, Volume 26)
ISBN 3-540-08685-4

Contents:
M. Cardona, L. Ley: Introduction. – *W. L. Schaich:* Theory of Photoemission: Independent Particle Model. – *S. T. Manson:* The Calculation of Photoionization Cross Sections: An Atomic View. – *D. A. Shirley:* Many-Electron and Final-State Effects: Beyond the One-Electron Picture. – *G. K. Wertheim, P. H. Citrin:* Fermi Surface Excitations in X-Ray Photoemission Line Shapes from Metals. – *N. V. Smith:* Angular Dependent Photoemission. – Appendix.

Photoemission in Solids II

Case Studies

Editors: L. Ley, M. Cardona
1979. 214 figures, 26 tables. XVIII, 401 pages
(Topics in Applied Physics, Volume 27)
ISBN 3-540-09202-1

Contents:
L. Ley, M. Cardona: Introduction. – *L. Ley, M. Cardona, R. A. Pollak:* Photoemission in Semiconductors. – *S. Hüfner:* Unfilled Inner Shells: Transition Metals and Compounds. – *M. Campagna, G. K. Wertheim, Y. Baer:* Unfilled Inner Shells: Rare Earths and Their Compounds. – *W. D. Grobman, E. E. Koch:* Photoemission from Organic Molecular Crystals. – *C. Kunz:* Synchrotron Radiation: Overview. – *P. Steiner, H. Höchst, S. Hüfner:* Simple Metals. – Appendix: Table of Core-Level Binding Energies. – Additional References with Titles. – Subject Index.

X-Ray Optics

Applications to Solids

Editor: H.-J. Queisser
1977. 133 figures, 14 tables. XI, 227 pages
(Topics in Applied Physics, Volume 22)
ISBN 3-540-08462-2

Contents:
H.-J. Queisser: Introduction: Structure and Structuring of Solids. – *M. Yoshimatsu, S. Kozaki:* High Brilliance X-Ray Sources. – *E. Spiller, R. Feder:* X-Ray Lithography. – *U. Bonse, W. Graeff:* X-Ray and Neutron Interferometry. – *A. Authier:* Section Topography. – *W. Hartmann:* Live Topography.

A. H. Eschenfelder

Magnetic Bubble Technology

1980. 271 figures, 8 tables. XIV, 317 pages
(Springer Series in Solid-State Sciences, Volume 14)
ISBN 3-540-09822-4

Contents:
Introduction to Magnetic Bubbles. – Static Properties of Magnetic Bubbles. – Dynamic Properties of Magnetic Bubbles. – Basic Permalloy-Bar Bubble Devices. – Other Bubble Device Forms. – Bubble Materials. – Device Chip Fabrication. – Chip Packaging. – Applications. – Future Prospects. – References. – Subject Index.

Springer-Verlag
Berlin
Heidelberg
New York

Physics of Superionic Conductors

Editor: M. B. Salamon

1979. 101 figures, 13 tables. XII, 255 pages
(Topics in Current Physics, Volume 15)
ISBN 3-540-09333-8

Contents:
M. B. Salamon: Introduction. – *J. B. Boyce, T. M. Hayes:* Structure and Its Influence on Superionic Conduction EXAFS Studies. – *S. M. Shapiro, F. Reidinger:* Neutron Scattering Studies of Superionic Conductors. – *H. U. Beyeler, P. Brüesch, L. Pietronero, W. R. Schneider, S. Strässler, H. R. Zeller:* Statics and Dynamics of Lattice Gas Models. – *M. J. Delaney, S. Ushioda:* Light Scattering in Superionic Conductors. – *P. M. Richards:* Magnetic Resonance in Superionic Conductors. – *M. B. Salamon:* Phase Transitions in Ionic Conductors. – *T. Geisel:* Continuous Stochastic Models. – Additional References with Titles. – Subject Index.

Solid Electrolytes

Editor: S. Geller

1977. 85 figures, 24 tables. XI, 229 pages
(Topics in Applied Physics, Volume 21)
ISBN 3-540-08338-3

Contents:
S. Geller: Introduction. – *H. Sato:* Some Theoretical Aspects of Solid Electrolytes. – *S. Geller:* Halogenide Solid Electrolytes. – *B. B. Owens, J. E. Oxley, A. F. Sammells*– Applications of Halogenide Solid Electrolytes. –*J. H. Kennedy:* The β-Aluminas. – *W. L. Worrell:* Oxide Solid Electrolytes. – *L. Heyne:* Electrochemistry of Mixed Ionic-Electronic Conductors.

X-Ray Optics

Applications to Solids

Editor: H.-J. Queisser

1977. 133 figures, 14 tables. XI, 227 pages
(Topics in Applied Physics, Volume 22)
ISBN 3-540-08462-2

Contents:
H.-J. Queisser: Introduction: Structure and Structuring of Solids. – *M. Yoshimatsu, S. Kozaki:* High Brilliance X-Ray Sources. – *E. Spiller, R. Feder:* X-Ray Lithography. – *U. Bonse, W. Graeff:* X-Ray and Neutron Interferometry. – *A. Authier:* Section Topography. – *W. Hartmann:* Live Topography.

MIX
Papier aus verantwortungsvollen Quellen
Paper from responsible sources
FSC® C105338

If you have any concerns about our products,
you can contact us on
ProductSafety@springernature.com

In case Publisher is established outside the EU,
the EU authorized representative is:
Springer Nature Customer Service Center GmbH
Europaplatz 3, 69115 Heidelberg, Germany

Printed by Libri Plureos GmbH
in Hamburg, Germany